STATISTICAL ANALYSIS FOR
BUSINESS DECISIONS

STATISTICAL ANALYSIS FOR BUSINESS DECISIONS

WILLIAM A. SPURR, Ph.D.

Professor of Business Statistics

and

CHARLES P. BONINI, Ph.D.

Associate Professor of Management Science

Graduate School of Business
Stanford University

 Revised Edition · 1973

RICHARD D. IRWIN, INC. Homewood, Illinois 60430
IRWIN-DORSEY INTERNATIONAL London, England WC2H 9NJ
IRWIN-DORSEY LIMITED Georgetown, Ontario L7G 4B3

Revised Edition
First Printing, May 1973

ISBN 0-256-00491-9
Library of Congress Catalog Card No. 72–95393
Printed in the United States of America

PREFACE

THE ROLE of quantitative analysis in business and economics has expanded tremendously in recent years with advances in statistical theory, electronic computers, and the growing appreciation of the scientific method in general, as opposed to intuitive methods of reasoning. New analytic techniques have sprung from probability theory, operations research, and decision theory, while computers have provided an effective catalyst to their widespread adoption. The basic university courses in statistics reflect this wide diversity in subject matter, as well as the varying goals of different schools and differing levels of students.

It is with this great diversity in mind that we have planned this text. A broad range of topics is included, from the traditional tools of analysis to the modern concepts of simulation and Bayesian decision theory; from simple graphic techniques to sophisticated topics such as survey sampling and analysis of variance. The instructor can structure his course by selecting subjects appropriate to the background and abilities of his students.

Since the book is planned for the general student who needs to use statistics in his chosen field of work, the principal emphasis is placed on the use of statistical methods as scientific tools in the analysis of practical business and economic problems, rather than on theory or mathematical derivations. The material has been presented as simply as possible, with a minimum of statistical jargon.

The main text requires no knowledge of mathematics beyond elementary algebra. The more advanced topics are marked by asterisks in the Contents, so that the instructor in the elementary course can easily omit them if desired. Optional material—some of it involving calculus or matrix algebra—appears in the appendixes of several chapters. Over 400 problems have been included to allow flexibility in assignments and a broad range of practical applications for class discussion, home study, or laboratory work. Almost all of the text and problems have been tested in the basic statistics courses at the Stanford Graduate School of Business, and revised on the basis of student evaluation.

In this edition, our main purpose has been to meet the changing needs of the business statistics course. To this end, we have omitted certain descriptive topics that appeared in the earlier edition and expanded the treatment of statistical inference and decision theory throughout. In particular, Chapter 11 on advanced test procedures (t, χ^2, and F distributions and nonparametric methods) has been added, since these topics are now becoming a part of many basic courses. Also, Chapter 15 presents new applications of Monte Carlo methods to decision problems, reflecting the increased importance of this topic. Other chapters have been reorganized for better readability. And finally, all material has been updated as needed, and numerous problems have been added.

The book is divided into six parts:

1. An introduction to the basic tools of analysis, such as ratios, frequency distributions, averages, and dispersion measures, in Chapters 1–4.

2. The elements of probability theory and the principal probability distributions are described and applied to decision making in Chapters 5–8. Probabilities of events, payoff tables, expected values, the value of information, and decision trees are all elements of a rational procedure for making decisions under uncertainty.

3. In order to draw inferences about sample information, it is desirable to set confidence limits or test hypotheses, as described in Chapters 9–11. In practical surveys, however, simple random sampling will not usually suffice, so Chapter 12 explores a variety of other sample designs that are more efficient or practicable. This topic is too often ignored in elementary books.

4. Probabilities and sample evidence are combined through Bayes' Theorem in Chapters 13 and 14 to improve the decision-making process. Here, as in Chapters 7 and 8, economic costs and profits are explicitly included in the analysis. This topic represents an important extension of the traditional interpretation of sample information. Simple methods of simulation and risk analysis are applied to business decision problems in Chapter 15.

5. Regression and correlation techniques are widely used and misused. The reader may well wish to be content with simple regression, but multiple regression is a more powerful tool, and is easily manage-

able in the new computer programs, so the entire treatment in Chapters 16–17 is recommended, if time permits.

6. Statistical analysis in business and economics requires considerable emphasis on time series, since the economist is vitally concerned with measuring and projecting economic growth, seasonal movements or business cycles. We therefore survey index numbers and time series analysis and forecasting, together with computer applications, in Chapters 18–20.

The book contains enough material for a two-semester course in statistics—say Chapters 1–12 for the first term and Chapters 13–20 for the second term. It may also be used for either a one-semester course or a more advanced course, by appropriate selection of topics. For example, a course along classical lines might be fashioned from Chapters 1–4, 9–11, and 16–20. In addition, Chapters 7, 8, and 13 might be included (or substituted for other chapters) if an introduction to Bayesian decision theory is desired.

An advanced course might include Chapters 5–8, 12–15, and 17. Other combinations of chapters may be selected to meet the requirements of specific schools and groups of students.

The authors are much indebted to Lester S. Kellogg and John H. Smith, whose major contributions to Spurr, Kellogg, and Smith, *Business and Economic Statistics* (1st ed. 1954, rev. ed. 1961; Homewood, Ill.: Richard D. Irwin, Inc.), provided the basis from which the present Chapters 1–4 and 18–19 have evolved. The general treatment of decision theory given in Chapters 7–8 and 13–14 follows in the tradition of the excellent pioneering work of Robert Schlaifer, *Probability and Statistics for Business Decisions* (New York: McGraw-Hill Book Co., Inc., 1959). The authors are also indebted to the following professors who contributed valuable ideas: William C. Dunkelberg, Karl A. Fox, Roy W. Jastram, Charles A. Holloway, James R. Miller, Donald G. Morrison, and Howard Raiffa. Finally, we wish to acknowledge the generous support of the Stanford Graduate School of Business in providing both time and facilities for us to complete this task.

April 1973   WILLIAM A. SPURR
                                                   CHARLES P. BONINI

CONTENTS

---

* Indicates sections that contain more advanced or optional material.

---

* Indicates sections that contain more advanced or optional material.

* Indicates sections that contain more advanced or optional material.

1. STATISTICS IN BUSINESS AND ECONOMICS

STATISTICS in today's business and economics includes: (1) statistical data, (2) statistical analysis, and (3) decision making. One is valueless without the others. Numerical data and methods of analysis and decision making are becoming increasingly important in business management and in every field of economics.

But what are statistical data? Not all numbers are statistical; logarithms, for instance, are merely abstract numbers. Statistical data are concrete numbers which represent objects—their counts or measurement. Statistics deals with numbers not merely as such but as expressions of significant relationships. It is not enough to collect and present the data, therefore; they must be carefully analyzed and interpreted as well, in order to make the best possible decisions based on the data. As Lord Kelvin put it:

> When you can measure what you are speaking about and express it in numbers you know something about it; but when you cannot measure it, when you cannot express it in numbers, your knowledge is of a meager and unsatisfactory kind: it may be the beginning of knowledge, but you have scarcely, in your thoughts, advanced to the stage of *science,* whatever the matter may be.

STATISTICAL ANALYSIS AS A SCIENTIFIC METHOD

When masses of numerical information are to be analyzed, some means of summarization must be found which will reveal their major characteristics. Statistical analysis meets this need. Hence, in a broad sense, statistical analysis is a scientific method of studying quantitative data. It is a means of summarizing the essential features and relationships of the data and then generalizing from these observations to determine broad patterns of behavior or future tendencies. Statistical

1

analysis therefore is useful in any field of knowledge in which extensive numerical information is needed.

The social and biological sciences, in particular, require masses of facts in order to determine general behavior, because of the wide variation in individuals. In the physical sciences, on the other hand, precisely controlled laboratory experiments can be used instead, to a large extent. The physicist can estimate the speed of light by repeated trials, with a small error of measurement, whereas the market analyst who wishes to determine consumer preferences toward compact cars must deal with a sample of consumers who vary widely in their preferences. He must design a questionnaire, select an unbiased sample, and estimate the sampling error. Human and biological groups are more variable in behavior than are most physical phenomena, so their study requires a statistical approach even more than in the physical sciences. Statistical analysis is therefore the fundamental method of quantitative reasoning not only in business and economics but also in sociology, anthropology, psychology, education, medicine, public health, and biology.

Statistical theory is founded on the mathematics of probability, which provides the basis for determining not only general tendencies but also the reliability of each generalization. The whole process of reasoning from the specific to the general may be called *statistical inference,* as well as generalization or induction. The field of statistical analysis itself is also called statistical methods or merely statistics. The latter term is used here in the singular sense, as opposed to "statistics" in the plural sense, which refers only to the observed data themselves.[1] Applications of statistical analysis in a particular field may be known under other names connoting the idea of *measurement* or *research,* such as econometrics, biometrics, psychometric methods, or forest mensuration—also business research, economic research, or marketing research methods. Finally, statistics plays an important part in the newer fields of operations research, management science, decision science, and systems analysis.

The importance of the statistical approach to the solution of practical problems has gradually come to be realized during recent times. The progress in this direction is explained by several developments. Fundamentally, the tremendous growth of population, large-scale production, and trade that followed the Industrial Revolution has required the production and use of a vast volume of statistics in every sphere of social activity. Statistical knowledge has increased in quantity, quality, and

---

[1] Note that the word "data" is plural; the singular is "datum."

frequency. The expanding needs of government have accelerated this growth. As a result, fact finding has become an integral part of economic progress.

Increasing public interest in and demand for social statistics rests, then, on the basic premise that the problems of society, as well as of natural science and technology, can be solved by the increase and diffusion of this especially matter-of-fact type of matter-of-fact knowledge. The whole world now seems to hold that statistics can be useful in understanding, assessing, and controlling the operations of society.[2]

Statisticians, too, have discovered new analytical techniques which have increased the value of statistical methods of planning and control. In particular, with the advent of the electronic computer in recent years, the statistician has acquired a means of dealing quickly with vast quantities of data. Electronic computers can perform a wide variety of functions in data processing: they can sort the information as desired, convert it into a different form, store it for future use, transfer it to other locations in the system, perform all types of arithmetic computations, and print the final results in readable form. All of this is done at high speeds in a completely integrated operation, with no human intervention. The versatility and speed of electronic data processing systems are therefore revolutionizing large-scale data handling and decision making in modern business.

The applied statisticians have also helped to dispel the aura of mystery which formerly surrounded the subject. This has been accomplished through a shift in teaching emphasis toward the applied side and through the publishing of textbooks and reference books which stress the simplicity of statistical application and avoid perpetuating the impression that one must be a master of advanced mathematics in order to do statistical work.

THE ROLE OF STATISTICS IN DECISION MAKING

Statistical data are collected and analyzed not only for the purpose of adding to scientific knowledge in general but also for the purpose of helping the rational man to make decisions. One of the most important functions of the business executive, the government official, or the administrator in any field is to make decisions. The function of statistics is to help decide what data are needed and how the data shall be collected, tabulated, analyzed, and interpreted in such a way as to lead to the best possible decision. Unfortunately, the complete facts are not

---

[2] Solomon Fabricant, "Factors in the Accumulation of Social Statistics," *Journal of the American Statistical Association*, June 1952, p. 259.

usually available, so incomplete data, or samples, must be used. Statistics then provides methods that help the executive make the best decision on the basis of these incomplete facts. Hence, statistics has come to be defined as a group of methods for making wise decisions in the face of uncertainty.

Of course, statistical methods do not provide the only basis for decision making. There are many intangible factors—the business climate, prospective government action, technological developments, or personnel relationships, for example—which make management an intuitive art rather than a science. Nevertheless, statistics provides the primary factual basis for reaching good decisions. As an IBM ad puts it, "No one can take the ultimate weight of decision making off your shoulders. But the more you know about how things really are, the lighter the burden will be."

And again:

All branches of statistics . . . deal with the same basic problem, namely, the problem of decision making in the face of uncertainty. All decisions rules . . . must be evaluated by their consequences. These consequences are expressible in terms of risks, or more intrinsically, in terms of the probabilities of taking the various permissible actions which are induced by the experiment, decision rule, and the possible states of the system. In brief . . . not facts from figures but rather decisions from observations should become the main emphasis in elementary statistical observations.[3]

In order to learn the logic of decision making, it is necessary first to study the laws of probability that govern uncertain events. Then, faced with a business problem involving uncertainty, one can set up a "payoff table" (see Chapter 7), listing down the column the future events that may occur and the probability of each. If there are no hard facts on which to base the probabilities, the executive can use his own judgment to estimate the odds that a given event will occur, thus incorporating personal judgment into the beginning of the decision process. Then list across the top the various acts or decisions that may be taken. In the body of the table, list the profit that would occur for each combination of a given act and a resulting event. Profits can be expressed either in dollars or in "utility units" representing one's subjective values. Finally, for each act, multiply the profit for each event by its probability and add the results for all possible events to obtain the "expected" profit that would ensue, on the average, from that act. The best act is then the one with the highest expected profit.

---

[3] M. A. Girshick, in *Journal of the American Statistical Association,* September 1953, p. 646.

This decision model can be extended to a series of decisions by means of a "decision tree" (see Chapter 7). Furthermore, if a decision is of doubtful validity, one can determine whether it is preferable to wait and obtain additional information, and how much information is needed, before acting. The probability that an event will occur can also be revised in the light of new facts by means of Bayes' theorem (see Chapter 13). Hence the name "Bayesian statistics" for this whole new method of decision making. Various business problems can also be clarified by setting up other probability models (see Chapter 15) which can be solved by similar methods. Thus, Bayesian decision theory enables the executive to quantify his ideas, narrow the range of his uncertainties about the future, and thus improve the likelihood of his making a correct decision.

The role of the electronic computer is becoming increasingly important in the decision-making process. The computer can be programmed to make simple decisions itself (as in inventory control) or else perform extensive analyses to aid the executive in making a more complex decision. Statistical methods provide not only the data but also the techniques used by the computer in decision making.

STATISTICS IN BUSINESS

The employment of statistical methods in the solution of business problems belongs almost exclusively to the 20th century. At an earlier date, when practically all business enterprises were small, management was able to comprehend its problems in detail by personal contact. The increased size of concerns in the present period has required more planning and greater regimentation of operations. At the same time, management has found it impossible to maintain personal contact with its problems. The alternative is control through the interpretation of numerical information. This chain of circumstances has led to the development of statistical methods of investigation as a primary aid in the performance of management functions.

According to a study made by the Pacific Telephone and Telegraph Company:

Today, management at all levels is guided quite generally by facts obtained through analysis of records rather than upon knowledge obtained merely through personal observation and experience. . . . Through application of appropriate statistical methods, current performance may be measured, significant relationships may be studied, past experience may be analyzed and probable future trends appraised. . . .

The use of statistical methods and the performance of analytical work which is largely statistical in character—whether or not it happens to be carried on under the distinctive label of "statistics"—occupy a conspicuous place in the work of all departments of the company.

Statistical analysis is thus used as a basis for the control of many operations in a company and for planning or forecasting its activities. Through the aid of statistical reports the executive can gain a summary picture of current operations which improves his factual basis for making valid decisions affecting future operations.

The principal statistical activities of a typical large and progressive firm are as follows:

1. A central economic research or statistical department operates under the guidance of an "economist" or "chief statistician." This department analyzes general business trends and forecasts business activity, commodity prices, and other economic factors. It may coordinate the internal company statistics compiled by other departments and issue summary reports of operations to top executives. It also makes periodic comparisons of the company's performance with that of its competitors.

2. A marketing research staff makes surveys of consumer preferences and purchasing power and forecasts probable future trends in sales. It may prepare a detailed sales budget for the coming year, broken down by individual products and by months. Finally, it has the responsibility for setting salesmen's quotas by territories and products, based on past performance, income studies, and salesmen's estimates.

3. The production department maintains a "quality control" staff that minimizes defective output by means of statistical checks, as described in Chapter 10. It prepares forecasts of production based on sales forecasts and other criteria and checks actual production against these estimates. It also maintains an inventory control system and makes time and motion studies.

4. The controller's department combines statistical and accounting methods in making the overall budget for the coming year—including sales; material, labor, and other costs; and net profits and capital requirements. It may maintain a standard cost system for controlling costs and setting prices of products.

5. The personnel department makes statistical studies of wage rates, incentive systems, the cost of living, employment trends, labor turnover rates, accident rates, and results of employee selection procedures.

6. The investment department maintains security analysts who study individual stocks and bonds and the general outlook for the securities markets.

7. The credit department performs statistical analyses to determine

how much credit to extend to each potential customer. Characteristics of those customers who have paid and those who have defaulted in the past are used for selecting future credit risks.

8. The executive department may include an "operations research" staff. This group consists of specialists, such as statisticians, mathematicians, and physicists, who apply scientific methods to the study of complex operation throughout the organization.

Some of the men and women who perform these functions are professional statisticians, but most of them have developed their knowledge of statistical analysis as an adjunct to their major specialties. In all departments of a business, personnel are concerned with the collection, classification, and presentation of statistics, even if their work requires no analysis. The general executive, too, must know some statistics as well as the basic principles of accounting, finance, business law, marketing, production management, and industrial relations in handling the various aspects of his job. He cannot depend entirely on specialists for his knowledge.

STATISTICS IN ECONOMICS

Economists and other social scientists are more concerned with conditions in the economy as a whole than with those in an individual concern, but they depend on statistics just as the business analyst does. Indeed, many of the statistical problems in economics are similar to, or identical with, those in business. Economists today are no longer content to theorize in abstract terms, citing statistics only as needed to buttress their arguments. Instead, they utilize the excellent data now available to build a sound factual foundation for their reasoning. Some of the uses of statistics in economics are as follows:

1. Extensive statistical studies of business cycles, long-term growth, and seasonal fluctuations serve to expand our knowledge of economic instability and to modify older theories.

2. Measures of gross national product and input-output analysis have greatly advanced overall economic knowledge and opened up entirely new fields of study.

3. Statistical surveys of prices are essential in studying the theories of prices, pricing policy, and price trends, as well as their relationships to the general problem of inflation.

4. Financial statistics are basic in the fields of money and banking, short-term credit, consumer finance, and public finance.

5. Operational studies of public utilities, including the transportation and communication industries, require both statistical and legal tools of

analysis. Such studies are necessary in connection with the federal and state regulation of these industries.

6. Analyses of population, land economics, and economic geography are basically statistical and geographic in their approach.

7. Studies of competition, oligopoly, and monopoly require statistical comparisons of market prices, costs, and profits of individual firms.

Statistical analysis is therefore carried on in every field of inductive economics—by individual professors, university economic research bureaus, chambers of commerce, trade associations, and such well-known research agencies as the National Bureau of Economic Research, the National Industrial Conference Board, the Twentieth Century Fund, and the Brookings Institution, to mention a few.

The most spectacular development of statistical analysis in economic research during recent years, however, has been in the federal government. As it has grown in size, the government has greatly expanded the scope of its statistical activities in every field of applied economics. Some agencies collect and publish statistics for their informational value to the public, while others compile data as a by-product of administrative or regulatory activities. Under the Full Employment Act of 1946 the President's Council of Economic Advisers and the congressional Joint Economic Committee employ many statistical indexes as guides in recommending to the President and Congress control measures designed to allay depression, inflation, or unemployment. Statistics has become as much a major tool of economic guidance and control by the federal government as it is an operational tool for individual concerns.

To conclude this introduction, we quote from M. J. Moroney's *Facts from Figures:*

If you are young, then I say: Learn something about statistics as soon as you can. Don't dismiss it through ignorance or because it calls for thought. . . . If you are older and already crowned with the laurels of success, see to it that those under your wing who look to you for advice are encouraged to look into this subject. In this way you will show that your arteries are not yet hardened, and you will be able to reap the benefits without doing overmuch work yourself. Whoever you are, if your work calls for the interpretation of data, you may be able to do without statistics, but you won't do so well.

Finally, the study of statistics helps in avoiding the common misuses of data. It is said: "There are three kinds of lies—lies, damn lies, and statistics," and conversely, "Figures don't lie, but liars figure." "Many people use statistics as a drunkard uses a street lamp—for support rather than for illumination." One can hardly pick up a newspaper without seeing some sensational headline based on scanty or doubtful data.

Many of the misuses that appear in statistical reports arise from the failure of the authors to maintain a critical attitude toward their work. Even facts and statements that are true in some sense can be quoted out of context or presented in such a way that they are bound to be misinterpreted by most readers. The scientific attitude toward evidence is skeptical rather than either cynical or uncritically enthusiastic. The investigator must seek the *truth* above all. It is not enough to avoid outright falsehood; he must be on the alert to detect possible distortion of truth.

THE ACCURACY OF ECONOMIC DATA

Not only are statistics often misused, but the basic data themselves vary widely in their accuracy, even though they may appear to be exact. Thus we read that: "The Census Bureau counted 22,580,289 blacks in the United States in its 1970 survey." "The thirteen regional Shippers Advisory Boards estimated yesterday that railroad freight loadings . . . in the current quarter would be 8,146,723 cars." "A State Industrial Commission study found that a bachelor girl can live a 'single, healthy, and moral' life on a minimum of $2,422.59 a year." (If she fails to receive that last $2.59, does her health or morals suffer, or both?) Certainly, none of these figures is correct to the last digit. Such detailed figures are misleading and suggest a degree of accuracy that does not exist by any means. In fact, most economic data should be rounded off to three or four significant figures for simplicity in tabulation, computation, and interpretation.[4] Additional figures are neither valid nor necessary in decision making (though they may be needed for accounting consistency).

On the other hand, many reported figures are subject to much wider errors than three or four significant figures would indicate. Therefore, it is important to estimate the size and type of errors inherent in the basic data. This may be done by studying the nature of the original data, the collection process, and the purpose for which the figures were gathered. For example, the *Survey of Current Business* reported that

---

[4] The following rules are recommended for rounding numbers: (*a*) When a number greater than five is dropped, increase the preceding digit by one. (*b*) When a number less than five is dropped, leave the preceding digit unchanged. (*c*) When the exact number five is dropped, increase the preceding digit by one if it is an odd number but leave it unchanged if it is an even number. That is, the rounded number is always even. This rule prevents cumulative errors in addition.

the value of new construction put in place in October 1972 amounted to $11,298 million. This might appear to be an exact figure, but actually it represents estimates by more than a dozen collection agencies derived from hundreds of different sources of varying reliability. Construction takes place on widely scattered sites and is carried on by tens of thousands of small contractors and by persons doing their own building and repair work, so that the above figure may be considerably in error. In order to understand the nature and limitations of basic statistics, therefore, one should study the text and footnotes accompanying a report, check other sources, and write the original collection agency, if necessary, for a description of its methods.

Sometimes the errors in data are estimated by the collection agency itself. For example, in "Income in 1970 of Families . . . ," the Bureau of the Census says: "Since the estimates in this report are based on a sample, they . . . are subject to errors of response and nonreporting and to sampling variability."[5] This is followed by a discussion of the errors and a table of the "Standard Error of Estimated Percentage of Families" (explained in Chapter 9) as a measure of sampling variability.

The U.S. Bureau of Labor Statistics, too, warns that unemployment figures for small subgroups of the population for one month are unreliable. Yet when it reported that Negro unemployment had risen from 8.4 percent in June 1965 to 9.1 percent in July, at the time of the Watts riots in Los Angeles, a number of writers cited these figures to prove that the expansion of the economy had left the Negro behind. Later though, the August figure was reported as 7.6 percent, and subsequent months were even lower. The July figure was a statistical blooper.

It is an excellent rule for the business analyst, therefore, to estimate the error in any figures he prepares or uses, so that he may avoid being misled by unreliable data.

Significant Figures in Computation

Two rules should be observed in performing basic calculations with approximate numbers:

1. In addition or subtraction, the result should contain no more decimal places than the least accurate of the numbers themselves. Thus, the *World Almanac* reported the area of Europe as 3,769,107 square miles, and that of Asia as 17,300,000 square miles (i.e., estimated to

---

[5] *Current Population Reports,* Series P-60, No. 80, October 4, 1971, p. 11.

the nearest 100,000). The total for Eurasia should then be stated as 21,100,000, not 21,069,107, square miles.

When applied to subtraction, however, this rule reveals a pitfall: A relatively small error in two large figures may produce a large percentage error in the difference. To illustrate, the number of unemployed persons in the nation is sometimes estimated by subtracting the number employed from the total labor force of those available for jobs. Suppose employment and labor force are each subject to an error of one million, or about 1 percent, in either direction. Then the resulting estimate of unemployment may be off two million, or 100 percent, as shown below.

Estimates of	Millions of Persons	Possible Error
Labor force.............	90 ± 1	1.1%
Employment............	88 ± 1	1.1%
Unemployment..........	2 ± 2	100%

This simple arithmetic accounts for the wide errors that frequently occur in estimates of unemployment, the federal deficit, personal savings, net profits of corporations, and similar values obtained by subtraction.

2. In multiplication and division (as well as in squares and square roots), the result has no more significant figures than the *least* number of significant figures in the numbers themselves. For example, suppose that in November the controller of the Apex Company estimates the calendar year's net profits at $2,736,000, based on indicated sales of $34,200,000 and an estimated net-profits-to-sales ratio of 8 percent. Then only one figure in the net profits estimate is really significant, since the 8 percent estimate means somewhere in the 7½ to 8½ percent range, and these end values multiplied by sales gives a profit range between $2,565,000 and $2,907,000.

In more extended calculations, however, the figures should not be rounded off until the final result is stated. This is to avoid cumulating the errors of rounding in subsequent operations of multiplication or subtraction.

SAMPLE SURVEYS

Original data may be collected from a complete *population* or from a *sample* selected from that population. The term population (or universe) here refers either to human populations (e.g., consumers, voters, college students), or to objects, such as manufactured products being tested for defective items.

Examples of a complete enumeration or *census* include the U.S. censuses of manufacturing and housing and the statistics of income and gasoline consumption, which are by-products of the tax-collecting function of the government. A poll of all employees in a plant is also a population census.

In contrast to these complete censuses are the great majority of surveys which depend upon obtaining a sample which will be typical of the whole population. For example, the Bureau of the Census has estimated the number of cars and other durable goods that American consumers plan to buy during the coming year from a sample of only 17,000 households out of the 53 million households in the country— only 1/30 of 1 percent of the total.[6] Similarly, the U.S. Department of Agriculture uses a sample of two quarts of grain in a carload (57,600 quarts) to determine the grade of the grain, and the U.S. Bureau of Labor Statistics Consumer Price Index is based on prices of a few hundred commodities and services obtained from a relatively small number of stores and other respondents.

There are three basic reasons for the widespread use of sampling:

1. Sampling usually saves a great deal of time and money. Often, when the cost of a complete census would be prohibitive, the necessary information can be obtained from a sample. The results of a survey need only be accurate enough to provide an adequate basis for decision making. Beyond a certain point, the increase in information from additional data is not worth the increase in cost.

2. In many cases, a complete census is impossible, as, for example, in making a quick check of consumer preferences for an entirely new product, or in the destructive testing required to determine the breaking strength of steel rods, or in measuring the effectiveness of a new antibiotic.

3. Finally, sampling may actually yield more accurate results than a complete survey. A small group of interviewers can be selected and trained more rigorously to reduce the biases in a survey than a very large staff. Similarly, in testing materials, a few careful measurements may be preferable to a larger number of crude measurements. Improvements in sampling techniques too, have led to many advances in modern survey methods.

Surveys can be conducted either by personal interviews or through questionnaires sent by mail. If personal (or telephone) interviewers

---

[6] *Federal Reserve Bulletin*, September 1960, pp. 977–1003.

are used, they can canvass the entire group to be sampled; they can explain questions carefully and evaluate the replies, thereby securing more reliable results than is possible by mail questionnaires. On the other hand, mail questionnaires are generally more economical, particularly if a wide area must be covered; so they are ordinarily used if the results can be made reliable. Sometimes the two methods may be combined, by first sending out a questionnaire and then making personal calls to selected nonrespondents. In any case, a definite sampling plan must be followed to assure that the responses are typical of the whole population (as described in Chapter 12).

SUMMARY

Statistical analysis is a scientific method of interpreting quantitative data. It is used to draw general inferences from the behavior of variable data. Statistical methods have become more important in recent times because of the growth of large-scale production and trade, the increasing scope of government, and improvements in statistical techniques themselves.

Statistical analysis is used in all branches of larger business organizations as a tool of planning and control. The principal statistical activities in business include general business analysis, marketing research, production control, budgeting, personnel and investment studies, credit analysis, and operations research.

Statistical analysis is also widely used in economics and social science generally, particularly in the study of economic fluctuations, social accounting, prices, finance, public utilities, regional analyses, and related subjects. The growth of government activities, too, has required more and better statistics for central planning and administrative purposes.

The basic steps in statistical analysis include (1) collecting the data from available sources or sample surveys, (2) analyzing and interpreting the figures by means of statistical techniques, and (3) using the results in making decisions, with the aid of probabilities and economic costs or profits.

The true meaning of facts is easily distorted. The statistical investigator therefore must be on guard to avoid misrepresenting the facts and to detect misuses of statistics by others. A critical attitude is essential.

The accuracy of figures must always be considered. Economic data are seldom accurate to more than three or four significant figures, so longer numbers should ordinarily be rounded off. The accuracy of any figure can be estimated by studying the method of collection.

The number of significant figures in computations is governed by the minimum number of significant figures in the data being processed. In subtraction, however, small errors in the original figures may produce a much larger error in the difference.

If the necessary figures cannot be found in published sources or in the internal records of a business, a sample survey must be made. Such a survey need not be a complete census but can be restricted to a limited group if the respondents represent a typical cross section of the entire population under study.

PROBLEMS

1. *a*) Explain the meaning of the term "statistics" when used in the singular sense as opposed to its use in the plural sense.
   b) Cite the application of statistical methods in some area or topic with which you are familiar.
   c) Name three other areas of quantitative methods that are closely related to statistics in your college, university, or other organization.

2. Describe the principal statistical activities of a typical large and progressive firm, citing any specific cases known to you.

3. Locate in the library and give the names of three major statistical journals, together with the associations that publish them, and briefly describe the type of subject matter contained therein.

4. Visit an economic research agency or one of the eight types of statistical departments in a business organization mentioned in the text, and hand in a two- to three-page outline of its statistical activities.

5. Outline one of the principal uses of statistics in economics with which you have had experience.

6. Hand in a clipping or photocopy from a newspaper or magazine illustrating a significant use of statistical analysis in business, economics, or other social science.
   a) What steps in analysis are illustrated: Collection of data from available sources or from original surveys? Analysis and interpretation of data?
   b) What inference or conclusion can you draw from this report?

7. Find the value of a wheat crop estimated at 3,500 bushels at a probable price of \$2.16⅞ per bushel. Express the result to the correct number of significant figures.

8. For the year ended January 31, 1972, Sears, Roebuck and Co. reported income before income taxes of \$949,965,971, less provision for these taxes of \$399,100,000, equaling net income of \$550,865,971 or \$3.56 per share

of stock. Express to the correct number of significant figures: (*a*) the net income and (*b*) the estimated number of shares outstanding.

9. At the beginning of 1972, the controller of X Company prepared a budget for the year which included the following estimates:

Sales........................	$50,000,000
Cost of sales....................	47,000,000
Net profits....................	$ 3,000,000

He believed that the error in his sales and cost-of-sales estimates would not exceed $1,500,000 in each case. Based on these statements:

a) What is the percentage of possible error in the net profits estimate?

b) If his sales estimate proves to be too high by $1,500,000 and his cost-of-sales estimate too low by that amount, what would net profits be?

10. A credit manager of a department store wished to estimate the number of credit transactions in the past month. There were exactly 2,842 credit accounts. By taking a random sample of accounts, it was estimated that there were an average of 2.4 transactions per account. The total number of transactions was, therefore, estimated at $2.4 \times 2,842 = 6,821$.

a) How many digits of this estimate are significant?

b) Bearing in mind that the estimate of 2.4 transactions per account, if accurate to one decimal place, might represent an exact value anywhere between 2.35 and 2.45, give the possible range in the total number of transactions.

11. State in each of the following examples of data collection whether a complete census or a sample should be taken. Give reasons for the answer in each case.

a) A retail dry goods association wished to study the distribution of operating expenses of its 61 members.

b) A marketing research agency wished to inquire from the owners of a certain make of refrigerator whether they would purchase the same make again.

c) A corporation president wanted information concerning how many of its 15,400 employees were homeowners, the value of their homes, the amount of mortgage, the interest rate paid, and the monthly payment on the mortgage.

12. The U.S. Fish and Wildlife Service engaged the firm of Crossley, S-D Surveys, Inc., to conduct a national survey of hunters and sport fishermen, to obtain data on the numbers of persons who hunt and fish, and the number of days and amount of money spent on such activities for a recent year. Assume that you are employed by Crossley to direct this survey.

a) In what types of business might the results of this survey be valuable?

b) Would you take a census (complete enumeration) or a sample of hunters and sport fishermen? Why?

SELECTED READINGS

FERBER, ROBERT, and VERDOON, P. J. *Research Methods in Economics and Business*. New York: Macmillan, 1962.
    Provides a broad perspective on means of solving research problems.

GOLDE, ROGER A. *Thinking with Figures in Business*. Reading, Mass.: Addison-Wesley, 1966.
    A primer on "techniques for improving your number sense."

HUFF, DARRELL. *How to Lie with Statistics*. New York: W. W. Norton, 1954.
    An amusing compendium of statistical misuses.

KENDALL, M. G., and BUCKLAND, W. R. *A Dictionary of Statistical Terms*. 3d ed. New York: Hafner, 1971.
    A comprehensive glossary, in English, French, German, Italian, and Spanish.

MANSFIELD, E. (ed.). *Elementary Statistics for Economics and Business:— Selected Readings*. New York: W. W. Norton, 1970 (paperback).
    Nineteen articles illustrating applications of statistics in various fields.

MORGENSTERN, OSKAR. *On the Accuracy of Economic Observations*. 2d ed. Princeton, New Jersey: Princeton University Press, 1963.
    A penetrating analysis of the many *in*accuracies of economic statistics. A condensed version appears as "Qui Numerare Incipit Errare Incipit" in *Fortune*, October 1963.

RIGBY, PAUL H. *Conceptual Foundations of Business Research*. New York: John Wiley, 1965.
    Describes the functions of scientific business research as providing the techniques for problem solving and decision making, as well as developing new concepts, testing hypotheses, and building models.

ROBERTS, HARRY V. "The New Business Statistics," *Journal of Business of the University of Chicago*, January 1960, pp. 21–30.
    Outlines the development of the decision-theory orientation of statistics.

SIELAFF, THEODORE J. *Statistics in Action*. San Jose, California: Lansford Press, 1963.
    Twenty-five articles by different authors show how the tools of statistics are used in dealing with business and economic problems.

WALLIS, W. A. and ROBERTS, H. V. *The Nature of Statistics*. New York: The Free Press, 1962 (paperback).
    A good introduction to the basic ideas of statistics, including uses and misuses, measurement, and applications.

2. ANALYSIS OF DATA: RATIOS AND FREQUENCY DISTRIBUTIONS

STATISTICAL METHODS deal with the collection, analysis, and interpretation of data. Sample survey methods of collecting data are discussed in Chapters 1, 9, and 12. Other data are available in published form or company records. Beginning in this chapter, we take up the principal methods of analyzing and interpreting data.

METHODS OF CLASSIFYING DATA

The first step in analysis is to classify the necessary figures into a table that will provide meaningful comparisons. Such data may be classified in three ways: by qualitative characteristics, by size, and by time. These classifications are illustrated in Table 2–1, which compares unemployment rates by sex, age, and race for 1970, 1971, and 1972.

Classification based on *qualitative* differences is illustrated by the breakdowns by sex and race. The distinction is one of kind rather than of amount. Other qualitative classifications could be made by marital status or occupation. Geographical classifications are also qualitative. Thus, unemployment rates could be reported by states or by metropolitan areas.

The criteria used in classifying qualitative data are often called *attributes*. An attribute is a characteristic than can be divided into two or more categories, such as the yes or no responses on a questionnaire; "defective" or "good" in describing the quality of a product; or a classification of employees as executives, office workers, or factory workers. However, attributes usually refer to only two categories (e.g., employed and unemployed), and ratios are used to compare just two

17

Table 2–1

UNEMPLOYMENT RATES IN THE U.S., 1970–72

As Percentage of Labor Force

	1970 Ave.	1971 Ave.	1972 Nov.
All civilian workers................	4.9	5.9	5.2
Men, 20 years and over...........	3.5	4.4	3.6
Women, 20 years and over........	4.8	5.7	5.0
Both sexes, 16–19 years...........	15.3	16.9	15.4
White workers.....................	4.5	5.4	4.6
Negro and other races.............	8.2	9.9	9.8

Source: *Survey of Current Business*, December 1972.

categories, such as the ratio of unemployed to total labor force. Ratios are discussed in the next section.

Data classified by *size* or *time,* on the other hand, are called *variables.* Thus, a size classification might be the number of unemployed classified by age of workers, where age is the variable. Similarly, the unemployed could be classified by years of education or by number of weeks out of work. Variables classified by size may be grouped into frequency distributions, and averages and measures of dispersion may be computed to summarize their characteristics, as described in the latter part of this chapter and in Chapters 3 and 4.

The columns showing 1970, 1971, and 1972 in Table 2–1 represent a time classification or time series. Time series may be further divided into (*a*) measurements taken at different points of time, like population or prices, and (*b*) cumulative data that build up from zero in a given period, like monthly steel production or weekly retail sales. Methods specially designed for studying time series are presented in Chapters 19 and 20.

RATIOS

A ratio or proportion is an extremely useful and simple device for comparing two attributes or qualitative characteristics. Thus, it is usually more significant to report the unemployment *rate,* as in Table 2–1, than simply to give the total number of unemployed. Ratios are also useful in comparing groups of variables classified by size, such as citing the percentage of factory workers who earn less than $5 an hour, even though the basic data are classified by size of hourly earnings.

Ratios are computed from a numerator and a base (denominator), which are usually expressed in the same units (e.g., dollars in a company's ratio of net profits to sales). Various terms are used for ratios in which the units differ. Thus, the birth *rate* is the number of births per thousand population; *density* of population is the number of persons in a region divided by its area; *per capita* national debt is the ratio of total debt to the number of persons in the country.

Selecting the Numerator and Base

For a ratio to be meaningful, the numerator and base may have to be adjusted or refined so as to exclude any extraneous factors that would obscure the direct relationship between them. For example, take the trend of deaths in automobile accidents from 1950 to 1971. The number of deaths increased 57 percent, as shown in Table 2–2. These

Table 2–2

FATALITIES IN MOTOR VEHICLE ACCIDENTS, 1950 AND 1971

	1950	1971	Percent Change
1. Persons killed in motor accidents.........34,763		54,700	+57
2. Deaths per 100,000 population............	23.0	26.5	+15
3. Deaths per 10,000 motor vehicles.........	7.1	4.76	−33
4. Deaths per 100,000,000 vehicle-miles.......	7.6	4.68	−38

SOURCE: National Safety Council, *Accident Facts*, 1972, pp. 58–59.

figures suggest that the "automobile menace" is increasing. The increase may be due to the growth of population, however, so the number of deaths per 100,000 population has been computed, as shown in line 2. This ratio has increased by only 15 percent. However, accidents are related more directly to the number of motor vehicles, which has increased more rapidly than the population. The number of deaths per 10,000 motor vehicles, therefore, is shown in line 3. Now we see a 33 percent *decrease* in this refined ratio. Finally, traffic deaths are related still more specifically to the number of vehicle-miles driven, and the average car was driven more miles in 1971 than in 1950. The number of deaths per 100,000,000 vehicle-miles appears in line 4. The decrease is now 38 percent. The more refined ratio therefore shows a substantial gain in safety, when the increased number of cars and mileage driven are taken into account, whereas the actual fatalities and the crude per capita ratio (lines 1 and 2) indicate just the opposite conclusion.

Which Item to Choose as Base

The base or denominator of a statistical ratio is always a standard with which the numerator is being compared. The following rules may be useful in selecting the base:

1. In comparing a part and the whole, the whole is always the base. Example: net profits to sales ratio = net profits ÷ sales.

2. In time comparisons of like items, the prior event is taken as the base. Example: this year's sales as a percentage of last year's.

3. In comparing a cause and effect or an independent event with one at least partly dependent on it, the cause or the independent item is nearly always the base. Example: price-earnings ratio of a common stock = price ÷ earnings. (Exception: stock yield = dividend ÷ price.)

When either of two items is equally logical as a base, custom often determines the choice. Example: rate of inventory turnover = sales ÷ inventory.

The Number of Units in the Base. The base may be expressed as a single unit, 100 units, or some other multiple of 10, depending on which is customary or most effective. Thus, the national debt of $1,806 per capita in 1970 is expressed in terms of *one* denominator unit, or one person; an interest rate of 5 percent means $5 for every $100 deposited, whereas the death rate may be reported as 9.0 per *thousand.* Furthermore, as shown in Table 2–2, the National Safety Council reports motor vehicle deaths per 10,000 motor vehicles, per 100,000 population, and per 100 million vehicle-miles. The larger numbers are used as a base so that the ratio can be reported mainly as a whole number rather than as a decimal fraction.

Cautions in the Use of Ratios

Ratios and percentages seem quite simple, but they are frequently miscalculated through using the wrong base, failing to subtract 100 percent in figuring increases, or misunderstanding the nature of the comparison. A textbook in office management states that "window envelopes cost around $1.00 less than regular envelopes, or $3.25, which represents a saving of 76.5 percent." This should be 23.5 percent—or 24 percent to avoid spurious accuracy. A newspaper article headlined "Bike sales up 300% over 1960" cites 1971 sales of 850,000 bicycles in California versus 300,000 in 1960. True, 850,000 is nearly 300 percent of 300,000, but the *increase* should have been reported as "nearly 200%."

The Ways and Means Committee of the House of Representatives in

1951 considered raising personal income tax rates 3 percentage points "across the board." The tax scale, then graduated from 20 percent up to 91 percent, would be made to run from 23 to 94 percent. Some critics attacked this as a "soak-the-poor measure," since a 3-point increase on the poor man's 20 percent represented a 15 percent jump, while 3 points on the rich man's 91 percent was a mere nudge of 3.3 percent. But other critics claimed that this was a "soak-the-rich" measure, since the poor man's take-home pay would be reduced from 80 to 77 cents on his dollar of income, or only 3¾ percent, while the rich man's take-home pay would be cut from 9 to 6 cents, or 33⅓ percent! The committee compromised by increasing taxes 12½ percent across the board. This expedient increased the minimum rate from 20 to 22½ percent, reasonably enough, but unfortunately boosted the maximum rate from 91 to 102.4 percent! (It was subsequently cut to 94½ percent.)[1] This controversy illustrates the importance of the careful use of percentages.

A further error in the use of percentages should be noted. The difference between two percentages, often called "percentage points," must not be interpreted as the percentage of change. Thus, it is incorrectly stated that "the production index for electric utilities increased from 130 in 1970 to 153 in November 1972, a 23 percent gain." The indexes are both percentages of the same 1967 base period, but the percentage change is the increase of 23 percentage points divided by the base level of 130, or only 18 percent.

Finally, ratios should not be used if the original number used as a base is very small. The report that 25 percent of the bank tellers in a town have been indicted for embezzlement would be misleading if there were only four tellers to begin with. Similarly, a 1,000 percent increase in profits over last year would hardly be significant if last year's profits totaled $1.

RATIO CHARTS

A ratio chart is one that shows ratios in their true proportion; that is, equal ratios or percentages cover equal spaces on the vertical scale. This is illustrated in Chart 2–1. The ratio scale is preferable to the arithmetic scale in comparing the *relative* changes in two curves, especially in time series.

Although arithmetic scales are satisfactory for showing absolute changes in the data, they fail to reveal clearly what is often of more importance—the relative or percentage changes. For example, it is

---

[1] National City Bank of New York, *Monthly Letter on Economic Conditions,* June 1951, pp. 66–67.

ordinarily not so significant that a company's sales increased more in dollars over a given period than those of its smaller competitor as that its percentage of increase was greater.

The ratio chart is also called a "semilogarithmic" or "semilog" chart because the natural numbers are plotted on the vertical scale at distances from the "1" bottom line proportional to their logarithms, while the horizontal axis shows time on the usual arithmetic scale. Thus, in Chart 2–1, the scale number "1" is at the bottom (since $\log 1 = 0$) and the top number 10 is one unit above (since $\log 10 = 1$), the unit

Chart 2–1

ONE-CYCLE RATIO OR SEMILOGARITHMIC CHART
WITH PERCENT MEASURING SCALE

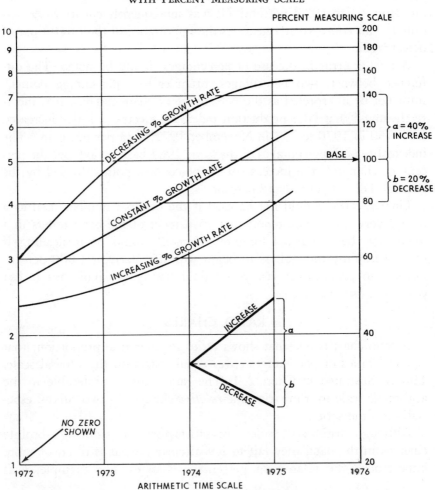

being 5 inches in this diagram. The "2" is marked .301 of the way up the graph (since log 2 = .301 in Appendix B), or 1.5 inches up; "3" is marked .477 of the way up; and so on. However, since only natural values are plotted, it is no more necessary to know logarithms in using a ratio chart than in using a slide rule. In fact, the ratio scale on a chart is the same as that on a slide rule.

A ratio chart should be so labeled, but if not, it may be identified in a publication by the fact that the vertical scale numbers get closer together as the scale rises. In particular, the vertical distances between 1 and 2, 3 and 6, and 5 and 10 are all the same, since these distances all represent the same ratio of 1 to 2 irrespective of their position on the chart.

In the ratio chart (as the term is generally used) only one scale is logarithmic. The double logarithmic chart, in which both scales are logarithmic, will be discussed in Chapter 16 in connection with regression analysis. Many types of logarithmic grids are available.

A log scale is said to have one *cycle* if the scale numbers extend only from 1 to 10; two cycles if the scale is divided into two equal parts covering the ranges 1 to 10 and 10 to 100, respectively; and so on. The scale can also be extended downward indefinitely to 0.1, 0.01, etc., but can never reach zero. Hence, the log scale cannot be used for a series that includes zero or negative values.

How to Plot

The choice between one-, two-, and three-cycle paper depends on the range of data. One-cycle paper is preferable, if the range permits, because this has the largest scale.

In order to plot data most easily, mark the "1" bottom line with one of the numbers 1, 2, 4, or 5, followed or preceded by any number of zeros, such as .01 million persons, 20 dollars, 4,000 tons, or 5 percent. If some other value were placed at the bottom, it would complicate plotting. Once the bottom value is selected—say $20—multiply this by the printed scale figures 1, 2, 3, . . . and mark them accordingly (20, 40, 60, . . .) until the top of the cycle is marked with a value ten times the bottom (200). This is a *must*. If the printed figures 1, 2, 3, were labeled 20, 30, 40, for example, the logarithmic proportions would be lost and the graph would be meaningless as a ratio chart.

Different scales can be used to compare series of disparate size or those expressed in different units. For example, the relative growth of a large and a small company, or of coal production in tons and oil in barrels, may be fairly gauged because the slopes of the curves register percentage changes, which are comparable even if the original units are not. Thus, the incompatible are made compatible. The choice of scale affects only the height of a curve above the bottom line, which is not significant; it does not affect the shape of the curve in any way.

Uses of the Ratio Chart

The slope of a line on a ratio chart indicates the percentage change between two points of time. A continuing line of the same slope or two parallel lines therefore represent the same relative movement. A given vertical distance corresponds to the same percentage difference anywhere on the chart. These characteristics give ratio charts the unique advantages described below.

Constant Rate of Growth as a Straight Line.   A series growing or declining by the same percentage each year, such as a sum of money at compound interest, or sales increasing 10 percent a year, appears on the ratio chart as a straight line. (This "logarithmic straight line" is also called an exponential curve or compound interest curve). If the series curves away from the straight line, it denotes a corresponding change in the rate of growth or the rate of decline, as shown in Chart 2–1. Many young industries expand at about a constant percentage rate each year until they mature, when the rate of growth tends to taper off, as in the top curve of the chart.

By watching a company's production curve on a ratio chart, therefore, the analyst can determine whether or not it is maintaining its past rate of gain. He can also project past trends in order to forecast future output, as described in Chapter 19.

Comparison between Two Curves.   The relative growth or decline of two or more curves can be seen at a glance by comparing their slopes on a ratio chart, irrespective of the size of the two series or the units in which they are measured. An arithmetic graph of two series on a single scale always emphasizes the growth of the larger one. Or, if two different scales are used to bring the curves together, the relationship is arbitrarily distorted. Even index numbers only afford easy comparison with one base level: if a different period is taken as a base, the relative changes in the indexes will differ. The ratio chart affords true relative comparisons between any two points on the grid, and yet absolute values can be read from the scale, unlike the case of index numbers.

Performing Calculations on a Ratio Chart.   Percentages or ratios may be read directly from a log scale in this way:

1. Mark a percentage measuring scale as shown in the right column

of Chart 2–1, or use a separate strip of graph paper. That is, mark the center line "100 percent" and the other percentages in proportion to the printed scale numbers.

2. Mark the *vertical* distance between any two points on the edge of a strip of paper (e.g., the increase *a* or decrease *b* between 1974 and 1975 on the lower part of the chart).

3. Lay off the increase upward, or the decrease downward, from the 100 percent base point of the measuring scale, and read the value of the second point as a percentage in terms of the first point as 100 percent. The percentage *change* is this figure minus 100. Thus, in Chart 2–1, the 1974–75 increase *a* is read off as 40 percent, while the decrease *b* is 20 percent.

Limitations of Ratio Charts

Ratio charts have certain limitations which restrict their use accordingly: (1) They do not give a visual idea of absolute magnitude as a distance above the base line, although these magnitudes can be read from the scale. (2) They are difficult for the layman to understand and so should not be used for simple illustrations where an arithmetic chart will do as well. (3) They cannot show zero or negative values. (4) Finally, they are sometimes mistakenly used to contract a wide range of absolute values into a small space. This is legitimate if relative movements are of interest, but if a picture of absolute changes is needed, an arithmetic scale should be used.

FREQUENCY DISTRIBUTIONS

Many types of data are classified according to size. Examples are rents paid for houses and wages of workers at a given time. In each case the original data are values of a *variable* (e.g., rent, which varies from house to house) which will be called *X*. These variables may be grouped by size into a *frequency distribution,* which lists only *class intervals* and the number or *frequency* (*f*) of values of *X* in each interval. A frequency distribution is a valuable device for summarizing unwieldly figures, so that a maximum of information can be presented with a minimum of detail.

Variables may represent either discrete or continuous data. Discrete data have distinct values, with no intermediate values. Thus, the number of children in a family can be two or three, but not 2.7. Continuous data can have any values over a range, such as the exact heights of men. However, continuous data are often treated as being discrete, as when heights are rounded to the nearest inch, and a man's height is reported at either 5 ft. 10 in. or 5 ft. 11 in. but not at any intervening value.

In order that the analysis of data may be meaningful, it is necessary that they be *homogeneous,* that is, sufficiently alike to be comparable for the purposes of the study. Thus, in Table 2–1, the more homogeneous subgroups, such as youths aged 16 to 19, reveal important differences in unemployment rates which are concealed in the total unemployment figures. The totals are *heterogeneous,* since they lump together male and female, different age groups, and different races.

The Array

Sometimes it is convenient to arrange the values of the variable in an array, as a preliminary step. An array is a listing of values arranged in order of size—either from smallest to largest or vice versa. Table 2–3, for example, shows the overall dimensions of 63 gears, taken from a quality control measurement. The raw data in panel A are too awk-

Table 2–3

RAW DATA AND ARRAY

DIMENSIONS OF 63 GEARS AS ILLUSTRATED, INCHES

SOURCE: Marchant Calculators, Inc., *Statistical Quality Control.*

ward to handle directly, so they have been combined in an *array* in panel B by means of a tally sheet.

This array not only shows the data in simpler form than in panel A but reveals at a glance certain salient characteristics—the range and the most common size (.4250 in.). Also, in this simple case where no further grouping of values is needed, the array is already in the form of a usable frequency distribution, with class intervals .0005 in. wide— the number of marks opposite each dimension indicating the frequency with which this measurement occurred.

Grouping Data into Classes

Most types of data, however, have so many different values that an array is excessively detailed. The figures must then be grouped into a manageable number of classes. The methods for doing this are illustrated below with data adapted from a survey of straight-time hourly earnings of 214 apprentice machine tool operators in machinery manufacturing plants in an eastern city. Studies of this type are needed for industrial relations analysis, labor-union wage negotiations, and many aspects of welfare economics.

Table 2–4 presents an array of these hourly earnings in the form of a tally sheet, with the number of operators at each earnings level noted in the column headed f (for frequency). This table still has too many separate values for easy analysis, so the data are grouped as shown in Table 2–5. For this purpose, class intervals 10 cents wide were chosen, beginning with the interval "$2.25 and under $2.35."

The reasons for this choice of intervals are as follows: The number of classes (eight) is large enough to show the general distribution of earnings and small enough to simplify analysis. The class limits ($2.25, $2.35, . . .) are multiples of 5 cents, which are simple round numbers, while the midpoints ($2.30, $2.40, . . .) are at the popular rates at multiples of 10 cents. This permits easy interpretation and minimized errors of grouping. Finally, the intervals ($2.25 and under $2.35, . . .) are defined clearly and unambiguously. These principles are discussed below.

Number and Width of Class Intervals

In general, it is advisable to divide the data into from 6 to 15 classes.[2] If the number of classes is too small, important characteristics of the

---

[2] Some writers, however, suggest from 6 to 15 classes for presentation but from 15 to 25 classes for accuracy in computations.

Table 2–4

MORE DETAILED ARRAY

STRAIGHT-TIME HOURLY EARNINGS OF 214 APPRENTICE MACHINE TOOL OPERATORS
IN MACHINERY MANUFACTURING PLANTS IN AN EASTERN CITY
(In Dollars per Hour)

EARN-INGS	OPERATORS		EARN-INGS	OPERATORS		EARN-INGS	OPERATORS	
	Tally	f		Tally	f		Tally	f
2.30	|	1	2.55	ⅢⅡ	5	2.80	ⅢⅡ	5
2.31			2.56	ⅢⅡ |	6	2.81	|	1
2.32	|	1	2.57	ⅢⅡ	3	2.82		
2.33			2.58	ⅢⅡ	4	2.83		
2.34			2.59	ⅢⅡ	5	2.84		
2.35	‖	2	2.60	ⅢⅡ ⅢⅡ |	11	2.85	|	1
2.36	‖	2	2.61	ⅢⅡ	4	2.86	|	1
2.37			2.62	‖‖	3	2.87	|	1
2.38	‖‖	3	2.63	ⅢⅡ ⅢⅡ ⅢⅡ ⅢⅡ	20	2.88		
2.39	‖	2	2.64	‖	2	2.89		
2.40	ⅢⅡ ‖	7	2.65	ⅢⅡ ‖‖	9	2.90		
2.41	|	1	2.66	‖	2	2.91		
2.42			2.67	‖‖	3	2.92		
2.43	|	1	2.68	‖	2	2.93		
2.44	ⅢⅡ	5	2.69	‖‖	3	2.94		
2.45	‖‖	4	2.70	ⅢⅡ ⅢⅡ ‖‖	13	2.95		
2.46	‖‖	3	2.71	‖‖	3	2.96		
2.47	ⅢⅡ	5	2.72	ⅢⅡ |	6	2.97	|	1
2.48	‖‖	3	2.73	|	1	2.98	|	1
2.49	‖	2	2.74	‖‖	3	2.99		
2.50	ⅢⅡ ⅢⅡ ‖	12	2.75	ⅢⅡ ⅢⅡ |	11	3.00		
2.51	ⅢⅡ	5	2.76	ⅢⅡ	5	3.01		
2.52	|	1	2.77	|	1	3.02	|	1
2.53	ⅢⅡ ⅢⅡ ‖	12	2.78			3.03		
2.54	‖	2	2.79	‖	2	3.04	|	1

data may be concealed. The use of too many classes may show unnecessary detail, as well as a confusing zigzag of frequencies, and
blanks in some classes. (This is the case in Table 2–4, which lists 75
one-cent intervals.)

Within these limits, the exact number of classes is determined by
the width of the interval. This interval is usually selected as a convenient round number located so that clusters of data occur at its
midpoints, as described in the next section. Thus, in Table 2–4, earnings tend to cluster at multiples of 10 cents, so we have used $2.30,
$2.40, and so on as class midpoints, and the 10-cent interval gives us
eight classes. (There are also minor clusters at odd multiples of five
cents, however, so we could have used five-cent intervals centered at
those points; but it is doubtful if the slight increase in accuracy justifies
the extra detail.)

Choice of Class Limits and Midpoints

The midpoint of a class interval (the point halfway between its limits) represents the average value of all the items in the class. This usage involves *errors of grouping,* which are similar to errors of rounding off numbers in general. To minimize errors of grouping, locate the midpoints of the intervals at any points of concentration around which values tend to cluster. Otherwise, any averages or other computed measures would be biased. Thus, if monthly salaries paid college graduates were set by a company at multiples of $50—say $900, $950, etc.—and they were reported in a frequency distribution with classes such as "$900 and under $950," so that the midpoint of

Table 2–5

FREQUENCY DISTRIBUTION

HOURLY EARNINGS OF 214 APPRENTICE TOOL OPERATORS

Hourly Earnings	Midpoint	Number of Operators *f*	Percent of Operators
$2.25 and under $2.35	$2.30	2	1
$2.35 and under $2.45	2.40	23	11
$2.45 and under $2.55	2.50	49	23
$2.55 and under $2.65	2.60	63	29
$2.65 and under $2.75	2.70	45	21
$2.75 and under $2.85	2.80	25	12
$2.85 and under $2.95	2.90	3	1
$2.95 and under $3.05	3.00	4	2
Total		214	100

$925 was used to represent salaries that were all actually $900, a computed average would overstate the true value by $25.

The class limits should be stated precisely to avoid ambiguity. For example, in Table 2–5, it would be clear to say either "$2.25 and under $2.35" or "$2.25–$2.34" (for discrete data expressed to the nearest cent), but not "$2.25–$2.35," since then the value $2.35 could fall into either of the first two classes.

Uniformity in Width of Class Intervals

All intervals used in a frequency distribution should have the same width, if possible, because frequencies are easier to interpret and computations are facilitated. Intervals of varying width are confusing and awkward to use in analysis. Unequal intervals are often necessary, however, in order to cover a wide range of data, as in the following grouping of annual incomes:

Under $2,000	$ 6,000–$ 9,999
$2,000–$3,999	$10,000–$19,999
$4,000–$5,999	$20,000 and over

In such cases, it is also rather common to have *open-end classes* at the extremes, with the extreme limits of the end classes not shown, e.g., "under $2,000" and "$20,000 and over." The sum or average of the values in such open-end classes should be indicated, if possible, to aid in computing averages and other summary measures.

Relative Frequency Distributions

It is often desirable to show each frequency as a relative part or percentage of the total, as shown in the last column of Table 2–5.

The use of percentages has four advantages: (1) It permits comparisons of the individual frequencies with each other and with the total on a common 100 percent base. (2) It facilitates comparisons between two frequency distributions having different numbers of items, provided they have identical class limits, as in Chart 2–4. (3) It permits one to make inferences from sample data regarding the population, provided the sample is carefully selected. For example, it might be inferred from Table 2–5 that about 29 percent of *all* Class A machine tool operators in the area earn from $2.55 to $2.65 an hour. (4) It provides a basis for estimating probabilities. Thus, if we take an operator at random, we can say that the probability is .29 that he will earn from $2.55 to $2.65 an hour. The use of relative frequencies to estimate probabilities is described in Chapter 5.

CHARTS OF FREQUENCY DISTRIBUTIONS

Frequency charts provide a quick and simple method of summarizing and presenting facts. To construct such a chart, measure the variable X along the horizontal scale and label either the class limits or midpoints. Then, at the midpoint, plot the frequency of the class on the vertical scale (assuming classes of equal width). The vertical scale must always begin at zero, but the horizontal scale need only include the range of X values and one extra interval at each end. The two most common frequency diagrams of sample data are the histogram—a vertical bar chart—and the frequency polygon—a line chart. The smooth frequency curve, used to describe the distribution of values in a population, is discussed later in this chapter.

The Histogram

A histogram is a set of vertical bars whose *areas* are proportional to the frequencies represented. When the class intervals, or bar widths, are equal, the *height* alone can be used to represent the frequency in that class. The height of the bar thus shows frequency *per unit width*.

In Chart 2–2, for example, the histogram represents the earnings of the 214 machine tool operators listed in Table 2–5. This chart shows at a glance how the earnings are distributed.

The tallest bar represents the *modal class,* which contains the greatest concentration of earnings figures. On either side, the bars taper off in height, showing that the farther the earnings are from the modal class, the fewer are the number of workers. Many types of economic data have a similar distribution.

If there are two separate modal classes or peaks in a histogram, the data may prove to be heterogeneous (e.g., foremen might have been included with operators). In this case, the figures should be separated into homogeneous groups before being analyzed.

The height of each bar of a histogram is equal to the frequency of the

Chart 2–2

HISTOGRAM

HOURLY EARNINGS OF 214 APPRENTICE MACHINE TOOL OPERATORS

NUMBER OF OPERATORS (f)

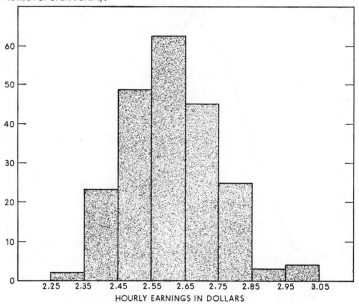

HOURLY EARNINGS IN DOLLARS

class when intervals are equal in width; but when the width varies, frequency is represented only by area rather than by height. Thus, in Chart 2–2, if the seven operators in the two classes $2.85 to $3.05 were combined into a single class, the height of this bar should be plotted as $7 \div 2 = 3\frac{1}{2}$, so that it would have the same area as the two right-hand bars shown. If the combined bar were drawn with a height of 7, it would double the apparent number of these highly paid workers.

The Frequency Polygon

The frequency polygon is a line chart plotted on the same scales as a histogram. To draw a polygon, plot each frequency on the vertical scale over the midpoint of the interval on the X axis (assuming classes of equal width). Then connect these points with straight lines, and extend them to an interval of zero frequency at each end.

Chart 2–3 shows the frequency polygon in comparison with the equivalent histogram (which is lightly blocked in as background). The frequency polygon (including the base line) encloses an area equal to

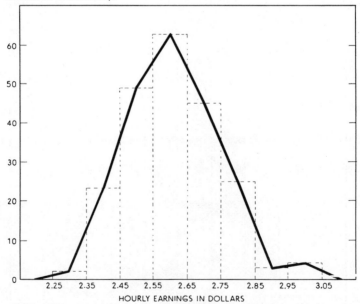

Chart 2–3

FREQUENCY POLYGON

HOURLY EARNINGS OF 214 APPRENTICE MACHINE TOOL OPERATORS

NUMBER OF OPERATORS (f)

HOURLY EARNINGS IN DOLLARS

that of the histogram,[3] although the areas in individual classes are shifted slightly from the classes to which the frequencies belong.

Histograms versus Frequency Polygons

The histogram has the following advantages over the frequency polygon: (1) the area within each bar represents the exact number of values in a class, (2) the individual classes stand out more clearly than in a frequency polygon, and (3) separated bars may be used to emphasize gaps in a discrete distribution.

Frequency polygons have these advantages: (1) they are simpler than bar charts, having fewer lines, (2) they resemble the smooth curve which describes a population of continuous data better than does the histogram, and (3) they are simpler for comparing two frequency diagrams.

Histograms are usually preferable when classes are few, frequency polygons when classes are numerous. Either type of chart, however, can ordinarily be used.

Comparison of Two Frequency Distributions

Two frequency distributions can best be compared by plotting their relative frequencies as polygons on the same scales. To illustrate, Chart 2–4 compares the earnings of our Class A apprentice machine tool operators with those of Class B apprentices. The frequencies are expressed as percentages of their respective totals. Comparison of the two curves shows that (1) Class A operators earn more than Class B operators for the most part; (2) the most common earnings rates are in the $2.25 to $2.35 bracket for the Class B workers, as compared with $2.55 to $2.65 for the Class A men; and (3) there is a much greater concentration of Class B earnings than Class A earnings in these modal classes, as shown by the relative heights of the two curves.

CUMULATIVE FREQUENCY DISTRIBUTIONS

Sometimes one needs to know the answers to questions such as "How many operators earn less than $2.75 an hour?" If so, it is convenient to add the frequencies cumulatively, beginning at either end, and list the resulting subtotals in a cumulative frequency distribution, as in Table 2–6, columns 3 and 4.

---

[3] This follows from the fact that each pair of adjoining triangles formed by the top lines of the polygon and the histogram in Chart 2–3 are equal in area. Similar areas are not equal, however, when intervals are of unequal width.

Chart 2–4

COMPARISON OF FREQUENCY DISTRIBUTIONS

HOURLY EARNINGS OF CLASS A AND CLASS B APPRENTICE
MACHINE TOOL OPERATORS

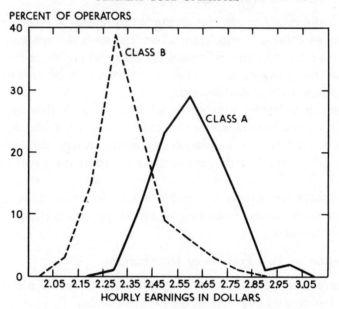

Table 2–6

CUMULATIVE FREQUENCY DISTRIBUTIONS

HOURLY EARNINGS OF 214 APPRENTICE MACHINE TOOL OPERATORS

(1) Hourly Earnings	(2) Number in Class with Lower Limit Shown	(3) Number Earning Less	(4) Number Earning as Much or More
$2.25	2	0	214
2.35	23	2	212
2.45	49	25	189
2.55	63	74	140
2.65	45	137	77
2.75	25	182	32
2.85	3	207	7
2.95	4	210	4
3.05	0	214	0
Total	214		

SOURCE: Table 2–5.

The table shows at a glance how many operators earn *less than* any amount listed, or that amount *or more*. Thus, 182 operators earn less than $2.75, while 32 earn $2.75 or more. Columns 3 and 4 could also be expressed as percentages of the total number of operators (214) for better comparability with other groups or for making inferences about a larger population.

The graph of a cumulative frequency distribution is called a cumulative frequency curve or an *ogive* (pronounced ō'jīve), because its shape resembles that of an ogive or rib of a Gothic arch. The data in Table 2–6 are graphed in Chart 2–5. The percentage scale at the right is

Chart 2–5

CUMULATIVE FREQUENCY CURVES

HOURLY EARNINGS OF 214 APPRENTICE MACHINE TOOL OPERATORS

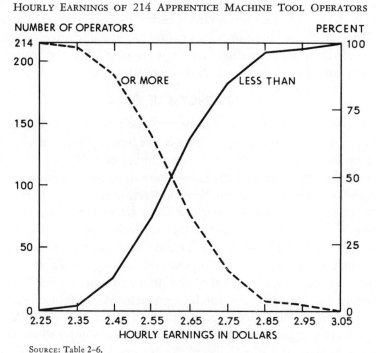

SOURCE: Table 2–6.

made so that 100 percent corresponds to 214 operators on the left-hand scale. The ogives then show graphically what number or percentage of the operators earn less than the amounts listed in Table 2–6, and what percentage earn those amounts or more.

In addition, the ogives permit easy interpolation for finding values between the plotted points. For example, the upward ogive shows that 25 percent, or about 53 operators, earn less than $2.51, while the downward ogive shows that 25 percent earn $2.70 or more. The intersection of the two curves at the 50 percent horizontal line indicates that about half the workers earn $2.60 or less, and half more. These three earnings figures are the quartiles and median, discussed in the next two chapters.

The same percentages can be used to make inferences about *all* comparable machine tool operators, provided the group of 214 is a good sample of the population. In this case, the sample was carefully selected, so it can be inferred that about 25 percent of all such operators earn less than $2.51, and so on.

An ogive can also be drawn as a smooth curve through the plotted points, with the aid of a French curve, rather than as a series of straight lines. The use of the curve implies *gradual change* in degree of concentration—often a more realistic assumption than that the values are uniformly distributed over each class interval.

FREQUENCY CURVES

A smooth curve can be drawn to portray the frequency distribution of a *population* of continuous data. This is the limiting form of either the histogram or frequency polygon as the number of values in the sample becomes infinitely large and the class intervals become infinitely small. A frequency curve smooths out sampling errors which are particularly evident in small samples—and provides a frequency value for *every* value of X, rather than just one value for each class interval. Smooth curves cannot be used, however, for data that cluster around certain values, such as the machine tool operators' earnings in Table 2–4.

Chart 2–6 shows a histogram of the prices charged by 3,395 dealers throughout the United States for laying mash. The height of each bar shows the number of dealers reporting prices within that price interval. A smooth curve has been drawn by Frederick V. Waugh to show "the general nature of the distribution." Such curves may be fitted either graphically on a judgment basis, or by mathematical methods. A careful study of the data is necessary in either case to assure a realistic fit. In the graphic method, the curve should be drawn in such a way that the area cut from each bar is approximately equal to the area added to that bar by the curve. Chart 2–6 deviates from this rule slightly in the case of the two tallest bars in order to follow a "normal curve." This type of curve is described below.

Chart 2–6

FREQUENCY CURVE FITTED TO SAMPLE DATA

Laying Mash: Prices Reported by Feed Dealers, September 1949

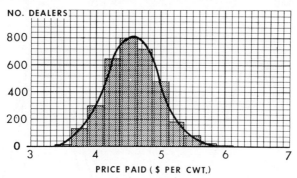

Source: Frederick V. Waugh, *Graphic Analysis in Economics*, U.S. Department of Agriculture, Agricultural Handbook 128 (1957), p. 3.

Types of Frequency Curves

Some common types of frequency curves are illustrated in Chart 2–7. The most important is the bell-shaped *normal curve* shown in Chart 2–6 and panel A of Chart 2–7. This curve describes the distribution of many kinds of measurement in the physical, biological, and social sciences. Thus, the prices of laying mash in Chart 2–6 vary with freight rates, differences in ingredients, dealers' markup, etc., but nevertheless form a nearly normal distribution. The normal curve is particularly important, moreover, because it reflects variations due to *chance,* such as the errors in random sampling. This curve will be used in later chapters in studying the reliability of sample measures and in making inferences about populations.

The two curves in panel B of Chart 2–7 are symmetrical like the normal curve, but one is more peaked, with longer tails; the other is more squat, and with shorter tails than the normal curve. The peaked curve might represent prices of gasoline in a city where most service stations charged about the same price, but a few prices were widely scattered. The squat curve would show that prices were distributed more evenly over a limited range, without being concentrated at one point.

Curves C and D represent distributions that also have a central tendency, as shown by the peak near the center of the curve, but the two branches of the curve are unequal or skewed. Curve C, with the longer branch to the left in the negative direction, is called "skewed to the left" or "negatively skewed." This type of curve commonly results

Chart 2-7
TYPES OF FREQUENCY CURVES

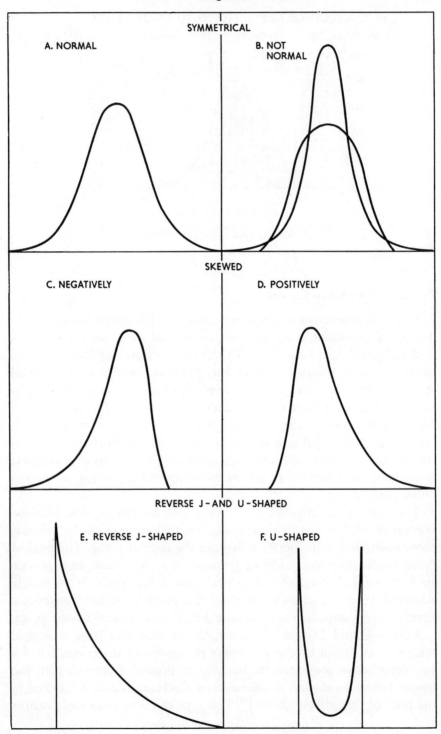

SYMMETRICAL

A. NORMAL

B. NOT NORMAL

SKEWED

C. NEGATIVELY

D. POSITIVELY

REVERSE J- AND U-SHAPED

E. REVERSE J-SHAPED

F. U-SHAPED

from a distribution having a fixed upper limit but a more remote lower limit, as in the case when test scores cluster closer to the perfect score than to zero. Curve D, which is skewed to the right, or positively skewed, is the most common type encountered in business and economic data. Distributions of personal earnings or assets of companies, for example, tend to cluster closer to the lower limit of zero than to the indefinite upper limit. An appropriate test given to a uniform group of job applicants might produce a symmetrical grade distribution; whereas a more difficult test would produce scores lower on the average and skewed to the right, and an easier test would produce scores higher on the average and skewed to the left.

Curves E and F are less common. The reverse J-shaped curve occurs in some distributions, such as income tax payments, where the smallest returns are most numerous and the number of returns (on the Y axis) drops off sharply at first and then more gradually as the size of payment (on the X axis) increases. The U curve may be illustrated by the number of houses classified by percentage of mortgage debt to house value, where many houses have no debt or a heavy debt, while relatively few have a middle-sized debt in relation to house value. The averages and measures of dispersion discussed in the next chapter apply especially to curve types A, B, C, and D, which have a pronounced central tendency; types E and F cannot be summarized so easily.

SUMMARY

Statistics may be classified by qualitative characteristics, by size, or by time. Data that are classified by qualitative characteristics, or attributes, may be summarized and compared by means of ratios. On the other hand, the values of a variable that are classified by size at a given point of time are grouped in a frequency distribution to facilitate analysis.

A ratio is the quotient of two related values. The base, or denominator, is the standard with which the numerator is compared. Ratios should be refined, if possible, by adjusting the numerator and denominator to eliminate any extraneous factors obscuring their relationship. The base may be expressed in any convenient multiple of 10 units, although the percentage form is most common. Ratios must be interperted with care, particularly in distinguishing percentage change from the difference between two percentages.

Ratio or semilogarithmic charts show relative comparisons by means of a vertical logarithmic scale, with an arithmetic time scale. A ratio scale is constructed by plotting natural numbers at distances from the bottom line proportional to their logarithms, as on a slide rule. The bottom of the scale should be marked 1, 2, 4, or 5 (with appropriate zeros and units) and this value multiplied by the printed scale figures to get the other values.

The ratio chart is useful for three types of comparison: (1) It shows a constant percentage rate of growth as a straight line, so changes in this rate are denoted by curvature of the line, and trend forecasts can sometimes be made. (2) The relative growth or fluctuations of two curves may be compared more accurately than in arithmetic charts, since parallel lines indicate the same percentage rates of change anywhere on the chart, and steeper slopes indicate higher rates. (3) Percentages or ratios may be read directly from the vertical scale and applied toward further graphic analysis.

In constructing a frequency distribution, the range of the variable is divided into intervals, and only the number of values of X in each class is shown, thus sacrificing some detail for conciseness. The values of X are first arrayed by listing them individually or marking them on a tally sheet in the order of their size. The figures are then grouped into from 6 to 15 classes so as to show the important characteristics of the data, but without undue detail. Class limits are chosen so that points of concentration, if any, are at midpoints or symmetrical about such points, in order that each midpoint will approximate the average value of X in the class interval. The intervals should be equal in size, if possible. The limits of the classes must be specified unambiguously. Frequencies may be expressed as percentages of the total number to facilitate comparisons or to make inferences from samples.

Frequency distributions may be charted by plotting frequencies on the Y scale above the class midpoints on the X axis. Either a histogram (bar chart) or a frequency polygon (line chart) may be used. Two frequency distributions may be conveniently compared by plotting the percentage frequencies as polygons on the same scales. Frequencies may also be added up from either end and plotted as a cumulative frequency curve or ogive to show the number or proportion of values less than or greater than a given amount.

A smooth curve drawn through a histogram or frequency polygon of a continuous distribution approximates the frequency curve for the population from which the sample was drawn, provided the sample is carefully selected and the data do not cluster at certain points.

Frequency distributions may assume a normal bell-shaped curve or some other symmetrical form; they may be skewed or asymmetrical either to the left or right; or, in extreme cases, they may assume the shape of a reverse **J** or a **U**.

PROBLEMS

1. *a*) Present a brief table, condensed from an available publication, that illustrates the classification of data by qualitative characteristics, by size, and by time.
   b) Compute ratios to compare the qualitative characteristics.
   c) What other type of classification of the raw data might have been used to clarify the findings?

2. What refinement would you recommend in the denominator of each of these ratios?
   a) Employees injured in shop accidents to total number of employees of airlines.
   b) The number employed in a community to the number of persons in the community.
   c) The number of Ford automobiles manufactured to the total number of motor vehicles sold in the United States.

3. What refinement would you recommend in the numerator or the denominator of these ratios?
   a) Dollars of bad debt to total sales.
   b) Freight revenues to number of trains in service.
   c) Airplane fatalities to population.

4. Given the following:

Month	Apparel Sales	Number of Days Store Was Open
February..........	$31,872	23
March...........	33,084	26

Find the percentage change in average daily sales from February to March.

5. Given the following information concerning federal credit unions:

Area	Number of C.U.'s	Members (thousands)	Loans Made During Year	
			Number (thousands)	Amount (millions)
United States............	8,350	4,502	3,300	$1,580
Pennsylvania............	843	433	300	129

   a) Compute whatever ratios you consider necessary to compare the state's operations with the nation's.
   b) Write a statement of your findings.

6. General Electric Company's earnings per share in 1970 were 90 percent of the 1967 level, and in 1971 were 130 percent of the same base. What is:
   a) The difference between the 1970 and 1971 figures in percentage points?
   b) Earnings per share in 1971 as a percentage of 1970?
   c) The percentage change from 1970 to 1971?

7. Xerox Corporation reports the following total operating revenues, which are also listed as indexes, with 1962 = 100 (percent).

Year	Millions of Dollars	Index (1962 = 100)
1962	115	100
1964	318	277
1966	753	655
1968	1,224	1,064
1970	1,719	1,495

Find:
   a) The percentage increase in 1964 revenues over 1962.
   b) The percentage increase in 1966 revenues over 1962 and over 1964.
   c) The increase in the index from 1968 to 1970 in percentage points, and in percentage.
   d) The 1970 revenues as a percentage of those in 1966.
   e) What percentage revenues would have to drop from the 1970 figure to reach the 1962 level.

8. a) Discuss the relative advantages of arithmetic and logarithmic vertical scales for time series charts.
   b) How would you label the bottom and top of a printed ratio sheet for data having the following ranges: 390 to 1,400 tons; 65 to 3,200 million passenger-miles; $0.16 to $55.50; 89 million to 180 million population? How many cycles does your ratio sheet have in each case—1, 2, or 3?

9. a) Draw a ratio chart of the data given below.
   b) Interpret the facts shown by your chart.

SELECTED FARM STATISTICS, 1940–1970

Year	Number of Farms (thousands)	Gross Farm Income (millions)	Number of Tractors on Farms (thousands)
1940	6,350	$11.0	1,545
1945	5,967	25.8	2,354
1950	5,648	32.3	3,394
1955	4,654	33.1	4,345
1960	3,962	38.1	4,685
1965	3,340	44.9	4,783
1970	2,924	56.2	4,790

SOURCE: *Statistical Abstract of the U.S.*, 1971.

10. *a*) Compare the growth of two industries or companies since 1960 by plotting their annual production or sales curves on a ratio chart.
    b) Compare the percentage rates of change in different years for one of the curves.
    c) Compare the relative growth of the two curves during this period.
    d) Mark a percentage measuring scale on the chart. Show the percentage change in each series between the first and last years by measuring the vertical difference on this scale.

11. Define and give the purpose of (*a*) an array, (*b*) relative frequency distribution, (*c*) frequency polygon, (*d*) ogive, and (*e*) normal curve.

12. Indicate which of the following are correct statements and amend any that are incorrect:
    a) Points of concentration are always present in an array and should be considered in preparing a frequency distribution.
    b) All frequency distributions should have at most 15 class intervals.
    c) Class intervals of unequal width should never be used.
    d) Class limits should be established so that the average value of the items in each interval is approximately equal to the midpoint of the interval.
    e) In presenting a distribution of continuous data, the best way to designate the classes is by listing the class midpoints.

13. State wherein each of the following meets or fails to meet the principles of constructing a frequency distribution.

(*a*)		(*b*)	
Income	Average Monthly Rent	Age (Years)	Persons (*thousands*)
Under $2,000	$62.70	All ages	5,390
$2,000–$2,900	65.40	Under 4	335
$2,900–$4,000	70.00	Under 2	87
$4,000–$4,900	81.10	4–9	602
$5,000–$6,500	93.50	10–15	721
etc.		16–25	1,358
		etc.	

14–16. A survey of typical starting salaries offered college men with bachelors' degrees by 191 companies in 1971 showed the results on the next page:

14. *a*) Plot histograms for two fields in the table below as assigned, using separate graphs.
    b) Plot frequency polygons for the same two fields, using either one or two graphs.
    c) Compare the merits of the histogram and the polygon in this case.

Monthly Starting Salary*	Field				
	Accounting	Sales Marketing	General Business Admin.	Prod. Mgt.	Economics-Finance
$601–$ 640	0	2	3	0	0
641– 680	3	11	14	3	0
681– 720	5	12	17	5	2
721– 760	16	26	34	10	7
761– 800	34	12	21	9	9
801– 840	20	1	3	6	9
841– 880	13	4	1	1	2
881– 920	5	2	0	2	0
921– 960	1	0	1	0	0
961– 1,000	2	1	0	0	1
Number of companies reporting	99	71	94	36	30

* Class limits in the end classes have been modified slightly in order to avoid open-end classes and to facilitate analysis.

NOTE: These data will be used also in Chapters 3 and 4.

SOURCE: Frank S. Endicott, *Trends in Employment of College and University Graduates in Business and Industry* (Evanston, Ill.: Northwestern University Press, 1971).

15. *a*) Compute a percentage frequency table for the two fields assigned in 14(*a*) above. Use these computations to construct two percentage frequency polygons on the same graph.
   b) What is the reason for using percentage frequencies in comparing two distributions?
   c) What conclusions can you draw about the relative salaries in the two fields from this graph?
   d) In what situation would percentage frequencies be unnecessary for comparing two distributions?

16. *a*) Construct a "more than" cumulative frequency table and ogive for one of the fields in the above table as assigned.
   b) Construct a "less than" table and ogive for the same field.
   c) How many companies offered starting salaries to college men in this field of more than $680? more than $800?
   d) How many companies offered starting salaries to college men in this field of $720 or less? $840 or less?

17. *a*) Make a frequency table, using the 112 items in the four columns assigned to you from the following table (see numbered assignments below table).
   b) Give reasons for your choice of class limits and width of class intervals.
   c) Draw a graph showing your frequency distribution.
   d) What information concerning earnings of women in this plant can be derived from your table and graph?

NOTE: This problem will be continued in Chapters 3 and 4.

DAILY EARNINGS OF 168 WOMEN IN AN ELECTRONIC ASSEMBLY PLANT

JANUARY 15, 1973

(In Dollars)

(a)	(b)	(c)	(d)	(e)	(f)
15.20	18.00	11.20	16.00	20.00	13.60
11.60	14.00	12.00	11.30	12.20	12.00
8.00	12.00	17.60	15.60	8.50	8.00
12.80	12.80	9.50	12.00	14.50	10.00
14.00	11.80	12.00	10.60	16.00	12.60
6.40	9.20	14.00	12.00	12.60	14.00
12.00	7.60	12.00	15.00	12.00	6.50
12.40	14.80	8.20	6.00	8.00	16.00
24.00	18.00	28.00	8.00	19.00	14.00
14.60	16.80	16.80	16.00	22.00	14.60
9.00	14.20	14.40	17.20	15.20	19.20
16.50	12.00	21.20	14.40	10.00	12.30
20.00	12.00	20.00	12.50	14.00	11.60
18.00	21.00	23.00	20.00	16.00	16.40
14.10	8.00	14.00	18.80	16.40	16.00
22.50	16.00	16.10	12.00	12.00	20.00
12.00	24.00	19.90	12.00	23.80	21.40
20.80	19.60	12.90	8.40	28.40	24.00
16.00	27.00	24.00	23.50	17.30	28.80
18.00	20.00	16.00	20.00	18.00	15.20
7.20	10.40	8.00	21.60	14.00	25.00
14.00	15.50	11.80	24.40	11.40	12.00
26.00	21.80	15.00	14.00	24.50	20.40
16.00	14.00	16.00	16.20	6.00	17.60
16.00	6.00	12.40	28.00	20.00	8.80
12.00	16.00	18.40	16.90	16.00	16.00
19.40	12.40	15.50	13.00	12.00	18.00
10.00	16.00	6.00	14.00	13.20	12.00

Assignments:

No.	Columns	No.	Columns	No.	Columns
1...........	a b c d	6..........	a b e f	11..........	b c d e
2...........	a b c e	7..........	a c d e	12..........	b c d f
3...........	a b c f	8..........	a c d f	13..........	b c e f
4...........	a b d e	9..........	a c e f	14..........	b d e f
5..........	a b d f	10..........	a d e f	15..........	c d e f

18. U.S. family incomes in 1971 were distributed as follows, according to the Census Bureau's *Consumer Income* for July 1972:

Income	Percent	Income	Percent
Under $1,000.....................	1.5	$ 7,000 to 7,999.................	6.2
1,000 to 1,999.....................	2.6	8,000 to 9,999.................	12.3
2,000 to 2,999.....................	4.2	10,000 to 14,999.................	26.9
3,000 to 3,999.....................	4.8	15,000 to 24,999.................	19.5
4,000 to 4,999.....................	5.4	$25,000 and over................	5.3
5,000 to 5,999.....................	5.7		
6,000 to 6,999.....................	5.5	Total families....................	100.0

a) Criticize the choice of class intervals and class limits.

b) Plot a histogram of this distribution. Then draw a smooth curve to approximate the true continuous distribution of incomes. What type of frequency curve is this—normal, negatively skewed, etc.?

19. You are employed by a concern that has just received a shipment of 200 sheets of ⅛-inch insulating board for use in the manufacture of power transformers. You are directed to check these boards for thickness, using a 0 to 1-inch micrometer. Thickness is a major characteristic affecting the quality of the board, and consequently, the quality of the transformer. The actual measurements are shown in thousandths of an inch (and rounded to the nearest thousandths).

Thickness	Number of Sheets	Thickness	Number of Sheets
118	2	125	51
119	8	126	14
120	5	127	23
121	9	128	14
122	8	129	10
123	23	130	5
124	27	131	1
		Total	200

NOTE: These figures will be used in problems of chapters 3 and 4.

a) You wish to prepare a chart summarizing the results of your inspection. Plot a frequency polygon (line chart) showing the distribution of thicknesses for the 200 sheets.

b) What essential characteristics of this shipment of insulating board can you determine by inspecting the chart?

c) Draw a smooth curve through your graph to iron out the zigzag sampling errors and to approximate the distribution of thicknesses for all future shipments of this insulating board. (The total frequencies under the two curves should be the same.) Mathematical curves are used for this purpose in more advanced analysis.

d) Would it be better to present the data as shown, in a report to the company executives, or to combine it into five classes of .003-inch width (118–120, 121–123, . . .) for simplicity? Why?

20. An automobile advertisement lists the following distribution of gas mileage reported by owners of its new cars:

Miles per Gallon	Percent	Miles per Gallon	Percent
15 and under 16*	6	19 and under 20	14
16 and under 17	10	20 and under 21	18
17 and under 18	16	21 and under 22*	12
18 and under 19	24	Total owners	100

* Open-end classes have been assigned arbitrary limits to facilitate later computations.

a) Plot a histogram of gas mileage, and draw a smooth curve through it to iron out sampling irregularities and approximate the continuous distribution of mileage performance for the whole population of car owners. What type of frequency distribution is this?

b) List a cumulative frequency distribution and draw an ogive showing the percentage of owners reporting a given gas mileage or more. From this curve, half the owners get what gas mileage or more? The most economical fourth of the owners get what gas mileage or more? (Give results to nearest tenth of a gallon.)

21. You are comparing two brands of a certain type of electron tube. You obtain the following frequency distributions for their life in hours.

Life (Hours)	Frequency		Relative Frequency	
	Brand A	Brand B	Brand A	Brand B
Under 50.....................	1	3	0.8%	3.8%
50 and under 100.............	8	8	6.7	10.0
100 and under 150.............	18	12	15.0	15.0
150 and under 200.............	40	14	33.3	17.5
200 and under 250.............	26	13	21.7	16.3
250 and under 300.............	12	10	10.0	12.5
300 and under 350.............	6	9	5.0	11.2
350 and under 400.............	3	6	2.5	7.5
400 and under 450.............	2	3	1.7	3.8
450 and under 500.............	1	1	0.8	1.2
500 and above................	3*	1*	2.5	1.2
Total........................	120	80	100.0%	100.0%

* The mean life for those tubes still burning after 500 hours was 700 for Brand A and 600 for Brand B.

a) Plot on the same chart the relative frequencies of the two brands. (For this purpose, omit the class 500 and above.) Why should you use percentages rather than the actual number of tubes?

b) Are these frequency distributions fairly normal, skewed to the left, skewed to the right, J-shaped, or U-shaped?

c) Use your chart to compare the two frequency distributions.

d) Calculate cumulative frequency distributions for the two brands of tubes. Then plot these distributions on a chart. At what life are approximately 50 percent of Brand A tubes still burning? 50 percent of Brand B tubes? (This can be obtained from your chart—where the cumulative frequency curves cross the 50 percent cumulative frequency line.) Using this result and your analysis in part (*c*) above, which tube do you think you should buy to obtain greater total life? Why?

e) Suppose your company had a policy of replacing all tubes after 150 hours. Would this change your answer to (*d*) above?

22. Given the life table for 1,000 cars shown on the next page:

a) Chart the number of cars scrapped (on X-axis) in a frequency polygon (with age on Y-axis).

b) What conclusions can you draw about car scrappage from this graph?

c) Plot ogives for the cumulative number of cars scrapped and the number surviving.

Age (Years)	Number Scrapped During Year	Cumulative No. Scrapped	Number Surviving
1–2	0	0	1,000
2–3	9	9	991
3–4	13	22	978
4–5	14	36	964
5–6	18	54	946
6–7	29	83	917
7–8	52	135	865
8–9	86	221	779
9–10	109	330	670
10–11	121	451	549
11–12	115	566	434
12–13	104	670	330
13–14	89	759	241
14–15	72	831	169
15–16	54	885	115

d) At the intersection point of the two curves, note the number of cars and the age in years. The latter is the median life.

e) At what age were 25 percent of the cars scrapped? 75 percent? These are the quartiles.

SELECTED READINGS

Selected readings for this chapter are included in the list that appears on page 93.

3. AVERAGES

A BASIC step in statistical analysis is to develop concise summary figures that will describe unwieldy masses of raw data. The initial stages in this analytic process have already been described—that is, appraising the accuracy of data, classifying facts, comparing them by ratios, and condensing them into a frequency distribution.

An important type of summary measure is the *average*. Averages are familiar to everyone in such examples as average prices of securities, a man of average income, and the usual rate of interest charged a bank's customers. Careful analysis of these examples shows that they involve several different concepts of "average" which should be distinguished from each other. No single average can be used indiscriminately.

The most common averages are (1) the arithmetic mean, (2) the median, and (3) the mode. The first is determined by calculation, the second by its position in an array, and the third by finding the point about which values of the variable cluster most closely. These will be described in turn.

THE ARITHMETIC MEAN

The most common average is the arithmetic mean or, more simply, the mean.[1] The term "average," when used alone, usually refers to the mean. The mean of any series of values is found by adding them and dividing their sum by the number of values.

---

[1] The arithmetic mean is distinguished from the quadratic mean, which averages the squares of numbers; the geometric mean, which averages their logarithms; and the harmonic mean, which averages reciprocals. The quadratic mean (of deviations from the arithmetic mean) appears in Chapter 4 as the "standard deviation." The other means, however, are seldom used and will not be considered here.

Ungrouped Data

The general method of computing the mean is the same whether the data are ungrouped or grouped in a frequency distribution, but the formulas look a little different. As an example of ungrouped data, consider a man working at piece rates who earns $4.80, $5.05, $5.00, and $5.15 in four successive hours. The mean of his hourly earnings is found by adding his earnings for the four hours and dividing by four. The earnings total $20.00, so the mean is $5.00. This process is generalized by the following formula:

$$\overline{X} = \frac{\Sigma X}{n}$$

where $\overline{X}$ (read "X bar") is the mean of the variable X (hourly earnings in dollars); Σ is the Greek letter capital sigma (corresponding to our S), which means "the sum of"; and n is the number of values.[2]

When a variable has a number of identical values, multiplication can be used as a short-cut for addition in totaling X. Thus, to find the average dimension of the 63 gears in Table 2–3, one could add the 63 figures in panel A; but it would be easier to multiply each dimension in panel B by its *frequency* and add the products as follows: $1(.4270) + 4(.4265) + 10(.4260) + \ldots$. Specifically, since there are 10 gears measuring .4260, it is simpler to multiply 10 by .4260 than to add .4260 ten times. The whole process is summarized by the formula:

$$\overline{X} = \frac{\Sigma f X}{n}$$

where f is the frequency, and $\Sigma f X$ means that each different value of X is multiplied by its frequency and the products (fX) are then added. Using either formula,

$$\overline{X} = \frac{26.7820}{63} = .4251, \text{ the mean dimension in inches}$$

The Weighted Mean. In many types of problems, the values to be averaged are of different degrees of importance. In such cases, each value is multiplied by a numerical weight based on its relative impor-

---

[2] Strictly speaking, the symbols $\overline{X}$ and n apply only to sample data. In later chapters, μ (the Greek letter mu) will be used to designate the mean of an entire population and N, the number of values in the population. Hence, $\mu = \Sigma X/N$.

tance, and the total is divided by the sum of the weights. The result is called a weighted mean. The weights are handled just as if they were frequencies. Hence, a weighted mean can be computed by the above formula—taking f as the weight and n as the sum of the weights.

Thus, an aptitude score may be based on an English test with weight 2 and a mathematics test with weight 1. The weights total 3. If a person makes 90 and 60, respectively, on these tests his combined aptitude score is

$$\bar{X} = \frac{\Sigma fX}{n} = \frac{2(90) + 1(60)}{3} = \frac{240}{3} = 80$$

Weighted means are used extensively in the construction of index numbers, to be described in Chapter 18.

All means can be regarded as weighted in some way, either explicitly or implicitly. From this point of view, the "unweighted" mean is one in which the weights are all equal. In computing any mean, therefore, it is important to use appropriate weights. In averaging the ratios of profits to sales for 30 retail grocers, for example, the total profits for all 30 grocers can be divided by their total sales to allow the larger firms more weight in the results, or the firms may be weighted equally by taking a simple average of the 30 ratios.

Grouped Data

The mean of data grouped in a frequency distribution is computed in the same way as described above. In a frequency distribution, however, the *midpoint* of each interval is used to represent all values of X in the interval. Accordingly, each midpoint is multiplied by the number of values in that class. The sum of these products is then divided by the total number of values of X to find the mean.

The formula for computing the arithmetic mean from a frequency distribution is therefore

$$\bar{X} = \frac{\Sigma fX}{n}$$

where fX is the frequency (number of values) in an interval times its midpoint X, and ΣfX is the sum of these products. The total number of values, n, is also the sum of the frequencies.

In calculating the arithmetic mean for the earnings of machine tool operators shown in Table 3–1, the midpoint of the first class ($2.30)

Table 3–1

DIRECT METHOD OF COMPUTING THE ARITHMETIC MEAN
FROM A FREQUENCY DISTRIBUTION

HOURLY EARNINGS OF 214 APPRENTICE MACHINE TOOL OPERATORS

Hourly Earnings, Dollars	(1) Class Midpoint X	(2) Number of Operators (Frequency) f	(3) Frequency × Midpoint fX
2.25 and under 2.35.........$2.30		2	$ 4.60
2.35 and under 2.45......... 2.40		23	55.20
2.45 and under 2.55......... 2.50		49	122.50
2.55 and under 2.65......... 2.60		63	163.80
2.65 and under 2.75......... 2.70		45	121.50
2.75 and under 2.85......... 2.80		25	70.00
2.85 and under 2.95......... 2.90		3	8.70
2.95 and under 3.05......... 3.00		4	12.00
Total.....................		214	$558.30

SOURCE: Table 2–5.

times the two operators in that class gives their combined earnings of $4.60. The total earnings for all classes is $558.30, and the mean is:

$$\bar{X} = \frac{\Sigma fX}{n} = \frac{558.30}{214}$$

$$= 2.609 \text{ dollars per hour}$$

The mean computed from a frequency distribution is subject to a slight error of grouping, since all values are rounded off to the nearest class midpoint, as noted in Chapter 2. The error can be minimized by placing the midpoints of class intervals at points around which the data tend to cluster or midway between such points within intervals. Grouping errors of opposite sign then tend to offset each other, so that the error in the grouped mean is usually negligible. Thus, the arithmetic mean of $2.609 per hour obtained from the frequency distribution is only $.003 greater than the exact mean of $2.606 per hour computed from the original figures.

Short-Cut Method. The direct method of computing the arithmetic mean from a frequency distribution is simple when all numbers involved are simple integers. However, it sometimes requires multiplication of many pairs of large numbers and laborious addition of their products. If class intervals are of equal width, the computations can be simplified by using a short-cut method in which the multipliers are

Table 3–2

SHORT-CUT METHOD OF COMPUTING THE ARITHMETIC MEAN
FROM A FREQUENCY DISTRIBUTION

HOURLY EARNINGS OF 214 APPRENTICE MACHINE TOOL OPERATORS

(1)	(2)	(3)	(4)	(5)
		Number of	Deviation from	
	Class	Operators	Assumed	Frequency ×
Hourly Earnings	Midpoint	(Frequency)	Mean	Deviation
(Dollars)	X	f	d	fd
2.25 and under 2.35.............$2.30		2	−3	− 6
2.35 and under 2.45..............2.40		23	−2	−46
2.45 and under 2.55............. 2.50		49	−1	−49
2.55 and under 2.65............. 2.60*		63	0	0
2.65 and under 2.75............. 2.70		45	1	45
2.75 and under 2.85............. 2.80		25	2	50
2.85 and under 2.95............. 2.90		3	3	9
2.95 and under 3.05............. 3.00		4	4	16
Total.........................		214		19

* Selected as $\bar{X}_a$ arbitrarily.

reduced to small whole numbers. This method is illustrated in Table 3–2.

The steps for computing the mean by the short-cut method are as follows:

1. List the class limits (if desired), the midpoints, and the frequencies, as shown in columns 1 to 3.
2. Select any midpoint as the assumed mean $(\bar{X}_a)$, preferably the midpoint of one of the middle intervals. In Table 3–2 the assumed mean is taken as $2.60.
3. List the deviation (d) of each class midpoint from the assumed mean in units of the class interval, as in column 4. Thus a zero is written opposite $2.60, the next larger midpoint is marked $+1$, the next smaller -1, and so on in whole numbers, 1, 2, 3, ... Be sure to mark the deviations of the larger midpoints plus and the smaller midpoints minus, irrespective of which end is listed first in the table. If there were a gap and then some values, say in the "3.15 and under 3.25" class, that class would have a deviation of 6, not 5, class units from the assumed mean.
4. Multiply the frequency in each class by its deviation and list the product (fd) in column 5, being sure to include the sign.
5. Total these products (Σfd).

The arithmetic mean computed by the short-cut method is then

$$\bar{X} = \bar{X}_a + \frac{i\Sigma fd}{n}$$

where i is the width of the class interval, Σfd is the sum of f times d for each class (not Σf times Σd), and the other symbols are defined above. In Table 3–2, therefore,

$$\bar{X} = \bar{X}_a + \frac{i\Sigma fd}{n}$$

$$= 2.60 + \frac{.10(19)}{214}$$

$$= 2.609 \text{ dollars per hour}$$

The short-cut method thus yields precisely the same result as the formula for the direct method. In case the intervals in a frequency distribution vary in width, however, the direct method, $\bar{X} = \Sigma fX/n$, should be used. The short-cut method might be used if the difference between each class midpoint and the assumed mean were expressed in units of some common factor (i), but this may be an awkward procedure.

Open-End Distributions. On some occasions it is necessary to compute the mean from a frequency distribution having open-end classes whose lower or upper limit is not indicated, such as a salary class "'$825 or less." Although open-end intervals should be avoided ordinarily, it is possible to compute the mean in such cases provided either the individual values, their average, or their total is available for each open-end class to supply the missing data. Simply use the average of the open-end interval as the X value for that interval in the computation of the overall arithmetic mean. If the values for the open-end interval are missing, the median or mode should be used in preference to the mean, since they do not depend on extreme values.

Attribute Data

When the data for analysis are attributes (i.e., classified into only two categories), the arithmetic mean has a particular interpretation. A ratio or proportion may be considered to be a special case of the arithmetic mean in which all the values are ones or zeros. Thus, if 20 out of 100 bolts inspected are defective, and we count the defectives as ones and the others as zeros, the *average* of the 20 ones and the 80 zeros is .20, which is the same as the *proportion* defective.

THE MEDIAN

The median of any set of data is the middle value in order of size if n is odd, or the mean of the two middle items if n is even. When there are a few very large or small values, the median is often superior to the mean as an average.[3] For example, the *Monthly Labor Review* reports median wages and salaries by occupations, and *Dun's* reports median operating ratios for samples of business firms because the median represents the typical middle item undistorted by large values that so greatly affect the mean, as illustrated below.

The median can sometimes be found when other averages are not defined because individuals are not measured quantitatively. For example, employees in a plant can be ranked in order of merit without assigning a numerical grade to each individual. To find the value of the median under these conditions, only one or two individuals need be measured or graded.

Ungrouped Data

In ungrouped data, the median is most easily found when the values are arranged in an array. Consider the price-earnings ratios 19.6, 17.3, 19.2, 14.0, and 29.9 (i.e., common stock prices divided by earnings per share) for five electronics companies. Arranged in order of size, the five ratios are

$$14.0, \; 17.3, \; 19.2, \; 19.6, \; 29.9$$

The median is then the middle value, or 19.2. If a sixth ratio, 30.0, were added, the median would be the mean of the two middle items 19.2 and 19.6, or 19.4. In general, the median in an array is not computed from a formula but is selected as the value whose rank or "order number" is $n/2 + 1/2$, counting from the lowest value. Thus, for the six ratios above, the order number of the median is $6/2 + \frac{1}{2} = 3\frac{1}{2}$, i.e., halfway between the third and fourth values.

This example illustrates an important advantage of the median over the mean. The ratio of the price of a stock to the earnings per share is sometimes very large when the earnings are abnormally small, as in the case of the 29.9 ratio above. Because of this figure, the mean (20.0) exceeds any of the other four ratios. The median is often more reliable than the mean in samples from populations in which such extreme

---

[3] A "modified mean" or "extended median" is sometimes used. This is the mean of a central group of values in an array or frequency distribution, omitting any very large and small values that are considered to be so extreme and atypical as to distort the overall mean. The modified mean, is, therefore, a compromise between the mean and the median, selected to combine the best features of both.

deviations occur, because the reliability of the mean is greatly affected by extreme deviations while the reliability of the median depends chiefly upon the degree of clustering about the median of the population.

Grouped Data

When data are grouped in a frequency distribution, the median falls in the class interval whose frequency is the first to make the cumulative frequency greater than $n/2$. It is convenient to call this the median class. The median (Md) may then be located within the median class by means of the interpolation formula

$$Md = L + \frac{i(n/2 - F)}{f}$$

where L is the lower limit of the median class, i its width, f its frequency, F the cumulative frequency below the median class, and n the total number of values of X.

In applying this formula to the earnings data of Table 3–1 above, the first step is to locate the class that contains the middle value, i.e., the one ranked $n/2 = 214/2 = 107$.[4] By cumulating the f column, the successive subtotals are found to be 2, 25, 74, 137, etc. The first subtotal to exceed $n/2$ is 137. Accordingly, the fourth class is the median class. Its lower limit is $L = 2.55$; its frequency is $f = 63$; the cumulative frequency for X less than L is $F = 74$; and the interval is $i = .10$. Substituting these values in the formula, the median is:

$$Md = L + \frac{i(n/2 - F)}{f}$$

$$= 2.55 + \frac{.10(107 - 74)}{63}$$

$$= 2.602 \text{ dollars per hour}$$

This value is only an approximation to the median of the original ungrouped data, since it is interpolated on the assumption that values of X in the median class are *evenly distributed* over that interval. In this case the true median, taken from the original data in Table 2–4, is exactly \$2.60, because the earnings around the median cluster at this point.

---

[4] The middle value interpolated over a continuous range is at the exact midpoint $n/2$ in rank, rather than $n/2 + 1/2$ as in discrete data.

About half of the 214 earnings are smaller than the median of $2.60 and about half are larger. The proportion on each side of the median is exactly one half when the median is between the two middle values. Nevertheless, the proportion of items on each side of the median may be more or less than one half. In ungrouped data, one or more values may be equal to the median, so that the proportion of values smaller (or greater) than the median may be considerably less than one half —it can never be greater. In grouped data, more than one half of the original values may be on one side of the interpolated median because of uneven distribution of values within the median class. For these reasons, it is better to say that the proportion of values on each side of the median is only approximately equal to one half.

The median can be determined in an open-end distribution exactly as above, since it is not affected by the size of extreme values.

In a frequency distribution, the median can also be read off graphically from a cumulative frequency curve or ogive, as described on page 36. The graphic method yields the same result as the interpolation formula of the preceding section, except for errors in plotting and reading the scale.

THE MODE

The mode in statistics means just what it does in the dictionary—the prevalent or most frequently encountered thing. More precisely, the mode is defined as the *value which occurs most often* or the value around which there is the greatest degree of clustering. The modal wage is the one received by the greatest number of workers. The modal interest rate for mortgages is the one that occurs more often than any other. If the most common or usual value is the one needed for a business decision, the mode is the appropriate type of average to use.

It is particularly important that the data used to determine the mode be homogeneous or enough alike to be comparable. Heterogeneous data, such as wages of both skilled and unskilled workers, may be *bimodal,* with two modes (or more) having equally high frequency. The mode is ordinarily meaningful only if there is a marked concentration of values about a single point.

Ungrouped Data

The mode can occasionally be determined directly from ungrouped data. When a large proportion of values are equal, no process of grouping could dislodge this value from its modal position. This is especially true of discrete data having only a limited number of possible distinct values. For example, if a bank charges the general run of its customers 8 percent interest on commercial loans, then 8 percent is the mode of interest rates, irrespective of what rates apply in special cases. Similarly, a survey indicates that more parents prefer to have three children than any other number. Thus, three is the modal family size preferred by parents.

Grouped Data

Most types of data, however, must be grouped in a frequency distribution in order to locate the mode. To illustrate, in the array of hourly earnings listed by cents in Table 2–4, the most frequently occurring rate is $2.63, but $2.70 is almost as popular; and there are other scattered points of concentration, such as $2.50 and $2.75, which cause doubts as to where the major area of concentration really is. By grouping the earnings as in Table 3–1, however, there appears only a single mode. This occurs in the $2.55 to $2.65 interval. The modal interval can be described by saying, "More earnings fall in the $2.55 to $2.65 class than in any other."

The value of the mode within this interval may be estimated graphically in a continuous distribution by drawing a smooth curve through the histogram so that the area cut from each bar is about equal to the area added to that bar by the curve. The mode is then the X value at the peak of the frequency curve. Thus, in Chart 2–6 the modal price of laying mash is about $4.57 per hundredweight.

Interpolation formulas are also used to locate a "single-valued" mode within the modal interval.[5] More simply, the midpoint of the modal interval could be taken as the mode, but this is recommended only if values cluster at this point. Ordinarily, a single-valued estimate of the mode is neither accurate nor necessary in practice; it is usually enough to cite the modal interval.

The modal interval itself is only a rough estimate, since it depends on the choice of class limits. Grouping the data in different class intervals may produce different values of the modal interval. In some types of data, therefore, the mode is practically indeterminate. Hence, the mode or modal interval should be used only if the problem specifically requires the most common value as an average rather than the middle or the mean value.

---

[5] See Spurr, Kellogg, and Smith, *Business and Economic Statistics* (1st ed., Homewood, Ill.: Richard D. Irwin, 1954), pp. 208–10, for a description of the most common method.

WHICH AVERAGE TO USE?

Much of the chapter thus far has been devoted to methods of computing the various types of averages. In the course of the several explanations, the distinctive features of the measures have been set forth in some detail but in incidental fashion. At this point, the reader may well ask, "Which of these various averages should I use?"

No single answer can be given to this question. The selection of the proper average depends upon three main factors:

1. The concept of the typical value required by the problem. Is a composite average of all values needed (arithmetic mean) or is a middle value wanted (median) or the most common value (mode)?

2. The type of data available. Are they badly skewed (avoid the mean)? Do they have a gap around the middle (avoid the median), or lack a major point of concentration (avoid the mode)? In particular, the choice between the arithmetic mean and the median of a sample depends on the shape of the frequency curve for the population. Refer to Chart 2–7. If the distribution is normal (panel A) or flat-topped with few extreme values (panel B, lower curve), the mean has a smaller sampling error than the median. That is, the mean of the sample is likely to be closer to the true mean of the population than the median of the sample is to the true median. On the other hand, if the distribution is sharply peaked around the median and includes some extreme values (panel B, higher curve), the median has the smaller sampling error. This is because the clustering around the population median makes the sample median more accurate, and extreme values make the sample mean erratic.

3. The peculiarities or characteristics of the averages themselves. These will be summarized below, under "Characteristics of Averages."

As a rule of thumb, the arithmetic mean should ordinarily be used as a simple, widely understood average which gives due weight to all values. The median is commonly preferred to the mean if a simpler, middle value is needed—particularly if the data are badly skewed, as is common in economic measurements. Finally, the mode may be used if the most usual or common value is wanted.

CHARACTERISTICS OF AVERAGES

The arithmetic mean, median, and mode have the same value in a symmetrical "normal" distribution. If the distribution is skewed, the mode remains under the highest point of the curve, the arithmetic

Chart 3–1

RELATIONSHIP OF ARITHMETIC MEAN, MEDIAN, MODE IN A
POSITIVELY SKEWED DISTRIBUTION

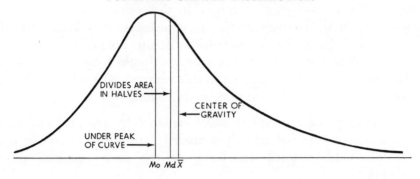

mean is pulled out in the direction of the extreme values, and the median, which is affected by the *number* of extreme items but not their *value,* tends to fall between the mean and the mode.[6]

Chart 3–1 shows the relation of the arithmetic mean, median, and mode in a positively skewed distribution—by far the most common type in business and economic data. Here the arithmetic mean is the largest value, and the mode is the smallest. Thus, the mean income of "unrelated individuals" in 1971 was $4,774, while the median was $3,316 and the mode only about $1,640, according to the Census Bureau's *Consumer Income* report of July 1972. The mean is the X value of the center of gravity. That is, if the area under the curve were a solid piece of metal, a fulcrum under $\overline{X}$ would balance it. The median divides the area under the curve (i.e., the total frequency) into two equal parts. The mode is the value of X under the highest point of the curve.

The characteristics of the individual averages are listed below.

Arithmetic Mean

1. The arithmetic mean is the most widely known and widely used average.
2. It is, nevertheless, an artificial concept, since it may not coincide with any actual value.
3. It is affected by the value of every item, but
4. It may be affected too much by extreme values.
5. It can be computed from the original data without forming an array or frequency distribution, or from the total value and number of items alone.

---

[6] The median falls roughly one third of the way from the mean toward the mode in a continuous distribution of only moderate skewness.

6. Being determined by a rigid formula, it lends itself to subsequent algebraic treatment better than the median or mode.
7. It is less affected by sampling errors than the median in a normal or flat-topped distribution.

Median

1. The median is a simple concept—easy to understand and easy to compute.
2. It is affected by the number but not the value of extreme items.
3. It is widely used in skewed distributions where the arithmetic mean would be distorted by extreme values.
4. It may be located in an open-end distribution or one where the data may be ranked but not measured quantitatively.
5. It is unreliable if the data do not cluster at the center of the distribution.
6. The median will have a smaller sampling error than the mean if the data *do* cluster markedly at the middle or if there are abnormally large or small values.

Mode

1. The mode can best be computed from a frequency distribution, unless one value predominates in an array.
2. It can be located in open-end distributions, since it is not affected by either the number or value of items in remote classes.
3. The mode is erratic if there are but few values or zigzag frequencies —particularly if there are several modes or peaks.
4. It is affected by the arbitrary selection of class limits and class intervals.

SUMMARY OF FORMULAS

Since the characteristics of the various averages have been summarized above, the chapter may be concluded by listing the principal formulas used:

Type of Average	*Ungrouped Data*	*Grouped Data*
Arithmetic mean	$\bar{X} = \dfrac{\Sigma X}{n}$	$\bar{X} = \dfrac{\Sigma f X}{n}$
		$= \bar{X}_a + \dfrac{i\Sigma fd}{n}$
Median	Value # $n/2 + \frac{1}{2}$ in an array	$Md = L + \dfrac{i(n/2 - F)}{f}$
Mode	Most common value	Same

PROBLEMS

1. One method of saving money regularly is to buy common stock at periodic intervals. Is it better policy, then, to buy the same number of shares in a company each year or to invest a constant number of dollars, irrespective of the price of the stock?

   To illustrate, Investor A buys 7 shares of DuPont and 25 shares of Dun and Bradstreet common stock at the average between the year's high and low prices (listed below) in each of the years 1966–70. Investor B invests $1,000, as nearly as possible, in each of these stocks at the same times and prices. His results are detailed in the table. DuPont declined and Dun and Bradstreet advanced in price over this period (prices are yearly averages).

COMMON STOCK PURCHASES BY INVESTOR B

Year	DuPont Share Price	DuPont Shares Bought	DuPont Total Cost	Dun & Bradstreet Share Price	Dun & Bradstreet Shares Bought	Dun & Bradstreet Total Cost
1966	$193	5	$ 965	$ 30	33	$ 990
1967	163	6	978	38	26	988
1968	163	6	978	44	23	1,012
1969	133	8	1,064	50	20	1,000
1970	113	9	1,017	51	20	1,020
Total	$765	34	$5,002	$213	122	$5,010

   a) Give the average cost per share for Investor A (constant shares) and Investor B (constant dollars), for each stock.
   b) Which investor achieved the lower average cost for DuPont? for Dun and Bradstreet?
   c) Explain these differences in terms of the weights used in computing the averages.

2. In the dollar-averaging method of investment, the same amount of money is invested each month in a variable number of shares of common stock. Thus, $50 will buy one share of a stock selling at $50 a share in one month, but two shares of that stock if it sells at $25 in another month. The three shares then cost $100, or an average of $33⅓ per share, as compared with the average market price of $37½ in the two months [(50 + 25) ÷ 2], irrespective of whether the market is rising or falling. Explain this apparent anomaly in terms of the two types of averages represented.

Stock	1972 Investment	1972 Dividend	1972 Yield	1974 Investment	1974 Dividend	1974 Yield
A	$ 8,000	$ 480	6%	$ 5,000	$300	6%
B	5,000	200	4	12,000	480	4
C	6,000	480	8	2,000	160	8
Total	$19,000	$1,160		$19,000	$940	
Average yield			6.11%			4.95%

3. An investor owns three stocks on which he receives the above dividends in 1972 and 1974:

   a) How are the average yields obtained?

   b) Inasmuch as none of the individual yields has changed, how do you explain the decrease in average yield?

4. A company has 200 executives receiving $500 a week and 800 workers receiving $200 a week. In time of recession, all salaries and wages are cut 20 percent and 600 of the 800 workers are laid off. Yet the public relations department publishes a statement that the average wage has increased. Explain.

5. From Chapter 2, Problem 17, on the earnings of women in an assembly plant:

   a) Compute the arithmetic mean from your frequency distribution. (Indicate all computations in this and following problems.) Discuss the grouping errors that affect this value.

   b) Find the median both from the original data and from your frequency distribution. If these values differ, explain why.

   c) What does the comparison of mean and median reveal about the shape of the distribution?

   d) State the modal interval. Which of the three averages is most meaningful in this case? Why?

6. a) Compute the mean starting salary offered to college men, shown in Chapter 2, Problem 14, in whichever of the five fields is assigned.

   b) Is this mean more or less accurate than one computed from the original ungrouped salary data? Why?

7. a) Find the median starting salary for whichever field was assigned in Problem 6 above.

   b) Give the modal interval for the same field.

   c) Explain the difference in the meaning of these two averages.

   d) If the last four classes had been grouped into one class and labeled "Over $840," which measure or measures would have been affected—the mean, median, or mode? Why?

8. The durations of 11 business cycles in the United States from March 1919 to November 1970, measured from trough to trough, were 28, 36, 40, 64, 63, 88, 48, 58, 44, 34, and 117 months, respectively, according to The National Bureau of Economic Research.

   a) List the mean and median of these periods.

   b) Which of these averages is preferable? Why?

   c) What is the difficulty in computing the mode for the figures listed above?

9. Under a wages-and-hours law it is considered desirable that the number of hours of work per week should be standardized for some 250 establish-

ments, all now operating under similar conditions except with respect to hours of work. What should be the standardized number of hours (*a*) if the object is to keep the total hours of work the same and (*b*) if the object is to change as few establishments as possible?

10. U-Fix Stores was a chain of 81 building supply and home repair stores in the northwestern United States. In a recent year the distribution of annual sales for these stores was:

Annual Sales (thousands of dollars)	Number of Stores
Less than 100	8
100 and under 200	32
200 and under 300	18
300 and under 500	16
500 and under 1,000	6
More than 1,000	1
Total	81

The smallest store had annual sales of about $50,000 and the largest had sales of about $1,600,000.

a) Estimate the total annual sales for all 81 retail stores. Give the mean sales per store.

b) Give the median sales per store.

11. Regarding the dimensions of 63 gears in Table 2–3, page 26:

a) Is this distribution discrete or continuous? Symmetrical or skewed to the right or left?

b) Find the mean and median to the nearest .0001 inch. (Express data as deviations from .4250 to simplify calculations.)

c) Which type of average is usually the best estimate of the corresponding population value for a distribution of this kind? Why?

12. Chapter 2, Problem 18, reports the distribution of family incomes in 1971. The mean income was stated by the Census Bureau to be $11,583.

a) Estimate the median income. What is its significance?

b) Give the modal interval.

c) Explain why the mean, median, and mode differ in value. Which is the best measure of typical family income? Why?

13. In your report on the thickness of 200 sheets of 1/8-inch insulating board used for power transformers, you wish to include a statement of the average width for the 200 sheets listed in Problem 19 of Chapter 2. (Micrometer readings were taken to the nearest .001 inch.)

a) Compute the arithmetic mean, using the shortest method possible. Explain your choice of formula.

b) Estimate the median thickness to the nearest hundred thousandth of an inch.

c) From the above measures, would you conclude that the average thickness of this shipment is above or below the supplier's specification of ⅛-inch or 125 thousandths?

14. In a study of the buying habits of a supermarket's customers, you have recorded the purchases of 15 customers over the past month. You compile the table below, showing the number and value of their purchases during July 1973.

Customer	Visits during Month (1)	Total Expenditure (2)	Average Exp. per Visit (3)
1	20	$ 62	$ 3.10
2	10	54	5.40
3	7	40	5.70
4	11	64	5.80
5	8	48	6.00
6	12	74	6.20
7	10	76	7.60
8	9	74	8.20
9	8	69	8.60
10	9	83	9.20
11	9	105	11.70
12	7	94	13.40
13	5	72	14.40
14	4	65	16.20
15	6	100	16.70
Total	135	$1,080	$138.20

In the process of analysis, several divergent views develop as to what is the correct "average" with which to describe these data. One analyst claims that the average size of purchase is $8.00 (i.e., $1,080/135); another contends that the appropriate figure is $9.21 (i.e., $138.20/15); still a third analyst claims the median is the appropriate measure and selects $8.20 (the middle value in column 3); a fourth analyst, also claiming to have the median, selects $6.20 (the middle or 68th visit in column 1, which falls in the sixth row with average purchase $6.20).
a) What is the meaning of each of these four numbers?
b) Which number do you think is appropriate? Why?

15. In Chapter 2, Problem 20:
a) Compute the mean mileage per gallon.
b) Interpolate to estimate the median mileage.
c) What does the difference between the mean and median indicate about the skewness of this distribution?

16. The age of 100 refrigerators turned in for new models in a recent survey is:

Years	No. of Refrigerators
0 and under 1	10
1 and under 2	19
2 and under 3	26
3 and under 4	18
4 and under 5	13
5 and under 6	8
6 and under 7	3
7 and over	3*
Total	100

* The average age of these three refrigerators is 10½ years.

a) What is the arithmetic mean of the ages of these 100 refrigerators?
b) Estimate the median age of the refrigerators.

17. A trucking concern kept statistics for several years on two makes of tires. It found the following results in miles of wear:

Tire	Median	Mean
A	25,000	27,000
B	27,000	25,000

Assuming that the two tires sell at the same price, which make would you advise the trucking concern to purchase? Why?

18. The U. B. Glad Company operates a small bulk plant which wholesales gasoline to independent retailers. Last week's sales are shown:

Gallons (000)	No. of Sales
0 and under 10	10
10 and under 20	20
20 and under 30	30
30 and under 40	25
40 and under 50	15
50 and under 60	10
60 and under 70	5
70 and under 80	5
Total	120

a) Compute from the above frequency distribution the total number of gallons sold last week.
b) Compute the average (mean) gallons per sale.
c) Is the mode above or below 25,000 gallons? How do you know?
d) Compute the median sale.

19. The president of a company states that the shares of the company are widely distributed. To illustrate his point, he presents the following frequency distribution:

Shares Held	Stockholders (Thousands)
1–10	10
11–20	18
21–50	20
51–100	12
101–500	4
501–1,000	2
Above 1,000*	1
	67

* The average number of shares for stockholders in this group is 2,500 shares.

a) Do you agree with the president's statement? Why?

b) What is the mean number of shares held? What is the median number of shares held?

SELECTED READINGS

Selected readings for this chapter are included in the list that appears on page 93.

4. DISPERSION

THE preceding chapters covered two basic methods of describing a set of data: first, the frequency distribution, which groups a large number of values into a few classes; second, the average, which summarizes the typical value. This chapter describes other measures that are needed to show how the data vary about the average, because this variation is sometimes as important as the average itself.

Four important characteristics of a distribution of values may be described by summary measures:

1. Average—typical size.
2. Dispersion—variation, spread, or scatter.
3. Skewness—asymmetry or lopsidedness.
4. Kurtosis—peakedness or relative influence of extreme deviations.

These four characteristics are illustrated in Chart 4–1 by smooth frequency curves. A frequency curve, as defined in Chapter 2, portrays the frequency distribution of a population of continuous data in which the area under any segment of the curve corresponds to the number of values in that interval. Chart 4–1 is drawn so that the total area under each curve is unity and the area within any interval is equal to the relative frequency for that interval.

Suppose these curves represent the distribution of wage rates in a large factory. Panel 1 then shows that wages in department A *average* lower than those in department B, although both have the same dispersion. In panel 2, department A has a wider variation or *dispersion* of wages than department B, although both have the same average. The curves in both panels are symmetrical and normal (as defined in Chapter 6). Panel 3 illustrates *skewness*. Here most of the wages in department *A* are near the minimum rate, although some are much higher (i.e., skewness is positive or to the right); while in department *B* most

Chart 4–1

FOUR SUMMARY MEASURES OF A FREQUENCY DISTRIBUTION

1. Average Is Small (*A*) or Large (*B*)

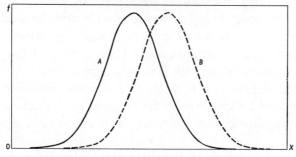

2. Dispersion Is Wide (*A*) or Narrow (*B*)

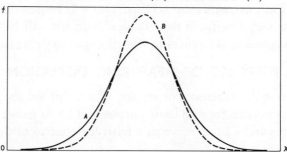

3. Skewness Is Positive (*A*) or Negative (*B*)

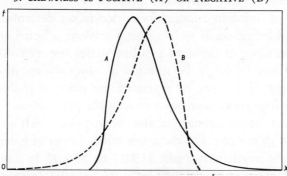

4. Kurtosis Is Peaked (*A*), Flat-Topped (*B*) or Normal (*C*)

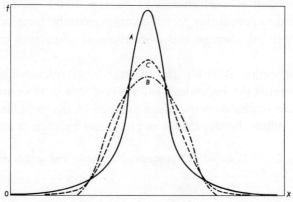

of the wages are near the maximum (skewness is negative or to the left). Finally, panel 4 shows different types of *kurtosis* in three symmetrical distributions having the same average and the same dispersion (as measured by the standard deviation, to be explained later). The distribution in department *A* is peaked, since most of the workers receive about the same wage with few very high or low wages; while the distribution in department *B* is flat-topped, indicating that the typical wages cover a wider spread with fewer extreme deviations; and in department C the distribution is normal, as if it had been determined by chance.[1]

Averages and measures of dispersion are the most important of these four kinds of summary measures. Dispersion will be described at length, and skewness very briefly, in this chapter. Kurtosis will be omitted, except for nontechnical references to the effects of extreme deviations.

PURPOSES OF MEASURING DISPERSION

Dispersion is the variation, or scatter, of a set of values. Measures of dispersion are needed for two basic purposes: (1) to gauge the reliability of averages and (2) to serve as a basis for control of the variability itself.

To illustrate the first purpose, suppose a company analyst is measuring the cost of living in a large city as one factor determining whether wages should be raised. If in five filling stations selected at random he finds that the price of standard gasoline varies between 40.9 and 41.9 cents per gallon, he might be justified in using the mean of as few as five prices, say 41.2 cents, to represent the price of gasoline. That is, the mean of five prices represents closely the price at each station, and it provides a reliable estimate of the mean price of all standard-grade gasoline sold in the city. On the other hand, prices of a certain type of woman's dress might vary from $19.95 to $34.95 in five stores. The mean of so few prices would then be highly unreliable as an estimate of the mean price of all such dresses in the city, but a measure of dispersion is needed to reveal this fact. To summarize the facts in most cases, therefore, both an average and a measure of dispersion must be presented.

When dispersion is small, the average is a typical value in that it closely represents the individual values, and it is reliable in that it is a good estimate of the corresponding average in the population. On the other hand, when the dispersion is great, the average is not so typical

---

[1] Curves *A, B,* and *C* are called leptokurtic, platykurtic, and mesokurtic, respectively.

and, unless the sample is very large, the average may be quite unreliable (see Chapter 9).

The second basic purpose of measuring dispersion is to determine the nature and causes of variation in order to control the variation itself. In matters of health, variations in body temperature, pulse beat, and blood pressure are basic guides to diagnosis. Prescribed treatment is designed to control their variation. In industrial production, efficient operation requires control of quality variation, the causes of which are sought through inspection and quality control programs. Thus, measurement of dispersion is basic to the control of causes of variation.

Measures of dispersion include: (1) the range, (2) the quartile deviation, (3) the mean deviation, and (4) the standard deviation. These measures are analogous to the averages described in Chapter 3, both in their characteristics and methods of calculation.

THE RANGE

The range is simply the difference between the largest and the smallest values of a variable. For the gasoline prices varying from 40.9 to 41.9 cents per gallon, the range is one cent. The range can be easily found in an array, but it can be determined in a frequency distribution only if the high and low values in the end classes are known.

Sometimes the range is indicated merely by citing the largest and smallest figures themselves. Quotations of stock prices include the high and low for the day. Weather reports state the maximum and minimum temperatures. If the high and low values are not widely separated from adjoining values, as in these cases, the range may be a fairly good measure of dispersion. In particular, the range is the basic measure of variation used in quality control, as described in Chapter 10.

However, if the two extremes are erratic, the range is unreliable and misleading because it gives no hint of the dispersion of the intervening values. In the distribution of prices paid for cars, for example, the range might extend from a Rolls-Royce at $20,000 to a used Jeep at $800; this would give little information about the variation in prices paid by the majority of consumers. In general, if the population contains a few extreme deviations, the range obtained from a random sample is more unreliable than any other measure of dispersion. For these reasons, the range is not recommended for general use.

The influence of extreme deviations on a measure of dispersion can be reduced by excluding a specified proportion of values at each end of the array and using the range of the remaining central values as the measure of dispersion. The simplest and most useful of these measures are based on the quartiles, as explained below.

THE QUARTILE DEVIATION

The quartiles are the three points which divide an array or frequency distribution into four roughly equal groups.[2] That is, the first or lower quartile, Q_1, separates the lowest valued quarter of the total number of values from the second quarter; the second quartile, Q_2 (almost always called the median), separates the second quarter from the third quarter; and the third or upper quartile, Q_3, separates the third quarter from the top quarter. Consequently, the quartile range, Q_3-Q_1, includes the middle half of the items. The quartile deviation, Q, is half this range. That is,

$$Q = \frac{(Q_3 - Q_1)}{2}$$

The quartiles are widely used as measures of dispersion. *Dun's*, for example, reports the medians and quartiles of 14 operating ratios in each of 32 types of wholesalers. Thus, the quartiles of net profits on net working capital of 199 grocery wholesalers in 1970 were 4.43 and 17.90 percent, compared with the median of 8.67 percent.[3] This means that while the "typical" grocery wholesaler earned 8.67 percent on net working capital, about one fourth of the companies earned less than 4.43 percent and one fourth earned over 17.90 percent, indicating a wide spread of profitability in this field. Similarly, the National Industrial Conference Board's *Management Record* reports the median and quartile salaries for various occupations by cities.

Ungrouped Data

The first and third quartiles are found in an array just as is the median, which is the second quartile. They are the values whose ranks or order numbers are $n/4 + 1/2$ and $3n/4 + 1/2$, respectively, counting from the lowest value. Fractional order numbers are interpolated between neighboring values in the array.

---

[2] The groups are rarely exactly equal, for reasons described under the median and because n is seldom a multiple of four.

The term "quartile" is sometimes applied to an entire range of values rather than to a point. Thus, a score might be said to fall "in the upper quartile" (i.e., between the top value and the upper quartile partition point). Such a range, however, should be called "quarter" to avoid confusion with "quartile," which should refer only to a point.

[3] *Dun's*, October 1971, pp. 64–65.

In the case of the hourly earnings of 214 machine tool operators listed in Table 2–4, the value of Q_1 is the earnings whose rank is $214/4 + 1/2$, or 54. This is the earnings of the 54th man,[4] the middle man of the lower paid half of the operators. Similarly, the value of Q_3 is the earnings of the man who is 161st from the bottom or 54th from the top, the middle man of the upper half. The values of Q_1 and Q_3 are found to be $2.50 and $2.70, respectively, from the original ungrouped data in Table 2–4. This means that about one fourth of the operators earn less than $2.50, one fourth exceed $2.70, and the middle half fall between these values. The quartile deviation is then $(2.70 — 2.50) \div 2$, or $.10.

Grouped Data

The quartiles can be estimated for a frequency distribution in the same way as the median by these analogous formulas:

$$Q_1 = L + \frac{i(n/4 - F)}{f} \qquad Q_3 = L + \frac{i(3n/4 - F)}{f}$$

where L is the lower limit of the class containing the quartile, i is the class width, f is the frequency or number in that class, F is the cumulative frequency below that class, and n is the total number of values. In these formulas, it is assumed that values of X are spread evenly over each interval, as explained in connection with the median.

For the machine tool operators' earnings grouped in Table 4–1, Q_1, the 54th value, falls in the third class ($L = \$2.45$, $f = 49$, $F = 25$); and Q_3, the 161st value, falls in the fifth class ($L = \$2.65$, $f = 45$, $F = 137$). Therefore,

$$Q_1 = 2.45 + .10(53.5 - 25) \div 49$$
$$= 2.45 + .10(.58)$$
$$= 2.508 \text{ dollars per hour}$$

$$Q_3 = 2.65 + .10(160.5 - 137) \div 45$$
$$= 2.65 + .10(.52)$$
$$= 2.702 \text{ dollars per hour}$$

The quartile range is then $2.702 — 2.508 = .194$ dollars per hour, and the quartile deviation is half this, or $.097. These estimates check

---

[4] If there were 215 operators, Q_1 would rank $215/4 + 1/2$, or $54\frac{1}{4}$, i.e., one fourth of the way from the earnings of the 54th man to that of the 55th man from the bottom.

Table 4–1

INTERPOLATION FOR QUARTILES
IN A FREQUENCY DISTRIBUTION

HOURLY EARNINGS OF 214 APPRENTICE MACHINE TOOL
OPERATORS

Lower Limit of Class (L)	Number in Class (f)	Number Earning Less (F)	Location of Quartiles
$2.25	2	0	
2.35	23	2	
2.45	49	25	$Q_1 = \#54$
2.55	63	74	
2.65	45	137	$Q_3 = \#161$
2.75	25	182	
2.85	3	207	
2.95	4	210	
3.05	0	214	
Total	214		

fairly closely with the exact values already obtained from the un-grouped data.

The quartiles can be located graphically from a cumulative frequency curve, or ogive, in the same manner as the median.

The quartiles are relatively unaffected by extreme deviations. On the other hand, their reliability depends on the degree of concentration at the quartiles of the population from which the sample is selected. In particular, if there are gaps in the population around the quartiles, the sample quartiles are unreliable.

Other positional measures of dispersion include the *deciles,* which divide the data into 10 equal groups, and the *percentiles,* which divide the data into 100 equal groups. These values are calculated and interpreted in the same way as the quartiles.

The measures of dispersion which follow differ from the quartile deviation in that they take into account the deviation of every value from the average.

THE MEAN DEVIATION

The mean deviation, sometimes called the average deviation, is exactly what its name implies. It is simply the mean of the absolute deviations of all the values from some central point, such as the arithmetic mean or median. The deviations must be averaged as if they were all positive, since the mean of plus and minus deviations would be zero (if measured from the mean), or nearly so. The mean deviation theoretically should be measured from the median since it is then smallest, but it is usually more convenient to measure the deviations from the mean, as described below. There is little difference in the results.

The mean deviation is a concise and simple measure of variability. Unlike the range and quartile deviation, it takes every item into account, and it is simpler and less affected by extreme deviations that the standard deviation, which will be described in the next section. It is therefore often used in small samples that include extreme values.

Ungrouped Data

The formula for the mean deviation (measured from the arithmetic mean) in a set of ungrouped data is

$$MD = \frac{\Sigma|X - \bar{X}|}{n}$$

where the blinkers | | mean that the signs are ignored. That is, the absolute deviations from the mean are added, and the sum (Σ) is divided by the number of values (n) to find the mean deviation (MD).

Table 4–2

COMPUTATION OF MEAN DEVIATION
FOR UNGROUPED DATA

PRICE-EARNINGS RATIOS OF FIVE ELECTRONICS STOCKS

Common Stock	Price-Earnings Ratio (X)	Deviation from Mean $\|X - \bar{X}\|$
A	19.6	0.4
B	17.3	2.7
C	19.2	0.8
D	14.0	6.0
E	29.9	9.9
Total	100.0	19.8
Mean	20.0 = $\bar{X}$	4.0 = MD

The mean deviation is computed in Table 4–2 for the price-earnings ratios of five electronics stocks, whose mean is 20.0. That is,

$$MD = \frac{\Sigma|X - \bar{X}|}{n} = \frac{19.8}{5} = 4.0$$

This means that while the five price-earnings ratios averaged 20.0, there was a wide variation among them, since the average departure from the mean was 4.0. Furthermore, the sample includes only five stocks. Therefore, the average ratio of 20.0 must be considered rather unreliable as an estimate of the typical price-earnings ratio for electronics stocks generally, assuming a large population of such stocks.

Grouped Data

The mean deviation can be computed from grouped data by the formula

$$\text{MD} = \frac{\Sigma f |X - \bar{X}|}{n}$$

where $|X - \bar{X}|$ is the absolute deviation of the class midpoint (X) from the arithmetic mean, ignoring signs, and f is the frequency in that class.[5] This formula will not be illustrated here, since its practical use is limited. The mean deviation has certain logical and mathematical limitations, such as disregarding plus and minus signs in averaging deviations. Consequently, the standard deviation is usually used instead for large distributions of grouped data.

THE STANDARD DEVIATION

The standard deviation is founded by (1) *squaring* the deviations of individual values from the arithmetic mean, (2) summing the squares, (3) dividing the sum by $(n - 1)$, and (4) extracting the square root. Like the mean deviation, the standard deviation is based on the deviations of all values, but it is better adapted to further statistical analysis. This is partly because squaring the deviations makes them all positive, so that the standard deviation is easier to handle algebraically than the mean deviation. The standard deviation is therefore of such importance that it is, in fact, the "standard" measure of dispersion.

Ungrouped Data

The basic formula for the standard deviation of ungrouped data is

$$s = \sqrt{\frac{\Sigma(X - \bar{X})^2}{n - 1}}$$

---

[5] For a short-cut method of computing the mean deviation for grouped data, see W. A. Spurr, L. S. Kellogg, and J. H. Smith, *Business and Economic Statistics* (Homewood, Ill.: Richard D. Irwin, 1954), pp. 227–28.

where s is the standard deviation; $(X - \bar{X})$ is the deviation of any value of X from the arithmetic mean $\bar{X}$; $\Sigma(X - \bar{X})^2$ is the sum of the squared deviations; and n is the number of items in the sample. The deviations may be squared most easily by referring to a table of squares, such as Appendix C or *Barlow's Tables*.

The square of the standard deviation (s^2) is called the *variance*. This is an important concept in statistical inference, to be considered later.

The above formula is now commonly used in statistics because it provides the best estimate of the standard deviation of the population from which the sample was drawn. An alternative formula for the standard deviation is $\sqrt{\Sigma(X - \bar{X})^2/n}$, which measures the dispersion of the sample itself but tends to understate the dispersion of the population. Since we usually take a sample in order to estimate population values, we will use $n - 1$ in our equations for s, the sample standard deviation, and will regard s as an estimate of σ (small sigma), the population standard deviation. (However, n may be substituted for $n - 1$ if desired; it makes little difference when n is large, as in most economic data.)

For the population, $\sigma = \sqrt{\Sigma(X - \mu)^2/N}$, where μ (small mu in Greek) is the population mean, and N is the number of values. Here, the variance (σ^2) is simply the average of the squared deviations from the mean.

In the sample of five price-earnings ratios listed in Table 4–3, col-

Table 4–3

COMPUTATION OF STANDARD DEVIATION
FOR UNGROUPED DATA

PRICE-EARNINGS RATIO OF FIVE ELECTRONICS STOCKS

(1)	(2)	(3) (4) Direct Method		(5)
Common Stock	Price-Earnings Ratio (X)	Deviation from Mean $(X - \bar{X})$	$(X - \bar{X})^2$	Short-Cut Method X^2
A....................	19.6	− .4	.16	384.16
B....................	17.3	−2.7	7.29	299.29
C....................	19.2	− .8	.64	368.64
D....................	14.0	−6.0	36.00	196.00
E....................	29.9	9.9	98.01	894.01
Total...............	100.0	0.0	142.10	2,142.10
Mean...............	20.0			

umn 2, the deviations from the mean of 20.0 are shown in column 3 and the squares in column 4. Their sum, $\Sigma(X - \bar{X})^2$, is 142.10, and $n = 5$ stocks. The standard deviation is then

$$s = \sqrt{\frac{\Sigma(X - \bar{X})^2}{n - 1}} = \sqrt{\frac{142.10}{4}} = 6.0$$

Short-Cut Method. While the above formula describes the standard deviation succinctly, it may be easier to compute its value directly from the original data, without finding the deviations from the mean. The following formula gives the same result as the one above:

$$s = \sqrt{\frac{\Sigma X^2 - (\Sigma X)^2/n}{n - 1}}$$

In Table 4–3, column 5 shows the original X values squared for use in this formula; columns 3 and 4 are not needed. Then,

$$s = \sqrt{\frac{2{,}142.10 - (100.0)^2/5}{4}} = \sqrt{35.52} = 6.0$$

The standard deviation is larger than the mean deviation of 4.0. This is always true because the squaring of the deviations puts more emphasis upon the extreme items.

Grouped Data

In a frequency distribution the midpoint of each class is used to represent every value in that class. The basic formula for the standard deviation therefore becomes

$$s = \sqrt{\frac{\Sigma f(X - \bar{X})^2}{n - 1}}$$

where $(X - \bar{X})^2$ is the deviation of the class midpoint (X) from the arithmetic mean and f is the frequency in that class.

A brief illustration is given in Table 4–4, which shows the prices of a transistor radio in six stores. The mean price is $26. Then:

$$s = \sqrt{\frac{\Sigma f(X - \bar{X})^2}{n = 1}} = \sqrt{\frac{6}{5}} = 1.10 \text{ dollars}$$

Table 4-4

COMPUTATION OF STANDARD DEVIATION
FOR GROUPED DATA

PRICES OF A TRANSISTOR RADIO IN SIX STORES

(1) Price in Dollars (Class Midpoint) X	(2) Number of Stores (Frequency) f	(3) Deviation from Mean (Dollars) $(X - \bar{X})$	(4) $(X - \bar{X})^2$	(5) $f(X - \bar{X})^2$
24	1	-2	4	4
25	0	-1	1	0
26	3	0	0	0
27	2	1	1	2
Total	6			6

Short-Cut Methods. An alternative short-cut formula uses the class midpoints (X) themselves rather than their deviations $(X - \bar{X})$ from the mean, as follows:

$$s = \sqrt{\frac{\Sigma f X^2 - (\Sigma f X)^2/n}{n - 1}}$$

These two formulas are the same as those for ungrouped data except for using X as the class midpoint and f as the class frequency. The short-cut formula will not be illustrated because in practice the standard deviation of grouped data is usually computed by a still shorter method, similar to that used for the arithmetic mean in Chapter 3.

The shortest method of computing the standard deviation of grouped data having class intervals of equal width is to use the formula:

$$s = i\sqrt{\frac{\Sigma f d^2 - (\Sigma f d)^2/n}{n - 1}}$$

where i is the width of the class interval, f is the frequency, d is the deviation of a class midpoint from the assumed mean in class interval units, $\Sigma f d^2$ is the sum of f times d^2 for each class (not Σf times Σd^2), and n is the total number of items.

The method is illustrated in Table 4–5. The first four columns of this table are identical with those used in Table 3–2 to find the arithmetic mean by the short-cut method. The steps are listed on pages 52–54. The last column $(f d^2)$ may be computed by multiplying d by $f d$, i.e., col. 3 × col. 4. [This is *not* $(f d)^2$.] Since the d's are small

Table 4–5

COMPUTATION OF STANDARD DEVIATION
FOR GROUPED DATA—SHORTEST METHOD

HOURLY EARNINGS OF 214 APPRENTICE MACHINE TOOL OPERATORS

(1) Class Midpoint (Dollars) X	(2) Frequency f	(3) Deviation from Assumed Mean in Classes d	(4) fd	(5) fd^2
2.30	2	−3	− 6	18
2.40	23	−2	−46	92
2.50	49	−1	−49	49
2.60	63	0	0	0
2.70	45	1	45	45
2.80	25	2	50	100
2.90	3	3	9	27
3.00	4	4	16	64
Total	214		19	395

integers, columns 4 and 5 can usually be computed mentally. Then the column totals are substituted in the formula as follows:

$$s = i\sqrt{\frac{\Sigma fd^2 - (\Sigma fd)^2/n}{n-1}}$$

$$= .10\sqrt{\frac{395 - (19)^2/214}{213}}$$

$$= .10\sqrt{1.85}$$

$$= .136 \text{ dollars per hour}$$

The result of this formula is the same as for the two other formulas for the standard deviation given, but the computations in columns 3, 4, and 5 are simpler. In any case, the standard deviation for grouped data is slightly less exact than that computed from the original data, since in formulas containing f the values in each class are rounded off to the class midpoint.[6]

---

[6] The three formulas for grouped data would be exact if every value of X were equal to its class midpoint. In case the concentration of values tapers off on both sides of the mean, as in a normal distribution, it is appropriate to adjust for grouping errors by subtracting $i^2 \div 12$ from the variance s^2. This is called *Sheppard's adjustment*. This adjustment is not generally recommended, however, because (1) when major points of concentration occur at midpoints the unadjusted formula is more nearly appropriate, (2) when values of X are evenly distributed over the intervals the one-twelfth adjustment should be *added*, not subtracted. Hence, the unadjusted formula is not only appropriate for one assumption but is also the mean of results obtained from two other assumptions. Finally, (3) errors of grouping are often small in comparison with other types of errors.

If the widths of class intervals in a frequency distribution are unequal, the class deviations must be adjusted to uniform units (such as the smallest interval or the highest common factor) in order to apply the short-cut formula. Otherwise one of the longer formulas should be used. If the distribution has an open end, neither the mean deviation nor the standard deviation can be computed unless the missing end values can be estimated.

RELATION BETWEEN MEASURES OF DISPERSION

In a normal distribution there is a fixed relationship between the three principal measures of dispersion. The quartile deviation is smallest, the mean deviation next, and the standard deviation σ is largest, in the following proportions:[7]

$$Q \approx 2/3\sigma$$
$$MD \approx 4/5\sigma$$

where the sign $\approx$ denotes approximate equality. These proportions are useful in estimating one measure of dispersion when another is known or in checking roughly the accuracy of a calculated value. Thus, if the computed standard deviation differs very widely from its value estimated as $3/2$ of Q, either an error has been made or the distribution differs considerably from normal.

Another comparison can be made of the proportion of items that are typically included within the interval of one Q, MD, or σ measured both above and below the population mean μ. In a normal distribution,

$\mu \pm Q$ includes 50 percent of the items
$\mu \pm MD$ includes 57.51 percent of the items
$\mu \pm \sigma$ includes 68.27 percent of the items

These relationships are shown graphically in Chart 4–2. Note that the standard deviation is the distance between the mean and the point of inflection on the normal curve, that is, the point where the curve changes from being concave downward to being concave upward, and where it is steepest.

For the machine tool operators, the interval around the sample mean $\overline{X} \pm Q$ is \$2.609 $\pm$ \$.097, or from \$2.512 to \$2.706 per hour. This interval actually includes about 50 percent of the workers, and so the distribution is nearly normal in this respect. The proportions within the intervals $\overline{X} \pm MD$ and $\overline{X} \pm s$ are also nearly normal for the hourly earnings, since they contain 55 and 67 percent of the workers, respectively.

---

[7] More precisely, $Q = .6745\sigma$ and $MD = .7979\sigma$.

Chart 4–2

PROPORTIONS OF AREA OF NORMAL CURVE INCLUDED IN INTERVALS
BASED ON COMMON MEASURES OF DISPERSION

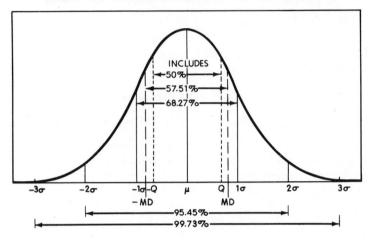

The proportions of items typically falling within 1, 2, and 3 standard deviations of the mean are even more widely used in statistical analysis. In a normal distribution,

$$\mu \pm \sigma \text{ includes } 68.27 \text{ percent of the items}$$
$$\mu \pm 2\sigma \text{ includes } 95.45 \text{ percent of the items}$$
$$\mu \pm 3\sigma \text{ includes } 99.73 \text{ percent of the items}$$

These relations are also shown graphically in Chart 4–2. The interval $\bar{X} \pm 2\sigma$ thus includes about 19 out of 20 of the items, while $\bar{X} \pm 3\sigma$ includes nearly all of them. In the case of the machine tool operators, the interval $2.609 \pm (3 \times \$.136)$, or from $2.201 to $3.017, includes 212 out of 214 workers (Table 2–4). In general, so long as the departure from symmetry is only moderate, an interval of 3σ on both sides of the average will give the practical limits of the distribution.

Which Measure of Dispersion to Use?

As in the case of averages, the selection of the proper measure of dispersion depends on three main factors:

1. The concept of dispersion required by the problem. Is a single pair of values adequate, such as the two extremes or the two quartiles (range or Q)? Or is a simple average of all absolute deviations from the mean or median needed (i.e., mean deviation)? Or an average (the standard deviation) that is better adapted for further calculations?

2. The type of data available. If they are few in number, or contain extreme values, avoid the standard deviation. If they are generally skewed, avoid the mean deviation as well. If they have gaps around the quartiles, the quartile deviation should be avoided.
3. The peculiarities of the dispersion measures themselves. These are summarized under "Characteristics of Measures of Dispersion," below.

As a rule of thumb, the median and quartiles may be used as simple, easily understandable summary values for rough or skewed data, as in a distribution of personal incomes, but the overall range should be avoided.[8] The mean deviation is commonly used to give equal weight to all deviations where n is small and in ungrouped data, even if the distribution is somewhat erratic, as in time series. But if n is large and the distribution is fairly symmetrical, and if more refined analysis is needed, such as the study of inference or correlation, the standard deviation should be used instead. A major reason for the widespread use of the standard deviation is that it has the smallest sampling error of any dispersion measure when the distribution is normal; that is, the sample value tends to deviate from the population value by the smallest percentage.

Characteristics of Measures of Dispersion

The characteristics of the individual measures of dispersion are summarized below:

Range:

1. The range is the easiest measure to compute and to understand, but
2. It is often unreliable, being based on two extreme values only.

Quartile Deviation:

1. The quartile deviation is also easy to calculate and to understand.
2. It depends on only two values, which include the middle half of the items.
3. It is usually superior to the range as a rough measure of dispersion.
4. It may be determined in an open-end distribution, or one in which the data may be ranked but not measured quantitatively.
5. It is also useful in badly skewed distributions or those in which

---

[8] An exception is the use of the range in quality control, discussed in Chapter 10.

other measures of dispersion would be warped by extreme values.
6. However, it is unreliable if there are gaps in the data around the quartiles.

Mean Deviation:

1. The mean deviation has the advantage of giving equal weight to the deviation of every value from the mean or median.
2. Therefore, it is a more sensitive measure of dispersion than those described above and ordinarily has a smaller sampling error.
3. It is also easier to compute and to understand and is less affected by extreme values than the standard deviation.
4. Unfortunately, it is difficult to handle algebraically, since minus signs must be ignored in its computation.

Standard Deviation:

1. The standard deviation is usually more useful and better adapted to further analysis than the mean deviation.
2. It is more reliable as an estimator of the population value than any other dispersion measure, provided the distribution is normal.
3. It is the most widely used measure of dispersion and the easiest to handle algebraically.
4. However, it is harder to compute and more difficult to understand, and
5. It is greatly affected by extreme values that may be due to skewness of data.

MEASURES OF RELATIVE DISPERSION

The measures of dispersion so far described are expressed in original units, such as dollars. These values may be used to compare the variation in two distributions provided the variables are expressed in the same units and are of about the same average size. In case the two sets of data are expressed in different units, however (such as tons of coal and cubic feet of gas), or if the average size is very different (such as executives' salaries versus laborers' wages), the absolute measures of dispersion are not comparable and measures of *relative* dispersion should be used instead.

A measure of relative dispersion is the ratio of a measure of absolute dispersion to an appropriate average and is usually expressed as a percentage. It is sometimes called a *coefficient of dispersion,* a "coefficient" being a ratio or pure number that is independent of the unit of measurement. A coefficient of dispersion may be computed from either the quartile or mean deviation[9] but is usually expressed as the ratio of the standard deviation to the mean, $s/\overline{X}$.

Thus, for the apprentice machine tool operators' earnings, the coefficient of dispersion is

$$s/\overline{X} = .136/2.609 = 5.2\%$$

That is, the standard deviation is 5.2 percent of the mean earnings. If a group of plumbers had a standard deviation of $\$.160$ and mean earnings of $\$8.00$ an hour, their earnings would vary more than those of the operators in dollars, to be sure ($\$.160$ versus $\$.136$), but they would vary less relative to their average earnings ($.160 \div 8.00 = 2.0$ percent versus 5.2 percent). The relative measure is the more significant comparison.

Standard Deviation Units

Individual deviations from the mean $(X - \overline{X})$ may also be reduced to comparable units by dividing them by the standard deviation (s). Thus, for a machine tool operator earning $\$2.80$ an hour, or $\$.191$ above the mean of $\$2.609$, $x/s = .191/.136 = 1.40$. His wage is, therefore, 1.40 standard deviations above the mean, a value which is comparable with, say, his output in units produced, which may be 2.20 standard deviations above the mean. Perhaps he rates a raise in pay! Or, in a college entrance test in which the mean is adjusted to 500 and the standard deviation is 100, a candidate who scores 700 knows that he is "two sigmas" above the mean. If the distribution is normal, then only about 2 percent of the candidates did better, since $\mu \pm 2\sigma$ includes 95.45 percent of the scores, leaving 4.55 percent above and below this range, or 2.27 percent above $\mu + 2\sigma$.

The values of x/s will vary from approximately $+3$ to -3 for any set of data, since this spread includes nearly all the items in a normal distribution. The interval $\overline{X} \pm 3s$ therefore provides the practical limits of variation used in quality control and other applications. Variation greater than these limits indicates the presence of abnormal forces that may have to be corrected.

SKEWNESS

Skewness means the lack of symmetry in the shape of a frequency curve. The extent of this lopsidedness is another important characteristic of a frequency distribution.

---

[9] The formulas are $(Q_3 - Q_1)/(Q_3 + Q_1)$ and $MD/\overline{X}$, respectively.

The simplest measure of skewness is based on the spread between the arithmetic mean and median. They are identical in a symmetrical distribution. In a skewed distribution, however, the mean is pulled out in the direction of the extreme values while the mode remains under the highest point of the curve, and the median, which is affected by the number of extreme values but not their value, tends to fall about one third of the way from the mean toward the mode, provided the skewness is moderate.

A *coefficient of skewness* may therefore be defined as follows:

$$Sk = \frac{3(\bar{X} - Md)}{s}$$

where $\bar{X}$ is the mean, Md is the median, and s is the standard deviation.

The numerator $3(\bar{X} - Md)$ is used instead of $(\bar{X} - \text{Mode})$ because the mode is often difficult to locate accurately. Dividing by s expresses the measure in standard deviation units, so that it is comparable between distributions that differ in unit of measurement or in average size. If the mean exceeds the median, the skewness is positive; otherwise it is negative.

The formula will not be illustrated here because of its limited practical use. The accurate measurement of skewness requires more advanced techniques. In elementary analysis, skewness is ordinarily treated in descriptive terms rather than being summarized by a single measure.

USES OF MEASURES OF DISPERSION

There are many other uses of dispersion measures than those which have just been described. The following summary briefly indicates these various applications.

Aid in Description

The simplest and most common use of a measure of dispersion is in the description of data. Averages are typical values, but measures of dispersion indicate the scatter of the data. The extent and direction of skewness should also be noted.

Comparison of Dispersion

The average values of two sets of data may be very similar, while the range and pattern of scatter differ greatly. If the data are generally alike, the measures of dispersion can be compared in absolute units to determine how the data differ in their variability. When several sets of data are expressed in different kinds of units or in similar units of widely different size, comparisons based on measures of relative dispersion are usually more appropriate.

Provision of a Standard

By the use of measures of dispersion, particularly the standard deviation, it is possible to compare the variation in a given group of data with that of the normal curve as a standard. It has been pointed out that approximately 68 percent of all the items in a normal distribution are included between one standard deviation above the mean and one standard deviation below the mean. When characteristics of a variable are expressed in standard deviation units, its distribution can be compared with a normal distribution. This use is at the very heart of studies of reliability of sample averages, quality control programs in industrial production, and other applications of statistical methods.

Measurement of Sampling Errors

Reliability of sample averages is an important part of statistical analysis. Averages vary by chance from sample to sample in the same population. In order to evaluate the reliability of the average in a single sample, we must know more about the variation of that average in all possible samples. The standard deviation is ordinarily used in this type of study, as explained in Chapter 9.

SUMMARY OF FORMULAS

Since the characteristics of the various measures of dispersion and skewness have been summarized above, the chapter may be concluded by listing the principal formulas used:

Measure	Ungrouped Data	Grouped Data
Range...............	Substract end values	Same
Quartile deviation....	$Q = \dfrac{Q_3 - Q_1}{2}$	Same
	Q_1 is #$n/4 + 1/2$*	$Q_1 = L + \dfrac{i(n/4 - F)}{f}$
	Q_3 is #$3n/4 + 1/2$*	$Q_3 = L + \dfrac{i(3n/4 - F)}{f}$
Mean deviation......	$MD = \dfrac{\Sigma\lvert X - \bar{X}\rvert}{n}$	$MD = \dfrac{\Sigma f\lvert X - \bar{X}\rvert}{n}$
Standard deviation....	$s = \sqrt{\dfrac{\Sigma(X - \bar{X})^2}{n - 1}}$	$s = \sqrt{\dfrac{\Sigma f(X - \bar{X})^2}{n - 1}}$

* In an array, counting from lowest value.

Short-cut method..... $s = \sqrt{\dfrac{\Sigma X^2 - (\Sigma X)^2/n}{n-1}}$ $s = \sqrt{\dfrac{\Sigma fX^2 - (\Sigma fX)^2/n}{n-1}}$

Shorter method, for
classes of equal width. $s = i\sqrt{\dfrac{\Sigma fd^2 - (\Sigma fd)^2/n}{n-1}}$

Relative dispersion...Divide measure of absolute dispersion by appropriate average, e.g., $s/\overline{X}$.

Skewness............ $Sk = \dfrac{3(\overline{X} - Md)}{s}$ Same

PROBLEMS

1. As market analyst for a drug manufacturer who is considering entering the Philadelphia market, you wish to study the retail price behavior of milk of magnesia, tincture of iodine, and other standard items. You compile the following data from a sample survey of Philadelphia stores:

DISTRIBUTION OF PRICES IN PHILADELPHIA DRUG STORES

Price in Cents (Standard Size)	Milk of Magnesia (Percent of All Stores Surveyed)	Tincture of Iodine (Percent of All Stores Surveyed)
15		25
16		
17		
18		
19	5	1
20		26
21	2	5
22		
23	3	5
24		
25	7	38
26		
27	3	
28		
29	41	
30	6	
31		
32		
33	5	
34	5	
35	8	
36	5	
37		
38	1	
39	9	
Total stores	100%	100%
Mean	30.18 cents	20.84 cents
Median	29.00 cents	20.00 cents
Standard deviation	4.90 cents	4.00 cents

SOURCE: *Retail Price Behavior*, University of Maryland, Studies in Business and Economics, Vol. 4, No. 2, p. 8.

Compare these two distributions as to their:
a) Averages.
b) Dispersion (both absolute and relative).
c) Skewness.

2. Cite actual or hypothetical illustrations, not given in the text, of each of the following:
   a) Two main purposes of measuring dispersion.
   b) Positive and negative skewness.
   c) Narrow dispersion and peaked kurtosis.

3. The following values show the number of hours of operation before repairs were required for eight power saws: 35, 27, 21, 29, 35, 29, 27, and 21; total 224 hours. Compute and explain briefly the meaning of:
   a) The third quartile.
   b) The mean deviation.
   c) The standard deviation and variance.
   d) A measure of relative dispersion, using the standard deviation.
   e) The largest value (35) expressed in standard deviation units above the mean.

4. The National Bureau of Economic Research computes mean deviations to show how different business cycles vary in duration and other respects.
   a) Find the mean deviation of the 11 cycle-duration periods in Chapter 3, Problem 8.
   b) Are cycles fairly uniform or variable in duration? Explain.
   c) Why would you say that the Bureau uses the mean deviation instead of the standard deviation to measure variability of business cycle behavior?

5. In Chapter 2, Problem 17, on earnings of women in an assembly plant:
   a) Find the range and quartile deviation from your original list of 112 items.
   b) Interpolate the quartiles and compute the quartile deviation from your frequency distribution of these data.
   c) Why do the quartile values differ in (a) and (b)?

6. Using your frequency distribution in the problem above:
   a) Compute the standard deviation.
   b) Explain the meaning of this measure in terms of electronic workers' earnings.
   c) Should this value of s differ from the following? Give reasons.
      (1) The s of the original ungrouped data.
      (2) The s for the other formulas containing f.

d) Estimate the mean deviation from the standard deviation, assuming a nearly normal distribution.

7. Answer the same questions as in Problem 6 above, for the starting salaries of college men in whichever of the five fields is assigned in Chapter 2, Problems 14–16.

8. A purchasing agent obtained samples of incandescent lamps from two suppliers. He had the samples tested in his own laboratory for length of life, with the following results:

| | SAMPLES FROM | |
LENGTH OF LIFE IN HOURS	Company A	Company B
700 and under 900......................10		3
900 and under 1,100.....................16		42
1,100 and under 1,300...................26		12
1,300 and under 1,500.................. 8		3
Total...............................60		60

a) Which company's lamps have the greater average length of life?
b) Which company's lamps are more uniform?

9. *a*) What ratio is MD to Q in a normal distribution?
   b) The interval $\mu \pm 3\sigma$ includes nearly all the items in a normal distribution. Express this range in Q units.
   c) If you compute the standard deviation to be .612 pounds and note as a rough check that the overall range is 36 pounds, what is the most obvious type of error you might have made?
   d) In a normal distribution of test scores with $\mu = 60$, $\sigma = 9$, what percentage of scores exceeds 33? 51? 78?

10. If a test of 100 pieces of cotton thread shows a mean breaking strength of 15 pounds and a median breaking strength of 14.8 pounds, with a standard deviation of 3 pounds, about what number of pieces of thread in the lot should have a breaking strength between 12 and 21 pounds?

11. Regarding the dimensions of 63 gears in Table 2–3, page 26:
   a) Estimate the standard deviation of the whole lot from which this sample was drawn.
   b) Check your result against the rough estimate of σ as one sixth of the range (since the interval $\overline{X} \pm 3\sigma$ includes practically all items in a normal distribution).

c) How much does the largest gear (.4270) differ from the mean in standard deviation units?

12. Refer to Problem 10 in Chapter 3:
    a) Calculate the standard deviation of sales per store.
    b) Estimate the quartiles and the quartile range.

13. To conclude your report on the thickness of 200 sheets of ⅛-inch insulating board used for power transformers, you need a measure of variation for the 200 sheets listed in Chapter 2, Problem 19. This measure is to be used in quality control, so it should reflect the variability of every sheet. The distribution is nearly normal; hence the standard deviation is appropriate.
    a) Compute the standard deviation by the shortest possible method.
    b) Compute a coefficient of dispersion. What is the advantage of this measure of relative dispersion as compared with the corresponding measure of absolute dispersion?
    c) Another sheet, measuring 116 thousandths of an inch in width, is received from the same supplier. How many standard-deviation units is this below your mean? Is this sheet inside or outside the control limits $\bar{X} \pm 3s$ computed for the earlier shipment? (This subject is developed further in Chapter 10 on statistical quality control.)

14. In Chapter 2, Problem 18, on family income:
    a) Compute whatever measure of dispersion you think most appropriate and explains its significance.
    b) If there are any dispersion measures you cannot compute from these data, name them and indicate why you cannot.

15. In Chapter 2, Problem 20, on gasoline mileage:
    a) Compute the standard deviation.
    b) Find the estimated variance for all such cars. Explain its significance.
    c) If you get 14 miles per gallon with your car, how many standard deviations are you below the mean of 18.8 miles per gallon?

16. In Chapter 3, Problem 16:
    a) Estimate the quartile deviation of refrigerator ages to the nearest year.
    b) Is the distribution of refrigerator ages normal, negatively skewed, open-ended, or bimodal?

17. A firm which services household appliances for a national manufacturer is trying to determine where it should locate a service facility and its fleet of service trucks. The territory to be serviced lies along a straight highway and includes nine cities of roughly equal size. (See the sketch.) The manager decided to use the mean distance (counting the north end of the territory as zero) as the location for the facility and the truck fleet. Thus, he has decided upon City F for the facility (mean = 225/9 = 25).

MAP OF SERVICE TERRITORY	
Miles from City A	
0	City A
5	City B
10	City C
15	City D
20	City E
25	City F
40	City G
50	City H
60	City I
Total 225	

a) Compute the mean deviation of miles from the mean.

b) What does this figure tell the manager about the distance his service trucks will have to travel?

c) Before the manager has found a location, an assistant suggests that perhaps the median is a better measure to use here. Accordingly, the assistant suggests that City E, which is the middle city at 20 miles on the scale, be chosen as the site. Compute the mean deviation *about the median* (20).

d) By comparing this with the answer to (a) above, determine in which city the facility should be located. Why?

e) Do you think there is any better location? Explain.

18. As a further step in your analysis you wish to compare the dispersion of burning life for the two brands of electron tubes described in Chapter 2, Problem 21. The following calculations have been made from the raw data:

	Brand A	Brand B
ΣX	25,525	17,825
ΣX^2	6,888,125	4,999,375
n	120	80
$\bar{X}$	212.71	222.81

a) Calculate the standard deviation for each brand of tube.

b) Estimate the quartile deviation for each distribution from your cumulative frequency curve [Chapter 2, Problem 21 (d)].

c) Compare the dispersion of the two distributions using both measures. Which measure gives the best general description in this case? Why?

d) In Chapter 2, Problem 21 (d) you estimated the medians graphically. Using this estimate and the means above, what can you say about the skewness of these distributions?

19. Percentiles are like quartiles except that they divide the number of items in a distribution into 100 equal groups instead of four groups. Find the 10th percentile of gasoline sales in Chapter 2, Problem 20; i.e. the number of gallons of gasoline which exceeds 10 percent of the sales but which is exceeded by 90 percent of the sales. Use an interpolation formula patterned after that given for the quartiles on page 73.

SELECTED READINGS

CROXTON, FREDERICK E.; COWDEN, DUDLEY J.; and BOLCH, BEN W. *Practical Business Statistics.* 4th ed. Englewood Cliffs, N.J.: Prentice-Hall, 1969.
Chapters 2 to 5 provide a detailed treatment of ratios, frequency distributions, averages, and dispersion.

FREUND, JOHN E., and WILLIAMS, FRANK J. *Modern Business Statistics,* rev. by B. Perles and C. Sullivan. Englewood Cliffs, N.J.: Prentice-Hall, 1969.
A clear discussion of frequency distributions, measures of "location," and variation appears in chapters 2–5.

LEABO, DICK A. *Basic Statistics.* 3d ed. Homewood, Ill.: Richard D. Irwin, 1968.
Chapters 3 to 5 contain a careful account of basic statistical measures.

MEYERS, CECIL H. *Elementary Business and Economic Statistics,* chaps. 2 to 4. Belmont, Ca.: Wadsworth Publishing, 1966.
A detailed discussion of frequency distributions, averages and dispersion.

NETER, JOHN; WASSERMAN, WILLIAM; and WHITMORE, G. A. *Fundamental Statistics for Business and Economics.* 4th ed. Boston: Allyn & Bacon, 1973.
Includes analysis of relationships by cross classification of data, as well as ratios and frequency distribution analysis.

U.S. GOVERNMENT, SMALL BUSINESS ADMINISTRATION. *Ratio Analysis for Small Business.* Washington, D.C.: Superintendent of Documents, 1957.
A popular survey of principal business ratios, sources of published ratios, and their analysis and evaluation.

YULE, G. UDNY, and KENDALL, M. G. *An Introduction to the Theory of Statistics.* 14th ed. London: Charles Griffin, 1950.
Chapters 5 to 7 provide a comprehensive treatment of frequency distributions, averages, dispersion, skewness, and kurtosis.

5. AN INTRODUCTION TO PROBABILITY THEORY

PROBABILITY THEORY is a branch of mathematics that is eminently useful to the businessman. To a great extent, statistics is built upon the foundations of probability. The evaluation of information obtained from samples depends upon probability theory for its interpretation. Also, the businessman—like the poker player or military strategist—must make decisions in the face of uncertainty as to the future. He can express his judgment by attaching a numerical probability to each possible event that might affect the outcome of his decisions, and he can use these probabilities, together with economic information, to improve his decision-making process.

BASIC CONCEPTS

A *probability* is a number between 0 and 1, inclusive, representing the chance or likelihood that an event will occur. A probability of zero ($P = 0$) means the event is impossible; if $P = .50$, there is "half a chance" that it will occur; if $P = 1$, the event is certain to occur. The value of P cannot be negative or greater than one.

A probability may be thought of as the relative frequency of "successes" (i.e., the occurrence of a certain event) in a random process over a great number of trials. Relative frequency is the number of successes divided by the number of trials. Suppose we roll dice, and define a success as throwing an ace (1). If the dice are "fair," the six faces 1 through 6 are equally likely, and the ratio of aces to total throws will approach $1/6$ in the long run. We then define the probability of throwing an ace as $1/6$. The process of shooting dice is a random one because we do not know in advance the outcome of any given roll. In general, if r is the number of successes in n trials, then the limit of r/n

for larger and larger values of n is defined as the *probability of success in a single trial.*

Sources of Probabilities

The theoretical concept given above is difficult to apply in practice, but we can *estimate* probabilities in any of three ways:

1. *Relative Frequency of Past Events.* Probabalities can be estimated from relative frequencies either in a controlled experiment or in a sample survey of a large, finite population. To illustrate an experiment, suppose we set up a machine to turn out a new part and conduct an extended test run in which 5 percent of the parts prove to be defective. Then, if the process is controlled so that there is no change in quality of output, we can say that the probability is .05 that the next part will be defective. Of course, this part will in fact be either defective or good; our prior probability is derived from the long-run experience with many parts.

The probabilities for more complicated events can be determined from the probabilities for much simpler events by means of *simulation*—using an experimental model designed to approximate actual conditions. In studying an inventory system, for example, the orders of customers, the stock available, and the time necessary to replenish stocks are incorporated in the model. A customer order is initiated and its effect is traced upon the inventory system. This is repeated for other orders and the behavior of the inventory system determined (e.g., the probabilities that demand will exceed supply by 0, 1, 2, . . . items, respectively). Simulation is described in Chapter 15.

Probabilities can also be estimated from the relative frequency with which an event occurs in a sample survey of a large finite population. Thus, in Table 2–5, the survey of machine tool operators reveals that 29 percent of the total earn about $2.60 an hour. Then, the estimated probability is .29 that an operator drawn at random from the whole group of such operators would earn about $2.60. Similarly, the probabilities for men and women buyers in the next section are based on their relative frequencies in the sample survey cited.

2. *Theoretical Distributions.* In some situations, probabilities can be determined without recourse to relative frequencies. Thus, in rolling dice, we can state the probability of an ace as $1/6$ without actually rolling a die, simply because the six faces are equally likely to turn up. The probabilities for complicated events, too, can be derived from simple assumptions. For example, in tossing a fair coin four times, the probabilities of from 0 to 4 heads may be derived from the fact that the

probability of a head on one flip is $1/2$. The probability is $1/16$ for no heads, $1/4$ for one head, etc., as listed in Table 5–8 later in this chapter. Such probabilities can be determined from the binomial distribution described in Chapter 6 without recourse to experiments or surveys based on past experience. The validity of such theoretical distributions depends upon how closely the assumptions match the real-world situation. (For example, the probabilities in Table 5–8 do not apply if, in fact, our coin is bent.)

3. *Subjective Judgment.* If none of these methods can be used, the decision maker must estimate probabilities on the basis of his judgment and experience. An automobile manufacturer may judge the chances to be two out of three that customers will prefer one body style over another. The weatherman may say: "The chances are 6 out of 10 for rain." Most betting odds on athletic events are set by personal judgment. To include these situations, we enlarge the definition to include *subjective probability.* A subjective probability is an evaluation by a decision maker of the relative likelihood of unknown events.[1] It is his betting odds on the occurrence of the event. Since it is personal to the decision maker, two individuals may attach different subjective probabilities to the same event. Even so, these subjective probabilities can be used in decision making in the same manner as the more objective probabilities described above. Appendix B at the end of this chapter describes a procedure that has been found helpful in assessing subjective probabilities.

Joint, Marginal, and Conditional Probabilities

Before proceeding, it is necessary to establish certain definitions. This can be done best by illustration. In studying the buying behavior of customers of a certain product, suppose you have taken the random sample of 1,000 customers entering a department store shown in Table 5–1.

Suppose we are going to pick a customer from this group by chance. Then:

1. *Simple Probability.* Probability of drawing a man: $P(M) = .30$. The symbol $P(A)$ is used to denote the probability of an event A. The event "not-A" is represented by $\sim A$. Thus, the simple probability of drawing a woman is $P(\sim M) = .70$.

---

[1] We could be more precise and define subjective probability in terms of decision makers' preferences for hypothetical lotteries. For our purposes, the intuitive definition above will suffice. For more detail, see Howard Raiffa, *Decision Analysis* (Reading, Mass.: Addison-Wesley, 1968), chap. 5.

Table 5–1

BUYING BEHAVIOR OF 1,000 MEN AND WOMEN

(Percent of Total)

	Men (M)	Women (~M)	Total
Buyer (B)...............	3	17	20
Nonbuyer (~B).........	27	53	80
Total..................	30	70	100

2. *Joint Probability.* The probability of getting a customer with two (or more) specific characteristics. For example, the probability of drawing a customer who is both a buyer and a man is $P(B, M) = .03$, and the probability of drawing a customer who is a woman nonbuyer is $P(\sim M, \sim B) = .53$.

3. *Marginal Probability (on the margin of the table).* The total probability of drawing a man, made up of the probability of men buyers plus the probability of men nonbuyers, is:

$$P(M) = P(M, B) + P(M, \sim B) = .03 + .27 = .30$$

Marginal probability is no more than simple probability viewed in a different light. That is, simple probability is a singular concept, whereas the marginal probability is essentially a sum of joint probabilities.

4. *Conditional Probability.* Suppose that we know that the customer drawn was a man. Given this information, what is the probability that he is also a buyer? This is the conditional probability $P(B \mid M)$. The symbol $P(B \mid M)$ is read as the probability of a buyer *given* a man. Since 30 percent of the customers are men and 3 percent are buyers, $P(B \mid M) = .03/.30 = .10$. From the above illustration, we can determine the general rule or mathematical definition of conditional probability:

Conditional probability of B given M:

$$P(B \mid M) = \frac{P(B, M)}{P(M)} = \frac{\text{Joint probability of } B \text{ and } M}{\text{Marginal probability of } M}$$

From this definition we can find, for example, the probability of a buyer, given that the customer is a woman:

$$P(B \mid \sim M) = \frac{P(B, \sim M)}{P(\sim M)} = \frac{.17}{.70} = .243$$

On the other hand, consider $P(M \mid B)$, the probability of the customer being a man, given that he is a buyer:

$$P(M \mid B) = \frac{P(B, M)}{P(B)} = \frac{.03}{.20} = .15$$

Note that this is not equal to $P(B \mid M)$ above.

As another illustration, suppose that we had an ordinary deck of cards. The cards can be classified as shown in Table 5–2. From the table, we see that:

Table 5–2
PROBABILITIES IN DRAWING CARDS

	Red Card, R	Black (Nonred), $\sim R$	Total
Honor (A, K, Q, J, 10)..H	10/52	10/52	20/52
Nonhonor.......$\sim H$	16/52	16/52	32/52
Total...................$26/52 = 1/2$		$26/52 = 1/2$	1

Simple Probability. The probability of drawing a red card is $P(R) = 1/2$.

Joint Probability. The probability of drawing a black honor is $P(H, \smile R) = 10/52$.

Marginal Probability. The probability of drawing a red card, viewed as the sum of the probabilities of red honors and red nonhonors, is:

$$P(R) = P(H, R) + P(\smile H, R) = 10/52 + 16/52 = 1/2$$

Conditional Probability. The probability of an honor, given that we have drawn a red card, is:

$$P(H \mid R) = \frac{P(H, R)}{P(R)} = \frac{10/52}{26/52} = 10/26$$

Note that the simple probability of drawing an honor is also the same, that is, $P(H) = 10/26$. Hence, our knowledge that the card was red gave us no additional information about whether or not it was an honor, since the probabilities were exactly the same. This property is known as *statistical independence.*

Definition of Statistical Independence

When $P(H \mid R) = P(H)$, we say that the events H and R are statistically independent. That is, the event H is just as likely to occur when event R occurs as it is when event $\sim R$ occurs. (There is the same fraction of red honors as black honors.) Statistical independence implies that knowledge of one event is of no value in predicting the occurrence of the other event.

To illustrate the notion of statistical independence, let us carry on the example of the buying behavior of customers and classify customers by age as well as sex. We could have the information in Table 5–3.

Table 5–3

BUYING BEHAVIOR OF 1,000 MEN AND WOMEN, BY AGE

(Percent of Total)

	Men (M)		Women ($\sim$M)		
	Young (Y)	Older ($\sim$Y)	Young (Y)	Older ($\sim$Y)	Total
Buyer (B).........1		2	4	13	20
Nonbuyer ($\sim B$).....5		22	15	38	80
Total..............6		24	19	51	100

From it, the reader can easily verify that:

Total men   = 30%   Total young = 25%
Total women = 70%   Total older   = 75%

Now, the simple probability of a buyer is $P(B) = .20$. The marginal probability of a young person is:

$$P(Y) = P(B, M, Y) + P(\sim B, M, Y) + P(B, \sim M, Y)$$
$$+ P(\sim B, \sim M, Y)$$
$$= .01 + .05 + .04 + .15 = .25$$

The conditional probability of a buyer, given a young person, is:

$$P(B \mid Y) = \frac{P(B, Y)}{P(Y)} = \frac{.01 + .04}{.25} = .20$$

Note that this conditional probability equals the simple probability of a buyer, $P(B)$. Hence, age and buying behavior are statistically independent. Knowledge of age is of no value in predicting whether or not a person is a buyer. The fact that age and buying behavior are independent also implies that

$$P(\sim B \mid Y) = P(\sim B); \; P(B \mid \sim Y) = P(B);$$
$$\text{and } P(\sim B \mid \sim Y) = P(\sim B)$$

Buying behavior and sex are not independent, however. Recall that the probability of buyer, given a man, is $P(B \mid M) = .10$. But the probability of a buyer is $P(B) = .20$. Hence, B and M are not independent. Knowledge of the sex of a customer gives us a better probability estimate as to whether the person will be a buyer. (Men are less likly to buy than women.)

RULES FOR DEALING WITH PROBABILITIES

Addition of Probabilities

A set of events are said to be *mutually exclusive* if the occurrence of one excludes the occurrence of any of the others. For example, in drawing cards from a deck, the occurrence of the event "draw of a king" eliminates the possibility of the event "draw of a queen." Hence, the events are mutually exclusive.

If the events in a set are mutually exclusive, the probability of one or another of the events occurring is the sum of the probabilities of the events occurring individually. Thus, if events A and B are mutually exclusive,

$$P(A \text{ or } B) = P(A) + P(B)$$

This is known as the *addition rule* for probabilities. Actually, the rule is fairly obvious; we have used it several times without stating it. For example, the probability of drawing a spade from a deck of cards is $1/4$. The probability of drawing a spade *or* a heart is $1/4$ plus $1/4$ or $1/2$.

If two events A and B are not mutually exclusive, then there is some probability that both can occur. The area of overlap is precisely the joint probability $P(A, B)$ illustrated in Chart 5–1. This area is counted

Chart 5–1

PROBABILITY OF NONMUTUALLY EXCLUSIVE EVENTS

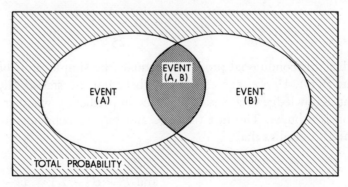

twice in the addition formula used above for mutually exclusive events. We can modify the formula to obtain the *addition rule for events that are not mutually exclusive:*

$$P(A \text{ or } B) = P(A) + P(B) - P(A, B)$$

In the example illustrated in Table 5–1, the events "buyer" and "man" are not mutually exclusive, since there are male buyers; that is, the event "buyer" does not rule out the possibility of the event "man." Hence, the probability of a man or buyer is:

$$P(M \text{ or } B) = P(M) + P(B) - P(M, B)$$
$$= .30 \ + \ .20 \ - \ .03 \ = .47$$

A set of events is said to be *collectively exhaustive* if all possible occurrences are included. For example, the set of events "drawing a red card" and "drawing a black card" are collectively exhaustive; there are no other possibilities. The set of events "man," "buyer," and "woman nonbuyer" are collectively exhaustive (though not mutually exclusive).

The sum of the probabilities for a set of mutually exclusive and collectively exhaustive events equals one. This follows from the addition rule and from the fact that some event must occur.

Multiplication of Probabilities

The rule for multiplication of probabilities is merely an extension of the definition of conditional probability. The joint probability that *both* events A and B will occur equals the probability of A times the conditional probability of B, given A. In symbols,

$$P(A, B) = P(A) \, P(B \mid A)$$

As examples, consider the following:

If we know that the probability of a man customer is $P(M) = .30$, and the probability that a man customer will be a buyer is $P(B \mid M) = .10$, the probability that a customer will be both a man and a buyer is

$$P(M, B) = P(M) \, P(B \mid M) = .30 \times .10 = .03$$

Suppose there were three balls in an urn, two white and one black. What is the probability of drawing both of the white balls in two draws (without putting the first ball back)?

Probability of white on first draw $= P(W_1) = 2/3$

Probability of second white, given first white $= P(W_2 \mid W_1) = 1/2$

Hence, the probability of a first white and a second white is

$$P(W_1, W_2) = P(W_1) \, P(W_2 \mid W_1) = 2/3 \times 1/2 = 1/3$$

Multiplication of Probabilities for Independent Events. When events are independent, $P(B|A) = P(B)$ and hence the rule becomes $P(A, B) = P(A) \, P(B)$. That is, the probability that two or more independent events will occur is the product of the simple probabilities. Consider, as an example, the tossing of a fair coin: $P(\text{head}) = 1/2$. The probability of two heads in a row is $1/2 \times 1/2 = 1/4$, since the results of the two tosses are independent.

Consider the urn with the three balls, two white and one black, discussed above. But now suppose we replace the first ball after it is drawn. (This is known as sampling with replacement.) The draws are then independent, and the probability of two white balls in two draws is:

$$P(W_1, W_2) = P(W_1) \, P(W_2) = 2/3 \times 2/3 = 4/9$$

PROBABILITY TREES

It is sometimes useful in working with probabilities to express the problems in the form of a probability tree. This is a simple graphic device to help keep various parts of the problem straight, and is best explained by an example.

Suppose a ball is drawn from an urn containing 3 red and 6 black balls. If a red ball is drawn, another ball is drawn from a second urn containing 7 green and 3 orange balls. If the first ball drawn is black, a second ball is drawn from yet a third urn containing 4 green and 6 orange balls. What is the probability of a green ball on the second

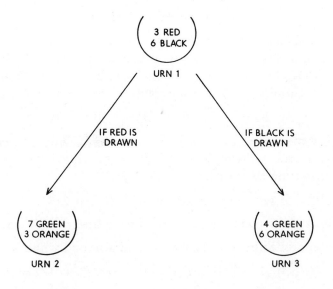

Chart 5–2

PROBABILITY TREE

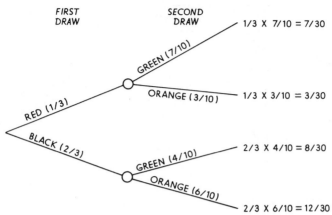

FIRST DRAW

SECOND DRAW

GREEN (7/10)   1/3 X 7/10 = 7/30

RED (1/3)

ORANGE (3/10)   1/3 X 3/10 = 3/30

BLACK (2/3)

GREEN (4/10)   2/3 X 4/10 = 8/30

ORANGE (6/10)   2/3 X 6/10 = 12/30

draw? The probability tree for this problem is shown in Chart 5–2. The probabilities shown at the ends of the tree are the joint probabilities of the various events along the branches leading to that end. For example, the probability of 7/30 is the joint probability of a red and then a green ball. The calculation $1/3 \times 7/10 = 7/30$ is an application of the multiplication rule. The probability tree is simply a graph to make clear the order of events and the probabilities.

The original question called for the probability of a green ball on the second draw. Note that this occurs at the first and third branches. Using the addition rule, the probability of a green ball is $7/30 + 8/30 = 1/2$.

EXAMPLES IN THE USE OF PROBABILITIES

Example 1—Rolling Dice

Two dice are rolled. Assuming that each die is fair, what is the probability of rolling a seven? The six different ways that a seven can appear are listed in Table 5–4.

Since the two dice are independent, the probability of obtaining a seven by any one of the ways in Table 5–4 is $1/6 \times 1/6 = 1/36$ (using the multiplication rule). The six different ways are mutually exclusive (we cannot obtain a seven two different ways at the same time). Using the addition rule, the total probability of obtaining a seven is 1/36, taken six times $= 6/36 = 1/6$.

Table 5–4

DIFFERENT WAYS OF ROLLING A SEVEN

First Die	Second Die	Probability
1	6	1/36
2	5	1/36
3	4	1/36
4	3	1/36
5	2	1/36
6	1	1/36
Total		1/6

Example 2—Sampling

Of 50 loan accounts at a local bank, 8 are known to be behind on their payments. If 5 accounts are selected at random from the 50 accounts, what is the probability that at least one of the accounts selected will be behind in payments?

Note that the probability that at least one account selected is behind is one minus the probability that all accounts are current. So we first find the probability that none of the five accounts is behind (i.e., that all accounts selected are current). The probability that the first account selected is current is $P(C_1) = 42/50$. For the second account, the conditional probability of a current account, given a current account on the first selection, is $P(C_2 \mid C_1) = 41/49$ (of the 49 remaining accounts, 41 are current). Hence, the probability of two current accounts is:

$$P(C_1, C_2) = P(C_1)\,P(C_2 \mid C_1) = (42/50)(41/49)$$

by use of the multiplication rule. For the third account, the conditional probability of a current account, given current accounts for the first two selected, is $P(C_3 \mid C_1, C_2) = 40/48$. Hence,

$$P(C_1, C_2, C_3) = P(C_1)P(C_2 \mid C_1)P(C_3 \mid C_1, C_2) = (42/50)(41/49)(40/48)$$

Continuing in this fashion, we have the probability that all five accounts selected are current:

$$P(C_1, C_2, C_3, C_4, C_5) = (42/50)(41/49)(40/48)(39/47)(38/46) = .40$$

Then the probability that at least one account selected is behind is one minus the probability that all are current:

$$1 - .40 = .60$$

Example 3—Probability Tree

The ABC Varnish Company currently has 20 percent of the varnish market in a certain region. Its major competitor, XYZ Varnish Company, has the remaining 80 percent. The research and development department reports substantial progress on a much improved product. R.&D. chemists estimate that there is an 80 percent chance of developing the improved varnish.

If the new varnish is developed and marketed by ABC, there is a .60 chance that XYZ will also develop a similar product. If this happens, the chances are .20 that ABC will have an 80 percent market share, a .30 chance that ABC will have a 60 percent market share, and a .50 chance for a 40 percent market share. If XYZ is not able to also develop a new varnish, then ABC has a .70 chance at an 80 percent market share and a .30 chance at a 50 percent market share. In the event ABC is not able to develop the new varnish, it will retain its current 20 percent market share.

What is the probability that ABC will gain a 60 percent or better market share? The above problem description is indeed somewhat confusing. But displaying it in a probability tree makes it much clearer. From the tree in Chart 5–3, we see that the probability of at least a 60 percent market share is obtained by adding the probabilities at the ends of the first, second, and fourth branches:

Chart 5–3

PROBABILITY TREE FOR NEW VARNISH

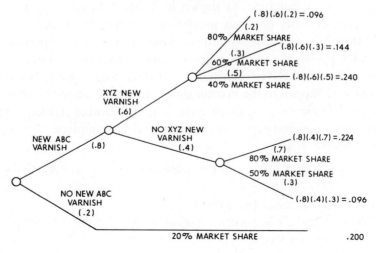

Probability of at least 60 percent market share $= .096 + .144 + .244 = .484$

Example 4—Brand Loyalty

Marketing analysts are concerned with the loyalty of a customer to a particular brand and with the effect of this loyalty on the brand's share of the market. There are two brands of a given product, A and B. Let us suppose that a customer who purchases Brand A in a given period (t) has a .50 probability of purchasing A again in the next period $(t + 1)$, and a .50 probability of purchasing Brand B. Those who buy Brand B in period t, however, have a .70 probability of repeating a Brand B purchase (they are more loyal than Brand A customers) and a .30 probability of switching to Brand A in period $t + 1$. This is shown in Table 5–5.

Table 5–5

PROBABILITIES OF REPEAT PURCHASES AND
BRAND SWITCHES

Brand Purchased in Period (t)	Brand Purchased in Period $(t + 1)$	
	Brand A	Brand B
Brand A.............	.50	.50
Brand B.............	.30	.70

Assume that brand-buying behavior is dependent only on the immediately preceding purchase, as shown in Table 5–5, and is statistically independent of other previous purchases. Assume also that the probabilities shown in the table remain the same from period to period.

Let us suppose, at a given point in time t, that each brand has 50 percent of the market (as many customers buy A as buy B). We might ask what will happen to the market share of each brand after one period has elapsed (time $t + 1$). During the period, Brand A has kept .50 of its own customers and captured .30 of Brand B customers. That is, the shares at time $t + 1$ are:

Brand $A = (.50)(A\text{'s } 50 \text{ percent market share}) + (.30)(B\text{'s } 50 \text{ percent market share})$
   $= 40$ percent of the market
Brand $B = (.70)(B\text{'s } 50 \text{ percent market share}) + (.50)(A\text{'s } 50 \text{ percent market share})$
   $= 60$ percent of the market

At the end of the first period, Brand B has increased its share to 60 percent of the market. The process is repeated during the second period, so that the shares at time $t + 2$ are

Brand A = $(.50)(A$'s 40 percent market share$)$ + $(.30)(B$'s 60 percent market share$)$
          = 38 percent of the market
Brand B = $(.70)(B$'s 60 percent market share$)$ + $(.50)(A$'s 40 percent market share$)$
          = 62 percent of the market

Again, Brand B's share increases, but only slightly. If the process is repeated over many periods, an equilibrium is reached with Brand A having three eighths of the market and Brand B having five eighths of the market. At this point, the number of customers leaving Brand A is exactly balanced by those switching from B to A.

Many marketing strategies (such as pricing, advertising, and merchandise deals) are aimed at influencing brand loyalty (i.e., influencing such probabilities as those shown in Table 5–5). The above probability analysis traces the effects of these strategies on market share.

Example 5—Project Scheduling

Construction or research and development projects require the scheduling and coordination of large numbers of tasks. It is usually important to complete the project by a scheduled date. When the times to complete some of the tasks are uncertain, the project completion time itself

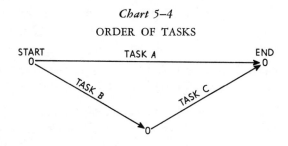

Chart 5–4

ORDER OF TASKS

is uncertain. However, we can determine the probability for completion at any time.

Consider the following simplified example. A project is made up of three tasks, A, B, and C. Task B must be completed before C can start. Task A is not dependent upon B and C (it is done in parallel) but both A and C must be completed before the project is considered finished. This arrangement, with lines indicating tasks, is illustrated in Chart 5–4.

The time needed to complete each task is uncertain, owing to weather conditions and other unpredictable factors. However, probabilities are assigned to task completion times as shown in Table 5–6.

Table 5–6

PROBABILITIES AND TIMES TO COMPLETE
TASKS *A, B,* AND *C*

Task	Completion Time, Weeks	Probability
A	4	.50
	6	.50
		1.00
B	1	.25
	3	.75
		1.00
C	2	.80
	4	.20
		1.00

Let us denote the event "Task *A* takes four weeks to complete" by the symbol *A*-4. Similarly, we have *A*-6, *B*-1, etc. Assume that task completion times are independent—the time taken to complete *B,* for example, does not influence the time for *C.*

We wish to determine the probabilities associated with total project completion time. If the events *A*-4, *B*-1, and *C*-2 all occur, the total project will take four weeks (this is the four weeks required for *A*; the *B* and *C* tasks take only a total of three weeks). Hence, the probability of the event *T*-4 (total project time equals four weeks) is:

$$P(T\text{-}4) = P(A\text{-}4, B\text{-}1, C\text{-}2) = P(A\text{-}4)P(B\text{-}1)P(C\text{-}2)$$
$$= (.50)(.25)(.80) = .10$$

using the multiplication rule for independent events.

The event *T*-5 can be obtained either by the set of events *A*-4, *B*-1, *C*-4 or by the set *A*-4, *B*-3, *C*-2. These sets are mutually exclusive—either one or the other happens, not both; and

$$P(A\text{-}4, B\text{-}1, C\text{-}4) = (.50)(.25)(.20) = .025$$
$$P(A\text{-}4, B\text{-}3, C\text{-}2) = (.50)(.75)(.80) = .300$$

Hence, the probability of *T*-5 is the sum:...... .325

The probabilities for the values of *T*-6 and *T*-7 can be determined in a similar manner and are shown in Table 5–7. From simple proba-

Table 5–7

PROBABILITIES AND TIMES TO
COMPLETE TOTAL PROJECT

Project Completion Time, Weeks	Probability
4	.10
5	.325
6	.425
7	.15
	1.000

bility information about the time to complete individual tasks, we have determined a complete set of probabilities for total project time.

PROBABILITY DISTRIBUTIONS

Consider an example of tossing four coins. The probabilities for various numbers of heads (r) are shown in Table 5–8 and are graphed

Table 5–8

PROBABILITIES OF VARIOUS NUM-
BERS OF HEADS IN FOUR TOSSES OF
A FAIR COIN

Number of Heads, r	Probability, $P(r)$
0	1/16
1	1/4
2	3/8
3	1/4
4	1/16
	1

in Chart 5–5. Note that this table simply expresses a functional relationship between values of a variable r and another set of values $P(r)$. This type of function is called a *probability distribution.* We call the variable r (number of heads) a *random variable.* It is random in the sense that we cannot predetermine the exact value that the variable will take on any trial; only the probabilities that it will take certain values are known. Each probability $P(r)$ applies to a given value of r. As noted above, each value of $P(r)$ must be between 0 and 1, and the total probabilities of mutually exclusive and collectively exhaustive events (e.g., for 0, 1, 2, 3, and 4 heads) must equal 1.

Chart 5–5

GRAPH OF PROBABILITY FUNCTION OF TABLE 5–8

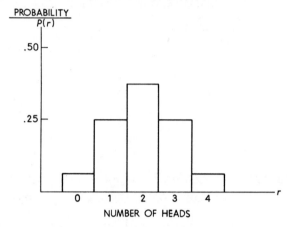

Discrete and Continuous Distributions

A probability distribution is continuous or discrete depending on whether the random variable can take on any real number in a specified interval or is restricted to specific values (often integers).

The distribution above is discrete, since the random variable r can take on only specific integer values. There are 0 or 1 or 2 or 3 or 4 heads in four flips of a coin. It is not possible to get $1\frac{1}{2}$ heads or 1.648 heads. On the other hand, the distribution of diameters of ball bearings is continuous, since the random variable can take on any value (if we have fine enough measuring instruments).

In the probability distributions in Tables 5–7 and 5–8, the relationship between the random variable and the probability function is defined by the table itself. Other probability distributions may be defined by mathematical equations. For example, the function $P(X) = .25X - .05X^2$ may define a discrete probability distribution in which the random variable X can take on the integer values 1, 2, 3, or 4. Similarly, the continuous function $P(X) = .06X - .006X^2$ may define a continuous probability distribution in which the random variable can take on any value between 0 and 10 (i.e., $0 < X < 10$). The graphs of these functions are shown in Chart 5–6. Four specific probability distributions, defined by mathematical equations, are studied in detail in Chapter 6.

Chart 5–6

EXAMPLES OF PROBABILITY DISTRIBUTIONS
DEFINED BY MATHEMATICAL EQUATIONS

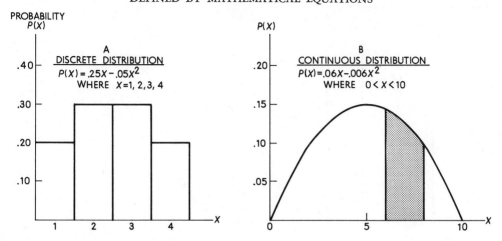

Graphs of Probability Distributions

Graphs of discrete probability distributions are illustrated in Charts 5–5 and 5–6A. The values of the random variable are shown on the X axis and the associated probabilities on the Y axis. These histograms are the same as those in Chapter 2, except that the vertical scale shows probability rather than frequency.

Continuous probability distributions are represented by a smooth curve, such as in Chart 5–6B. However the values of $P(X)$ represent only the height of the curve at any point X and are *not* probabilities. In a continuous distribution, the probability of the random variable having any exact value is infinitely small. We can only speak of the probability of a random variable being in a specified range. For example, the probability that X falls between 6 and 8, or $P\ (6 < X < 8)$, is represented by the shaded area in Chart 5–6B. The total area under the curve (i.e., the probability for all values of X) is taken as 1. Thus, probability is associated with an area under the curve for continuous distributions.

It is sometimes convenient to have graphs of the probability that a random variable is less than (or greater than) a given value. These graphs of cumulative distributions (see Chart 5–7) are like the ogives of Chapter 2, except that probabilities are cumulated and plotted instead of frequencies. (Note the use of the cumulative distribution in the probability assessment procedure described in Appendix B.)

Chart 5–7

CUMULATIVE DISTRIBUTIONS

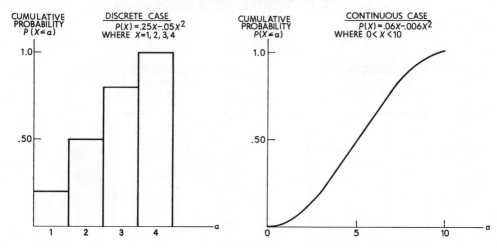

EXPECTED VALUE AND VARIANCE OF PROBABILITY DISTRIBUTIONS

The *expected value* of a discrete random variable X is defined as follows:

$$E(X) = \Sigma[X \cdot P(X)]$$

where $P(X)$ is the probability for each value of X.

Note that we multiply each value of X by its probability and sum the products. The concept of expected value then corresponds to that of a weighted mean, $\overline{X} = \Sigma fX/n$, where the probability $P(X)$ is equivalent to the relative frequency f, and $n = 1$, since the sum of the probabilities equals 1.

Consider a new car agency that sells from 0 to 6 cars (X) a day. In a typical period, the agency makes no sales on 20 percent of the days; it sells 1 car on 25 percent of the days, and so on, as shown in Table 5–9. These relative frequencies might be used as estimates of probabilities $P(X)$ for future sales.

To find the expected value, multiply X by $P(X)$ and sum the products (column 3):

$$E(X) = \Sigma[X \cdot P(X)] = 2.00$$

That is, average or expected sales are 2 cars a day. The expected value is called the *first moment* of a probability distribution.

Table 5–9

PROBABILITY DISTRIBUTION OF CAR SALES
EXPECTED VALUE AND VARIANCE

Cars Sold X	Probability P(X)	X · P(X)	X − E(X)	[X − E(X)]²	[X − E(X)]² · P(X)
0	.20	0	−2	4	.80
1	.25	.25	−1	1	.25
2	.25	.50	0	0	.0
3	.10	.30	1	1	.10
4	.10	.40	2	4	.40
5	.05	.25	3	9	.45
6	.05	.30	4	16	.80
Total	1.00	2.00			2.80

The principal measure of dispersion for a probability distribution is the *variance* (the square of the standard deviation or σ^2) which is defined as:

Variance = $\Sigma\{[X - E(X)]^2 \cdot P(X)\}$ in a discrete distribution

This is equivalent to the formula $s^2 = \Sigma f(X - \overline{X})^2/n$ (Chapter 4)[2] where $P(X)$ is used in place of f; $X - E(X) = X - \overline{X}$; and $n = 1$. To compute the variance, take the deviation from the mean, that is, $X - E(X)$, square it, multiply it by the probability $P(X)$, and then sum the products (columns 4 to 6).

For the car sales,

Variance = 2.80 (column 6, bottom)
Standard deviation = $\sqrt{2.80} = 1.67$ cars

The variance is called the *second moment about the mean*. The further the individual values of X are from the mean, the larger the second moment.

We could define the third moment about the mean (a measure of skewness) and fourth moment (a measure of kurtosis) and so on. These, however, have limited usefulness.

The calculation of the expected value and variance for continuous distributions requires the use of the calculus. (See Appendix A at the end of this chapter.) However, the basic notions apply equally well to continuous distributions.

The expected value, standard deviation, and variance of a probability

---

[2] The denominator $n-1$ does not apply here.

distribution are useful in themselves as measures of central tendency and dispersion, as are similar measures for frequency distributions (described in Chapters 3 and 4). These measures also will be found useful in subsequent chapters in summarizing distributions, in decision problems, and in sampling analysis.

SUMMARY

A probability is a number between zero and one describing the relative likelihood of a possible event. Probabilities are often thought of as the limit of the ratio of "successes" to total trials in a long-run experiment. However, probabilities may be estimated from any of three sources: (1) the relative frequency of past events, based on either an experiment or survey; (2) theoretical distributions; or (3) the subjective judgment of the decision maker.

A simple probability is the probability of the occurrence of a single event. A joint probability is the probability that two or more events will both occur. A conditional probability is the probability of the occurrence of one event, given that some other event has occurred. A marginal probability is the probability of the occurrence of a single event, determined as the sum of the joint probabilities involving that event.

Two events are statistically independent if the conditional probability of one, given the other, is equal to the simple probability of the first; that is, if $P(A \mid B) = P(A)$. Independence implies that knowledge of one event is of no value in predicting the other event.

If two events are mutually exclusive, the probability that one or the other will occur is the sum of the respective simple probabilities; that is, $P(A \text{ or } B) = P(A) + P(B)$. If the events are not mutually exclusive, the probability that one or the other will occur is the sum of the respective simple probabilities minus the joint probability of the two events: $P(A \text{ or } B) = P(A) + P(B) - P(A, B)$.

The joint probability that two events (A and B) will both occur is the simple probability of one times the conditional probability of the second given the first; that is, $P(A, B) = P(A) P(B|A)$. When the events are independent, $P(B|A) = P(B)$, so the joint probability is merely the product of the simple probabilities: $P(A, B) = P(A)P(B)$.

A probability tree is a diagram displaying logical order in probability problems involving several steps. Each branch represents a possible event and its probability, so that the joint probability of any combination of events can be easily found.

A probability distribution is a functional relationship between a random variable r and a set of probabilities $P(r)$. Probability distributions may be discrete or continuous, depending on whether the random variable can take on only a restricted set of values (e.g., only integers) or can take on any value within an interval. Probabilities may be graphed in the same way as are frequencies in Chapter 2.

The expected value of a discrete probability distribution is the weighted average of the random variable, the weights being the respective probabilities, that is, $E(X) = \Sigma X \cdot P(X)$. The variance of a discrete probability distribution is the sum of the squared deviations from the mean times the respective probabilities:

$$\sigma^2 = \Sigma\{[X - E(X)]^2 P(X)\}$$

The standard deviation is the square root of the variance. These general concepts will be applied to four specific probability distributions in the next chapter.

APPENDIX A: EXPECTED VALUE AND VARIANCE OF CONTINUOUS DISTRIBUTIONS

Definition. A continuous distribution $f(X)$ with random variable X is a function such that:

$$f(X) \geq 0 \text{ for all } X, \text{ and}$$
$$\int_{\text{all } X} f(X)\, dX = 1$$

Expected Value. The expected value of the random variable X is defined to be:

$$E(X) = \int_{\text{all } X} Xf(X)\, dX$$

Thus, for the function $f(X) = .06X - .006X^2,\ 0 < X < 10$:

$$E(X) = \int_0^{10} X(.06X - .006X^2)\, dX = \left. \frac{.06X^3}{3} - \frac{.006X^4}{4} \right|_0^{10}$$
$$= 20 - 15 = 5$$

In general, the expected value of any expression involving X, say $g(X)$, is:

$$E[g(X)] = \int_{\text{all } X} g(X)f(X)\, dX$$

Variance. The variance (σ^2) is the expected value of the function $[X - E(X)]^2$.

$$\sigma^2 = E\{[X - E(X)]^2\} = \int_{\text{all } X} [X - E(X)]^2 f(X) \, dX$$

In our example, $E(X) = 5$, and

$$\sigma^2 = \int_0^{10} (X - 5)^2 (.06X - .006X^2) \, dX$$

$$= \int_0^{10} (X^2 - 10X + 25)(.06X - .006X^2) \, dX$$

$$= \int_0^{10} X^2(.06X - .006X^2) \, dX$$

$$- 10 \int_0^{10} X(.06X - .006X^2) \, dX$$

$$+ 25 \int_0^{10} (.06X - .006X^2) \, dX$$

$$= \left(\frac{.06X^4}{4} - \frac{.006X^5}{5} \right) \Big|_0^{10} - 10(5) + 25(1)$$

$$= (150 - 120) - 50 + 25 = 5$$

and the standard deviation $\sigma = \sqrt{5} = 2.24$

Evaluation of Probabilities. The integration operation can be used to measure areas under curves and hence to evaluate probabilities for continuous distributions. For example, the probability that X is between 5 and 7 in our example is

$$P(5 < X < 7) = \int_5^7 (.06X - .006X^2) \, dX$$

$$= .03X^2 - .002X^3 \Big|_5^7$$

$$= .284$$

APPENDIX B: ESTIMATING SUBJECTIVE PROBABILITY DISTRIBUTIONS

Probability estimates of unknown events are necessarily subjective. Consider the example of a manager who must make an estimate of the cost per unit for producing a new product that his company is considering. He would, of course, examine the information available, including engineering estimates of production rates and labor and material usage. He might examine historical records of similar products. All this information, while useful, still leaves substantial uncertainty about what the actual cost will be. To express this uncertainty, the manager might assess a subjective probability distribution for the unknown factor, cost per unit.

This appendix describes a method that has been found useful in assessing such probability distributions. It is called the *bet-yourself* method. You will see why in a minute.

Estimating the Median

The first step is to pick a middle value—the median. Recall that the median divides the area of a probability distribution into two equal parts. Suppose our manager made a first guess at $3.25 per case. He then formulates two bets about the true value of the unknown factor (cost per unit):

Bet 1: True value   is below   estimated median ($3.25 per case)
Bet 2: True value   is above   estimated median

The manager now imagines that some important prize (for example, a vacation to Hawaii) depends upon winning one of the above bets. And he has a choice about which bet he would rather have. If the bets appear about even to him, the estimate of the median is satisfactory. That is, each bet has a 50 percent chance of winning. On the other hand, if the manager has a strong preference for one or the other bet, then the estimate of the median needs to be revised. For example, if he preferred Bet 2 (implying that it had a better than 50 percent chance of winning), he might revise his estimate of the median to $3.30 per case and reformulate the bets. If after this, he felt the bets were even, then $3.30 would be the estimate of the median.

Estimating the Quartiles

The quartiles divide the probability area into four equal parts. The quartiles are the lower quartile, Q_1; the median, which is the middle quartile; and the upper quartile, Q_3 (see Chapter 4 for details).

The procedure for estimating Q_1 and Q_3 is similar to that for estimating the median described above. To estimate Q_1, for example, the manager would start by picking a value of the unknown factor (cost per unit) somewhere below the median estimate. Let us say he picks $3.20 per case. He then formulates bets as:

Bet 1: True value   will be below   estimated Q_1 ($3.20)
Bet 2: True value   will be between   estimated Q_1 and estimated Md
                                        ($3.20 to $3.30)

If these bets seem about even, then the estimated quartile is satisfactory. Otherwise, it must be revised up or down until the manager is indifferent between the two bets. Let us suppose that the $3.20 estimate for Q_1 is satisfactory.

The upper quartile, Q_3, is estimated in the same manner, by picking a value above the median, formulating bets, and revising until the bets are even. Let us suppose the manager estimated Q_3 at $3.50 using this procedure.

A Consistency Check: The Quartile Range

Once the manager has made these estimates, it is useful to check for consistency by formulating the following bets:

Bet 1: True value   is inside   the quartile range, i.e., between Q_1 and
                                   Q_3 ($3.20 and $3.50)

Bet 2: True value   is outside   the quartile range, i.e., below Q_1 or
                                   above Q_3 (below $3.20 or above $3.50)

These bets, based upon the previous estimates, should be about the same. If the manager has a preference for one or the other, he should revise some of his estimates. Since estimating probabilities is a difficult task, it is not unusual for these bets not to seem fair, and for a second round of assessment to be needed.

In our example, let us suppose that the manager is indifferent about the two bets above, and does not need to revise his estimates.

Estimates of Extreme Values

The manager now must pick values near the extremes of his probability distribution. Suppose he considers a low estimate first. He tries to imagine what factors would have to be present to have a really low cost per case. Suppose he finally settles on an estimate of $3.00 per case, assuming only one chance in 10 that the cost could be this low. This will be an estimate of the lower decile, abbreviated LD, or 10 percent point of the probability distribution. The manager then formulates the following bets:

Bet 1: Prize is won if   a red ball is drawn from an urn containing one red
                            and nine white balls

Bet 2: Prize is won if   true value will be less than estimated LD ($3.00
                            per case)

As before, if these bets seem about the same, the estimate is satisfactory; if not, the estimated LD must be revised. Let us suppose that the bets are considered even by the manager.

A similar procedure is used to estimate the upper decile or UD. Suppose that the manager settles on $4.00 for this estimate.

The Cumulative Probability Distribution

The five estimates obtained by the process described above are enough to describe the cumulative probability distribution for the unknown factor (cost per case in our example). The five points are plotted and a smooth curve drawn connecting them in Chart 5–8. This curve describes the subjective probability distribution for the unknown cost per

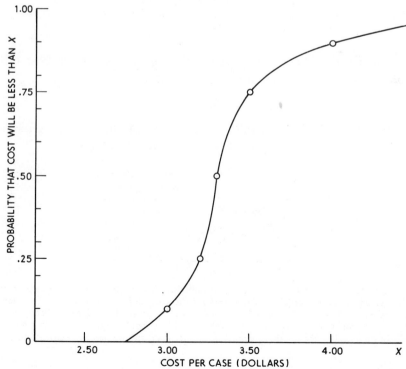

Chart 5–8

CUMULATIVE PROBABILITY DISTRIBUTION
COST PER CASE FOR NEW PRODUCT

unit, and can be used in decision problems such as those described in Chapter 7.

A probability density function can also be drawn and is shown in Chart 5–9. This is obtained from the cumulative distribution.[3] Note

Chart 5–9

SUBJECTIVE PROBABILITY DENSITY FUNCTION
Cost per Case for New Product

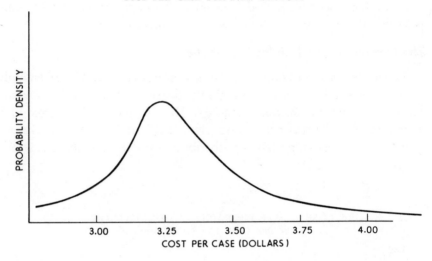

that the density function is skewed to the right, indicating a substantial chance that the actual cost might be well above the median value.

PROBLEMS

1. An automobile dealer classified his car sales over the last year according to method of payment (figures are percentages of total sales):

Type of Car	Method of Payment	
	Cash	Credit
New	6%	18%
Used	30%	46%

---

[3] The smoothed density function can be obtained by breaking up the range of the unknown value into intervals and determining the cumulative probability at the boundaries of each interval. The probability in each interval is the difference between the cumulative values at the boundaries. Finally, a histogram can be plotted using these probabilities, and a smooth curve drawn to approximate the probability density function.

a) In selecting a purchaser at random, what is the simple probability of new car purchase?

b) What is the joint probability of selling a used car on credit?

c) What is the conditional probability that a used car purchaser will pay cash?

d) Is the type of car purchased independent (in the statistical sense) of the method of payment? Why?

2. Suppose businessmen read periodicals as follows:

Fortune	5%
U.S. News	15
Wall Street Journal	15
None of the above	15
Fortune and U.S. News	5
Fortune and *Wall Street Journal*	15
U.S. News and *Wall Street Journal*	10
All three	20
Total	100%

a) If a certain businessman reads *Fortune* and the *Wall Street Journal,* what is the probability that he also reads *U.S. News?*

b) What proportion of businessmen read *Fortune?*

c) Are the events "reader of *Fortune*" and "reader of the *Wall Street Journal*" independent events?

d) Are the events "reader of *U.S. News*" and "reader of the *Wall Street Journal*" independent?

3. An investor classified the 80 stocks in his portfolio in the following manner:

	Industrial Stocks Percent	Utility Stocks Percent
Large companies (in top 100 of assets):		
Price increased (in past year)	4	1
Price decreased	8	7
Total	12	8
Small companies:		
Price increased	17	3
Price decreased	55	5
Total	72	8
Total (100%)	84	16

In this portfolio:

a) If a stock were drawn at random, what would the probability be that it was one that had increased in price? What kind of probability is this (simple, joint, marginal, or conditional)?

b) What is the probability of a stock having increased in price if it was a large company industrial stock? What kind of probability is this?

c) Is size of company independent of price behavior in this portfolio? Why?

d) Is the type of stock (industrial versus utility) independent of the price behavior in this portfolio? Why?

e) Is price behavior independent of both size and type of stock? Explain.

4. Suppose 70 percent of the corporations in a certain industry have a lawyer on the board of directors and suppose 40 percent have a banker on the board. What proportion of the corporations have neither a banker nor a lawyer on the board?

5. In analyzing sales of a certain product in a retail store over the past year you discover that 10 percent of the purchases were made by men and 20 percent of the purchases were over $10 in value. If you know that 80 percent of male customers make purchases over $10:
   a) What percentage of purchases over $10 are made by men?
   b) What percentage of purchases are made by men or are over $10?

6. If 30 percent of the households in a certain city have electric dryers and 40 percent have electric stoves, and if 25 percent of those who have electric stoves also have electric dryers, what proportion of those who have electric dryers also have electric stoves?

7. A market research firm is interested in surveying certain attitudes in a small community. There are 125 households, broken down according to annual income, having telephone service, and ownership of a television set.

	Households with Annual Income of $8,000 or Less		Households with Annual Income above $8,000	
	Phone	No Phone	Phone	No Phone
Own TV set.	27	20	18	10
No TV set.	18	10	12	10

   a) What is the probability of obtaining a TV owner in drawing at random?
   b) If a household has income in excess of $8,000 and is a telephone subscriber, what is the probability that it has a TV?
   c) What is the conditional probability of drawing a household that owns a TV given that the household is a telephone subscriber?
   d) Are the events "ownership of a TV" and "telephone subscriber" statistically independent?
   e) Are the events "income of $8,000 or less" and "ownership of TV" independent events?

8. As a bond salesman, you are considering using a list of stockholders for direct mail advertising. You know that 40 percent of investors hold stocks only and 10 percent hold bonds only, another 20 percent hold both stocks and bonds, and the other 30 percent hold neither. Then, if an investor is a stockholder, what is the probability that he is also a bondholder?

9. A piece of electronic equipment has three essential parts. In the past Part A has failed 20% of the time; Part B, 40% of the time; and Part C, 30% of the time. Part A operates independently of Parts B and C. Parts B and C are interconnected, however, so that failure of either part affects the other. In those instances when Part C failed, the chances were two out of three that Part B would also fail.

Assume that at least two of the three parts must operate to enable the equipment to function. What is the probability that the equipment will function?

10. The police chief of a metropolitan area was reviewing the statistics on the number of pedestrian fatalities in the past year. Of a total of 12 deaths, he noted that 6 were killed while crossing with the proper light and 6 were killed in crossing the street against a red light. Could the police chief conclude that it was just as dangerous to obey traffic signals in crossing the street as it was to disregard them? Explain.

11. If an employee shirks his work 30 percent of the time, what is the probability that he will be caught if his boss checks on him four times at random?

12. As manager at a crucial point in a ball game, you feel that your pitcher has a 70 percent chance of getting the next batter out. You could replace him with a relief pitcher who has a 90 percent chance of getting the batter out if he is at his best, but only a 40 percent chance if he is not at his best. Your pitching coach in the bullpen informs you that, on the basis of watching his warm-up, he feels that the relief pitcher has about a 70 percent chance of being at his best. Do you change pitchers?

13. Which of the following functions are probability distributions? Explain.
   a) $P(X) = X/10$ for $X = 1, 2, 3, 4$.
   b) $P(X) = X^2/10$ for $X = 1, 2, 3, 4$.
   c) $P(X) = .40 - .02X^2$ for $X = 1, 2, 3, 4$.

14. Find the expected value and variance of the distribution shown in Table 5–7.

15. Find the expected value and variance of the distribution shown in Table 5–8.

16. Find the expected value and variance of the probability distribution:

$$P(X) = .25X - .05X^2 \text{ for } X = 1, 2, 3, 4.$$

17. The following represents a probability distribution for the number of orchids (Z) demanded by customers in a certain florist shop. Calculate the expected value and variance of Z.

Z	$P(Z)$
0	.05
1	.10
2	.25
3	.30
4	.20
5	.10
6 and up	0
	1.00

18. Consider the probability distribution given by the following table:

X	$P(X)$
0.............	.18
1.............	.32
2.............	.20
3.............	.12
4.............	.08
5.............	.06
6.............	.03
7.............	.01
	1.00

a) What is the expected value of X?
b) What is the variance of X?
c) What is the conditional probability that $X = 2$, given that X is an even number or zero?

19. An executive is in the process of hiring a new production manager. There are four candidates for the job. He will interview each individually on successive days. Because of certain factors outside his control, the manager must decide after interviewing each whether he will hire him or not (that is, he cannot wait until he has talked to all of them before deciding). Only one will be hired.

The executive has decided to rate each, after his interview, as either excellent, good, average, or poor. With his present knowledge, the manager thinks the candidates are about the same—each having a .2 chance of being rated excellent on the basis of the interview, a .5 chance of being rated good, a .2 chance of being rated average, and a .1 chance of being rated poor.

The executive has decided to adopt the following strategy: If the first candidate is rated excellent, hire him; otherwise go to the second. If the second is rated excellent, hire him; otherwise go to the third. If the third is rated excellent or good, hire him; otherwise go to the fourth. He would of course, be stuck with whatever rating the fourth merits.

What is the probability that this strategy will result in hiring a manager rated excellent or good? What is the probability of hiring a poorly rated manager?

20. Refer to Problem 19 above. Suppose the manager could wait until he had interviewed all four and then hire the best. How much does this improve his chances of hiring a manager rated excellent or good?

21. One of the most famous probability problems started with the bets of the Chevalier de Méré, a 17th century Frenchman. The Chevalier was betting (and winning) that he could roll at least one six in four rolls of a single die. He then switched to betting that he could roll one 12 in 24 rolls of a pair of dice. He reasoned that, since the odds were one in six that the second die would be a six, he should have the same odds of winning his second bet as his first. But experience apparently did not bear this out,

and he turned to the famous mathematician Blaise Pascal. This led to the beginnings of probability theory.

What is wrong with the Chevalier's reasoning? Calculate the odds of each bet (note that the evaluation of the second bet requires a lot of arithmetic or the use of logarithms.)

22. Refer to Example 3 on page 105. Derive the complete probability distribution for the resultant market share. Calculate the mean and standard deviation of this distribution.

23. The game of craps is played by a player rolling two dice. If a 7 or an 11 total appears on the first roll, he wins. If a 2, 3, or 12 total appears on the first roll, he loses immediately. If any other total (4, 5, 6, 8, 9, 10) appears on the first roll (called the player's point) the player rolls again. In fact he continues to roll until he matches his point (in which case he wins), or until he rolls a 7 (he loses).

Calculate the odds of winning. Hint: The probability of winning on any roll after the first can be determined by considering only the two relevant long-run possibilities—making his point or rolling a 7.

24. Consider Example 4 on page 106. Suppose the following numbers represent the probabilities of repeat purchases or switches:

Brand Purchased in Period (t)	Brand Purchased in Period ($t + 1$)	
	Brand A	Brand B
Brand A....................	.40	.60
Brand B....................	.40	.60

Show that Brand A—40 percent, Brand B—60 percent, is an equilibrium distribution of market shares; i.e., market shares are the same in period ($t + 1$) as they were in (t).

25. Carry through the illustration of Example 5, page 107, on the assumption that there is a .3 probability of Task A taking four weeks, and a .7 probability of its taking six weeks.

26. A company has two warehouses, A and B. Each warehouse carries a normal stock of three units of a certain product. Daily demand (requests) for units of this product at *each* warehouse has the following probability distribution:

Demand (units)	Probability
1......................	.30
2......................	.40
3......................	.20
4......................	.10
	1.00

a) What is the probability that warehouse A will have more demand than stock on a given day?

b) What is the probability that A or B (not both) will have more demand than stock on a given day?

c) What is the probability that both warehouses will have more demand than stock available on a given day?

27. Suppose that the company in Problem 26 consolidated warehouses A and B into a central warehouse C. A normal stock of six units is to be carried at the central warehouse C.

a) Determine the probability distribution of demand for warehouse C from the individual distributions for A and B. [Hint: The probability of a demand for three units at C is (probability of one demand at A times the probability of a two demand at B) plus (probability of a 2 demand at A times the probability of a 1 demand at B) etc.]

b) From the distribution determined in (a), what is the probability of having one more demanded than stock available? Of having two more demanded than stock available? Compare these with the answers to Parts (b) and (c) of Problem 26. If the answers are different, why are they so?

28. Management of the Alzo Company is considering marketing a new product. Market research indicates that there is a .40 probability that the total market for the product is 10,000 units; a .40 probability for an 8,000-unit total market; and a .20 probability for a 6,000-unit market.

It is not known whether Alzo's competitor, Barden, will offer a similar product. Chances are about 50/50 that Barden will. If Barden does not offer a competitive product, then Alzo will capture the entire market. If Barden does enter the market, it will capture part of the market, depending upon the price charged. If Barden sets a competitive price, Alzo management feels that Barden will have .20 chance of taking 60 percent of the market, a .50 chance of taking 40 percent of the market, and a .30 chance of taking 20 percent of the market. On the other hand, if Barden resorts to price cutting, Barden has a .70 chance of taking 60 percent of the market and a .30 chance of taking 40 percent of the market.

Based upon past experience, Alzo feels that the chances are three out of four that Barden will set a competitive price.

Determine the probability distribution for number of units sold. What are expected sales?

29. Suppose that in Problem 28 Barden's pricing strategy depended upon the size of the market, so that if the market was 10,000 or 8,000 units, the chances were 8/10 that Barden would set a competitive price. But if the market was only 6,000 units, the chances were 6/10 that Barden would resort to price cutting. Determine the probability distribution for sales (units) and expected sales.

30. A project is composed of five tasks, A, B, C, D, and E. The order in which the tasks must be performed is shown in the network diagram (lines

represent tasks). That is, Task *A* must be done before either *B* or *E* can start; both *C* and *E* must be done before *D* can start; and both *B* and *D* must be done before the project is considered finished. Thus, there are three sequences of tasks (called paths through the network) that can hold up total project completion time: *A-B, C-D,* and *A-E-D.* The total project

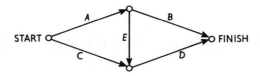

completion time is the time taken to complete the longest of these sets of tasks. For example, if *A* takes 5 weeks; *B,* 6 weeks; *E,* 2 weeks; *C,* 9 weeks; and *D,* 4 weeks; then *A-B* is 11 weeks, *C-D* is 13 weeks, and *A-E-D* is 11 weeks. The total project time is 13 weeks, determined by the *C-D* set of tasks. The table below lists the times and probabilities to complete each of the tasks.

Task	Weeks	Probability
A.	5	.50
	7	.50
B.	6	.80
	9	.20
C.	5	.40
	9	.60
D.	4	.50
	6	.50
E.	2	1.00

Determine the probability distribution for project completion time. Calculate the expected completion time.

SELECTED READINGS

Selected readings for this chapter are included in the list that appears on page 155.

6. PROBABILITY DISTRIBUTIONS

THIS CHAPTER describes four probability distributions that govern the behavior of many business processes. These probability distributions will be used in Chapter 7, together with the economic consequences of business actions, to develop a rational procedure for decision making under uncertainty. In addition, the distributions will serve as a basis for evaluating sample evidence (Chapter 9).

In Chapter 2 we classified statistical data into two categories: *attributes,* which are classified into two or more discrete groups (e.g., heads or tails), and *variables,* which can be measured along a scale. The binomial and Poisson distributions describe the behavior of attributes, while the normal and exponential distributions describe the behavior of variables.

THE BINOMIAL DISTRIBUTION

We shall first discuss a few examples of the binomial distribution to illustrate the points involved. Consider the following kinds of problems:

1. What is the probability of getting four heads in 10 flips of a coin?
2. If a certain district is 60 percent Republican, what is the probability of getting fewer than 30 Democrats in a sample of 100 voters?
3. If a certain process produces transistors, 4 percent of which (on the average) are defective, what is the probability of getting more than four defectives out of 50 items?

Bent Coin Example

A coin is bent so that it turns up heads 60 percent of the time. We can ask the following question: "What is the probability of 5 heads in 5 flips?"

The events are independent; using the multiplication rule:

$$\text{Probability of 5 heads} = P(5 \text{ heads})$$
$$= .6 \times .6 \times .6 \times .6 \times .6$$
$$= .078$$

Now, what is the probability of 3 heads in 5 flips? If the order is specified (e.g., HHHTT) we can answer the question exactly as above:

$$P(3 \text{ heads in the order HHHTT}) = .6 \times .6 \times .6 \times .4 \times .4$$
$$= .6^3 \times .4^2$$
$$= .034$$

In general, this probability is $p^r q^{(n-r)}$; the symbols being described below. In any other order, the answer is still the same, thus:

$$P(3 \text{ heads in order TTHHH}) = .4 \times .4 \times .6 \times .6 \times .6$$
$$= .034$$

Hence, the order is unimportant, so we need to know in how many ways (that is, how many arrangements) 3 heads can occur in 5 flips.

This is the number of *combinations* of 5 things taken 3 at a time; that is, there are two groupings (heads and tails), and we wish to know how many ways we can arrange the 5 flips into the 2 groupings. It can be shown that the number of combinations in which r successes can occur in n trials is

$$_nC_r = \frac{n!}{r!(n-r)!}$$

where n factorial is $n! = 1 \times 2 \times 3 \times \ldots \times n$ and $0! = 1$ by definition.

The number of combinations in which 3 heads can occur in 5 trials is, therefore,

$$_5C_3 = \frac{5!}{3!2!} = \frac{1 \times 2 \times 3 \times 4 \times 5}{1 \times 2 \times 3 \times 1 \times 2} = 10$$

(There are 10 ways in which 3 heads can occur in 5 flips of a coin.) Let us now return to our original question (the probability of 3 heads in 5 flips of the bent coin). We must multiply the number of combinations of 3 heads in 5 flips by the probability of 3 heads in 5 flips occurring in some specific order:

$$P(3 \text{ heads in 5 flips}) = 10 \times .034 = .34$$

The Binomial Probability Formula

In general, the probability of r successes in n trials is:

$$P(r) = {_nC_r}p^r q^{(n-r)}$$

where r is the number of successes (i.e., heads); n is the size of the sample (i.e., number of flips); p is the probability of a success (i.e., a head); $q = (1 - p)$ is the probability of a failure (i.e., a tail); and $P(r) =$ probability of exactly r successes (i.e., r heads).

Example. Probability of 3 heads and 2 tails with our bent coin:

$$n = 5 \text{ flips}$$
$$r = 3 \text{ heads}$$
$$n - r = 2$$
$$p = .6, \text{ the probability of a head}$$
$$q = 1 - p = .4$$
$$P(r) = {_nC_r}p^r q^{(n-r)} = \frac{5!}{3!2!}(.6)^3(.4)^2 = 10 \times .034 = .34$$

If we carried this procedure out, we could find the probability of any number of heads in 5 flips of our bent coin. The results would be:

$$
\begin{array}{lll}
\text{Probability of 0 heads} = & P(0) = & .01 \\
\text{Probability of 1 head} = & P(1) = & .08 \\
\text{Probability of 2 heads} = & P(2) = & .23 \\
\text{Probability of 3 heads} = & P(3) = & .34 \\
\text{Probability of 4 heads} = & P(4) = & .26 \\
\text{Probability of 5 heads} = & P(5) = & \underline{.08} \\
\text{Total} = & & 1.00
\end{array}
$$

These results are plotted in Chart 6–1, with the random variable (heads) on the X axis and the probability on the Y axis.

This is one example of the *binomial distribution*. Note that for each flip of the coin (i.e., each trial), there were only two possible outcomes—heads or tails. We can use the same kind of analysis whenever we count only two outcomes to each trial (subject to the assumptions below), for example, when we are sampling to determine party affiliation (Democrat or Republican), or in determining if a manufactured product is good or defective, or in any case where there is only a yes or no answer.

The formula for $P(r)$ defines a whole family of distributions of r, one for each combination of the values of n and p. The quantities n and p are called the *parameters* of the binomial distribution, since they

Chart 6–1

BINOMIAL DISTRIBUTION ($p = .6$, $n = 5$)

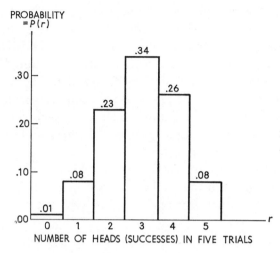

NUMBER OF HEADS (SUCCESSES) IN FIVE TRIALS

determine the probabilities for all values of r. We will use the symbol $P(r|n,p)$ to denote the probability of r successes given n and p.

The *expected value* or mean number of successes $E(r)$ in a binomial distribution is np, and the *variance* is npq. Thus, in the bent coin example ($n = 5$, $p = .6$),

$E(r) = np = 5 \times .6 = 3$ heads (the expected or average number of heads in 5 tosses)

Variance $= npq = 5 \times .6 \times .4 = 1.2$
Standard deviation $= \sqrt{1.2} = 1.1$ heads

Assumptions Underlying the Binomial Distribution

1. *For each trial, the random variable can take on only one of two values*—success or failure.

2. *The trials are independent.* What happens on the first trial does not affect the second, and so on. If we are flipping a coin, this means that heads will occur with the same probability, regardless of whether the previous flip was a head or a tail.

This assumption implies that we are sampling from an infinite population. Flipping a coin can be considered an infinite process, for we can conceive flipping the coin forever. Likewise, if we inspect items from a lot of manufactured parts, and if we replace each item after it is inspected, we can again consider this an infinite population since we would never exhaust it. This latter process is called *sampling with re-placement.*

Oftentimes in actual practice, we do not replace items in sampling from a large lot (i.e., *sampling without replacement*), and we violate the assumptions of the binomial distribution. Theoretically, we should use the *hypergeometric distribution* instead, when sampling without replacement from a finite population. This will not be described here since, in the great majority of practical applications, it can be approximated by the binomial distribution. This is because the binomial is approximately equal to the hypergeometric if the sample size (i.e., the number of trials) is small relative to the number of items in the population. A good rule of thumb is 20 percent. That is, if the sample size is less than 20 percent of the total number of items in the whole population, then the binomial distribution can be used even when sampling without replacement.

3. *The value of p, the probability of success, remains the same from trial to trial.* The assumption implies that, for example, the coin does not become more and more bent as the trials proceed, or that a machine does not wear and produce a higher proportion of defective items over time.

Mathematically, we can derive the binomial distribution from these three assumptions. If a process in the real world satisfies these assumptions, then we use the binomial probabilities to represent the real-world probabilities.

Tables of the Binomial Distribution

Calculating the binomial probabilities from the formula

$$P(r) = {}_nC_r p^r q^{n-r}$$

would be quite time consuming if n were very large. Hence we resort to tables for obtaining the values. Several comprehensive tables are available.[1] We have included a shorter set of tables in Appendixes F and G at the end of the book. Appendix F lists the individual probabilities in the binomial distribution for values of n from 2 to 25, and for various values of p from .01 to .5. Values for p greater than .5 can be read from this table by reversing the definition of "success" and "failure."

Appendix G is a table of the cumulative binomial distribution. That is, it shows the probability of r or more successes for any given value of

---

[1] See, for example, *Tables of the Binomial Probability Distribution,* U.S. Department of Commerce, National Bureau of Standards, Applied Mathematics Series No. 6 (Washington, D.C.: U.S. Government Printing Office, 1949).

r, and for the same values of n and p as above. Examples of the use of these tables follow.

Examples of the Binomial Distribution

1. A large lot of a certain manufactured part is known to contain 5 percent defective parts. If a sample of three parts is drawn at random, what is the probability that none of the parts is defective?

First let us check the binomial assumptions. The first assumption says that each part can take on only two possible values. Here we have only good or defective, so we are all right on that account.

The second assumption implies that the trials (i.e., drawings) are independent. If we were to replace each part before the next is drawn, this assumption would be strictly true. However, our sample size, three items, is quite small relative to the size of the large lot, so that any error introduced on this account will be small.

The third assumption implies that the value of p remains the same as we continue to sample. Since we are sampling from a fixed lot of items which does not change, the assumption is valid.

Having satisfied ourselves that the binomial distribution is appropriate (or a close approximation), we proceed to calculate the required probability. In our example, $p = .05$, $n = 3$, and $r = 0$. The probability of zero defectives is:

$$P(r = 0) = {_3}C_0 p^0 q^3 = \frac{3!}{3!0!}(.05)^0(.95)^3 = .857$$

2. Suppose, for our second example, we use the same circumstances as above, namely, a large lot of manufactured parts which is known to contain 5 percent defective parts. Let us now, however, take a sample of 20 items, and ask the following three questions: (a) What is the probability of exactly 2 defective items out of the 20 sampled? (b) What is the probability of 2 or more defective items? and (c) What is the probability of 2 or less defectives?

The evaluation of the required probabilities would involve considerable calculation, so we shall look up the values in the table instead.

 a) The probability of exactly two defectives: This value can be found directly in Appendix F for $n = 20$, $p = .05$, and $r = 2$. The value is $P(r = 2 | n = 20, p = .05) = .189$.

b) The probability of 2 or more defectives: This value can be found directly in Appendix G for $n = 20$, $p = .05$, and $r = 2$. The value is $P(r \geq 2 | n = 20, p = .05) = .264$.

c) The probability of 2 or less defectives: This cannot be read directly from either of our tables. Instead, we recognize the fact that the probability of 2 or less defectives plus the probability of 3 or more defectives must be one. In symbols,

$$P(r \leq 2) + P(r \geq 3) = 1 \qquad \text{or} \qquad P(r \leq 2) = 1 - P(r \geq 3)$$

Now, the probability of three or more defectives is read easily from the table: $P(r \geq 3) = .075$. Hence:

$$P(r \leq 2) = 1 - .075 = .925$$

That is, the probability of two or less defectives is equal to 1 minus the probability of 3 or more defectives.

3. Exactly 60 percent of the workers in a certain plant belong to a union. If management drew a sample of 15 workers at random from the plant: (a) What is the probability that exactly 8 will belong to the union? (b) What is the probability that 8 or more will belong?

Again, we cannot answer these questions by direct reference to the table, since the table extends only to $p = .50$. Hence, we must rephrase the question as follows: 40 percent of the workers are nonunion. (a) What is the probability of obtaining exactly 7 nonunion members in the sample (i.e., 8 union members $+$ 7 nonunion members $= 15$ men in sample). This is:

$$P(r = 7 | n = 15, p = .40) = .177$$

The probability of 7 nonunion members is identical to the probability of 8 union members, since they represent the same situation.

Similarly (b), the probability of 8 or more union members is equivalent to the probability of 7 or fewer nonunion members (i.e., fewer than 8). As in example 2:

$$P(r \leq 7 | n = 15, p = .40) = 1 - P(r \geq 8 | n = 15, p = .40)$$
$$= 1 - .213 = .787$$

(It is suggested that the student work several exercises to be sure he understands how to evaluate binomial probabilities.)

Uses of the Binomial Distribution

Although the binomial distribution is restricted to samples from two-valued populations, the number of applications is quite large. Industrial quality control is a major application, as cited above. Here, very often, items are classified as defective or good or as having failed a test or not. In public opinion polls voters may be for or against a candidate or in favor of or against some proposition. In market research a consumer may prefer "our product" or "Brand X"; he may buy or fail to buy a product; or he may recall a brand name or fail to recall it. Medical research (a drug cures a patient or not) and economic surveys (employed or unemployed) provide two more of the many applications of the binomial distribution in describing attributes.

THE POISSON DISTRIBUTION

Another discrete distribution of some practical importance is the Poisson distribution. The Poisson is like the binomial except that we conceive of a very large number of trials and a very small probability of a success on any trial. This may best be explained by an example. If we were to inspect an enameled refrigerator door of a standard size, we might find zero, one, or two blemishes, or even more, in a given square foot of enameling. We can count the blemish spots. It is impossible to count the number of nonblemished spots (they are practically infinite). We cannot use the binomial distribution in this case because we do not know the value of n, the total number of possible spots. Or putting it another way, the binomial is defined in terms of a particular attribute which has values of zero or one, whereas the Poisson is defined with respect to *some unit of measurement* and there may be zero, one, two, or more outcomes (e.g., blemishes) within a given measurement unit (e.g., a square foot of enameling). In statistical quality control, therefore, the Poisson distribution is applied to the number of defects per unit (X), whereas the binomial is applied to the number of defective units (r).

Formula and Assumptions of the Poisson Distribution

The Poisson probability function is

$$P(X) = \frac{e^{-m}m^X}{X!} \quad \text{for} \quad X = 0, 1, 2, \ldots$$

where X is the random variable, the number of occurrences per unit of measurement; m is the mean or average number of occurrences of X per unit of measurment; and e is a constant (the base of natural logarithms) with value of 2.718. . . .

In the example of the enameling process, the random variable X is the number of blemishes in a square foot. X is an integer, since 0, 1, 2, 3, etc., blemishes only—not 1.25—can occur in a square foot. The value m need not be an integer, since the average number of blemishes can take on any value. Note that m is the only parameter of the Poisson distribution; that is, if we know only the average, we can find the probability that any specified number of blemishes will occur.

It is curious to note that the variance of the Poisson distribution is equal to m. Hence, the variance equals the mean, and the standard deviation is $\sqrt{m}$ — a very simple situation indeed!

The assumptions underlying the Poisson distribution are similar to those for the binomial:

1. Within any unit of measurement (area of opportunity) there are a large number of possible points for an occurrence, and the probability of an occurrence in any one point is very small. Further, the random variable X must be an integer within the unit of measurement.

2. *Independence:* any number of occurrences can happen in one unit of measurement and this will not affect the number of occurrences in other units of measurements. In our enameling example, this assumption implies that five blemishes in one particular square foot do not influence the probabilities for any other square foot.

3. *Stability:* the value of m (the average or mean) must remain constant. Thus about the same number of blemishes, on the average, must occur at all points of the refrigerator doors inspected.

Examples of the Poisson Distribution

1. Suppose in our example that enameling blemishes occurred on the average of one per square foot of refrigerator door (and the assumptions of stability and independence are valid). The probability that a square foot will have no blemishes is:

$$P(X = 0 | m = 1) = \frac{e^{-1}1^0}{0!} = e^{-1} = .368$$

The probabilities of one, two, and three blemishes in a square foot are:

$$P(X = 1 | m = 1) = \frac{e^{-1}1^1}{1!} = e^{-1} = .368$$

$$P(X = 2 | m = 1) = \frac{e^{-1}1^2}{2!} = \frac{e^{-1}}{2} = .184$$

$$P(X = 3 | m = 1) = \frac{e^{-1}1^3}{3!} = \frac{e^{-1}}{6} = .061$$

2. Consider a telephone switchboard. Suppose calls arrive at random. What would this mean? Let us look at each second of time. In most seconds there would be no calls arriving; in some seconds one call would arrive. If this were all, we could treat the process as a binomial distribution. However, in some seconds two or three or even more calls may arrive. The Poisson distribution deals with this kind of process. Note,

however, that the assumption of stability would be violated if more persons, on the average, called the switchboard at certain times during the day than at other times.[2]

3. A certain part in a machine breaks at random. We can use the Poisson distribution to evaluate the probabilities of no breakages on a certain day, of one breakage, or two breakages, or more. Note, however, that if breakage was a function of how long the part had been in operation (i.e., wear), the assumption of stability would be violated.

Tables of the Poisson Distribution

Appendix H at the end of the book is a table of individual probabilities of the Poisson distribution for selected values of m from .001 to 10.[3] Appendix I is a table of the cumulative Poisson distribution for X or more occurrences. The use of these tables is very similar to that of the binomial tables. An example is given below.

A certain part breaks, on the average, twice a month. What is the probability (a) that exactly 3 breakages will occur in a given month, (b) that 3 or more breakages will occur, and (c) that fewer than 3 will occur?

(a) $P(X = 3|m = 2) = .180$ Appendix H
(b) $P(X \geq 3|m = 2) = .323$ Appendix I
(c) $P(X < 3|m = 2) = 1 - P(X \geq 3|m = 2) = 1 - .323 = .677$

Poisson Approximation to the Binomial

Another important use of the Poisson distribution is as an approximation to the binomial. Indeed, we can think of the Poisson as the limiting distribution to the binomial as n becomes large and p becomes small. Thus, when n is large and p is small, we can use the Poisson to evaluate binomial probabilities.

How large must n be and how small must p be? As a rule of thumb, we can use the Poisson to approximate the binomial if

$n \geq 10$ and $p \leq .01$ or $n \geq 20$ and $p \leq .03$ or
$n \geq 50$ and $p \leq .05$ or $n \geq 100$ and $p \leq .08$

These requirements achieve a moderate degree of accuracy in the approximation. If very fine precision is required, larger sample sizes would be required.

---

[2] We could treat this by breaking the day up into parts such that m was stable over each part.

[3] If $m > 10$, use the normal curve as an approximation, with $\mu = m$ and $\sigma = \sqrt{m}$. Make the same correction for discrete values as shown on page 143.

To approximate a binomial probability, we simply set $np = m$ and look the values up in the Poisson table. As an example: Suppose we are sampling 1,000 items of which .001 are defective on the average; that is, $n = 1,000$, $p = .001$, and $np = m = 1$ (an average of one defective per 1,000). We can estimate the probability of getting any number of defects in our sample by using the Poisson table, as follows:

$$P \text{ (0 defectives)} = .368$$
$$P \text{ (1 defective)} = .368, \text{ etc.}$$

Uses of the Poisson Distribution

The Poisson distribution, like the binomial, is widely used in industrial quality control. In describing the number of defects per unit, the Poisson applies particularly when (1) a natural unit does not exist, as in defects per 100 square yards of cloth, the unit of area being arbitrary; or (2) where the unit is quite complex (e.g., aircraft instruments), so that all units have some defects. The distribution is also used to predict chance or rare events such as accidents, fires, and breakdowns. In fact, it was first developed from a study of the number of Prussian soldiers killed each year by kicks from horses.

Finally, the Poisson distribution is widely used in waiting line or queuing problems. The arrivals of cars at toll booths, customers at supermarket checkout counters, telephone calls at switchboards, and airplanes at airports are examples of situations that can be described by the Poisson distribution. The management in these cases must have adequate facilities so that customers do not have long waits.

THE NORMAL DISTRIBUTION

By far the most important distribution in statistics is the no. nal distribution. This function was described in Chapter 2 as a continuous distribution represented by a symmetrical, bell-shaped curve (see Charts 2–6, 2–7, 4–1, and 4–2). The equation for the normal distribution is:

$$f(X) = \frac{1}{\sqrt{2\pi}\sigma} e^{-\frac{1}{2}\frac{(X-\mu)^2}{\sigma^2}}$$

where X is the random variable and μ and σ are the parameters. The constant π is 3.14159 . . . and e is 2.718. . . . For the normal distribution, the expected value or mean is $E(X) = \mu$ and the variance is σ^2. Normal distributions can take on many different shapes, depending on the values of these two parameters. Consider, for example, Chart 4–1, panels 1 and 2. Since the normal curve is a continuous distribution, the random variable X can take on any value, rather than only discrete values, as in the binomial and Poisson distributions.

It would be difficult to measure the probabilities under the normal curve were it not for a simple transformation which makes it possible to use only a single table. The trick is simply that we discuss normal distributions and associated probabilities in terms of standard deviation (σ) units from the mean (μ) of the distribution.

It was pointed out in Chart 4–2 that in a normal distribution:

$$\mu \pm \sigma \quad \text{includes 68.27\% of the values}$$
$$\mu \pm 2\sigma \quad \text{includes 95.45\% of the values}$$
$$\mu \pm 3\sigma \quad \text{includes 99.73\% of the values}$$

That is, if we draw a single item from this distribution, the probability is .6827 (about two chances out of three) that it will fall within the interval $\mu \pm \sigma$, the probability is .9545 that it will fall within the interval $\mu \pm 2\sigma$, and so on. These probabilities hold for all normal distributions, regardless of the mean or standard deviation. Furthermore, we can similarly evaluate probabilities for any number of standard deviations difference from the mean.

Table of Areas under the Normal Curve

We can determine these probabilities from a table of areas under the normal curve. Appendix D shows the proportion of the total area which lies between the mean and any other point X along the horizontal axis. To use the table, first take $X - \mu$ and divide by σ as follows:

$$z = \frac{X - \mu}{\sigma}$$

The value z is called the *standard normal deviate* and represents the number of standard deviation units the random variable X is above or below the mean. The whole table then represents a *standardized normal distribution* with mean $\mu = 0$ and standard deviation $\sigma = 1$. (For a sample, substitute $\overline{X}$ for μ and s for σ.)

The left-hand stub and the heading of Appendix D show the values of these deviations (z) from zero (the mean itself) to five, a point far out under the tail of the normal curve. The body of the table shows the proportion of the total area between the mean and any given value of z. Since the normal curve is symmetrical about the mean, the table can be used for points on either side of the mean.[4]

---

[4] Theoretically, the curve extends indefinitely on each side of the mean without touching the X axis. However, only a negligible part of the area lies more than four or five standard deviations from the mean, so the infinite tails can be ignored.

To illustrate, suppose a large number of job applicants take an aptitude test given by the personnel department of a company. The scores on the test form a normal distribution[5] with an arithmetic mean of 80 and standard deviation of 4. Now consider the following cases. These are illustrated in Chart 6–2, panels A to D, respectively.

Chart 6–2

FINDING AREA UNDER A NORMAL CURVE
IN APPENDIX D

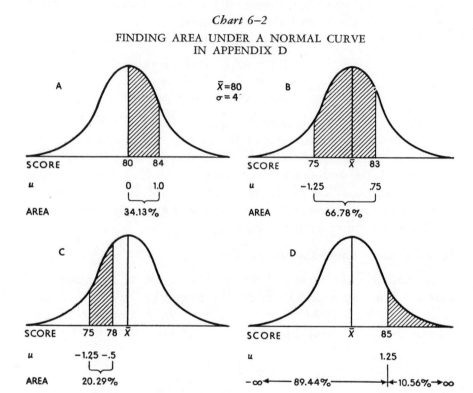

A. What proportion of applicants should score between 80 and 84? The deviation of the point 84 from the mean 80 is 4, so in standard deviation units, $z = 4/4 = 1.0$. Looking in Appendix D opposite $z = 1.0$, the proportion of the total area in this segment is .3413, or 34.13 percent. The table shows probabilities, while the chart shows relative areas. The two are equivalent, since the area under any segment of the curve is proportional to the probability. The proportion of scores that fall between the mean and one standard deviation on both sides of the mean is twice 34.13 percent, or 68.26 percent—the same value that was given for $\mu \pm \sigma$ previously (except for a slight error in rounding).

[5] The distribution of scores may be treated as continuous, since differences between successive scores are small.

Many intervals do not terminate at the mean. These may be broken down, however, into intervals that do terminate at the mean, as shown below. Hence, Appendix D can be used for any interval.

B. What proportion of scores should fall between 75 and 83? Since these points fall on both sides of the mean, the areas between the mean and each point must be added. For the score 83, $z = (83 - 80)/4 = .75$. In Appendix D, look down the z column to .7 and across to the column headed .05; the area is .2734. Similarly, for 75, $z = (75 - 80)/4 = -1.25$, and the area is .3944. The combined area is then $.2734 + .3944 = .6678$ or 66.78 percent.

C. What proportion of scores should fall between 75 and 78? Since both points are on the same side of the mean, the areas between each point and the mean must be subtracted to get the area between them. For 75, the area is .3944 as above. For 78, $z = -.5$ and the area is .1915. The area between 75 and 78 is then $.3944 - .1915 = .2029$, or 20.29 percent of the total area.

D. What proportion of scores should exceed 85? This is 50 percent—the entire segment above the mean—minus the proportion of scores between the mean and 85, or 39.44 percent (for $z = 1.25$). The answer is then 10.56 percent. Similarly, the proportion of scores below 85 (the unshaded part of panel D) is $50 + 39.44 = 89.44$ percent.

The table of areas under a normal curve thus serves to show the probabilities for any segment of the curve. When in doubt as to how to apply this table, draw a rough diagram, as in Chart 6–2, to picture the areas needed.

Normal Approximation to the Binomial

We noted before that when n is large and p is near zero or one, we can use the Poisson distribution to approximate the binomial. On the other hand, when n is large and p is *not* close to zero or one, we can use the normal distribution to approximate the binomial. How large must n be and how large must p be?

The influence of sample size and value of p on the shape of the binomial distribution is illustrated in Chart 6–3. The chart represents the distributions of r, the number of successes for various combinations of values on n and p. The polygons show that the distribution of r is discrete rather than continuous. They also show how skewness depends on n, the size of the sample, and the population value of the proportion p.

Effect of p on the Distribution. In panel A of Chart 6–3, probability distributions of number of successes are shown for samples of a

Chart 6–3

BINOMIAL DISTRIBUTIONS OF NUMBER OF SUCCESSES
A. Fixed Size of Sample, $n = 10$,
and Different Values of p

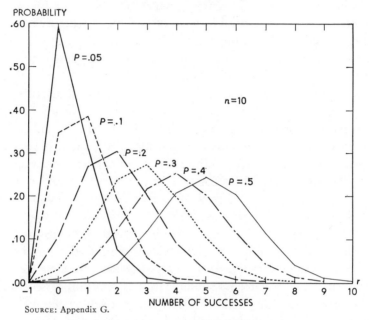

SOURCE: Appendix G.

B. Fixed Value of Proportion, $p = .1$,
and Different Sizes of Sample

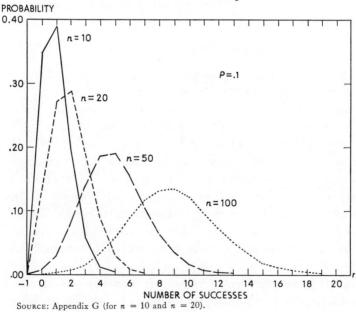

SOURCE: Appendix G (for $n = 10$ and $n = 20$).

fixed size, $n = 10$, but for varying values of p from .05 to .5. When $p = .05$, the distribution has a high degree of positive skewness. As the value of p approaches one half (.5), the skewness approaches zero, so that when $p = .5$ the distribution is perfectly symmetrical and nearly normal.

Effect of Sample Size. In panel B of Chart 6–3, probability distributions are shown for a fixed value of a proportion ($p = .1$), but for varying sizes of sample from 10 to 100. For small values of n the skewness is large and positive; as n increases, the approach to the symmetrical normal curve is rather striking. The same curves apply to q as to p, substituting "number of failures" for "number of successes."

The curves illustrate the fact that n should be large, or else p should be not too close to zero or one, to justify the use of the methods presented below, since they are based on the assumption that the distribution of the number of successes is approximately normal. As a rule of thumb, both np and nq should be about five or more for this assumption to be valid. Thus, if $n = 10$, p would have to be .5 to make $np = 5$, as in the right-hand curve of panel A. On the other hand, if $p = .1$, n would have to be as large as 50 (panel B) for the distribution to be roughly normal. The assumption of normality is useful both because it is valid for most practical problems involving large samples and because it is simpler than using the binomial distribution.

How can we make the approximation? Proceed as follows:

1. Set np equal to μ and $\sqrt{npq}$ equal to σ.
2. Remember that the binomial is a discrete distribution. To allow for this we have to use a factor of $+\frac{1}{2}$ or $-\frac{1}{2}$ added to X, depending upon the circumstances. To find the probability of r or *less* successes, *add* $\frac{1}{2}$ to the value of X in calculating the normal deviate $z;$ to find the probability of r or *more* successes, *subtract* $\frac{1}{2}$ from the value of X in determining z.
3. Look up the probabilities in the normal table (Appendix D).

Example. The probability of a defective item is $p = .20$. We take a sample of 400 items from a very large lot.

a. What is the probability of 90 or more defectives?

$$\mu = np = 80$$
$$\sigma_r = \sqrt{npq} = \sqrt{400 \times .2 \times .8} = 8$$

Now, the dividing line between 90 or more and the rest of the distribution in $89\frac{1}{2}$. That is, the probability of being greater than $89\frac{1}{2}$ for the continuous normal distribution is approximately the same as the probability of 90 or more in the discrete binomial.

$$z = \frac{X - \mu}{\sigma} = \frac{89\frac{1}{2} - 80}{8} = 1.19$$
$$P(z > 1.19) = .1170$$

b. What is the probability of exactly 90 defectives? The probability of more than 90 defectives in the binomial distribution is equivalent to the probability of more than $90\frac{1}{2}$ defectives in the normal distribution. For $X = 90\frac{1}{2}$,

$$z = \frac{90\frac{1}{2} - 80}{8} = 1.31$$
$$P(z > 1.31) = .0951$$
$$P(\text{exactly } 90) = P(1.19 < z < 1.31) = .1170 - .0951 = .0219$$

The shaded area in Chart 6–4 illustrates this probability.

Chart 6–4

NORMAL APPROXIMATION TO BINOMIAL DISTRIBUTION

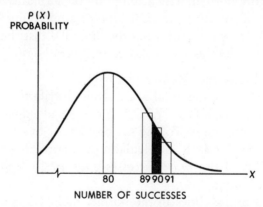

NUMBER OF SUCCESSES

Normal Probability Paper

Normal probability paper is special graph paper with a scale such that the *cumulative* normal distribution plots as a straight line (see Chart 6–5).

The major use of this paper is in testing whether a particular distribution is normal. For example, you have samples from some population (e.g., scores of employees on a manual dexterity text) and you wish to

Chart 6–5

CUMULATIVE HOURLY EARNINGS OF 214 APPRENTICE
MACHINE TOOL OPERATORS PLOTTED AS PERCENT OF
TOTAL ON NORMAL PROBABILITY PAPER

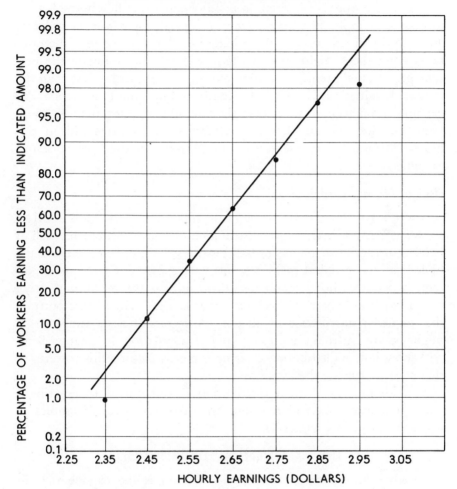

know if the distribution is normal. Simply plot the cumulative distribution on normal probability paper. If the distribution is normal, the
points should lie close to a straight line (there will be some chance
variation about the line).

The cumulative hourly earnings of 214 apprentice machine tool
operators (see Table 2–6) are plotted on normal probability paper in
Chart 6–5. A straight line has been drawn by inspection through these
points. The five points between $2.45 and $2.85 lie nearly on the line,

indicating that the distribution of earnings is roughly normal over this middle range. The two end points, however, are out of line; hence, the distribution is not normal near its extremes.

A second purpose of normal probability paper is to fit a normal curve to a set of sample data drawn from a normal population in order to estimate the distribution of the population. Thus, if we read the ordinates from the straight line in Chart 6–5, we can estimate the percentages of all apprentice machine tool operators earning less than the indicated values of X. This device irons out sampling errors. For example, 85 percent of workers in the sample earned less than $2.75 an hour, but we estimate that 87 percent of all workers fall in this group (assuming a representative sample from a normal population of earnings).

Uses of Normal Distribution

The normal distribution is the most important distribution in all of statistics. First, it describes the distribution of many phenomena such as heights of people, diameters of ball bearings, IQ's, in fact a great many biological and physical measurements (see Chart 2–6). More important, it describes how certain measures, such as the mean, vary from sample to sample because of chance; that is, the normal curve portrays the frequency distribution of all possible means of large samples that might be drawn from almost any kind of population. In Chapter 9 we will show how a distribution of sample means follows this pattern, so that we can estimate the sampling error.

In addition, the normal distribution is often used to describe a decision maker's uncertainty about an unknown factor. Examples of this are shown in Chapter 14.

THE EXPONENTIAL DISTRIBUTION

Another important continuous distribution is the exponential. Its probability function is:

$$f(t) = \lambda e^{-\lambda t}$$

where t is a random variable representing the time between successive arrivals (e.g., arrivals at a service booth); λ (lambda) is the average rate of arrivals (the same as m in the Poisson process), and the reciprocal $1/\lambda$ is the average time between arrivals; and e is the constant 2.718, the base of natural logarithms. λ is the sole parameter; it determines the entire distribution. Both t and λ must be positive.

The exponential distribution has a reverse J-shape, as shown in

Chart 6–6. The mean of this distribution is $1/\lambda$ and the variance is simply $1/\lambda^2$.

The cumulative probabilities for the exponential distribution can be evaluated from the following relationship:

$$P(t > a) = e^{-\lambda a}$$

where a is any given value of t. The table in Appendix J at the end of the book lists powers of e^{-x} (i.e., $e^{-\lambda a}$) for values of x from 0 to 10.

Example 1. The life of a certain type of electronic part is known to have an exponential distribution with a mean life of one week. What is the probability that a given part will have a life in excess of three weeks? Here $a = 3$ and the average life $= 1/\lambda = 1$ week. Hence $\lambda = 1$ and

$$P(t > 3) = e^{-\lambda a} = e^{-1(3)} = .05, \text{ from Appendix J}$$

This is the unshaded area to the right of a under the curve in Chart 6–6, taking the whole area as the probability of one.

What is the probability of a life of between one and three weeks?

$$P(t > 1) = e^{-1(1)} = .368$$

and
$$P(1 < t < 3) = P(t > 1) - P(t > 3)$$
$$= .368 - .050 = .318$$

This is the shaded area in Chart 6–6. [$P(t = 3)$ is infinitesimal.]

Chart 6–6

EXPONENTIAL DISTRIBUTION FOR $\lambda = 1$

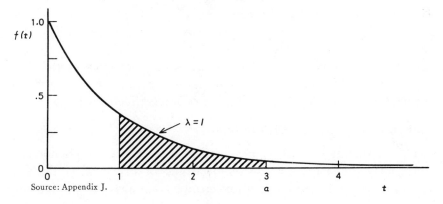

Source: Appendix J.

Example 2. Suppose the time between arrivals of customers at a teller's window in a bank is known to be exponential with a mean of .25 minutes. What is the probability of a gap of less than 6 seconds

($a = .10$ minute) between a given arrival and the next? Note that the mean $.25 = 1/\lambda$, so $\lambda = 4$. Then:

$$P(t < .10) = 1 - P(t > .10)$$
$$= 1 - e^{-(4)(.10)} = 1 - e^{-.4}$$
$$= 1 - .670 = .330, \text{ from Appendix J}$$

or about one chance in three. (Note that $P(t) = .10$ is infinitesimal.)

Properties of the Exponential Distribution

The exponential distribution has the property of being *memoryless*. For example, if the burning time of light bulbs is exponentially distributed, it would mean that a light bulb that has already burnt 100 hours has the same probability of burning an additional 200 hours as a bulb that has not yet been tested. In a sense, the bulb has no memory of its previous history.

To see this, continue Example 1 above and calculate the probability of an additional two weeks, given that one week of life has already expired:

$$P(t > 3 | t > 1) = \frac{P(t > 3)}{P(t > 1)} = \frac{.050}{.368} = .135$$

Note that this is identical to $P(t > 2) = e^{-1(2)} = .135$.

The exponential distribution also bears a special relationship to the Poisson distribution. The Poisson describes the number of occurrences per unit of measure (e.g., the number of telephone calls per minute), whereas the exponential describes the value of the measure per occurrence (e.g., the time between successive telephone calls). The time between successive arrivals is called the *interarrival time.* Thus the two distributions may be used to describe the same phenomenon, the Poisson describing the number of occurrences per unit of time and the exponential describing the distribution of interarrival times.

Uses of the Exponential Distribution

The exponential distribution is used to represent many phenomena, particularly the operating life of electronic or industrial equipment. The time between failures of an electronic computer and the burning life of electric light bulbs are examples.

Because of the relationship between the Poisson and exponential distributions, the exponential is also used to describe the interarrival times whenever the Poisson is applicable. Thus the exponential is used quite extensively in queuing or waiting line theory to describe the interarrival times of customers at a facility and also the time needed to service a customer.

SUMMARY

This chapter describes four specific probability distributions: the binomial, the Poisson, the normal, and the exponential.

The binomial distribution characterizes situations in which we are sampling from a population of attributes having only two values (yes or no, success or failure, etc.). It describes the number of successes (r) achieved in a fixed number of trials (n). The binomial is a discrete distribution.

The assumptions underlying the binomial are: (1) the random variable can take on only one of two values—success or failure; (2) the trials are independent; and (3) the probability of a success remains the same from trial to trial.

The Poisson distribution, like the binomial, is a discrete distribution. The random variable X can take on the value of zero or any positive integer. The Poisson distribution is used to represent random occurrences in some unit of measurement, such as the number of telephone calls per unit of time or the number of defects per foot of wire.

The assumptions underlying the Poisson distribution are: (1) there is a very large number of possible occurrences in any unit of measurement; (2) there is independence from one unit of measurement to another; and (3) the average number of occurrences per unit remains the same.

If the number of trials (n) is large and the probability of success (p) is small, the Poisson distribution is a close approximation to the binomial.

The normal distribution is a continuous distribution represented by the familiar bell-shaped curve. The standardized normal distribution has a mean of zero and a standard deviation of one. Using this standard distribution and Appendix D, we can evaluate probabilities for any normal distribution.

If the number of trials (n) is large and the probability of success (p) is not close to either zero or one, the normal distribution is a good approximation to the binomial.

Normal probability paper may be used to test if a given set of data follows the normal distribution, or to estimate the distribution of a normal population from sample data.

The exponential distribution is a continuous J-shaped distribution. It is used to represent certain continuous phenomena such as the time between arrivals at a service booth or the life of electronic parts. It is also a complementary distribution to the Poisson, representing the interarrival times between occurrences, whereas the Poisson represents the number of arrivals per unit of time.

The four distributions studied in this chapter, together with their parameters, means, variances, and standard deviations are shown in the following table:

Distribution	Parameters	Mean	Variance	Standard Deviation
Binomial.................	n, p	np	npq	$\sqrt{npq}$
Poisson.................	m	m	m	$\sqrt{m}$
Normal.................	μ, σ	μ	σ^2	σ
Exponential.............	λ	$1/\lambda$	$1/\lambda^2$	$1/\lambda$

PROBLEMS

In problems 1 through 5 below, evaluate the binomial probabilities by using the binomial probability formula.

1. What is the probability of three heads in four flips of a fair coin?

2. What is the probability of drawing (with replacement) two red chips and one yellow chip in three drawings from a bag of chips containing 20 percent red and 80 percent yellow chips?

3. What is the probability of drawing three aces out of five cards from a deck of cards in which the card drawn is replaced and the deck shuffled before each draw?

4. What is the probability of drawing four defective parts from a large lot which is known to contain exactly 10 percent defective parts?

5. If 60 percent of television viewers are watching a certain program, what is the probability that more than half of those selected in a random sample of five will be watching the specified program?

6. Evaluate the following binomial probabilities, using Appendixes F and G.

   a) $P(r = 6 | n = 15, p = .35)$
   b) $P(r \geq 5 | n = 12, p = .25)$
   c) $P(r < 11 | n = 20, p = .45)$
   d) $P(r \leq 2 | n = 16, p = .06)$
   e) $P(r = 18 | n = 20, p = .95)$
   f) $P(r \geq 9 | n = 18, p = .60)$
   g) $P(r < 6 | n = 14, p = .70)$
   h) $P(5 \leq r \leq 13 | n = 20, p = .40)$
   i) $P(1 < r < 5 | n = 20, p = .12)$

7. Evaluate the following binomial probabilities, using Appendixes F and G.

   a) $P(r = 1|n = 8, p = .01)$ *f)* $P(r \geq 12|n = 20, p = .75)$
   b) $P(r \geq 2|n = 13, p = .15)$ *g)* $P(r < 5|n = 15, p = .60)$
   c) $P(r < 15|n = 20, p = .50)$ *h)* $P(7 \leq r \leq 10|n = 24, p = .55)$
   d) $P(r \leq 6|n = 20, p = .20)$ *i)* $P(2 < r < 5|n = 18, p = .30)$
   e) $P(r = 15|n = 25, p = .70)$

8. Evaluate the following Poisson probabilities, using Appendixes H and I.

   a) $P(X = 2|m = .20)$ *c)* $P(X < 5|m = 5)$
   b) $P(X \geq 3|m = .80)$ *d)* $P(2 < X \leq 6|m = 2.4)$

9. Evaluate the following Poisson probabilities, using Appendixes H and I.

   a) $P(X = 4|m = 2.6)$ *c)* $P(X < 2|m = 1)$
   b) $P(X \geq 1|m = .40)$ *d)* $P(10 \geq X \geq 5|m = 6.5)$

10. A part to a certain machine is known to break randomly on the average of once in five days. How many parts must be available so that there is less than one chance in 100 of having more breakages than parts available on a given day?

11. Ships are known to arrive randomly at a port on an average of two days apart. What is the probability of two or more ships arriving on the same day?

12. The Speedo Computer averages .05 breakdowns requiring service per hour of operating time. What is the probability of no breakdowns in an eight-hour day? In a 40-hour week? Assume a Poisson distribution of break-downs.

13. Evaluate the following exponential probabilities using Appendix J.

    a) $P(t \geq 5|\lambda = .5)$ *c)* $P(10 \leq t \leq 30|\lambda = .05)$
    b) $P(t \leq .2|\lambda = 2)$ *d)* $P(t \geq 100|\lambda = .03)$

14. Suppose that the time needed for a bank clerk to service a customer has an exponential distribution with a mean of 30 seconds.
    a) What is the probability that the time needed for a given customer will be longer than four minutes?
    b) What is the probability that the time needed will be between one and two minutes?

15. Refer to Problem 11 above. What is the probability that the time between two successive ship arrivals is more than two days?

16. Refer to Problem 12 above. Answer the same questions (probability of no breakdown in eight hours? in 40 hours?) assuming that the times be-

tween breakdowns are exponentially distributed. Are the answers the same as in Problem 12? Why or why not?

17. A commuter knows that the time between successive buses at a certain bus stop has an exponential distribution with a mean of five minutes.

   a) Suppose he arrives at the bus stop just in time to miss one bus. What is his expected waiting time until the next bus? What is the probability that he will wait longer than 10 minutes?

   b) Suppose when he arrives he finds out that the last bus left exactly five minutes ago. What is his expected waiting time?

   c) Suppose our commuter arrives at the bus stop but doesn't know how long ago the last bus departed. What is his expected waiting time?

18. The random variable X is normally distributed with mean 50 and standard deviation 20. Evaluate the following probabilities:

   a) $P(X \geq 75)$
   b) $P(X \leq 55)$
   c) $P(25 \leq X \leq 45)$
   d) $P(35 \leq X \leq 80)$

19. The random variable X is normally distributed with mean 18 and standard deviation 10. Evaluate the following probabilities:

   a) $P(X \geq 28)$
   b) $P(X \leq 17)$
   c) $P(12 \leq X \leq 16)$
   d) $P(15 \leq X \leq 24)$

20. Suppose the haddock catch in Boston over the past 10 years has averaged 100 million pounds annually, with a standard deviation of 5 million pounds. For Gloucester over the same period, the mean has been 10 million pounds, with a standard deviation of 2 million pounds. If in one year the Boston catch is 108 million pounds, how large must the Gloucester catch be that year to be just as exceptional? (Assume normal distributions.)

21. The average grade on an examination taken by a large number of students is 80. The standard deviation of the grades is 6. The instructor wishes to award A's to 10 percent of the class. Assuming grades are approximately normally distributed, above what numerical grade would he give an A?

22. A firm estimates that 3 percent of its accounts receivable cannot be collected. What is the probability that out of its 200 current accounts receivable, eight or more will be uncollectible?

23. A sales manager believes that 60 percent of consumers prefer his product over his competitor's. Under this assumption, what is the probability of obtaining fewer than 54 who prefer his product out of a random sample of 100 consumers?

24. The number of misprints on a page of a daily newspaper has a Poisson distribution. You are told that the average number of misprints is 1½ per

page. You examine three pages at random and find no misprints. What is the prior probability of this sample result?

25. It is estimated that weekly demand for gasoline at a new filling station will be approximately normally distributed, with an average of 1000 and a standard deviation of 50 gallons. The station will be supplied with gasoline once a week. What must the capacity of its tank be if the probability that its supply will be exhausted in a given week is to be no more than .01?

26. In a recent survey, 85 of 100 firms surveyed reported an increase in sales over the same month last year. If in fact 80 percent of all firms had such a sales increase, what is the probability of obtaining exactly the sample result observed? What is the probability of 85 or more firms out of 100 reporting sales increases?

27. Show that the binomial probability distribution has a mean $= np$ and a variance $= npq$ by computing the mean and variance for the probability distribution of the number of heads in four tosses of a fair coin.

Heads	Probability
0	1/16
1	1/4
2	3/8
3	1/4
4	1/16

28. A committee of 15 is chosen at random from a large company's employees, of whom 60 percent are women. What is the probability that women will find themselves a minority of the committee?

29. The union in your plant claims that only 20 percent of the workers are opposed to a strike. To investigate this claim, you take a random sample of 225 workers and question them. If the union claims were correct, what is the probability of obtaining more than 54 opponents of the strike in the sample?

30. An insurance company finds that half of 1 percent of the population die from a certain kind of accident each year. What is the probability that the company must pay off on more than:
 a) Three of 500 insured risks against such accidents in a given year?
 b) Three of 10,000 insured risks against such accidents in a given year?

31. At an airline reservation desk, incoming calls average about one every five minutes. Assuming that calls are independent and come at random, what is the probability of more than one call in any given minute?

32. The charge accounts at a certain department store have an average balance of $120 and a standard deviation of $40. Assuming that the account balances are normally distributed:

a) What proportion of the accounts is over $150?

b) What proportion of the accounts is between $100 and $150?

c) What proportion of the accounts is between $60 and $90?

33. The World Series is to be played between two teams, the Nationals and the Americans. The victor must win four out of seven games. Suppose that the Nationals have a superior team, so that the probability of their winning in any single game is .60. Assume that this probability remains the same from game to game, and that games are statistically independent.

a) What is the probability that the Nationals will win the series (i.e., will win the necessary four games)?

b) What is the probability of the Nationals winning in four games?

c) What is the probability of the series going exactly five games and the Nationals winning?

d) What is the probability of a seven-game series (the maximum possible number)?

34. A company purchases large lots of a certain electronic component. The decision to accept these purchased lots or to reject them (return them to the supplier) is based upon a sample of 20 items. If any of the 20 items are defective, the lot is rejected; otherwise, it is accepted.

a) What is the probability of rejecting a lot that has 1 percent defectives? What is the probability of accepting such a lot?

b) What is the probability of accepting a lot containing 10 percent defectives?

35. Suppose that the company in Problem 34 was considering using a sample of 50 items rather than the 20 items used previously. Assuming that a lot is accepted if fewer than two defectives are found and rejected if two or more defectives are found in the sample:

a) What is the probability of rejecting a lot with 1 percent defectives?

b) What is the probability of accepting a lot with 10 percent defectives? (*Hint:* Use Poisson approximation to the binomial.)

36. Calculate the probabilities of accepting a lot for each of the sampling plans in Problems 34 and 35 for the intermediate values of .02, .05, and .08 for the fraction defective in the lot. Plot these values and the ones calculated in Problems 34 and 35 on a chart. (The Y axis is the probability of accepting the lot; the X axis is the fraction defective in the lot). Connect the points for each plan by a smooth curve. These are the *operating characteristic* (or OC) curves for each sampling plan. Use the OC curves to compare the two sampling plans.

37. An auditor wishes to determine a rule to use in evaluating the accounts payable of a certain firm. There are 5,000 such accounts. The auditor considers the accounts as satisfactory if there are mistakes in only 1 percent of them. On the other hand, if 5 percent or more are in error, the auditor would require a thorough investigation. Since there are a large number of

accounts, the auditor plans to take a sample of 25 accounts and investigate these. His decision to certify the accounts payable or to require further investigation will depend upon the outcome of the sample. The auditor decides to certify the accounts if none or only one account of the 25 sampled is found in error and to require further investigation if two or more accounts prove in error.

a) If, in fact, 1 percent or 50 accounts are in error, what is the probability that the auditor will certify the accounts? What is the probability that he will decide upon further investigation?

b) If, in fact, 5 percent or 250 accounts are in error, what is the probability that the auditor will require further investigation? What is the probability that he will certify the accounts?

SELECTED READINGS

DRAKE, ALVIN W. *Fundamentals of Applied Probability Theory*. New York: McGraw-Hill, 1967.

A good, slightly more advanced treatment of probability and probability distributions is contained in Chapters 1, 2, and 4.

GOLDBERG, SAMUEL. *Probability, An Introduction.* Englewood Cliffs, N.J.: Prentice-Hall, 1960.

A detailed and systematic treatment of discrete probability.

HUFF, DARRELL. *How to Take a Chance*. New York: W. W. Norton, 1959.

A short and readable book dealing with probability with a humorous approach.

LEVINSON, HORACE C. *Chance, Luck, and Statistics*. New York: Dover Publications, 1963. (Paperback.)

Part I of this book deals with probability in an easily understandable way.

MOSTELLER, FREDERICK; ROURKE, ROBERT E.; and THOMAS, GEORGE B., JR. *Probability with Statistical Applications*. 2d ed. Reading, Mass.: Addison-Wesley, 1970.

A detailed treatment of probability at the elementary level.

NATIONAL BUREAU OF STANDARDS. *Tables of the Binomial Probability Distribution*. Washington, D.C.: U.S. Government Printing Office, Applied Mathematics Series No. 6, 1949.

Detailed tables of the binomial distribution.

RAIFFA, HOWARD. *Decision Analysis*. Reading, Mass.: Addison-Wesley, 1968.

The early chapters discuss assumptions underlying subjective probability. Chapter 5 is a specific treatment of estimating subjective probability.

SCHLAIFER, ROBERT. *Analysis of Decisions under Uncertainty*. New York: McGraw-Hill, 1969.

Part 2 of this book (Chapters 5 through 9) gives a detailed treatment of probability, including the assessment of subjective distributions.

STAEL VON HOLSTEIN, CARL-AXEL. *Assessment and Evaluation of Subjective Probability Distributions*. Stockholm: Economic Research Institute, Stockholm School of Economics, 1971.

A good summary of the work that has been done on subjective probability assessment, including several experiments by the author.

7. DECISION MAKING UNDER UNCERTAINTY

THIS CHAPTER is concerned with using probabilities and economic consequences of future events in a logical procedure for making decisions.

THE DECISION MAKING PROCESS

Any decision making problem has certain essential elements. First, there are *alternative* ways of solving the problem, that is, two or more possible actions or alternatives; otherwise, there is no decision problem. Second, there must be *goals* or objectives that the decision maker is trying to achieve. For example, a manager may have the goals of profitability and growth for his firm. Third, there must be a *process of analysis* whereby the alternatives are *evaluated* in terms of the goals. The decision maker can then pick that alternative which best achieves his goals. This is diagramed in Chart 7–1.

Models

Note that, as shown in Chart 7–1, the process of analysis makes use of a *model* or artificial representation of reality. Models have long been useful in scientific analysis. Engineers build scale replicas of aircraft and test them in wind tunnels, or construct replicas of dams before deciding to build them. Often, an equation may be used to represent some phase of reality, as with the laws in physics. For example, the equation

$$d = \tfrac{1}{2} g t^2$$

predicts the distance (d) that a freely falling object will travel as a function of the time (t) it has been falling. (The g is a constant.) This

Chart 7–1

THE DECISION MAKING PROCESS

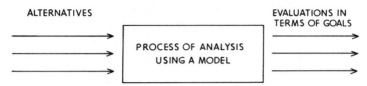

model is a very useful one for describing a particular aspect of the real world.

When making a simple decision, we sometimes use an intuitive model. In making important decisions, we use more formal models, specifying carefully the important variables and the relationships between these variables. Rarely does a model conform exactly with reality—it would have to include far too many factors and be very complex. For example, the physical law described above does not include the air resistance to the falling object. However, for a model to be useful, it need represent only the important variables affecting the decision at hand.

Certainty versus Uncertainty

In some business decisions, all the facts relevant to the decision are known in advance; that is, there is no uncertainty about future costs or profits. The decision problem is to select the best of the known alternatives. Consider the following as an example of this type of decision situation: A firm has several factories that ship goods to its warehouses. The factories and warehouses are scattered geographically around the country. The shipping costs from each factory to each warehouse are known with certainty. The capacities of the factories and the requirements of the warehouses are also known in advance. Despite the fact that all this information is known without error, the determination of the optimum (least cost) shipping schedule (i.e., which factories should ship to which warehouses) is not a trivial problem and often requires a complex mathematical model called linear programming.[1] Note again that all relevant information is known in advance; the solution to the problem involves a search through all

---

[1] This is the transportation problem in linear programming. For a discussion of this problem, refer to Daniel Teichroew, *Introduction to Management Science: Deterministic Models* (New York: John Wiley, 1964) or another text on operations research or linear programming.

alternatives to find the optimum one. These are the characteristics of *decision making under certainty.*

Contrast the problem faced by the buyer for a department store to the above illustration. The buyer must purchase in advance the merchandise needed by his store for a particular season. The cost of the merchandise and the price at which it will be sold may be known. The amount to order is what must be decided. If he orders too much merchandise, it may have to be sold at clearance prices, thereby reducing the profit for the store. Similarly, if too little merchandise is ordered, sales may be lost and the opportunity for additional profits may be forgone. To make this decision, the buyer must estimate the future demand for merchandise. Generally, he cannot know this beforehand; there is some uncertainty about the demand that will materialize owing to the appeal of the particular products, the trends in style, general economic conditions, and other factors. The buying decision is thus a *decision under uncertainty.* Such decisions are characterized by the fact that the value of one or more variables is not known to the decision maker at the time the decision is to be made. This is not to say that no information about the value of the uncertain variable is known. The department store buyer certainly has some estimate of future demand based upon his past experience, his evaluation of the merchandise, and his knowledge of economic conditions. Therefore, he may feel that certain levels of demand are more likely than others.

In business decision making under uncertainty, it is helpful to use models or representatives of reality based upon probabilities and probability distributions. For example, a manufacturer may have a production process that turns out parts classified as good or defective. The binomial probability distribution may serve as a model for this process if the assumptions of this distribution are approximately satisfied. Similarly, subjective probabilities are used to represent a decision maker's judgment about the likelihood of uncertain events. For example, estimated probabilities about market demand are helpful in decisions about plant size and location.

THE DECISION MAKING CRITERION

In decision making under uncertainty, alternative *actions* must be available to the decision maker. Similarly, there must be two or more events or values that can be taken on by the unknown variable. Such possible events are sometimes called *states of the world,* since they represent different happenings that can occur. The decision maker is

notue

uncertain because he does not know which event will happen (i.e., which state of the world will materialize).

The problem is to decide which action to select. Under uncertainty this choice is not easy because, generally, some actions will be better if certain events occur and different actions will be preferred if other events occur. Thus we need a *decision criterion*, or a rule for determining which action to select.

Consider these concepts in the following example. The Zip Car Rental Company rents cars at a rate of $10 per day. (The customer pays for his own gasoline and oil.) Cars are rented for one day only. Zip Company does not own its own cars but leases them on a daily basis from a large leasing firm. The larger firm pays the maintenance cost for the cars. Zip must specify the number of cars it intends to lease on a given day at least one week in advance. The daily lease fee paid to the leasing firm by Zip Company is $7 per day. (To avoid confusion, note that the word "lease" is used to denote the arrangement between Zip Company and the large leasing firm; the words "rent" and "rental" are used to denote relationships between Zip Company and its customers.)

Zip is faced with the decision of how many cars to lease for a given day one week hence. The demand for rental cars varies from day to day. If Zip Company leases more cars than are requested as rentals on a particular day, Zip Company will lose the lease fee of $7 for each car unrented. If demand for cars is greater than the number available, a profit of $3 per car (the $10 rent less the $7 lease fee) is forgone.

In this decision situation, the unknown factor is the number of cars wanted by customers for a given day. The possible events or states of the world are thus: "10 cars wanted"; "11 cars wanted"; etc. The actions available to the decision maker are: "lease 10 cars"; "lease 11 cars"; etc. We wish to decide which action is best.

In order to obtain some information, the manager of Zip Company recorded the number of requests for rental cars each day over a typical period of 100 days. This information is shown in Table 7–1. We can use these frequency data as a probability model or representation of the uncertainty facing the Zip Company. That is, we can use a relative frequency in Table 7–1 as an estimate of the *probability* that the specified number of rental cars will be requested on a given day. This implies that the probability is zero for 9 or fewer rental requests; the probability is .05 for exactly 10 rental requests; etc. Note that we are restricting the possible events to between 10 and 17 rentals requested.

The use of these frequencies as a probability distribution implies a

Table 7–1

REQUESTS FOR RENTAL CARS—ZIP CAR RENTAL COMPANY
SUMMARY FOR 100 DAYS

Cars Requested	Frequency: Days	Relative Frequency
9 or fewer	0	0
10	5	.05
11	5	.05
12	10	.10
13	15	.15
14	20	.20
15	25	.25
16	15	.15
17	5	.05
18 or more	0	0
	100	1.00

sort of "betting" model of reality. That is, we can conceive of a roulette wheel with 100 possible slots. Five of these slots are labeled "10"; five are labeled "11"; ten are labeled "12"; etc., corresponding to the frequencies, or estimated probabilities, in Table 7–1. Hence, the event "10" has only 5 chances in 100, or 1 chance in 20, of occurring, and so on. The use of these probabilities implies such a "betting" distribution about the real world.

To use the above probability distribution as a model of reality involves, of course, certain assumptions. We assume that the 100 days are a representative sample of past requests (i.e., there was no bias in the manner in which the sample was selected). We assume that the future will be the same as the past insofar as rental requests are concerned. We assume that the number of requests is independent from day to day and week to week. If these assumptions are valid, our model has some validity as a representation of the real-world situation.

Decisions Based upon Probabilities Only

When presented with the data in Table 7–1, you might be tempted to make the decision of how many cars to lease with this information alone. Some such decisions and rationalizations might be as follows:

a) Lease 10 cars. This would guarantee that all cars leased would be rented.

b) Lease 17 cars. This would guarantee that no rental customer would be turned away.

c) Lease 15 cars. This is the number most frequently requested (i.e., the mode).

d) Lease 14 cars. This is the mean or expected number requested, as shown in Table 7–2.

Table 7–2

CALCULATION OF EXPECTED NUMBER OF REQUESTS
FOR RENTAL CARS

X Number Requested	P(X) Probability	X · P(X)
10	.05	0.50
11	.05	0.55
12	.10	1.20
13	.15	1.95
14	.20	2.80
15	.25	3.75
16	.15	2.40
17	.05	0.85
	1.00	14.00

$$E(X) = \Sigma[X \cdot P(X)] = 14.00$$

Expected value   Sum total

The objection to all of the criteria (*a* to *d*) is that they make no use of the economic information available to the decision maker. To see why the decision must depend upon the costs of leasing a car and the rental price, consider the following illustrations:

1. If the cost of leasing a car were zero, then the *b* criterion above (lease 17 cars) would yield the most profitable decision.

2. If the cost of leasing a car were equal to the rental price, then the *a* criterion (or the alternative of going out of business) would be the least costly alternative. It would involve zero profit, which would be preferable to the other alternatives, since they would involve losses.

From these illustrations, it appears that the economic factors such as prices and costs very much influence the correct (or most profitable) decision.

Decisions Based upon Economic Factors Only

It is possible to go to the other extreme and rely entirely upon economic factors, thereby ignoring the probability information. Let us consider this approach.

First, we arrange in a table the economic consequence for each event and for each possible action. Such a table is called a payout or *payoff table.* In construction of payoff tables, it is important to include only costs or profits which result from the actions and events under consideration. Thus, only out-of-pocket costs and revenues are relevant. Overhead charges and depreciation should be excluded, since they do not represent actual flows of funds. Table 7–3 is a payoff table for this problem.

Table 7–3

PAYOFF TABLE

PROFITS (IN DOLLARS) FOR ZIP CAR RENTALS

Events: Number of Rental Cars Requested	Actions: Number of Cars Leased							
	10	11	12	13	14	15	16	17
10	30	23	16	9	2	−5	−12	−19
11	30	33	26	19	12	5	−2	−9
12	30	33	36	29	22	15	8	1
13	30	33	36	39	32	25	18	11
14	30	33	36	39	42	35	28	21
15	30	33	36	39	42	45	38	31
16	30	33	36	39	42	45	48	41
17	30	33	36	39	42	45	48	51

Recall that Zip Company leased cars for $7 per day and rented them in turn for $10 per day. From this we can derive the profit (or loss) in the table for each combination of action and event. Thus, if Zip Company leased 13 cars and rented 11 to customers, the profit would be 11 × $10 (i.e., $110 revenue) − 13 × $7 (i.e., $91 cost), or $19. We assume that there is no penalty cost (except for lost profit) when a customer requests a rental car and one is not available. The customer can be served by a competing rental agency.

Table 7–3 shows that the actions the Zip Company can take vary somewhat as to risk. The action "lease 10 cars" guarantees a profit of $30 regardless of what happens. In this sense, it is the least risky or most conservative action available.[2] In contrast, the action "lease 17 cars" is the most risky alternative in the sense that the possible profits

---

[2] The choice of the alternative with the highest minimal profit level is called a maximin strategy (maximizing the minimum profit). If the table is expressed in losses (negative profits), then the criterion is called minimax (i.e., select the alternative with the least [minimum] maximum loss). See references to Luce and Raiffa and others on pp. 215–16 for a discussion of these types of decision strategies.

range from a loss of $19 (when only 10 cars are rented) to a profit of $51 (when all 17 cars are rented).

Most decision makers would balk at the prospect of making a decision with only the information shown in Table 7–3. They would insist on knowing something about how likely the occurrence was of each possible event. The alternative "lease 10 cars" would generally be preferred if there were only a slight chance (say 1 in 100) that more than 10 rentals would be requested. Similarly, the alternative "lease 17 cars" would generally be preferred if requests were only rarely fewer than 17 rentals.

A person's preference or aversion to risky alternatives may depend upon how much he subjectively values the dollar amounts shown in Table 7–3. If a loss of $10 or more may cut his working capital seriously, the decision maker would avoid the alternatives "lease 16 cars" and "lease 17 cars," even though it might be very unlikely that the number of rental requests could be as low as 10 or 11. On the other hand, if profits of at least $40 were needed to satisfy a certain goal (e.g., to pay off a pressing debt), the decision maker might consider only leasing upward of 13 cars. Factors that affect the subjective worth of a gain (or loss) of a certain amount of money do influence the decision process. We shall consider such effects in detail in a later section. For now, the assumption is that no factors would subjectively change the value of money to the decision maker; that is, a gain of $20 is worth twice as much to the decision maker as a gain of $10.

Expected Monetary Value as a Decision Criterion

Both the probability information and the economic information are necessary for rational decision making under uncertainty. The procedure for incorporating both sets of information is the subject of this section. We begin by computing the *expected monetary value* for each alternative decision. Table 7–4 illustrates this computation for the action "lease 15 cars."

The column labeled "Profit" in Table 7–4 is the profit that would result for various numbers of rental requests if 15 cars were leased (see Table 7–3). The maximum profit is $45 when all 15 cars (or more) are requested for rental. If only 10 rentals are requested, there will be a loss (negative profit) of $5.

The expected monetary value (abbreviated EMV) or expected profit is interpreted in the same manner as the expected value of a random variable, $E(X)$. It is the average profit that would result if this decision were repeated many times, and each time the decision maker chose the

Table 7–4

CALCULATION OF EXPECTED MONETARY VALUE
FOR ACTION "LEASE 15 CARS"

Event: No. of Rental Cars Requested (X)	Probability $P(X)$	Profit π	Expected Profit $\pi \cdot P(X)$
10	.05	−$ 5	−$ 0.25
11	.05	5	0.25
12	.10	15	1.50
13	.15	25	3.75
14	.20	35	7.00
15	.25	45	11.25
16	.15	45	6.75
17	.05	45	2.25
	1.00		$32.50

$$\text{Expected Profit} = \text{EMV} = \Sigma[\pi \cdot P(X)] = \$32.50$$

expected monetary value

same alternative (in this case, "lease 15 cars"). It is the profit that is to
be expected in the long run even though the decision is to be made
only once. It is simply a weighted average profit, the weights being the
probabilities of the various events. Note that a profit of $32.50 can
never occur on any day, even though the EMV is $32.50. The actual
profit that will result will be one of the values in the "Profit" column of
Table 7–4.

The expected monetary value for each alternative can be computed
by the procedure illustrated in Table 7–4. These values are shown in
Table 7–5. The alternative "lease 13 cars" has the highest EMV. *Our*

Table 7–5

EXPECTED MONETARY VALUE (EXPECTED PROFIT)
FOR ALL ALTERNATIVES

Action: Number of Cars Leased	Expected Monetary Value (Expected Profit)
10	$30.00
11	32.50
12	34.50
13	35.50
14	35.00
15	32.50 *about*
16	27.50
17	21.00

8 actions

criterion for decision making under uncertainty is to pick that action with the highest expected profit (i.e., highest EMV).[3]

A little reflection should convince even the skeptical reader that this criterion is reasonable. If the decision were to be repeated day after day, the action "lease 13 cars" would bring the highest average profit. Recall that the use of probabilities as a model of the real world implied a betting distribution for the decision maker, the odds on various events occurring being represented by the probabilities. The action which maximizes the expected value is simply the most sensible bet or gamble in the face of the stipulated odds or probabilities.

Note that the decision selected (lease 13 cars) is not the one suggested by any of the criteria using the probabilities by themselves or using the economic information alone. The number of cars to lease is neither the mean (which is 14) nor the mode (which is 15).

An Example Using Subjective Probabilities. Football Concessions, Inc. had the franchise to sell ice cream, soft drinks, and hot dogs at Siwash University home football games. In the past, this concession had returned a small but consistent profit to the concessionaire. Siwash had mediocre football teams with relatively small crowds attending the games. On the other hand, because of the location in California, weather conditions were fairly predictable; thus crowd size and food orders could be estimated accurately.

However, Saturday, November 17, 1973, posed a problem for the concessionaire. Siwash was scheduled to meet its arch-rival Califlower University. Both teams were undefeated, so the winner would be the league champion. Advance sales of tickets indicated that if the weather were nice a crowd of 80,000 persons could be expected. On the other hand, it was raining on Friday and the weather prediction called for continued rain. If the rain were heavy, a crowd of perhaps only 20,000 would attend.

The concessionaire had to order his food the day before a game. He generally ordered on the basis of a cost of $.50 per person attending, and this had proved reasonably accurate in the past. He had a markup of 50 percent (that is, selling price was double the cost). He could generally save about 20 percent of the cost of anything he had left over.

The concessionaire is faced with a real decision problem under uncertainty. As a first step, he might draw up a payoff table for the problem. To simplify things slightly, we will assume only four possible actions and four events. The payoff table is shown in Table 7–6.

---

[3] Later, we shall discuss maximization of expected utility, where utility is a measure of risk evaluation. For the present, we are assuming a linear utility function for money (i.e., no aversion or preference for risk).

Table 7–6

FOOTBALL CONCESSIONS, INC.

PAYOFF TABLE (THOUSANDS OF DOLLARS)

Event: Crowd Size	Action: Order Food For			
	20,000	40,000	60,000	80,000
20,000	$10	$20	$−6	$−14
40,000	10	20	12	4
60,000	10	20	30	22
80,000	10	20	30	40

If the action taken in Table 7–6 is "order food for 20,000," the concessionaire will make $.50 per person (or $10 thousand in total), with nothing left over, whatever happens. If he orders food for more people than show up, he will have to throw some away. For example, if he orders for 80,000 and 40,000 actually attend, the net profit is $4,000, calculated as follows:

Cost for amount ordered, $.50 × 80,000..............$40,000
Revenue from sales, $1 × 40,000.................... 40,000
Refund on unused food, 20% of ($.50 × 40,000)...... 4,000
Net profit, $40,000 − $40,000 + $4,000.............. 4,000

The other payoffs in Table 7–6 are calculated in similar fashion.

If our concessionaire is to follow the procedure adopted by the car rental manager in the previous example, he will need estimates of the likelihood of the different crowd sizes. But, unlike the previous example, no past history is comparable. The situation for this game is unique.

How then is the concessionaire to obtain probabilities for the events? He must use *subjective probabilities.* That is, he must estimate the relative likelihood of the different crowd sizes. The probabilities are subjective because they represent his own judgment about what might happen.

Of course, the concessionaire should obtain as much information as possible before assessing these probabilities. He could, for example, obtain the latest weather forecast. He could check with the ticket office for the latest ticket sales and the number of requests for ticket returns. He might check with other concessionaires to see if they have any experience that would be relevant. However, it is doubtful that all this will give him very much information. Predicting weather is not easy, and even if he knew the weather, he could not necessarily predict the attendance. For example, a large crowd might come despite rain.

Suppose the concessionaire, after sufficient reflection, assigned the probabilities shown in Table 7–7 to the different crowd size pos-

Table 7–7

FOOTBALL CONCESSIONS, INC.

SUBJECTIVE PROBABILITIES FOR EVENTS

Event: Crowd Size	Subjective Probability
20,000	.30
40,000	.20
60,000	.10
80,000	.40
	1.00

sibility. In assigning these probabilities, he might have felt, for example, that if it rained a small crowd would come; if it cleared up, a large crowd would come; and that there was a lesser chance that a middle-sized crowd would attend. In setting these probabilities, the concessionaire might think of them as betting odds. For example, the probabilities in Table 7–7 imply that it is an even bet (50/50 odds) between a crowd size of 20,000 to 40,000 or a crowd size of 60,000 to 80,000. The probabilities also imply that the odds are 4 out of 10 that about 80,000 will attend, 1 out of 10 that about 60,000 will attend, and so on. To aid in establishing these probabilities, the concessionaire might see if, in his mind, the bets seem fair at the betting odds that are implied by his set of probabilities. (The appendix to Chapter 5 suggests a procedure for estimating subjective probability distributions.)

Using the probabilities in Table 7–7, the expected profit can be calculated for each action. This is illustrated for the action "Order food for 40,000 people" in Table 7–8. The expected values for the other actions can be calculated similarly, and the results are shown in Table 7–9. If the concessionaire accepts as his criterion, "pick the action with the highest expected value," he will order food for 60,000 people.

Before proceeding, however, think about the decision suggested above. If you were the concessionaire, would you in fact order food for 60,000 people? Many people who think the decision criterion "pick the action with the highest expected value" is eminently sensible in the

Table 7–8

FOOTBALL CONCESSIONS, INC.

EXPECTED VALUE FOR THE ACTION
"ORDER FOOD FOR 40,000 PEOPLE"

Event: Crowd Size	Probability	Profit	Probability × Profit
20,000	.30	$ 2,000	$ 600
40,000	.20	20,000	4,000
60,000	.10	20,000	2,000
80,000	.40	20,000	8,000
Expected value			$14,600

Zip Car Rental problem begin to have some doubts about its application here. So some discussion is in order.

Table 7–9

FOOTBALL CONCESSIONS, INC.

EXPECTED VALUES FOR ALL ACTIONS

Action: Order Food For	Expected Profit
20,000	$10,000
40,000	14,600
60,000	15,600
80,000	14,800

Subjective Probabilities versus Historical Data.   There are three major differences between the Football Concessions problem and the Zip Car Rental one. The first involves the use of subjective probabilities versus the more objective historical data that were available in the Zip problem. Questions such as "But are the subjective probabilities right?" or "Won't the concessionaire get a different answer if he assigns different probabilities?" often come to the mind of the skeptical.

The answer is, of course, that there is no such thing as a "correct" subjective probability. Subjective probabilities are a state of mind, not a state of nature. They express the decision maker's judgment, and are a help in making a decision *consistent* with that judgment. Even in the Zip Car rental problem, the decision maker had to make a subjective judgment that the past data he had collected would be relevant for future decisions. Note that two decision makers with different judgments about what might occur (i.e., different subjective probabilities) could make different decisions. There is nothing inconsistent about this. The decision criterion of maximizing expected value does not guarantee "right" decisions—it can only guarantee that decisions are consistent with the judgment of the decision maker.

One-Shot versus Repeated Decisions. The second difference between the two examples is that the Football Concession case is a "one-shot" decision situation, whereas the car rental problem involved repeating the same situation day after day. Is the expected value criterion valid for this one-time decision?

To answer this, suppose that our concessionaire is involved in several business decisions in the course of a year involving roughly the same magnitudes of payoffs as in our example. These might be concessions at other games, real estate ventures, or investments in stocks. Each of these situations involves different alternatives, events, payoffs, and probabilities. If the decision maker follows the criterion of maximizing expected monetary value in each, he will be better off, on the average, than using any other decision criterion. In this context, maximizing expected value can be thought of as maximizing average payoff over a number of different decision situations. Using the expected value criterion amounts to consistently playing the best bet.

If this is the concessionaire's only venture, however, he may wish to consider the risk involved by using utility values, as described below.

Adjustment for Risk. The third difference between the concessionaire and the car rental manager problems involves the large negative amounts (−$14,000) in the concessionaire problem. If losing this amount would seriously impair his financial position, would he consider this alternative? Note that even the recommended decision of ordering food for 60,000 people involves a possible loss of $6,000. This brings in the attitude towards risk on the part of the concessionaire. If the amounts of money involved in the decision are large relative to those in his other ventures, he may wish to take account of the risk involved, and may in fact tend to avoid more risky alternatives.

Adjustment for risk involves the use of utility or preference functions for money. This topic is discussed later in the chapter. It means that the decision criterion is modified so that it maximizes expected utility (risk-adjusted money) rather than expected monetary value (EMV).

DECISION TREES

In some situations the decision maker has only a single decision to make; for example, a manufacturer must decide whether to build a large or a small plant. Subsequent market conditions will determine what profit he will make.

Suppose it is possible for the manufacturer to build a small plant and expand it at a later date when the market demand for the new product is known. The cost of such an expansion would be $3 million, but the expanded facilities would enable the firm to supply a high-level market demand and hence to obtain the same $10 million profits (excluding plant cost) that could be obtained by building a large factory now.

Note that in such an example, the manufacturer is making a sequence of decisions: first, the decision "large versus small plant"; and second, at a later date, the decision "expand or not expand the small plant" (if he chose the small plant for the first decision). In between these decisions, the manufacturer obtains new information; that is, he discovers whether the market demand will be high or low. The manufacturer may improve his first decision, therefore, by taking account of the possibilities offered in the second decision.

Sequential Decisions and Decision Trees

One method of analyzing problems which involve a sequence of decisions is to express the alternatives in the form of a decision tree. The decision tree for the problem faced by the manufacturer is shown in Chart 7–2.

Starting at the left, the first two lines or *branches* of the decision tree represent the alternative actions for the first decision—either build a large or a small plant. At the end of each of the decision (or action) branches comes a fork with two branches representing the events high and low market demand for the new product. It is unknown at the time the first decision (size of plant) must be made which of these event branches will actually occur.

For the "build large plant" action, the tree ends after the event branches. However, for the "build small plant" action a second decision point is reached *after* each of the events "high demand" or "low demand." The decision maker can choose between the actions "expand the plant" and "no expansion" after he knows the market demand level. These actions are represented as branches on the decision tree. Including both action branches after each of the forks at the second decision point may seem unnecessary at first. One would generally expect to expand the plant in response to high demand and not to expand if low demand materialized. But we cannot be sure of this until we include the economic information in the tree, which we shall do below. There always is the possibility, for example, that the expansion will cost more

Chart 7–2

DECISION TREE FOR DECISION ABOUT NEW PLANT

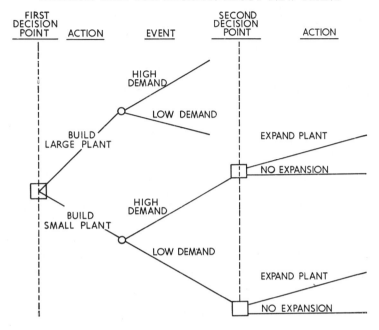

than the additional revenue even from high market demand. Hence, we should retain both alternative actions at each of the second decision points.

The decision tree as shown in Chart 7–2 represents the basic structure of this decision problem. The decision actions and the uncertain or chance events are shown; and the order in which various actions precede or follow events is indicated.

Analysis Using Decision Trees

Once we have set up a decision problem in the form of a tree, the next step is to analyze the problem and arrive at a solution.

Economic Information and Probabilities. The costs or profits of various actions and the likelihoods or probabilities of various events must be incorporated in the analysis just as was done with payoff tables in the earlier parts of this chapter. The probabilities for various events can be shown alongside each event branch, as is illustrated in Chart 7–3, where the probabilities are .6 that high demand will materialize and .4 for the low-demand possibility.

The economic consequences or payoffs are also determined as before.

They represent the net cash outflow or inflow for various action-event combinations. In Chart 7–3, the payoffs are represented at the ends of final branches of the tree. For a large plant and high demand, the net cash inflow is $6 million; and if demand is low, the payoff is $1 million. If a small plant is built initially and no expansion is made, the amounts are $4 million and $3 million. The payoff or net profit of $5 million related to expanding the plant with high demand is determined as follows:

```
Profit from high demand
   (with production ability to meet demand)......        $10 million
Less: Cost of building small plant...............$2 million
      Cost of expanding.........................  3 million
         Total cost...............................              5 million
Payoff.......................................              $ 5 million
```

Similarly, expanding in the face of low demand costs the $5 million as above and only gives $5 million in profit for a net payout of zero, as shown at the end of the "Small Plant—Low Demand—Expand" branch in Chart 7–3.

Chart 7–3

DECISION TREE FOR DECISION ABOUT NEW PLANT

(INCLUDING PROBABILITIES AND PAYOFFS)

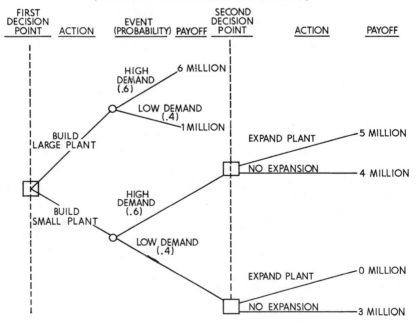

Working Backward on the Decision Tree. With the payoffs and probabilities shown on the decision tree, the next step is to begin the analysis with the aim of finding that decision (or sequence of decisions) which is best. To do this, we begin by working backward on the tree, from the final or end branches back toward the first decision point.

The second decision point is thus the first considered. At the end of the high demand branch is the fork shown in Chart 7–4, Panel A.

Chart 7–4

DECISIONS AT END BRANCHES

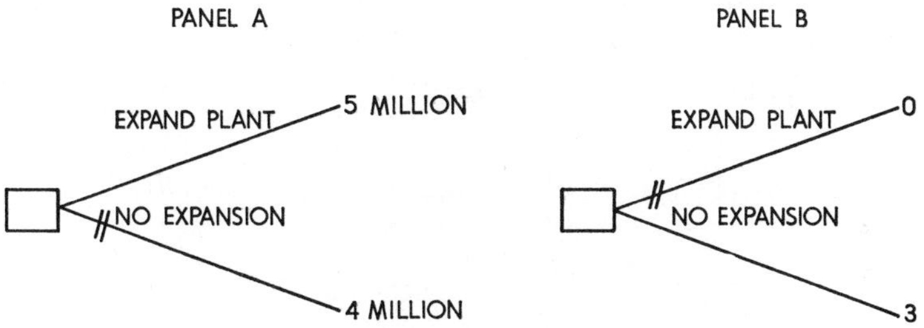

PANEL A   PANEL B

Since the action "expand the plant" leads to $5 million net profit as opposed to only $4 million for no expansion, that alternative is selected. The "no expansion" branch is removed from further consideration by drawing two lines through it, as shown. Similarly, for the decision at the end of the low demand branch, Chart 7–4, Panel B, the action "no expansion" is preferred (with net profit $3 million), and the action "expand plant" is eliminated. The reduced decision tree appears in Chart 7–5. This completes the analysis for the second decision point.

We now move backward to the "event" forks, with branches labeled "high demand" and "low demand," respectively. At each of these forks an expected value is taken, using the payoffs at the ends of the branches and the probabilities shown. For the fork at the end of the "build large plant" action, the expected value is $4 million ($6 million × .6 + $1 million × .4). For the fork at the end of the "build small plant" branch, the expected value is $4.2 million ($5 million × .6 + $3 million × .4). By replacing the event forks by their expected values, the final reduced form of the decision tree is obtained (Chart 7–6).

The best decision for the manufacturer, therefore, is to build the small plant now and to decide upon expansion later when market demand is known.

Discussion. The only immediate decision facing the manufacturer

Chart 7–5

REDUCED DECISION TREE

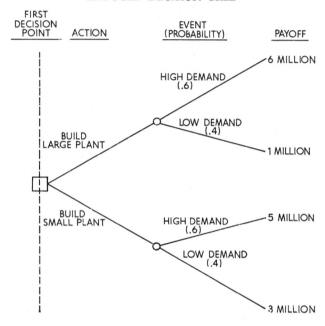

Chart 7–6

FINAL REDUCED DECISION TREE

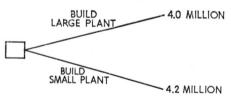

was the one involving the initial size of the plant. But in order to make this decision, he had to take account of the possibility of a subsequent decision on expansion. Thus, he makes a *sequence* of two decisions— (1) build a small plant and (2) expand if a large market potential materializes—rather than a single decision. Had the manufacturer considered only the single decision—large or small plant—without the subsequent expansion possibility, he would have arrived at exactly the opposite decision, namely to build the large plant.

A Further Example

To illustrate the use of the decision tree in a more complex situation, consider the following example: Artex Computers is interested in developing a tape drive for a proposed new computer. Artex does not have research people available to develop the new drive itself and so is going to subcontract the development to an independent research firm. Artex has offered a fee of $250,000 for developing the new tape drive and has asked for bids from various research firms. The bid is to be awarded not on the basis of price (set at $250,000) but on the basis of both the technical plan shown in the bid and the reputed technical competence of the firm submitting the bid.

Boro Research Institute is considering submitting a proposal (i.e., a bid) to Artex Computer to develop the new tape drive. Boro Research management estimated that it would cost about $50,000 to prepare a proposal; further, they estimated that the chances were about 50/50 that they would be awarded the contract.

However, Boro Research engineers were uncertain as to how they would develop the tape drive if they were awarded the contract. Three alternative approaches could be tried. One approach involved the use of certain electronic components. The engineers estimated that it would cost only $50,000 to develop a prototype (i.e., a test version) of the tape drive using the electronic approach, but that there was only a 50 percent chance that the prototype would be satisfactory. A second approach involved the use of certain magnetic apparatus. The cost of developing a prototype using this approach would be $80,000, with a 70 percent chance of success. Finally, there was a mechanical approach with a cost of $120,000, but the engineers were certain they could develop a successful prototype with this approach.

Boro Research would have sufficient time to try only two approaches. Thus, if either the magnetic or electronic approach were tried and it failed, the second attempt would have to use the mechanical approach in order to guarantee a successful prototype.

The management of Boro Research was uncertain how to take all this information into account in making the immediate decision—whether to spend $50,000 to develop a proposal to send to Artex Computers.

Since this decision problem seems complex, let us build the decision tree in steps. The first decision facing Boro Research involves the actions "prepare a proposal" and "do not prepare a proposal." If a proposal is developed and submitted to Artex Computers, then either of the events "contract awarded to Boro Research" or "Boro Research loses contract" must occur. Each event has the probability .5. These choices are shown in Chart 7–7.

Chart 7–7

BORO RESEARCH INSTITUTE

DECISION ON PREPARATION OF PROPOSAL

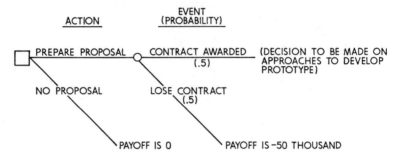

If Boro Research decides not to prepare a bid, the net payoff is zero. If a bid is prepared but the contract is lost, Boro Research loses the $50,000 cost of preparing the bid (i.e., the payoff is −$50,000). If the contract is awarded to Boro Research, then the next decision—the choice between alternative methods of developing a successful tape drive—must be made.

In the second decision, Boro Research must decide which of the three approaches—mechanical, electronic, or magnetic—to try first.[4] This decision is shown in Chart 7–8.

If the mechanical approach is selected, a successful prototype will be developed for sure and Boro Research will have a net return of $80,000 ($250,000 value of contract minus $50,000 proposal cost minus $120,000 to develop the mechanical prototype). If either of the other approaches is selected, it may succeed or fail. Failure means that the mechanical approach must be used in order to guarantee a successful

---

[4] Boro Research could possibly add a fourth alternative—develop both the electronic and magnetic prototypes simultaneously and follow with the mechanical only if both fail. This could be added as a branch of the tree. However, the cost of this would be at least $180,000 (more if neither approach produced a success), and this is greater than the cost of a mechanical prototype ($170,000).

Chart 7–8

BORO RESEARCH INSTITUTE
DECISION ON WHICH APPROACH TO TRY FIRST

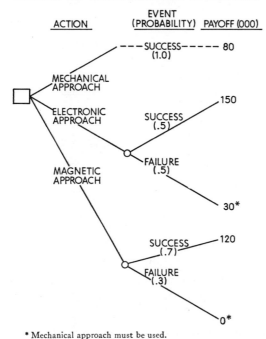

| EVENT |
| ACTION | (PROBABILITY) | PAYOFF (000) |

* Mechanical approach must be used.

prototype within the time available. The payoffs shown in Chart 7–8 are calculated as follows:

| | | Payoff (Thousands of Dollars) | | | | |
| | | | Cost of Other Proto- | Cost of Mech. Proto- | | |
End of Branch	Fee	Cost of Pro- posal	type	type		Profit
Electronic Approach						
Success	250	— 50	— 50		=	150
Failure	250	— 50	— 50	— 120	=	30
Magnetic Approach						
Success	250	— 50	— 80		=	120
Failure	250	— 50	— 80	— 120	=	0

The complete decision tree is shown as Chart 7–9. It is obtained by joining Charts 7–7 and 7–8.

Chart 7–9

COMPLETE DECISION TREE FOR BORO RESEARCH INSTITUTE

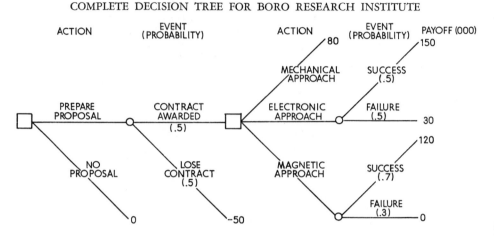

Working Backward. The expected values are calculated for each of the event forks in the far right part of the tree. Thus, the expected payoff associated with the electronic approach is $90,000 (.5 × 150 plus .5 × 30 = 90) and for the magnetic approach is $84,000 (.7 × 120 plus .3 × 0 = 84). These expected payoffs are inserted in circles beside the appropriate forks in Chart 7–10.

Moving left to the decision point, we see that the electronic approach offers the highest expected payoff ($90,000) and is the best choice. The value $90,000 is written (circled) beside the decision point and the nonpreferred approaches are crossed off by drawing || on the branches.

The tree now has a payoff of +$90,000 if the contract is awarded

Chart 7–10

BORO RESEARCH INSTITUTE ANALYSIS OF DECISION TREE

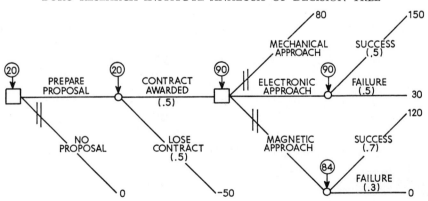

and —$50,000 if not. The expected value of preparing a proposal is $20,000 (.5 × 90 plus .5 × (—50) = 20). This is written in a circle beside the event fork.

Finally, the choice must be made between the expected payoff of $20,000 for preparing the proposal and zero if the proposal is not prepared. The first, of course, is selected, and the mark || drawn through the "no proposal" branch.

In summary, Boro Research should prepare the proposal, anticipating $20,000 as the expected value of this decision. If the contract is awarded, the electronic approach should be tried first; but if this fails, the mechanical approach must be used.

RISK IN DECISION MAKING: THE UTILITY OF MONEY

Expected monetary value is not always the best criterion to use in decision making. If you were offered your choice of one of two alternatives: either (*a*) a 50/50 chance of $250 or zero or (*b*) $100 for sure, you would probably take the $100. Most people would, despite the fact that the expected monetary value of the 50/50 gamble is $125. Is this evidence in conflict with the decision criterion which we expressed in an earlier section—the criterion that one should pick the decision alternative with the highest expected monetary value? Yes, it is! And we are now in a position to extend or elaborate upon our measure of value. The problem arises because the value of money to people is not always a linear function of the amount of money. Generally, $200 is not worth twice as much to a person of modest means as $100. It would matter a great deal to you whether I gave you zero or $100; but it probably would not matter a great deal if the choice were between $1,000,000 and $1,000,100. This is because money has diminishing utility to most of us; the first $100 we receive is most important, while successive increments of $100 have less and less subjective value.

We see the same phenomenon at work when people buy insurance. For most people, insurance is bound to be a "bad bet" from a purely monetary point of view, since the insurance company must pay its expenses and make a profit in addition to covering the risk. That is, the expected monetary value of insurance is negative, from the buyer's viewpoint. However, most of us are willing to pay a small amount (the insurance premium) to guard against a disastrous occurrence, even though the chance of such an event happening may be quite small.

In order to make decisions under uncertainty, we must have some way to measure a decision maker's attitude toward risk and express this in quantitative terms. The appendix at the end of this chapter gives a brief discussion of how this can be done. The result is a function relating dollar amounts to a measure of *utility*.[5] A typical function is shown in Chart 7–11.

For a person who has an aversion to risk (e.g., one who would prefer $100 for sure to a 50/50 chance at zero or $250), the shape of the function would reflect the diminishing utility of money to him, as shown. A person who is willing to use expected monetary value would

Chart 7–11

TYPICAL UTILITY FUNCTIONS

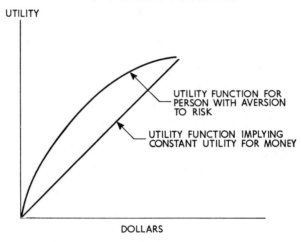

have a linear utility function. (He'd be indifferent between the alternatives of a certain $125 and a 50/50 chance of zero or $250.)

In many decision situations, the amounts of money involved are small relative to the resources of the decision maker. Thus, for inventory decisions that involve only a few thousand dollars, a large corporation would use expected monetary value. Over this range (plus or minus a few thousand dollars), the utility function for the company is approximately linear. For more important decisions (e.g., the decision to build a new factory or to enter a new market), monetary value alone is generally not appropriate. In such situations, the decision maker should determine the utility for money for him (as shown in the Appendix at the end of this chapter). The decision criterion is then to pick the alternative with the highest expected utility, rather than the highest expected monetary value.

---

[5] The word "utility" is somewhat misleading. It is merely a risk equivalence measure and bears no direct relationship to "utility" as commonly used in economic theory. The utility scale (the ordinate in Chart 7–11) is not unique. (The scale can be multiplied by a constant or shifted up or down without changing the function in any real sense.)

SUMMARY

This chapter described a procedure for making decisions in an uncertain environment. The procedure, in skeletal form, involves these steps:

1. Define the possible events that can occur.
2. Define the actions that can be taken.
3. Determine the value (in dollars or utility) of each action-event combination.
4. Describe the decision maker's uncertainty about the events by a set of probabilities.
5. Find the expected value of each alternative action by multiplying its value for each event by the probability and summing.
6. Select that alternative with the highest expected profit (or utility).

To specify this decision procedure is merely to organize the decision making process in a systematic and logical fashion. No one making a decision under uncertainty can avoid these steps—though he might do some steps in an intuitive manner. Our procedure is no more than a completely specified logical framework.

Subjective probabilities may be needed if objective historical data are not available. Such probabilities represent the personal judgment of the decision maker about the likelihood of events.

Decision trees may be used to analyze problems that involve a sequence of decisions. The various actions that may be taken are shown on the tree as branches emanating from a fork, and the various events that may occur are similarly represented. Hence, the tree diagram ties together a sequence of decisions and events.

The payoffs for various sequences of actions and events are shown at the end branches of the tree, and the probabilities for the various events are listed below each event.

The decision tree is analyzed by working backward from the final action or event to the first action to be chosen. At each stage an expected value is calculated over possible events, or else a choice is made among alternative actions, selecting the one with the highest expected value.

Utility values may be substituted for monetary values, for those whose subjective value of money is not linear, by methods described in the appendix of the chapter.

In the chapters that follow, we shall extend this analysis. We shall examine, first, the possibility of postponing the decision while additional information is collected (Chapter 8). Subsequently (Chapters 13 and 14), we shall consider obtaining information by sampling.

APPENDIX: DERIVATION OF UTILITY CURVES FOR DECISION MAKING UNDER UNCERTAINTY

Suppose a businessman had a choice of one of two contracts. The profit resulting from either contract is uncertain. The contracts and their probabilities and payoffs are:

Contract I				Contract II		
Event	Prob.	Payoff		Event	Prob.	Payoff
A	.30	+$9,000		Q	.25	+$7,500
B	.45	+ 6,000		R	.60	+ 2,000
C	.25	− 9,000		S	.15	− 5,000
	EMV =	+$3,150			EMV =	+$2,325

It is easy enough to calculate the expected monetary value of each contract shown above. In order to decide which contract the businessman prefers, however, we intend to ask him a series of questions. The questions are intended to measure his preferences in risk situations simpler than the above contracts.

We begin by selecting two reference points. One should be larger than the largest positive money value in the real decision problem. For this upper reference point, let us arbitrarily choose $10,000. The other reference point should be less than the lowest money value in the real problem; let us select −$10,000 for this reference point. We arbitrarily assign utility values of one and zero to these reference points.[6] That is,

$$u(+10,000) = 1$$
$$u(-10,000) = 0$$

We then give the decision maker a choice of the following kind: What is the maximum amount you would pay to be released from a contract that gives you a 1/2 chance at +$10,000 and a 1/2 chance at −$10,000?[7]

The answer to such a question would be a personal matter, depending upon the resources and the propensity for risk of the decision maker. Let us suppose that the decision maker said that he would be willing to pay up to $2,000 to be released from the gamble (i.e., from the contract giving a 1/2 chance at +$10,000 and a 1/2 chance at −$10,000). In other words, the decision maker is indifferent between a sure amount of −$2,000 and the gamble (or contract). We postulate

---

[6] The choice of scale is arbitrary. We could have chosen $u(+$10,000) = 502.6$ and $u(-$10,000) = -29$ if we wished. The use of a scale between 1 and 0 is convenient.

[7] The contract may have positive value, in which case the question should be: What is the minimum amount (positive) that you would accept to sell the contract to someone else?

that the utility of —$2,000 is equivalent to the expected utility of the contract:

$$u(-\$2,000) = 1/2u(+\$10,000) + 1/2u(-\$10,000)$$
$$= 1/2(1) + 1/2(0) = .5$$

Hence, our utility index for —$2,000 is .5. Using this figure, we can proceed to ask further questions. We might ask: What is the minimum amount the decision maker would accept for a contract that gave him a 1/2 chance for +$10,000 and a 1/2 chance for a —$2,000?[8] Suppose the answer is +$2,000. We then determine the utility index for +$2,000 as

$$u(+\$2,000) = 1/2u(+\$10,000) + 1/2u(-\$2,000)$$
$$= 1/2(1) + 1/2(.5) = .75$$

We can continue asking similar questions:[9] At what amount is the decision maker indifferent to a contract with a 1/2 chance of —$2,000 and a 1/2 chance at —$10,000? Suppose the answer is —$4,000. Then,

$$u(-\$4,000) = 1/2u(-\$10,000) + 1/2u(-\$2,000)$$
$$= 1/2(0) + 1/2(.5) = .25$$

Suppose we continued and determined more answers. These are shown, together with the ones discussed above, in the table:

Chance	Gamble	Indifference Amount	Utility Value
1/2 1/2	+$10,000 −$10,000	−$2,000	$u(-\$2,000) = .5$
1/2 1/2	+$10,000 −$ 2,000	+$2,000	$u(+\$2,000) = .75$
1/2 1/2	−$10,000 −$ 2,000	−$4,000	$u(-\$4,000) = .25$
1/2 1/2	+$ 2,000 −$ 2,000	−$ 500	$u(-\$ 500) = .625$
1/2 1/2	+$ 2,000 +$10,000	+$5,000	$u(+\$5,000) = .875$
1/2 1/2	−$10,000 −$ 4,000	−$5,000	$u(-\$5,000) = .125$

[8] The question would be worded, "How much would he pay to get out of a contract . . ." if the contract had negative value (less than zero dollars).

[9] An alternative procedure is to hold the amounts in the question constant (i.e., keep the +$10,000 and —$10,000) but change the odds for each question. The utility index is determined in the same manner.

The utility function is shown in Chart 7–12. A smooth curve has been drawn connecting the points determined above.

Chart 7–12

UTILITY CURVE FOR DECISION MAKER
CONSIDERING TWO CONTRACTS

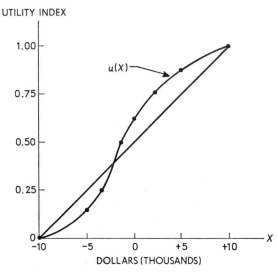

We can now return to the original situation with which we started this appendix. The two contracts are shown below, together with the corresponding utility index values. The utility values are read from Chart 7–12.

Contract I					Contract II			
Event	_Prob._	_Monetary Outcome_	_Utility Value_		_Event_	_Prob._	_Monetary Outcome_	_Utility Value_
A	.30	+$9,000	.98		Q	.25	+$7,500	.95
B	.45	+ 6,000	.90		R	.60	+ 2,000	.75
C	.25	− 9,000	.02		S	.15	− 5,000	.125
		EMV = +$3,150					EMV = +$2,325	
		Expected Utility = .704					Expected Utility = .706	

Contract II now has a slightly greater utility value, though Contract I has a much greater monetary value. Hence, this particular businessman should choose Contract II. Note that both contracts would be preferred to doing nothing, since $u(\$0) = .66$.

PROBLEMS

1. Characterize each of the following as decision making under certainty or uncertainty. Give your reason in one or two sentences.

    a) Decision about whether or not to develop a new type of product (e.g., a new type of drug).

    b) Decision about what price to put on a bid for a construction contract.

    c) The price to set for a product.

    d) Scheduling of production orders through a machine shop.

    e) Inventory decisions.

2. In each of the decision situations below, indicate in a general way what events might occur. From what sources would management obtain the probabilities of these events? To what extent are the probabilities subjective or objective?

    a) The number of clerks to staff a tool crib in a factory and the effect upon time spent by mechanics waiting for tools.

    b) The marketing of a new product.

    c) Company sales forecast 10 years in the future.

    d) The size of a new factory.

    e) How many items to stock in inventory.

3. Consider the following payoff table, which shows dollars of profit:

		Actions				
Event	Probability	A	B	C	D	E
I.........................	.05	100	120	210	140	180
II.........................	.05	110	160	190	140	180
III.........................	.10	130	200	170	140	100
IV.........................	.30	150	180	120	140	180
V.........................	.40	180	150	100	140	120
VI.........................	.10	250	100	100	140	120

Calculate the expected monetary value of each action. Which action gives the highest expected profit?

4. Suppose, in the payoff table in Problem 3, that the probabilities for events I through VI are

Event	Probability
I.........................	.10
II.........................	.40
III.........................	.30
IV.........................	.10
V.........................	.05
VI.........................	.05

Determine the expected value for each action. Which action gives the highest expected profit?

5. A merchant carries a perishable good in his inventory. Each item costs $5 and sells for $9. At the end of the day, any unsold items must be thrown away (no value). Assuming that demand for the item follows a Poisson distribution with mean $m = 3$ per day, how many items should the merchant stock on any given day? What is the expected profit?

6. Suppose in Problem 5 that the demand for the item followed this distribution:

Demand	Probability
0	0
1	.4
2	.3
3	.2
4	.1
5 or more	0
	1.0

How many items should the merchant stock? What is the expected profit?

7. A company is trying to decide what size plant to build in a certain area. Three alternatives are being considered: plants with capacities of 10,000, 15,000, and 20,000 units, respectively. Demand for the product is uncertain, but management has assigned the probabilities listed below to five levels of demand. The payoff table below also shows the profit (in millions of dollars) for each alternative and each possible level of demand. (Output may exceed rated capacity.)

Demand in Units Z	Probability $P(Z)$	Actions: Build Plant with Capacity Of		
		10,000 Units	15,000 Units	20,000 Units
5,000	.2	−4.0	−6.0	−8.0
10,000	.3	+1.0	0.0	−2.0
15,000	.2	+1.5	+6.0	+5.0
20,000	.2	+2.0	+7.5	+11.0
25,000	.1	+2.0	+8.0	+12.0

What size plant should be built?

8. Suppose your plant is having a new cylindrical extruder made to order by Farrell-Birmingham, a company that specializes in the manufacture of large, custom-made machinery such as this. One of the key parts in the extruder is a double-toothed pinion gear, which incurs a great deal of strain in the extruding process and is apt to break down.

Farrell-Birmingham will include extra gears, at a cost of $2,000 each, when they ship the extruder to you. If, on the other hand, you do not order enough extra gears initially and have to place a new order at some later date, Farrell-Birmingham will have to prepare a new mold and will charge you a flat fee of $14,000 for five extra gears.

Your plant foreman estimates that no more than five breakdowns of the pinion gear will occur during the life of the extruder and attaches the following probabilities to the number of failures to be expected:

No. of Breakdowns	Probability
0	.1
1	.2
2	.3
3	.2
4	.1
5	.1

Draw up a payoff table. How many extra gears should you order now? What is the expected cost? (Hint: remember that if you order two extra gears and have three breakdowns, you will have to place a second order.)

9. The Gusher Oil Company is considering leasing a particular parcel of land in a recently discovered oil area. The cost of the lease is $40,000. The cost of drilling an oil well on the site is $80,000. If oil is discovered, the net profit from the well (excluding drilling costs and the cost of the lease) will be $360,000.

Draw up a payoff table. Assuming that Gusher maximizes expected monetary value, what is the minimal probability of finding oil necessary for Gusher to take the lease and start drilling?

10. The LMN Company produces novelty items for the Christmas season. A particular item is sold for $1. Management assigns the following probabilities to various levels of sales:

Sales, Units	Probability
1,000	.1
1,500	.4
2,000	.3
2,500	.1
3,000	.1

The cost of manufacturing this novelty item varies with the number produced, as shown below:

Units Produced	Average Cost per Unit
1,000	60 ¢
1,500	46⅔
2,000	38¾
2,500	33⅖
3,000	29½

If more items are produced than sold, up to 1,000 units of the excess may be disposed of at a price of 10 cents each. Any additional excess items have no value. Items may be produced only in blocks of 500 units. Draw up a payoff table. How many units should be produced? What is the expected profit?

11. The credit manager of IJK Industrial Products considered extending a line of credit to Lastco Construction Company. Lastco was a new company and was definitely considered a credit risk. Based upon IJK's experience, approximately 30 percent of firms like Lastco failed within a year with a severe loss to creditors. Another 25 percent had serious financial troubles. Of the remaining 45 percent, 25 percent became sporadic customers and only 20 percent became good customers over a period of time.

Those customers that failed completely averaged sales of $1,500 each before failing and left an average unpaid balance of $800 which was totally lost.

Those that had severe financial troubles usually lost their credit, but only after they had made purchases of $2,000 and had unpaid balances of $1,000 of which half ($500) was ultimately collected.

Firms that were sporadic customers averaged sales of only $500 (with no credit losses). The good customers, however, averaged sales of approximately $6,000.

IJK was concerned about granting credit to Lastco. On the one hand, if credit was not extended to a potential customer, his business was lost. On the other hand there were substantial risks of nonpayment (as described above), and since IJK made an average contribution (price minus variable cost) of only 20 percent of sales, this exaggerated the problem. In addition, there were collection costs of $100 per customer for those that failed or were in financial trouble.

Draw up a payoff table for this decision problem. Should IJK grant credit to Lastco?

12. An oil company is about to drill 10 wells in an isolated part of the Middle East. A certain piece of equipment is used on each well and is subject to accidental breakage. The question arises as to how many spare parts (if any) the company should transport to the drilling site.

This particular part costs $50. If the parts are shipped with the original expedition, they will cost an additional $50 each to ship, or a total of $100. If parts are needed later, they will have to be shipped by air at a cost of $500 each to transport, for a total of $550, including the cost of the part itself. At the end of the drilling operation, all parts are to be abandoned.

The drilling company knows from past experience that, on the average, .30 parts break per well drilled. Breakage is accidental (i.e., random) and does not depend upon how long the part has been used.

How many spare parts should the company transport in the original expedition? Assume that breakage follows a Poisson distribution.

13. Suppose, for the example of Boro Research Institute described in the text, that Boro Research was not under a time constraint to produce the prototype. In this case, the firm could possibly try both the uncertain approaches (electronic and magnetic) before using the certain mechanical approach.

Draw up the decision tree for this case. How should Boro Research proceed to develop the prototype?

14. In which of the decision situations below do you think maximization of expected monetary value (as opposed to expected utility) is a satisfactory decision criterion?

 a) Building a new factory.
 b) Entering a new market.
 c) Buying out another company.
 d) Production schedules.
 e) Warehouse location.
 f) What quantities to order for inventory.

15. The Pearson Company is considering the purchase of a new machine which will be used exclusively in the production of a certain product. Two machines which are on the market would be satisfactory. Machine *A* has a purchase cost of $10,000 and will save $1 per item over the manufacturing process now used. Machine *B,* on the other hand, will cost $60,000 but will effect cost savings of $3 per item over the current cost. Both machines have a life of five years.

The future market is somewhat uncertain. Management expressed the following probabilities for total sales (units) over the five-year period.

Total Sales (Units)	Probability
10,000	.1
20,000	.3
30,000	.4
40,000	.2

Ignore all discounting in your calculations. Which machine should Pearson purchase? What are the expected savings for each action?

16. The Lockjaw Company is about to bid on a contract to manufacture a large electric generator for a municipal utility company. Lockjaw has two competitors, *A* and *B,* who will be submitting competitive bids. The lowest bidder will win. If two or more bid the same lowest price, the winner will be determined by random draw.

In order to obtain some feel for how Lockjaw had fared against its competitors in the past, the company statistician prepared the following tables concerning past bids by *A* and *B:*

A's Bid (Above Lockjaw Cost)	Relative Frequency	B's Bid (Above Lockjaw Cost)	Relative Frequency
$2,400	1/3	$2,400	1/4
1,200	1/3	1,200	1/2
600	1/3	600	1/4

Furthermore, there was no consistent pattern between the bids of *A* and *B* (i.e., they were statistically independent). Assume that Lockjaw has only three possible bids: (1) cost + $2,400; (2) cost + $1,200; (3) cost + $600. Which bid should be chosen? What is the expected profit?

Hint: Calculate the probability, for each alternative, of (1) winning outright, (2) tying with one competitor, (3) tying with both competitors, and (4) losing. Then set up a profit (payoff) table and calculate the expected profit for each strategy.

17. The Lark Company is considering replacing its No. 1 deplaning machine which is in need of considerable repair. There are two machines with which to replace it. Machine A is a completely automatic machine and could save Lark a considerable amount by eliminating the work that is now done manually. Machine A costs $75,000.

Machine B, on the other hand, costs only $20,000 and can turn out a product of equal quality. It is only slightly more mechanized than the current machine and hence would have considerably higher labor operating costs than Machine A.

The decision about which machine to purchase hinges to a large extent upon the projected sales. But the sales manager is very uncertain about what future sales will be. At the moment, Lark is the dominant firm in the industry. However, the sales manager thinks it is quite possible that several large manufacturers will enter the market soon. When questioned further, the sales manager stated that he believed that there was a 30 percent chance that Lark could maintain its dominant position, a 50 percent chance that it could keep a moderate share of the market, and a 20 percent chance that it would slip to a small share of the market.

Earnings were then projected showing the discounted contribution of the product (excluding initial cost of the machine) under each of these possibilities, in the following table:

	Share of Market		
	Dominant	Moderate	Small
Machine A $225,000		$125,000	$55,000
Machine B.................... 120,000		80,000	45,000

Which machine should Lark buy? Why?

18. Hony Pharmaceutics is a manufacturer engaged in the development and marketing of new drugs. The chief research chemist at Hony, Dr. Bing, informed the president, Mr. Hony, that recent research results have indicated a possible breakthrough to a new drug with wide medical use. Dr. Bing urged an extensive research program to develop the new drug. He estimated that with expenditures of $100,000 the new drug could be developed at the end of a year's work. When queried by Mr. Hony, Dr. Bing stated that he thought the chances were excellent, "9 or 10 to 1 odds," that the research group could in fact develop the drug.

Mr. Hony, worried about the sales prospects of a drug so costly to develop, talked to his marketing manager, Mr. Margin, who said that the market for the potential new drug depended upon the acceptance of the drug by the medical profession. Margin also stated that he had heard rumors that several other firms had been considering developing such a drug. If several firms developed competing drugs, they would have to split the market among them. Hony asked Margin to make future market estimates for different situations, including estimates of future profits. Margin made the estimates shown in the table:

Market Condition	Likelihood	Present Value of Profits
Large market potential	.1	$500,000
Moderate market potential	.6	250,000
Low market potential	.3	80,000
	1.0	

Margin pointed out that the profit figures did not include the costs of research and development or the cost of introducing the product ($50,000). This latter cost would be incurred only if the firm decided to enter the market after the drug was developed.

Mr. Hony was somewhat concerned about spending $100,000 for development of the drug in the face of such an uncertain market. He returned to Dr. Bing and asked if there was some way to develop the drug more cheaply or to postpone development until the market position was clearer. Dr. Bing said that he would prefer his previous suggestion— an orderly research program costing $100,000—but that an alternate was indeed possible. The alternate plan called for a low-level research program for eight months and then a crash program for four months. The cost of this would be $40,000 for the low-level part plus $110,000 for the crash program. Dr. Bing did not think this program would change the chances of a successful product development. One advantage of this approach, Dr. Bing added, was that the question of whether the drug could be developed successfully would be known at the end of the eight-month period. The decision could then be made at the end of eight months on whether to undertake the crash program. When consulted, Mr. Margin stated that at the end of eight months he would be able to estimate the market potential fairly accurately.

Mr. Hony inquired about the possibility of waiting until other drugs were on the market and then developing a drug on the basis of a chemical analysis of the competitive drug. Dr. Bing said that this was indeed possible and that such a drug could be developed for $50,000. Mr. Margin was dubious of the value of such an approach. He said that the first drugs out usually got the greater share of the market. He estimated that returns would only be about 40 percent of those given in the table. In addition, he indicated that there was a good chance, say one out of three, that no equivalent competitive drug would be marketed—in which case Hony would have nothing upon which to develop a drug.

a) Draw a decision tree for this problem.

b) Which action should Mr. Hony take in order to maximize his expected profit?

SELECTED READINGS

Selected readings for this chapter are included in the list that appears on pp. 215–16.

8. DECISION MAKING UNDER UNCERTAINTY: THE VALUE OF ADDITIONAL INFORMATION

CHAPTER 7 introduced a logical structure for decision making in an uncertain environment. In this chapter, we wish to elaborate upon these procedures from a different point of view. This will lead to the question of whether the decision maker should act now with the information available or whether he should postpone the decision and gather additional information.

OPPORTUNITY LOSS

In order to introduce the concept of opportunity loss, let us return to the example of the previous chapter. Recall that the Zip Car Rental Company leased cars for $7 per day and rented them in turn for $10 per day. The payoff table for the decision, including the probabilities and expected values, is shown in Table 8–1. In constructing such a table, it was important to include only real cash or out-of-pocket expenses and revenues. We explicitly excluded all fixed costs, as well as profits or costs from missed opportunities.[1] But these missed opportunity costs give us important insights into the decision problem.

Consider the action "lease 12 cars." If we lease 12 cars and receive only 10 rental requests, our profit is $16. This is not the best we could have done with 10 requests, since, had we leased 10 cars, we would have made $30. We had an opportunity to make an additional $14, if only we had known the true number of requests. The amount $14, then, is the *opportunity loss* associated with the decision "lease 12 cars" and

---

[1] Such concepts are implicitly included in the table, as we shall see immediately.

192

Table 8–1

PAYOFF TABLE FOR ZIP CAR RENTALS
(Dollars Profit)

Event: Number of Rental Requests	Probability	Actions: Number of Cars Leased							
		10	11	12	13	14	15	16	17
10	.05	30*	23	16	9	2	−5	−12	−19
11	.05	30	33*	26	19	12	5	−2	−9
12	.10	30	33	36*	29	22	15	8	1
13	.15	30	33	36	39*	32	25	18	11
14	.20	30	33	36	39	42*	35	28	21
15	.25	30	33	36	39	42	45*	38	31
16	.15	30	33	36	39	42	45	48*	41
17	.05	30	33	36	39	42	45	48	51*
	1.00								
Expected Profit		30.00	32.50	34.50	35.50†	35.00	32.50	27.50	21.00

* Figure represents maximum possible profit for each event.
† Maximum expected profit.

the event "10 rental requests." It is the amount we fall short of the optimal decision, given the event (in this case, 10 requests). The opportunity loss has also been designated by the term *regret,* since we regret having leased the two extra cars and thus having lost an extra $14 in profit.

There is an opportunity loss for each combination of event and action. We can draw up an opportunity loss table by subtracting each profit figure in a row from the *maximum* profit (asterisk) shown in that row. This is done in Table 8–2. Note that, in this decision situation, there are zeros on the diagonal of the table from the upper left to the lower right. This results because one can do no better than lease the exact number of cars that is requested; in each case this is the best action for the given event. There is no opportunity loss or regret. The values above the diagonal are in multiples of $7 (the daily lease rate), representing the opportunity losses of having leased more cars than were requested. Below the diagonal, the values are in multiples of $3, representing the profit that is forgone when there are more requests than leased cars available ($10 revenue less $7 cost per car).

It is important not to confuse opportunity loss with the accounting term "loss," which means a negative profit. Opportunity loss is always positive; it is measured relative to some optimal or best profit.

We can compute the *expected opportunity loss* in the same way we

Table 8–2

OPPORTUNITY LOSS TABLE FOR ZIP CAR RENTALS
(Dollars Regret)

Event: Number of Rental Requests	Probability	Actions: Number of Cars Leased							
		10	11	12	13	14	15	16	17
10	.05	0	7	14	21	28	35	42	49
11	.05	3	0	7	14	21	28	35	42
12	.10	6	3	0	7	14	21	28	35
13	.15	9	6	3	0	7	14	21	28
14	.20	12	9	6	3	0	7	14	21
15	.25	15	12	9	6	3	0	7	14
16	.15	18	15	12	9	6	3	0	7
17	.05	21	18	15	12	9	6	3	0
	1.00								
Expected Opportunity Loss		12.00	9.50	7.50	6.50*	7.00	9.50	14.50	21.00

* Minimum expected opportunity loss.

computed expected profit—by multiplying each opportunity loss in a given column by its probability and adding the products. This yields a weighted average of opportunity losses for each action—the loss we might expect in the long run if we consistently chose that action. Table 8–2 shows the expected opportunity loss (EOL) for each action. Note that the alternative "lease 13 cars" has the least EOL. That is, if we put in a firm order to lease 13 cars each day we would have less regret over lost opportunities than if we leased any other number of cars consistently. This must necessarily be the case. The use of opportunity losses is simply another way of looking at the same problem that was illustrated in Table 8–1, and that action with the highest expected profit must also have the least expected opportunity loss. That is, we can minimize EOL as our decision criterion as an alternative to maximizing expected profit.

EXPECTED VALUE OF PERFECT INFORMATION

We now turn to the problem of whether additional information should be collected before action is taken. More specifically, we would like to know how much additional profit would result from having more information. Thus, we can compare the value of this information with the cost of obtaining it.

While it often is not possible to assess the value of any specific amount of information, in terms of added profit, it is possible to put an upper limit on the value of additional information. In particular, we can determine the value of perfect information—that is, the exact knowledge of what event will occur.

Let us call the *expected value of perfect information* (or EVPI) the expected savings (or additional profit) from knowing the exact event that will occur. *Now, the expected value of perfect information is precisely the expected opportunity loss of the best action.* Recall that opportunity loss is the additional profit associated with picking the best decision. With perfect information about what will happen we could always make the best decision. Perfect information will save us precisely the amount of the opportunity loss. By multiplying the opportunity losses by the probabilities that each event will occur we obtain the expected opportunity loss and simultaneously the expected value of perfect information.

In the Zip Company case, the action "lease 13 cars" is the best action in the face of uncertainty about how many rentals will be needed. The opportunity losses (from Table 8–2) for this alternative are repeated in Table 8–3.

Table 8–3

OPPORTUNITY LOSSES FOR ACTION "LEASE 13 CARS"

Event: Number of Rental Requests	Probability	Opportunity Loss	Expected Value
10	.05	$21	$1.05
11	.05	14	.70
12	.10	7	.70
13	.15	0	0
14	.20	3	.60
15	.25	6	1.50
16	.15	9	1.35
17	.05	12	.60
	1.00		EOL = $6.50

When 10 rentals are requested, there is an opportunity loss of $21. If this event had been predicted beforehand, as it would with perfect information, the decision maker would have saved $21. Hence, perfect information is worth $21 in the event "10 rental requests" occurs. If 13 rentals are requested, perfect information is worth nothing because we would be making the best decision anyway. Perfect information is, in a sense, like a crystal ball, predicting accurately the future event. But before we have the crystal ball (i.e., perfect information) we do not know how much it will save us. It might save us $21 or $14 or any of the values in Table 8–3, column 3. The expected savings with the crystal ball (EVPI) is obtained by multiplying the probabilities by the savings (the opportunity loss) for each event and adding the products.

In most decision situations, it is not possible to obtain perfect predictions; accurate crystal balls just are not available. The EVPI puts an upper limit on what one would pay for additional information. In our example, EVPI = $6.50. A system for predicting future rental requests, no matter how accurate, would be worth no more than $6.50 per day.

Profit under Certainty: An Alternative Method for Determining EVPI

Another method for determining EVPI is to first determine the expected profit that would result if perfect information were available. Table 8–4 shows the optimal profits for each possible event. Even if

Table 8–4

PROFIT UNDER CERTAINTY

Event: Number of Rental Requests	Probability	Best Action	Profit from Best Action	Expected Value
10	.05	lease 10 cars	$30	$ 1.50
11	.05	lease 11 cars	33	1.65
12	.10	lease 12 cars	36	3.60
13	.15	lease 13 cars	39	5.85
14	.20	lease 14 cars	42	8.40
15	.25	lease 15 cars	45	11.25
16	.15	lease 16 cars	48	7.20
17	.05	lease 17 cars	51	2.55
	Expected Profit under Certainty			$42.00

we could make the best profit for each event, we do not know which will occur. Hence, we take the expected value. This is the *expected profit under certainty,* $42.00, and measures the profit level obtainable with a perfect predictor (i.e., knowing in advance the number of cars needed each day and leasing just that number). On the other hand, our best expected profit under *uncertainty* was $35.50, obtained by leasing 13 cars each day throughout the period. The difference between these numbers is $6.50; this is the expected value of the perfect information (EVPI).

An Example

A manufacturer must decide whether to build a new plant. The profitability of the plant will depend upon future economic conditions (either stability or growth). The payoffs for various actions and events and the subjective probabilities that the manufacturer assigns to stability and growth are shown in Table 8–5.

Table 8–5

PAYOFF TABLE
PROFITS FROM BUILDING NEW PLANT
(Millions of Dollars)

Event: Level of National Economy	Probability	Actions	
		Build	Do Not Build
Stability................ .2		3	5*
Growth................ .8		16*	12
	1.0		
Expected Profit		13.4	10.6

*Maximum possible profit for each event.

The opportunity loss table for this problem is shown as Table 8–6.

If the economy is stable, "do not build" is the better action and hence has an opportunity loss of zero. If instead the plant were to be built, it would reduce profit by $2 million, relative to the best alternative. Hence, the opportunity loss of "build," if stability occurs, is $2 million.

Similarly, if there is to be economic growth, "build" is the best alternative and has zero regret (opportunity loss). If the decision maker failed to build and there was growth, his opportunity loss would be $4 million, since his profit would be reduced by this much relative to the optimal decision.

Table 8–6

OPPORTUNITY LOSS TABLE
(Millions of Dollars)

Event: Level of National Economy	Probability	Actions	
		Build	Do Not Build
Stability................ .2		2	0
Growth................ .8		0	4
	1.0		
Expected Opportunity Loss		0.4	3.2

The expected value of perfect information is equal to the EOL of the best decision. In this case, the best decision is "build" and EVPI = 0.4 million or $400,000.

Alternatively, we can calculate the profit under certainty as shown in Table 8–7. EVPI is then determined as the expected profit under

Table 8–7
CALCULATION OF EXPECTED PROFIT UNDER CERTAINTY
(Millions of Dollars)

Event: Level of National Economy	Probability	Best Action	Profit from Best Action	Expected Value
Stability........... .2		Do not build	5	1.0
Growth........... .8		Build	16	12.8
Expected Profit under Certainty				13.8

certainty less the profit under uncertainty ($13.8 - 13.4 = 0.4$), yielding 0.4 million, as above.

Since this is a sizable amount, the decision maker might profitably seek more information about future economic trends before making his decision. This is not to say that one could ever get perfect information on future events. Perhaps the decision maker could hedge somewhat in this case by proceeding with the plans but still keeping alive the possibility that the project might be canceled if economic growth did not justify it.

LINEAR PROFIT FUNCTIONS

In the previous chapter and in the earlier sections of this chapter we presented a general framework for decision making under uncertainty. In this section we shall present a special case in which the analysis is considerably simpler than heretofore. This occurs when the profit for a given action can be represented as a linear function of the unknown variable. Let us illustrate this.

A manufacturer of children's toys has a new toy which he is considering marketing nationwide. The toy is a novelty item which would be discontinued after a single national selling campaign. The variable cost to manufacture the toy is 12 cents. The selling price to retail outlets is 57 cents, so the unit profit is $57\cancel{c} - 12\cancel{c} = 45\cancel{c}$. A national advertising campaign to sell the product would cost $2.7 million. Management is uncertain about how many of the toys will be sold. The probability distribution assigned to the unknown variable—number of units sold

—appears in Table 8–8. The possible actions are (1) market the new product or (2) abandon the product.

Table 8–8

PROBABILITIES AND EXPECTED VALUES OF
TOY SALES

Event: Number Sold X	Probability P(X)	Expected Value (Millions of Units) X · P(X)
4 million......................	.2	.8
6 million......................	.3	1.8
8 million......................	.4	3.2
10 million.....................	.1	1.0
	1.0	E(X) = 6.8

We could, of course, analyze this problem by drawing up a payoff table and proceeding as outlined in Chapter 7 and the first part of this chapter. Instead, let us find an equation that will relate profit to the unknown number of items sold (X). There is one equation for each action:

Market the product: Profit $\pi = -\$2{,}700{,}000 + \$.45X$
Abandon the product: Profit $= 0$

These equations are graphed in Chart 8–1.

Chart 8–1

PROFIT FUNCTIONS OF TWO ACTIONS
IN MARKETING NEW TOY

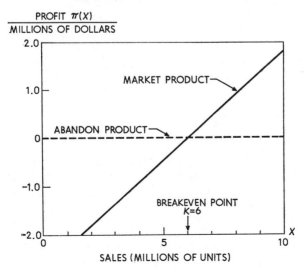

The first equation contains a negative $2.7 million (the cost of promotion campaign) and a variable contribution per unit of 45 cents times the number of units sold. Thus, if eight million were sold, profit would be:

$$\pi = -\$2,700,000 + (\$0.45)(8,000,000) = +\$900,000$$

Note that the profit equations are linear. That is, they are of the form

$$\pi = a + bX \tag{1}$$

where $\pi =$ profit; a and b are constants; and X is the unknown variable. When this is the case, the expected profit, $E(\pi)$, can be found by the following equation.[2]

$$E(\pi) = a + bE(X) \tag{2}$$

where $E(X)$ is the expected value of the unknown variable X.

For the decision "market the product," $a = -\$2,700,000$ and $b = \$0.45$. $E(X) = 6.8$ million unit sales, as in Table 8–8. Hence, the expected profit (using Equation 2) is

$$E(\pi) = -\$2,700,000 + (\$0.45)(6,800,000) = \$360,000$$

For the decision "abandon the product," both a and b are zero and $E(\pi) = 0$. If the toy manufacturer were to act now, therefore, he would market the product, since this action has a higher expected profit than the alternative (which has zero profit).

It is important to note that if the profit function is not linear, the expected profit cannot generally be obtained by substituting in the expected value of the unknown variable. This is a mistake which the unwary can easily make.

It is also instructive to calculate the break-even level of sales; that is, the volume of sales at which the decision maker is indifferent between the two alternatives. In this case it is the sales necessary to cover the advertising expenses. Let us denote this break-even value by K. Then

$$\$0.45 \, K = \$2,700,000$$
$$K = 6,000,000 \text{ units}$$

Once this value is known, the decision maker can simply compare the expected sales $E(X)$ with the break-even point K. If $E(X)$ is greater than K, marketing the product will be more profitable. If $E(X)$ is

---

[2] This can be shown as follows: $E(\pi) = \Sigma P(X)\pi = \Sigma P(X)[a + bX] = \Sigma aP(X) + \Sigma bXP(X) = a\Sigma P(X) + b\Sigma XP(X)$. But $\Sigma P(X) = 1$ because $P(X)$ is a probability function and $\Sigma XP(X)$ is defined to be $E(X)$. Hence, $E(\pi) = a + bE(X)$, as shown.

less than K, marketing the product would lead to negative profits, and it would be better to abandon the project.

Opportunity Loss Functions

When the profit function is linear, each function describing the possible opportunity losses from a given action can be described by two connected straight lines.[3] The loss functions for our illustration are shown in Chart 8–2. These functions are:

Chart 8–2

OPPORTUNITY LOSS FUNCTIONS FOR TWO ACTIONS
IN MARKETING NEW TOY

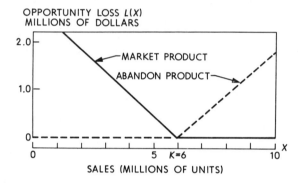

Action: Market the Product

$$\text{Opportunity loss} = L(X) = 0 \qquad \text{if } X \geq 6 \text{ million}$$

or,

$$L(X) = \$0.45\,(6{,}000{,}000 - X) \qquad \text{if } X < 6 \text{ million}$$

Action: Abandon the Product

$$\text{Opportunity loss} = L(X) = \$0.45\,(X - 6{,}000{,}000)$$
$$\text{if } X > 6 \text{ million}$$

or,

$$L(X) = 0 \qquad \text{if } X \leq 6 \text{ million}$$

Note that the break-even point, $K = 6$ million units, plays a key part in determining the loss functions. Their meaning is as follows: If we

---

[3] We are describing here the loss functions for two-action problems (i.e., only two actions are considered). For multiaction problems, each loss function still consists of connected straight lines, but the subsequent analysis is more complicated.

market the product and sales exceed the break-even value (six million), then there is no opportunity loss, since we have made the correct decision. If, on the other hand, sales are below six million, our regret (loss) is 45 cents for every unit that sales fall below six million, since, had we abandoned the project, we could have avoided this loss. Similarly, if we abandon the project and sales are at or below the break-even value, then our loss is zero, since we acted optimally. However, if sales are above six million, we suffer an opportunity loss of 45 cents for every unit above six million, since this is profit we could have obtained had we acted optimally.

Because these loss functions are broken rather than continuous straight lines, it is not generally possible to obtain a simple expression for the expected opportunity loss (EOL) and EVPI, except in the special case of the normal distribution considered in chapter 14.

However, we can compute the expected value of perfect information in our usual fashion. This is done in Table 8–9. The expected oppor-

Table 8–9

OPPORTUNITY LOSSES AND EXPECTED VALUE OF PERFECT INFORMATION

Event: Sales, Millions of Units, X	Probability $P(X)$	Opportunity Losses (Millions of Dollars)		Expected Value (Millions of Dollars)	
		Market Product	Abandon Product	Market Product	Abandon Product
4	.2	$0.9	$0	$0.18	$0
6	.3	0	0	0	0
8	.4	0	0.9	0	0.36
10	.1	0	1.8	0	0.18
	1.0			EOL = $0.18	$0.54

tunity loss for the best decision is $180,000. This is the expected value of perfect information.

THE VALUE OF IMPERFECT INFORMATION

The expected value of perfect information (EVPI) determines the upper limit on the value of additional information in a decision situation. However, in most cases, the information that we can obtain at a reasonable cost is imperfect in the sense that it will not predict exactly which event will occur. Such information can have value if, on the average, it improves the chances of making a correct decision and increases the expected profit.

In this section, we shall consider the possibility of conducting an experiment to obtain additional imperfect information.[4] The term "experiment" is intended here to be very broad. An experiment may be a study by economists to predict national economic activity, a consumer survey by a market research firm, an opinion poll conducted on behalf of a political candidate, a sample of production line items taken by an engineer to check on quality, or a seismic test to give an oil well drilling firm some indications of the presence of oil.

In general, we can evaluate the worth of a given experiment only if we can estimate the reliability of the resulting information. A market research study may be helpful in deciding upon whether or not to introduce a new product. However, only if the decision maker can say beforehand how closely the market research study can estimate the potential sales can he put a specific economic value on the experiment.

An example will make this clear. Let us suppose that the sales of a new product will be either at a high level or quite low (the product will be either a success or a flop). The payoff table for this decision is shown in Table 8–10. The value of $4 million is the net profit, over a

Table 8–10

PAYOFF TABLE FOR DECISION ON INTRODUCTION OF NEW PRODUCT
(Millions of Dollars)

Event	Probability	Actions	
		Introduce Product	No Introduction
High sales...............	.3	4.0	0
Low sales...............	.7	−2.0	0
Expected values............		−0.2	0

period of time, if the potential sales level is high. The −$2 million is the cost of an abortive introduction.

The indicated action is to abandon (i.e., not introduce) the product. However, the decision maker, being reluctant to give up a chance to make $4 million, may wonder if he should gather more information before action. As a first step, the EVPI can be obtained from the opportunity losses associated with the "no introduction" action ($4 million for high sales and zero for low sales). If these are multiplied by the respective probabilities, EVPI is determined as $1.2 million

---

[4] The authors are indebted to H. Bierman, Jr., C. P. Bonini, and W. H. Hausman, *Quantitative Analysis for Business Decisions* (3d ed.; Homewood, Ill.: Richard D. Irwin, 1969), pp. 80–85, for this example, originally prepared by C. P. Bonini.

$[(.3 \times 4) + (.7 \times 0) = 1.2]$. Thus, potentially at least, considerable value can be obtained through additional information.

The decision maker can perform an experiment in this situation. Let us suppose the experiment takes the form of a market survey conducted in two representative cities. Although, in the past, such a survey often predicted accurately the success or failure of a new product, occasionally success was predicted for a product that later failed, and vice versa. In addition, the results were often inconclusive.

If the marketing manager takes the survey before acting, he can base his decision upon the survey predictions. This problem can be expressed in terms of a decision tree, as shown in Chart 8–3. The upper part of the tree shows the decision process if no survey is taken. This is the same as Table 8–10, with probabilities of .3 and .7 for high and low sales, expected profit of −$0.2 million for introduction, and an indicated decision of no introduction.

The marketing manager attaches probabilities to the possible survey predictions as a function of the actual sales level, as specified in Table

Chart 8–3

DECISION TREE FOR PROBLEM ON INTRODUCTION OF NEW PRODUCT

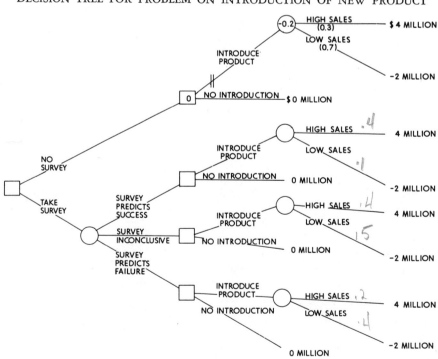

8–11. That is, he attaches probabilities to "success," "inconclusive," and "failure" predictions for the event "high sales" and different probabilities for the "low sales" event. Such probabilities would reflect past experience with surveys of this type, modified perhaps by the judgment of the marketing manager. Such probabilities may also be based upon sampling analysis, which will be introduced in the next chapter.

Table 8–11

CONDITIONAL PROBABILITIES OF SURVEY PREDICTIONS,
GIVEN ACTUAL SALES

Experimental Results (Survey Prediction)	Actual Level of Sales	
	High Sales (H)	Low Sales (L)
Survey predicts success (S) (i.e., high sales).........................	.4	.1
Survey inconclusive (I)......................	.4	.5
Survey predicts failure (F)...................	.2	.4
	1.0	1.0

The probabilities shown in Table 8–11 express the reliability or accuracy of the experiment. Only with these estimates can the marketing manager evaluate the economic worth of the survey.

The Revision of Probabilities: Bayes' Theorem

In order to complete the analysis of Chart 8–3, we need the probabilities of the various survey outcomes. (All we have available are the probabilities of the high and low sales level.) Similarly, we need the conditional probabilities of a high and low sales level given a prediction of success, etc.; whereas Table 8–11 gives the conditional probabilities in the reverse order, namely, the conditional probabilities of the various predictions given a high sales level, etc. To remedy this, the probabilities must be put in a different form.

We next construct a joint probability table, similar to those used in Chapter 5. In Table 8–12, the joint probability of both a high sales level (H) and a successful prediction (S) is obtained by multiplying the probability of a high sales level (.3) by the conditional probability of a successful prediction given a high sales level (which is .4 from Table 8–11):

$$P(H, S) = P(H)P(S|H) = (.3)(.4) = .12$$

Table 8–12

JOINT PROBABILITY TABLE

| Level of Sales | Survey Prediction | | | Total |
	Success (S)	Inconclusive (I)	Failure (F)	
High (H)......... .12		.12	.06	.30... $P(H)$
Low (L).......... .07		.35	.28	.70... $P(L)$
Total	.19	.47	.34	1.00
	$P(S)$	$P(I)$	$P(F)$	

Similarly:

$$P(L, S) = P(L)P(S|L) = (.7)(.1) = .07$$
$$P(H, I) = P(H)P(I|H) = (.3)(.4) = .12$$

and so on. Note that the marginal probabilities for "success," "inconclusive," and "failure" predictions are .19, .47, and .34, respectively. These are needed for our decision problem and are inserted in the proper places in Chart 8–4.

The decision tree also requires the conditional probabilities for the various levels of sales, given the survey prediction. These can be computed directly from the definition of conditional probability. For example, the probability of high sales, given a prediction of success, is:

$$P(H|S) = \frac{P(H, S)}{P(S)} = \frac{.12}{.19} = .632$$

And the probability of low sales, given a prediction of success, is:

$$P(L|S) = \frac{P(L, S)}{P(S)} = \frac{.07}{.19} = .368$$

Similarly:

$$P(H|I) = \frac{.12}{.47} = .255$$

$$P(L|I) = \frac{.35}{.47} = .745$$

and

$$P(H|F) = \frac{.06}{.34} = .176$$

$$P(L|F) = \frac{.28}{.34} = .824$$

Chart 8–4

DECISION TREE FOR INTRODUCTION OF NEW PRODUCT

(WITH PROBABILITIES)

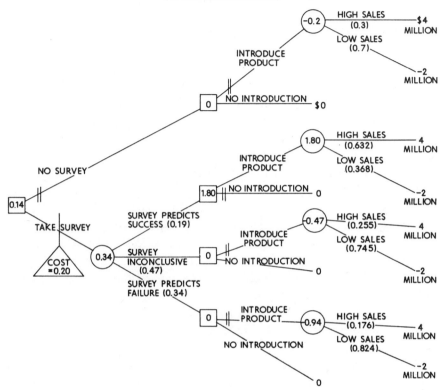

Although the calculation of the above probabilities is a simple exercise in the use of conditional probability, it is of sufficient importance to be noted carefully. In fact, it is called Bayes' theorem after a 17th-century clergyman who first emphasized this form of conditional probability.[5] Note that:

$$P(H|S) = \frac{P(H, S)}{P(S)} = \frac{P(H)P(S|H)}{P(H)P(S|H) + P(L)P(S|L)}$$

---

[5] A more general form of Bayes' theorem is as follows: given a set of mutually exclusive and collectively exhaustive events, $E_1, E_2 \ldots, E_n$, and an experimental outcome, e

$$P(E_j|e) = \frac{P(E_j) P(e|E_j)}{\sum_{i=1}^{n} P(E_i) P(e|E_i)} \qquad \text{for } j = 1, 2 \ldots, n$$

In this form, the conditional probability of a state of the world (high sales) given an experimental outcome (survey prediction of success) is expressed in terms of the conditional probabilities of the experimental outcome (success) given the various states of the world (high and low sales) and the simple probabilities of the states of the world.

We shall consider Bayes' theorem again in the evaluation of samples in Chapters 13 and 14. Bayes' theorem plays such a key role in the evaluation of experimental and sample evidence in decision making that the whole area is sometimes called Bayesian decision theory.

Returning to our example, the probabilities calculated above are listed in the appropriate places in Chart 8–4. All the necessary information is now available, and Chart 8–4 can be analyzed, starting from the right and working backward. The expected values are shown in the circles. Expected profit is positive for introducing the product only if a successful prediction is obtained from the market survey. Hence the product should not be introduced if an "inconclusive" or "failure" prediction is obtained. The expected profit from taking the survey is shown as $0.34 million. This is the expected profit associated with acting on the basis of the survey outcome. It does not include the cost of the survey, which is $0.2 million. When this is included, the net expected profit is $0.14 million. Since this is preferable to the zero profit from acting without obtaining the additional information, the survey should be taken.

Discussion. Taking a survey, in the above illustration, would be a means of obtaining additional information. The information would not be perfect since the survey could not tell exactly whether the sales would be high or low. The probabilities in Table 8–11 give the estimated reliability of the survey predictions. Estimates of this sort are necessary if the economic worth of taking the survey is to be determined.

In our example, the action of taking the survey gave an expected profit of $0.34 million above the best action without the survey (which was to abandon the product). Hence the value of the imperfect information was $0.34 million. Since this exceeded the cost ($0.2 million), the information was worth obtaining, and the survey should be taken.

Taking a sample represents a means of obtaining information. This information is imperfect, since the sample is not likely to represent exactly the population from which it is taken. Chapters 9 and 10 discuss the reliability of samples and Chapters 13 and 14 describe how sampling can be incorporated into the decision making process.

SUMMARY

The previous chapter introduced methods for decision making under uncertainty by which we could answer the following question: "If we must act now with the information available, what is the optimal act?" The first part of this chapter was directed at the question: "Should we act now or postpone the decision and collect additional information before acting?"

We first consider opportunity loss, which is part of the world of "might have been." It is the difference between the profit actually achieved and the profit that could have been obtained had the optimal action for a given event been selected. An opportunity loss table shows the opportunity loss for each combination of action and event. The expected opportunity loss (EOL) of any action is then the weighted average of the opportunity losses associated with that action, the weights being the probabilities of the various events.

The expected value of perfect information (EVPI) is the additional profit that could be made if the decision maker knew beforehand and could pick the optimal action for every possible event. The expected opportunity loss (EOL) of the best action is uniquely the expected value of perfect information (EVPI). The expected value of perfect information can also be obtained by calculating the expected profit under certainty and subtracting the highest expected profit under uncertainty. The expected value of perfect information is an important concept in the decision whether to act now or later. If EVPI is small, it means that our uncertainty is small when measured in economic terms; hence, there is little to be gained from additional information. On the other hand, if EVPI is large, then there is room for considerable improvement in the available information; possibly we should seek more information before acting.

When the profit for a given action can be expressed as a linear function of the unknown variable, then the expected profit for that action can be determined simply from the expected value of the unknown variable. The opportunity loss function is composed of two linear pieces.

Additional information that is obtained in the real world is generally imperfect. However, the economic value of the information can be measured if the reliability of the information can be determined. Bayes' theorem is used to calculate the probabilities used in evaluating the additional information.

PROBLEMS

1. Refer to Problem 3 of Chapter 7:
   a) Prepare an opportunity loss table for this decision situation.
   b) Calculate the expected opportunity loss for each action.
   c) What is EVPI?
   d) What is the expected profit under certainty?

2. Refer to Problem 6 of Chapter 7:
   a) Prepare an opportunity loss table.
   b) What is EVPI? Explain its meaning in this decision situation.

3. Refer to Problem 7 of Chapter 7:
   a) Prepare an opportunity loss table.
   b) What is the expected profit under certainty?
   c) What is EVPI?

4. Refer to Problem 10 of Chapter 7:
   a) What is the expected value of perfect information in this decision situation?
   b) How might the decision maker obtain additional information?

5. Refer to Problem 11 of Chapter 7:
   a) Determine the EOL of each action.
   b) Do you think IJK should obtain additional information about the financial position of new customers such as Lastco? Suppose, for example, a credit rating company could give an opinion of a potential customer for a $200 fee.
   c) Suppose the fee of the credit rating company was only $50. And, based upon past experience, the ratings (good, medium, poor) related to IJK experience as follows:

CREDIT RATINGS BY CUSTOMER CLASSIFICATION
(Percent of Total)

Credit Evaluation Rating	Event			
	Failed	Financial Troubles	Sporadic Customer	Good Customer
Good.........................	0%	10%	40%	40%
Medium.......................	40	50	50	50
Poor.........................	60	40	10	10
Total........................	100	100	100	100

Draw a decision tree for this problem. Would it be worthwhile to use the credit rating company to help screen customers?

6. Refer to Problem 17 of Chapter 7. The president of Lark suggests that the decision be postponed a year. He notes the great degree of uncertainty about

the future market position of the Lark Company, and he feels that the situation will be somewhat clearer in a year. It is determined that the present No. 1 deplaning machine can be repaired to last a year for $3,000. Should the decision be postponed? Why or why not?

7. Refer to Problem 15 of Chapter 7. Express the profits of each action as a linear function. Calculate expected sales. Use this value to determine the expected profit for each action.

8. The quality of a manufactured product varies from day to day due to weather conditions, machine settings, worker productivity, and other factors. Over the past 100 days the quality (fraction of the items which were defective) had the following frequency distribution:

Fraction Defective	Relative Frequency
.01	.20
.02	.40
.04	.20
.06	.10
.08	.05
.10	.03
.15	.02
	1.00

a) Using past relative frequencies as probabilities, what is the expected fraction defective?

b) Suppose that each defective item causes rework costs of $1.50 when the item is included in a final assembly operation. Express the rework costs for a lot of 1,000 items as a function of the fraction defective.

c) Use the answers to a and b to determine the expected rework cost.

9. The Zippy Razor Company makes a contribution (price minus variable cost) of 80 cents on each package of razor blades sold. Fixed costs of operating (costs independent of the sales level) are $180,000. The following probabilities are assigned to various sales levels (in thousands of packages) for next year.

Sales	Probability
100	.05
150	.05
200	.10
250	.40
300	.30
350	.10
	1.00

a) Express Zippy profits as a function of sales.

b) What is the break-even point?

c) Calculate expected sales and use this value to determine expected profit.

10. Wildcat Dynamics was an oil exploration company, founded in 1955. The company had been successful in bringing in wildcat wells in various parts

of the United States. By 1969 the company had reasonable financial reserves of its own, but also occasionally entered into partnership with a group of investors in Dallas. Hence, the firm usually did not have great difficulty in raising funds for a reasonably good wildcat prospect.

It was Wildcat Dynamic's policy to sell off the rights to produce the oil once a well was brought in. Activities were confined to locating possible sites, arranging for appropriate leases, and contracting for drilling operations.

In June of 1969, the company was trying to decide whether to drill on a parcel of offshore land in Louisiana. The lease had been taken out in 1967 but the company had been busy elsewhere. The lease would expire soon and the company had to decide whether or not to drill.

The cost of drilling at the site would be $70,000. This would all be lost if the well were dry. If the well were successful, the value would depend upon the extent of the reserves uncovered. For simplicity, management generally considered only two alternatives, described humorously as either a "wet" well or a "soaking" well. The revenue associated with selling the rights to a wet well were $220,000, or $150,000 in excess of the drilling cost. For a soaking well, revenues were expected to be $670,000, or $600,000 above the drilling cost.

William Cooper, the company geologist, was consulted about the chances of actually finding oil. He said that the chances depended upon whether or not a particular structure lay underneath the proposed drilling site. If the underlying lime-shale formation rose into a flat dome shape where Wildcat proposed to drill, there were substantially better chances of finding oil than if no such dome structure existed. Mr. Cooper estimated that there were roughly 6 chances in 10 of such a dome structure underneath the Wildcat site. He based this estimate upon experiences of other drillers in the area and his own knowledge of the geology of the area. Cooper also gave the following estimates of the probabilities of finding oil:

Drilling Result	Dome Structure	No Dome Structure
Dry well	.600	.850
Wet well	.250	.125
Soaking well	.150	.025
	1.000	1.000

Mr. Cooper's estimates represented his best judgment about the results of drilling. He indicated that another expert might come up with a different set of estimates and that in fact, there were no such things as "right" probability estimates in this business.

Cooper also suggested that Wildcat might consider the possibility of taking a seismic test on the site before drilling. This test would cost $10,000. The seismic test would give an estimate of the depth of the lime-shale formation, and hence give an indication of the existence or nonexistence of the dome structure. Cooper emphasized that the seismic test was not foolproof. Sometimes intermediate layers of rock reflect the seismic soundings sufficiently to give the impression of a dome when none is there, and sometimes the soundings are misinterpreted to say that no dome exists when in fact it does. Cooper gave the following estimates of the reliability of the seismic test:

From Seismic Soundings			Probability
Estimate of dome	given	dome exists.................	.90
Estimate of no dome	given	dome exists.................	.10
Estimate of dome	given	no dome exists..............	.20
Estimate of no dome	given	no dome exists..............	.80

a) Draw a decision tree for this problem.
b) What is the value of the additional information obtained from the seismic test?
c) What decision should Wildcat make?

11. Mr. John Fabian, the president of Wildcat Dynamics, was trying to decide about taking a seismic test or drilling on the Louisiana lease (see Problem 10 above). He asked his assistant, Frank Lindsay, to make an analysis of the problem that would incorporate the financial considerations and the probability estimates supplied by the geologist, William Cooper. However, Fabian was concerned with using the cash flows only in making the decision. He noted that the various alternatives involved different amounts of risk.

Lindsay reported that he thought that he could include an allowance for risk in the analysis if Mr. Fabian would supply answers to a few simple questions. These questions and Fabian's answers are given below:

1. *Lindsay's question:* You have a business venture that you assess has a 50/50 chance for a loss of $100,000 or a profit of $600,000. Someone offers to buy this venture from you. How much would he have to offer for you to be interested in selling? In other words, how much cash would you need to be indifferent about keeping or selling the venture?

Fabian's answer: I guess a price of $150,000 would be reasonable. If I were offered much above $150,000 I would rather have the cash. Much below $150,000, I would rather take my chances with the venture.

2. *Lindsay's question:* Now suppose you had a venture with a 50/50 chance for a profit of $150,000 or a profit of $600,000. At what price would you be indifferent to selling this venture?

Fabian's answer: $350,000.

3. *Lindsay's question:* Suppose you were committed to a venture that had a 50/50 chance for a loss of $100,000 and a profit of $150,000. How much would you pay to get out of this venture?

Fabian's answer: I sure wouldn't want to find myself in such a position. I guess I would be willing to spend about $20,000 to be relieved of such a risk.

a) Using the questions and answers above, assess Mr. Fabian's utility function for this decision situation.
b) Using this utility function, what decision should Mr. Fabian make?

12. Refer to Problems 10 and 11 above. Suppose that a syndicate in Dallas offers to provide part of the financing for the Louisiana drilling venture.

The syndicate offers to pay a percentage share of the drilling costs and seismic test costs (if taken) in return for the same percentage share in the revenues if the well is successful.

a) Suppose the proposed syndicate share is 50 percent. Should Mr. Fabian accept the offer?

b) Suppose the proposed syndicate share is 30 percent. Should Mr. Fabian accept this offer?

c) Suppose Mr. Fabian had a choice of either the 30 percent or the 50 percent offer, which should he accept?

13. Refer to the example on pages 202–8. Suppose the following probabilities described the reliability of the market survey, rather than those presented in Table 8–11.

	Actual Sales Level	
Experiment Result	High	Low
Success (S)	.5	.1
Inconclusive (I)	.4	.3
Failure (F)	.1	.6

a) What is the expected value of the additional information given by the survey in this case?

b) Should the survey be taken? What decision should be made if the survey outcome is inconclusive (I)?

14. The JFC Dynamics Company manufactured electronic components and marketed them to larger electronics firms. JFC was experimenting with a new product, which they hoped to sell for $139. On the basis of contacts with potential customers, JFC management estimated that they could sell about 2,000 units of the product annually. The equipment to manufacture the part had been purchased and installed, and the variable manufacturing cost was estimated to be $86.25 per unit.

One unanticipated problem developed in the manufacture of the item. A subassembly unit sometimes did not function properly when the whole item was tested. It was impossible to tell by simple inspection before the subassembly was put into the unit whether or not it would function properly. When the whole unit did not work properly, it had to be torn down completely, and the subassembly adjusted carefully using precision equipment. The cost of this tear-down, adjustment, and reassembly was $32. In the first production runs, approximately 25 percent of the items did not function properly because of the subassembly.

One suggested solution to the problem was to adjust each subassembly more accurately before it was assembled into the main unit. This would cost $10 for each such fine adjustment, and hence add $10 to the cost of the product.

The engineering supervisor had another suggestion. He knew of a piece of test equipment that could be rented that would give some indication of whether or not the subassembly would function properly before the subassembly was put in the final unit. When a subassembly was tested on this equipment it registered a reading of either "strong" or "weak." A sample of 60 subassemblies were carefully marked, tested on the test equipment, and then assembled in the final product. The performance in the final product is summarized below:

Reading on Test Equipment	Actual Performance of Final Unit		
	Satisfactory	Fail	Total
Strong	42	6	48
Weak	3	9	12
Total	45	15	60

The cost of the test would be $2 per unit tested.

a) Draw a decision tree for this problem.

b) What decision should JFC make?

15. Refer to the quotation from *The Wall Street Journal* contained in footnote 3 on page 398. Comment on the decision by the candy manufacturer to buy the insurance and pay the $10,000 premium from the point of view of:

a) The expected value of perfect information.

b) The decision maker's utility curve for money.

SELECTED READINGS

BIERMAN, H.; BONINI, C. P.; and HAUSMAN, W. *Quantitative Analysis for Business Decisions.* 4th ed. Homewood, Ill.: Richard D. Irwin, 1973.

Chapters 3, 4, and 5 treat decision making under uncertainty at about the same level as here. Chapter 17 deals with utility theory.

BROWN, R. V. "Do Managers Find Decision Theory Useful?" *Harvard Business Review* (May–June 1970).

A survey of some of the applications of decision theory and a discussion of the difficulties of implementing it.

HAMMOND, J. S. "Better Decisions with Preference Theory," *Harvard Business Review* (November–December 1967).

A readable introduction to utility theory.

HARLAN, N.; CHRISTENSON, C.; and VANCIL, R. *Managerial Economics: Text and Cases.* Homewood, Ill.: Richard D. Irwin, 1962.

Section III presents text and several short cases relevant to decision making under uncertainty.

HOWARD, R. A. (ed.). *IEEE Transactions on Systems Science and Cybernetics,* Special Issue on Decision Analysis (Vol. SSC-4, No. 3, September 1968).

Contains many articles, ranging from introductions to decision making under uncertainty and utility theory to advanced topics. Articles of particular relevance to the material in the previous two chapters are those by North, Howard, Wilson, Meyers and Pratt, and Spetzler.

LUCE, R. DUNCAN, and RAIFFA, HOWARD. *Games and Decisions.* New York: John Wiley, 1957.

Chapter 2 is a good presentation of the role of utility in decision making. Chapter 13 compares different decision criteria in the face of uncertainty.

MAGEE, JOHN F. "Decision Trees for Decision-Making," *Harvard Business Review* (July–August 1964) and "How to Use Decision Trees in Capital Investment," *Harvard Business Review* (September–October 1964).

These two articles describe the basic ideas about decision trees and show their application to several types of management decision problems.

RAIFFA, H. *Decision Analysis.* Reading, Mass.: Addison-Wesley, 1968.

An excellent, detailed treatment of decision making under uncertainty, without using sophisticated mathematics.

SCHLAIFER, R. *Analysis of Decisions Under Uncertainty.* New York: McGraw-Hill, 1969.

Part 1 deals with basic elements of decision analysis and diagramming of decision trees in detail. Part 2 deals with the assessment of preferences and probabilities in detail. A very practical reference source.

SWALM, R. O. "Utility Theory—Insights into Risk Taking," *Harvard Business Review* (November–December 1966).

Describes the derivation and application of utility functions in business firms.

9. INTRODUCTION TO STATISTICAL INFERENCE

THE ABILITY to make valid generalizations and predictions from sample data is an important step forward in scientific knowledge. The concept of sampling was introduced in Chapter 1. Chapters 2 to 4 presented the necessary tools of analysis—frequency distributions, averages, and measures of dispersion. Chapters 5 and 6 developed the fundamentals of probability theory. These basic concepts can now be brought together in the study of statistical inference.

Statistical inference is the process by which we draw a conclusion about some measure of a *population*[1] based on a sample value. The measure might represent either variables or attributes. Thus, we could estimate the average or *mean* amount of money that consumers plan to spend on a new car, or the *proportion* of consumers favoring foreign cars. The purpose of sampling is to estimate these same characteristics for the population from which the sample is selected.

The population measure is the *parameter,* while the sample measure is called a *statistic.* We will first consider the problem of estimating the arithmetic mean of a population from the mean of a sample. This is called a *point estimate,* since it endeavors to provide the best single estimated value of the parameter. An *interval estimate,* on the other hand, proceeds by specifying a range of values. Thus, after testing a sample of steel rods, we may make a point estimate that the mean breaking strength of all such rods is 10 pounds; but we might also make an interval estimate that the mean for all rods probably lies in the interval from 9 to 11 pounds, as described later.

---

[1] "Population" and "universe" are usually considered synonymous. The newer term "population" will be used in this discussion. These terms refer here to inanimate objects as well as to living beings.

Sample information may be used for either of two purposes—reporting or decision making. In a reporting role, the sample estimates (both point and interval estimates) are presented for the information of others. Government statistics (e.g., on unemployment) are a good example of this use of sample data. The sample information may be used also in this context to corroborate some point in exposition, as when a social scientist presents such information to help in drawing some conclusion. *Confidence intervals* will also be presented for the purpose of reporting sample evidence and drawing conclusions therefrom.

On the other hand, the sample information may be incorporated directly in a decision making procedure. *Tests of hypotheses* will be described in Chapter 10 as a means of decision making, as well as reporting sample findings. Or, to go a step further, the sample may be combined with the prior judgments of the decision maker and the economic consequences of various actions to arrive at the best decision. Chapters 13 and 14 incorporate samples in this decision-making context.

SAMPLING ERROR AND BIAS

A sample rarely produces, without error, the exact information needed for decision making. Some reasons for the deviation of sample results from the true population values are as follows.

Sampling Error

Sampling error is the random or chance error that occurs when we take a sample rather than testing the whole population. A sample is only partially representative of the larger population from which it is taken. Any two samples will differ from each other, since they will contain different elements of the population.

If a probability sample (see below) is taken properly, sampling error can be controlled and measured. This error depends in part on the type of sample chosen. Thus, a stratified sample generally has a smaller error, and a cluster sample a larger error, than a random sample of the same size, as described in Chapter 12. The error also depends on the size of sample—the smaller the sample the larger the error. But the sampling error does not include the effect of bias, which must be minimized in the design of the original survey. Nor can the error be precisely measured in nonprobability samples such as quota or judgment samples (Chapter 12).

Size of Sample. A basic error in statistical reasoning is to jump to a conclusion or generalization on the basis of too small a sample. As an example, a national magazine reported that a group of Colorado school-teachers had been given a test in history and failed with an average grade of 67, indicating that Colorado schoolteachers generally were deficient in history. An official of the Colorado Education Association retorted that only four teachers had been given the test, of whom three made the respectable average score of 83 and the fourth only 20, bringing the average of the four down to 67.

An extreme case of using too small a sample is that of generalizing from a sample of one, or citing only a single case. Thus, a typewriter manufacturer advertises that "Tests by leading educators prove that students who use typewriters get up to 38% better grades." Or, "All Indians walk single file; at least the one I saw did." In general, sampling error can be reduced by increasing the size of the sample. Since larger samples are more costly, a key element of sample design is balancing the cost of a sample with the value of the information provided. Sampling errors will be discussed at length in Chapters 9–12.

Bias

Conscious or unconscious bias is very common in statistical work. It is easy to detect the conscious bias in an advertisement that quotes statistics to "prove" the superiority of a given product, while a competitor's ad quotes other statistics to "prove" the superiority of his own product. But many compilers of statistics have an ax to grind. A jewelers' association quotes figures purporting to show that the double-ring wedding has become "an accepted national custom." A labor organization claims that a consumer price index, on which wages are based, should be revised upward because it understates real costs, while an employers' association defends the index, pointing out components that overstate real costs. The source of the data must be considered, as well as the conclusions themselves.

Unconscious bias in the selection of samples is more difficult to detect. It may arise in any of the three following ways.

Bias in the Manner in Which the Sample is Taken. If the sample is drawn in such a way that some elements of the population cannot be drawn at all, then some bias will arise. The classic example of this bias is the poll taken in 1936 by the *Literary Digest*. The *Literary Digest* mailed out 10 million cards and received about 2.3 million returns. On the basis of this sample, a victory by Alfred Landon for

President was predicted. Actually, F. D. Roosevelt won with about 60 percent of the vote. The trouble with the *Literary Digest* sample was that it was taken from lists of telephone subscribers and automobile registrants—in general, a higher income group not representative of the overall population of voters.

As another example, a feature article in *Advertising Age* is entitled "Obits Show 'Average' Adman is Dead at 62," based on obituaries of 300 advertising men who died during the previous year. Perhaps the advertising game *does* kill men off young, but there may be two defects in the sample used: (1) Since many young men have entered this field in recent years, those who died during the past year were relatively young; the surviving ones who will live to a riper old age of course are not counted. (2) If advertising is a young man's game, as reputed, older men go into other fields and are counted there when they die. As an analogy, the average age at death of college students is about 20 years, but this does not indicate that college graduates die young.

Sometimes in business research it is almost impossible to eliminate this kind of bias. Consider the firm that wishes to test a new advertising campaign. Very often it is economically feasible to select only one or two cities in which to test a new program. If the city selected is Atlanta, we obviously cannot measure the effect in Seattle. It is necessary to use business judgment to select an area that is "representative" of the nation as a whole. Experience in similar surveys and advertising programs would be useful in making this judgment.

Bias Due to Nonresponse.   In almost any survey a number of items are drawn in the sample for which no information is available. These may be people who do not mail back a questionnaire or who slam the door in the face of the interviewer. If these items are ignored, considerable bias may result, for the nonrespondents may be entirely different from the respondents. Thus, a significant part of the population may be ignored. For example, a business school alumni journal reported that the average graduate in an early class earned $87,049 in a recent year. This figure was based on 18 returns received from a questionnaire mailed to 62 class members. Unfortunately, the average income is not typical if a larger proportion of those with higher incomes return the questionnaire than do those with lower incomes or if some respondents exaggerate their incomes, as is sometimes the case. Furthermore, if a few alumni have very high incomes, these figures greatly inflate the average.[2]

---

[2] This example illustrates several misuses: (1) too small a sample, (2) nontypical sample, (3) spurious accuracy, and (4) use of mean instead of median (see Chapter 3).

Every effort should be made to reduce nonresponse. This can be partly done in the design stage of the survey by careful wording and pretesting of questionnaires and instruction sheets to those conducting the survey. Training of survey personnel is also helpful in reducing nonresponse. And finally, extensive searches and call-backs should be employed.

Measurement Bias.   Considerable bias can be introduced into a survey if the measurement device (e.g., questionnaire, interview, counting procedure) is not accurate, that is, does not measure what is intended. Consider the interviewer who found that most of those he interviewed said that they had never borrowed money from a loan company, despite the fact that the interviewer's list was drawn from a loan company's files.

Bias in questionnaires may occur in several ways. First, the choice of words or the phrasing of a question may suggest a certain answer. An example is "Did the frozen peas taste better to you than canned peas or dried peas?" This is the notorious "leading question." It would be much better to list the three types of prepared peas and request that the user number them in the order of preference. One market analyst reports that even such a seemingly innocuous wording as, "Have you read— [the latest novel]?" brought a much larger proportion of favorable replies than when a similar group of people was asked the question, "Do you happen to have read—[the same novel]?"

Second, estimates that are based on opinions rather than on actual figures may be biased. Suppose you were inquiring of a manufacturer of drugs whether his product was distributed at retail mainly through chain stores or independent stores. His direct contacts with the buyers of chain retailers might lead him to suppose that they were his chief customers, whereas a study of the sales records might well show the reverse. Questions should be objective rather than subjective.

Respondents may have unconscious biases about their own attitudes or actions. For this reason it is sometimes better to use indirect questions to obtain information. Thus, in a survey of consumer preferences, it was found that the question "What do you think *your neighbor* would like in his next automobile (chrome, space, economy, etc.)?" produced more unbiased replies than "What would *you* like in your next automobile?" The impartial investigator must check his words, as well as his figures, for possible bias.

Careful preparation of the questionnaire will help reduce these types of bias. In addition, a pretest and a follow-up check on the measuring device and the results of the survey are essential.

Control of nonsampling bias from the three sources just noted is of crucial importance in survey work. It is better to take a small sample which is relatively free of bias than to take a much larger sample with unknown bias. It is a common misconception to suppose that a large sample will iron out biases, as in the *Literary Digest* debacle. Finally, if biased data must be used in the absence of better information, the nature and probable direction of the bias must be considered in interpreting the results.

The discussion of sampling errors in Chapters 9–12 assumes that bias has been minimized in the sample design. Those planning an original survey should study one of the specialized books on the subject listed at the end of Chapter 12, both to minimize bias and to gauge its direction and extent. Too many surveys are based on some convenient "hunk" of the population, with sampling errors computed, but with little or no consideration for the bias that might far surpass the sampling error.

SIMPLE RANDOM SAMPLING

There are many effective methods of selecting samples, and these may be used in various combinations. The sample may be selected from the population as a whole, or it may be selected from certain parts or *strata* of the population. In either case, the sample may be selected at random, according to somebody's judgment, or by other methods. The individuals selected may be drawn one at a time or in clusters, such as the residents of selected city blocks. The clusters may be enumerated completely, or they may be subsampled by selecting, say, the head of every third household in the block. Thus, these procedures provide a great variety of sample designs. One distinction is made between *probability samples* and others. A probability sample is taken in such a manner that elements of the population have a specific probability of being included in the sample. A measure of sampling error can be estimated for most probability samples. Other methods rely on the judgment of the one selecting the sample or on other nonrandom procedures. While such samples may be quite useful, there is no accurate way of measuring their sampling error.

The basic concepts of statistical inference are applied to simple random samples in Chapters 9 to 11. While simple random sampling is not often used alone in business and economic research, it is important because it illustrates the fundamental principles of sampling and is a basic part of more complex types of sample design described in Chapter 12.

A *simple random sample* of *n* units is one selected from a population in such a way that each combination of *n* units has an equal chance of being selected. Thus, in selecting a simple random sample of five bolts from a shipment, every combination of five bolts in the shipment must have the same chance of selection. The bolts could not be picked only from certain boxes or just from the top of the pile.

This method is sometimes called "unrestricted" random sampling because units are selected from the population as a whole without any restriction, whereas procedures like stratification and clustering introduce restrictions (e.g., grouping the population before the sample is selected) designed to increase the precision of the sample or to reduce its cost.

Random sampling does not mean haphazard selection. Interviewing passers-by on a downtown street corner does not provide a random sample of a city's population because stay-at-homes have less chance of being interviewed than downtown shoppers or businessmen.

Random selection is determined objectively by some equivalent of a game of chance. For example, the residents of a city block might be numbered from 1 to 72 and a roulette wheel could be spun 10 times to determine the choice of 10 persons to be interviewed. However, selections are usually made from a *table of random numbers.* Such a table is just as efficient as operating a game of chance and is more convenient. In constructing a table of random numbers, the digits from 0 to 9 are drawn by some randomizing device so that each number is independent of any other. The RAND Corporation, for instance, programmed an electronic computer so as to produce the random numbers listed in its book *A Million Random Digits.* Table 9–1 is a section of another such table. (See Appendix L at the end of this book for a larger table.)

How to Use a Table of Random Numbers

To illustrate the use of this table, suppose you wish to select a random sample of 6 households from a city block of 78 households, as part of a market survey to determine brand preference for frozen foods. First, list all households by address and number them from 01 through 78. Second, take a page from a table of random numbers, and choose a starting point at any arbitrary point[3]—say, the 13th column, fifth row, in Table 9–1. This number is 43. Third, go down this column and the next columns to the right (or go in any predetermined direction) until you have selected six numbers between 01 and 78, with no repetitions.

---

[3] Ideally, the starting point should be selected by a game of chance. In practice, however, an arbitrary choice is generally considered satisfactory.

Table 9–1

RANDOM NUMBERS

03	47	43	73	86	36	96	47	36	61	46	98	63	71	62
97	74	24	67	62	42	81	14	57	20	42	53	32	37	32
16	76	62	27	66	56	50	26	71	07	32	90	79	78	53
12	56	85	99	26	96	96	68	27	31	05	03	72	93	15
55	59	56	35	64	38	54	82	46	22	31	62	43	09	90
16	22	77	94	39	49	54	43	54	82	17	37	93	23	78
84	42	17	53	31	57	24	55	06	88	77	04	74	47	67
63	01	63	78	59	16	95	55	67	19	98	10	50	71	75
33	21	12	34	29	78	64	56	07	82	52	42	07	44	38
57	60	86	32	44	09	47	27	96	54	49	17	46	09	62
18	18	07	92	46	44	17	16	58	09	79	83	86	19	62
26	62	38	97	75	84	16	07	44	99	83	11	46	32	24
23	42	40	64	74	82	97	77	77	81	07	45	32	14	08
52	36	28	19	95	50	92	26	11	97	00	56	76	31	38
37	85	94	35	12	83	39	50	08	30	42	34	07	96	88
70	29	17	12	13	40	33	20	38	26	13	89	51	03	74
56	62	18	37	35	96	83	50	87	75	97	12	25	93	47
99	49	57	22	77	88	42	95	45	72	16	64	36	16	00
16	08	15	04	72	33	27	14	34	09	45	59	34	68	49
31	16	93	32	43	50	27	89	87	19	20	15	37	00	49

SOURCE: R. A. Fisher and F. Yates, *Statistical Tables for Biological, Agricultural and Medical Research* (6th ed.; London: Oliver & Boyd, 1963), Table XXXIII, Random Numbers (1). This is part of a much larger table.

Beginning with 43, the next number down is 93, but it is ineligible, being larger than 78, so continue with 74, 50, 07, 46, 86 (ineligible), 46 (ineligible—already selected), and 32—a total of six eligible numbers. Thus, the numbers of the households to be surveyed are 7, 32, 43, 46, 50, and 74.

If there are exactly 100 items in the population, read "00" as 100. If there are more than 100 items, combine adjacent columns as necessary to form larger numbers. Thus, in the upper-left corner of Table 9–1, the columns beginning 034 could be used for three-digit numbers, or those beginning 0347 for four-digit numbers.

HOW SAMPLE MEANS ARE DISTRIBUTED

The use of the sample mean to make inferences about the population mean is a common problem in statistical inference. The following methods apply strictly to the *means* of simple random samples; they will be adapted to proportions and to other types of samples later. Therefore, the term "sample mean" in this chapter will refer to the arithmetic mean of a variable in a simple random sample.

The following symbols will be used:

	Sample Estimate	Population Value
Arithmetic mean.............................	$\bar{X}$	μ
Standard deviation..........................	s	σ
Standard error of the mean....................	$s_{\bar{x}}$	$\sigma_{\bar{x}}$
Number of items............................	n	N

If we are interested in estimating *totals* for a population, we simply multiply the estimate of the mean and standard error of the mean by the number of items in the population. Thus:

	Sample Estimate	Population Value
Population total.........................	$T = N\bar{X}$	$N\mu$
Standard error of population total.........	$s_T = Ns_{\bar{x}}$	$N\sigma_{\bar{x}}$

Inferences about a population are usually made from a single sample. This is only one of a large number of samples that might be drawn from the same population. By studying the variation of the means of all these samples, we can infer within what limits *our* sample mean is likely to fall. The means of all possible samples drawn from a given population may be grouped in a frequency distribution. This is called the *sampling distribution of the mean.* The mean and standard deviation of this distribution will describe the behavior of the sample means.

An Experiment

To illustrate the sampling distribution of the mean when the population is known, consider the following experiment.

A manufacturer of electrical equipment receives shipments of ball bearings from a steel company for use in electric fans. Specifications call for these balls to average one quarter of an inch in diameter, and none of them must deviate from the specification by more than a given tolerance. Since it is not feasible to measure every ball bearing, it is necessary to depend on sample inspection to avoid acceptance of unsatisfactory shipments.

The inspection supervisor wished to illustrate the sampling principles involved as part of the training program for inspectors. Accordingly, he selected one shipment of 565 ball bearings as the population. He then had the whole lot measured with automatic calipers. The results are shown in Table 9–2 columns 1 and 2. Thus, only one of the 565 balls was six thousandths of an inch below specification, four balls were five thousandths below, and so on; the average of all the balls (last row) was exactly equal to the specification.

Table 9–2

SAMPLING THE DIAMETERS OF 565 BALL BEARINGS

DIAMETER* (1)	NUMBER OF BALL BEARINGS IN						
	Popula-tion (2)	1st Sample (3)	2d Sample (4)	3d Sample (5)	4th Sample (6)	5th Sample (7)	All 5 Samples (8)
−6	1	...	...	1	...	1	2
−5	4	...	...	1	...	2	3
−4	15	...	2	1	1	...	4
−3	38	2	1	1	4	3	11
−2	70	8	7	5	3	10	33
−1	97	9	7	12	7	11	46
0	115	12	11	11	10	6	50
1	97	9	11	10	8	7	45
2	70	5	4	6	9	4	28
3	38	1	5	1	4	4	15
4	15	4	2	...	3	2	11
5	4	...	...	...	1	...	1
6	1	...	...	1	...	...	1
Number of ball bearings	565	50	50	50	50	50	250
Average diameter*	0	+.14	+.20	−.18	+.52	−.42	+.05

* Difference from specification (.250 inches) in thousandths of an inch.

Samples of 50 steel balls each were then selected at random from the bin containing the shipment, and their diameters were measured. After each 50 were selected, they were returned to the bin and thoroughly mixed so that the next sample could be selected from the same population as the first sample. In all, 100 samples of 50 balls each were selected.

The results of the first 5 of the 100 samples are shown in columns 3 to 7 of Table 9–2. Each of these samples differs from the others, and none of them is a perfect replica of the population. The mean diameter for each sample is shown in the last row.

The Three Distributions.   It is important to distinguish the three different distributions illustrated by this experiment. They are shown in Chart 9–1. First is the distribution of ball-bearing diameters (X) in the population itself—curve A. The figures are taken from Table 9–2, columns 1 and 2. Frequencies are plotted as percentages of the total on

Chart 9–1

THREE DISTRIBUTIONS INVOLVED IN ESTIMATING THE MEAN OF
BALL-BEARING DIAMETERS (TRUE MEAN = 0)

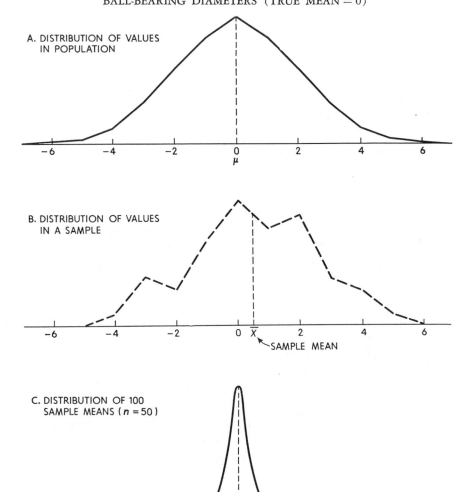

Unit: Thousandths of an inch differences from specification.
SOURCE: Table 9–2 and related data.

the Y axis, for comparability with curve B. (The curve would have been smooth if the ball bearings had been measured exactly rather than to the nearest .001 inch.) This population is normal, with its mean μ equal to zero. Other populations may be skewed or otherwise irregular.

Second is the distribution of the X values in a sample drawn from this population, such as the fourth sample in Table 9–2, shown in curve B. The sample distribution has somewhat the same general shape as the population, but it is more irregular, and its mean ($\overline{X}$) differs from the true mean (μ) because of sampling errors. As the sample size increases (e.g., Table 9–2, column 8), the shape of the sample distribution approaches more and more closely that of the population distribution, whether the latter be skewed or what not. Both the mean and the standard deviation of the sample also approach the population values.

Third is the sampling distribution of the *means* ($\overline{X}$) of a great many samples (curve C) of size $n = 50$ that can be drawn from this population. This curve shows the distribution of 100 sample means. It has been drawn with a smaller area than that under the other curves; otherwise it would be awkwardly tall. The five sample means shown in the bottom row of Table 9–2 fall well within the range of curve C. The mean of this distribution is very close to that of the population, and its dispersion or standard deviation is much less than that of curve A or B. If all possible samples of size 50 were drawn from this population, the distribution shown in curve C would be smoother, and nearly normal.

As the sample size increases, the distribution of sample means becomes still narrower in spread, and more normal in shape, as described below. Chart 9–2 shows how the sample means from a normal population tend to cluster more closely about the population mean as the sample size increases. The three curves in Chart 9–2 have the same area and are all normal, but they differ markedly in dispersion.

Sampling Concepts. The ball-bearing experiment illustrates several concepts in sampling:

1. Each of the means is approximately, but not exactly, equal to the population mean. Of the 100 samples selected in the larger study (not reported here in detail), only 5 exactly equaled the population in mean diameter, while 53 were above and 42 were below.

2. The sample means cluster much more closely about the population mean than do the original values. Thus, the means in the last row of the table vary only from $-.42$ to $+.52$, while the individual

Chart 9–2

SAMPLING DISTRIBUTIONS OF THE MEANS OF SAMPLES
OF SIZE $n = 4$ AND $n = 25$, COMPARED WITH
DISTRIBUTION OF A NORMAL POPULATION

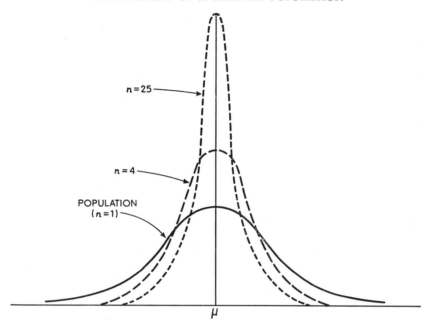

diameters (columns 1 and 2) range from -6 to $+6$. Hence, the standard deviation of the sample means is much smaller than the standard deviation of the original values.

3. If larger samples were taken, their means would cluster still more closely around the population mean, since the positive and negative errors of sampling tend to offset each other. This is illustrated by combining the five samples shown to obtain the larger sample of 250 balls listed in column 8. The mean of this larger sample is $+.05$, a result which is much closer to the population value (0) than is any of the means of the five samples of 50. The overall average of the 100 sample means proved to be $+.02$, which is closer yet to the population mean.

Thus, the larger the sample, the closer its mean is likely to be to the mean of the entire population, and the greater the precision of the sample mean. It can be shown that if all possible samples of a given size are drawn from a population, the arithmetic mean of the sample means will equal the population mean.

4. The distribution of sample means follows a normal curve. More precisely, if a number of random samples of size n are drawn from a given population, their means tend to form a normal distribution, provided (1) the size of sample is large[4] and (2) the population is not unduly skewed. If the population is skewed, the distribution of sample means will be much less skewed, in inverse proportion to the size of the sample. Thus, for samples of size 50 the distribution of means will only be $\frac{1}{50}$ as skewed as the population[5] (i.e., $n = 1$).

Central Limit Theorem.   The arithmetic mean therefore tends to be normally distributed as n increases in size, almost regardless of the shape of the original population. This principle is called the central limit theorem. It applies to the distribution of most other statistics as well, such as the median and standard deviation (but not the range). The central limit theorem gives the normal distribution its central place in the theory of sampling, since many important problems can be solved by this single pattern of sampling variability.

The distribution of sample means being normal, or nearly so, it can be completely described by its mean and its standard deviation. Furthermore, these values may be estimated from a single large random sample, as described under "The Standard Error of the Mean" below.

The Sample Mean as an Estimator of the True Mean

When we select a statistic such as the mean to estimate the population value, we ordinarily expect it to satisfy two criteria:

1. The statistic should, on the average, give the "correct" answer—the population value. That is, the mean of a distribution of all possible means for a given size of sample—the *expected value*—should equal the population value. Such an estimate is said to be *unbiased*. Means of random samples are unbiased estimators of the true means. Thus, in Table 9–2, the expected value is the overall mean of all possible sample means, each representing 50 ball bearings. This is zero, the same as the population mean. The mean of an individual sample, then, whatever its value, is said to be an unbiased estimator.

2. The second criterion states that the sampling distribution of the statistic be concentrated as closely as possible about the true population value. Such a statistic is said to be *efficient*. It can be shown that the sample mean is a more efficient estimator of the parameter than the

---

[4] In many cases a size of 30 is large enough but no exact number can be given; it depends in part on the population distribution.

[5] See F. E. Croxton and D. J. Cowden, *Applied General Statistics* (2d ed.; New York: Prentice-Hall, 1955), p. 627.

sample median in a normal population, since the sample values cluster more closely about the population value. In Chart 9–1, panel C, a distribution of sample medians would have a wider spread than that shown for the means.[6] (The median may be more efficient, however, for sharply peaked, long-tailed distributions, as noted in Chapter 3.)

THE STANDARD ERROR OF THE MEAN

The standard deviation of the distribution of sample means is called the standard error of the mean. (The word "error" is used here in place of "deviation" to emphasize that variation among sample means is due to sampling errors.) The standard error measures (inversely) the *precision* of the sample estimate, that is, how closely the sample value is likely to approach the true value. (The standard error does not, however, include errors of bias.) The smaller the standard error, the greater the precision.[7]

Where the population is large in relation to the sample size, the formula for the standard error of the mean is

$$\sigma_{\bar{x}} = \frac{\sigma}{\sqrt{n}}$$

where σ is the standard deviation of X in the population and n is the size of the sample.

Thus, in the ball-bearing example the standard deviation of the population (Table 9–2, column 2) is (unit = .001 inch):

$$\sigma = \sqrt{\frac{\Sigma f(X - \bar{X})^2}{N}} = \sqrt{\frac{2{,}190}{565}} = 1.969$$

Then, for samples of size 50, the standard error of the mean is:

$$\sigma_{\bar{x}} = \frac{\sigma}{\sqrt{n}} = \frac{1.969}{\sqrt{50}} = .278$$

and for samples of size 250,

$$\sigma_{\bar{x}} = \frac{1.969}{\sqrt{250}} = .124$$

---

[6] The standard error of the median is 1.25 times that of the mean in a normal population.

[7] "Precision" or "reliability" as used in statistics means how closely we can reproduce from a sample the results that would be obtained if we took a complete census, using the same methods of measurement, interview procedures, etc. The "accuracy" of a survey takes into account these sampling errors as well as nonsampling errors arising from bias due to methods of measurement, questionnaire design, etc. that would affect the census as well as the sample. We can only measure precision, but it is the overall accuracy that we attempt to maximize in designing surveys.

The standard error of the sample means, therefore, varies directly with the standard deviation of the population σ, and inversely with $\sqrt{n}$. By increasing the sample size, the standard error of the mean can be reduced to any desired level. However, the reduction is not *pro rata*. The sample size must be quadrupled to cut the standard error in half.

Finding the Standard Error of the Mean When σ Is Unknown

In practice, the standard deviation of the population (σ) is usually unknown, but it can be estimated as being equal to the standard deviation of a single large sample (s). That is, instead of $\sigma_{\bar{x}} = \sigma/\sqrt{n}$, we can say

$$s_{\bar{x}} = \frac{s}{\sqrt{n}}$$

where $s_{\bar{x}}$ is the standard error of the mean estimated from a single sample, s is the standard deviation of the sample, and n is the sample size.[8]

Thus, for the first sample in Table 9–2, the standard deviation is:

$$s = \sqrt{\frac{\Sigma f(X-\bar{X})^2}{n-1}} = \sqrt{\frac{161}{49}} = 1.81$$

and the standard error of the mean is:

$$s_{\bar{x}} = \frac{s}{\sqrt{n}} = \frac{1.81}{\sqrt{50}} = .256$$

This estimate of the standard error of the mean differs by 8 percent from the true $\sigma_{\bar{x}}$ of .278.

Again, for the combined sample of 250,

$$s = \sqrt{\frac{1,017}{249}} = 2.021$$

and

$$s_{\bar{x}} = \frac{2.021}{\sqrt{250}} = .127$$

For the larger sample, the estimated standard error of the mean differs by only 2 percent from the true $\sigma_{\bar{x}}$ of .124. This illustrates the princi-

---

[8] Sometimes n is used instead of $n-1$ in the formula for s, e.g., $s = \sqrt{\Sigma fx^2/n}$. In this case use $s_x = s/\sqrt{n-1}$ to achieve the same result as above. That is, by combining the two formulas, $s_x = \sqrt{\Sigma fx^2/n(n-1)}$ in either case. (Omit f in formulas for ungrouped data.)

ple that the standard error of the mean can usually be estimated satisfactorily from the standard deviation of a single large sample (the larger the better) when the standard deviation of the population is unknown.

Effect of Population Size. The above formulas for $\sigma_{\bar{x}}$ and $s_{\bar{x}}$ are correct if the population is infinitely large, or if the sampling is carried out with replacement, which amounts to the same thing. Sampling with replacement means that after each item is selected it is replaced and has a chance of being selected again. These formulas are also substantially correct when the sample is a small part—say less than 5 percent—of a finite population. Thus far, the ball bearing experiment has been treated as if its population were infinite.

Where the sample comprises a large part of the population and is done without replacement, the expression $\sigma/\sqrt{n}$ should be multiplied by $\sqrt{(N-n)/(N-1)}$, or approximately $\sqrt{1-n/N}$, where n is the sample size and N is the population size. That is,

$$\sigma_{\bar{x}} = \frac{\sigma}{\sqrt{n}} \sqrt{1 - \frac{n}{N}}$$

for finite populations. The term $1 - n/N$ is the proportion of the population not included in the sample. The adjustment is called the *finite population correction.*[9] Its use always reduces the standard error.

For example, since each sample of 50 ball bearings in Table 9–2, columns 3 to 7, was drawn without replacement of each individual ball from the finite population of 565 balls, we should have

$$\sigma_{\bar{x}} = \frac{1.969}{\sqrt{50}} \sqrt{1 - \frac{50}{565}}$$
$$= .278 \times .955$$
$$= .265$$

instead of .278 for sampling with replacement.

Thus, the precision of the sample estimate, measured by $\sigma_{\bar{x}}$, is determined not only by the absolute size of the sample but also to some extent by the proportion of the population sampled. This is in accordance with common sense. A 10 percent sample certainly seems more reliable than a 5 percent sample.

---

[9] See M. H. Hansen, W. N. Hurwitz, and W. G. Madow, *Sample Survey Methods and Theory* (New York: John Wiley, 1953, Vol. I, pp. 122–24; and W. A. Wallis and H. V. Roberts, *Statistics, A New Approach* (New York: The Free Press, 1956), pp. 368–71. The finite population correction is also called the finite population factor, finite multiplier, and finite sampling correction.

In most actual surveys, however, the sample is such a small part of the population that n/N is negligible, and $\sigma_{\bar{x}}$ is virtually equal to $\sigma/\sqrt{n}$. Hence, the reliability of a sample usually depends almost entirely on the absolute size of the sample and *not* on the percentage of the population sampled. In planning a market survey of consumers in a large city, one should ask questions like "Is a sample of 1,000 big enough?" and not "Is a 10 percent sample big enough?" The size of the city makes little difference.

How $\sigma_{\bar{x}}$ Is Used

The standard error of the mean in the ball-bearing example is .265 thousandths of an inch for samples of size 50. Since .265 is the standard deviation of all possible means of size 50, and the distribution of means of large samples is normal, we can say what proportion of the sample means lies within any given interval of the true (population) mean. In this case the true mean is known ($\mu = 0$). Then 68.27 percent of the sample means fall within one standard error ($\sigma_{\bar{x}}$) of the true mean, that is, from $+.265$ to $-.265$. As noted in Chapter 6, this means that there is a *probability* of about 68 percent—or 68 chances out of 100—that a *single* sample mean will fall within the interval of $\mu \pm \sigma_{\bar{x}}$, or $0 \pm .265$; and so on for any other degree of probability desired.

These figures also show just how much more closely the sample means cluster than do the individual ball-bearing diameters. While 68 percent of the means lie within $\sigma_{\bar{x}}$ or .265 thousandths of an inch from the true mean, the same percentage of individual ball bearings lie within σ or 1.969 thousandths of the true mean—a far wider spread.

If the distribution of the population is not normal, the above figures are still approximately correct for larger samples. In an experiment at the University of California, Berkeley, some 3,000 independent samples of 30 items each were drawn at random (using a table of random numbers) from a skewed population consisting of 200 weekly earnings figures for a group of wage earners and clerical workers in the San Francisco Bay Area. The population values ranged from $17.50 to $116.91 a week and averaged $57.95. The arithmetic mean, the standard deviation, and the approximate standard error of the mean $s_{\bar{x}}$ were computed for each sample. The question then arose: What percentage of the 3,000 sample means fell within various multiples of the standard error around the true population mean μ of $57.95? The results were as follows:

	$\mu \pm s_{\bar{x}}$	$\mu \pm 2s_{\bar{x}}$	$\mu \pm 3s_{\bar{x}}$
Theoretical expectancy....	68.27%	95.45%	99.73%
Experimental results......	68.4 %	95.2 %	99.6 %

This shows a remarkable agreement between fact and theory, despite the fact that (1) the sample size was but 30 items; (2) the sample standard deviation s was used, instead of the true population value σ; and (3) the population was not normally distributed. The theory therefore works well in practice. For smaller samples, however—say when n is under 30—the above values may have to be adjusted, as described in Chapter 11.

The corresponding results for any other probability or interval in the sampling distribution of means can be found in Appendix D, just as we previously did for individual values. For example, within what interval will exactly 95 percent of the sample means fall in the ball-bearing case ($n = 50$)? Since the proportion .95 lies on both sides of the population mean, look up half this amount, .475, for the proportion on one side of the mean, in the body of Appendix D. The interval is then $\pm 1.96\sigma_{\bar{x}}$ or $\pm .519$ thousandths of an inch.

It is customary to state probabilities in such round numbers as 95 or 99 percent, so the following relationships in a normal distribution are important:

Mean $\pm\ 1.96\sigma$ includes 95 percent of the area
Mean $\pm\ 2.58\sigma$ includes 99 percent of the area

These are often used instead of the statements that the mean $\pm 2\sigma$ includes 95.45 percent of the area and $\pm 3\sigma$ includes 99.73 percent.

When the population mean is *not* known and we use a sample mean to estimate it, we can only say that 68 percent of the sample means lie within one standard error of the true mean, wherever that may be, and similarly for other intervals. Nevertheless, we will see in the next section how this information about the spread of sample means around the unspecified true mean can be used to make satisfactory estimates of the true mean.

CONFIDENCE INTERVALS

It is often necessary to estimate the unknown mean (or other parameter) of a population. To do so, we need both the sample value and a measure of the margin of error to which this value is subject. This may be done as follows:

1. Find the mean $\bar{X}$ and its standard error $(s_{\bar{x}} = s/\sqrt{n})\sqrt{1 - n/N}$ from a large random sample as point estimates of the population values.

2. Specify a zone based on $\bar{X}$ and $s_{\bar{x}}$ within which we may be confident that the true population mean does lie. This is called a *confidence interval*. The end points of this interval are called *confidence limits*.

3. State the probability—say, 95 or 99 percent—that such a zone will include the population mean. This probability is called the *confidence coefficient* or *level of confidence*. It must be set in advance. Each confidence interval that may be chosen has an associated probability of including the population mean—the wider the interval, the greater the probability. Thus, the zone $\bar{X} \pm 1.96\sigma_{\bar{x}}$ is the 95 percent confidence interval. This relationship is based on the fact that 95 percent of all sample means tend to fall within $1.96\ \sigma_{\bar{x}}$ of the population mean, where $\sigma_{\bar{x}}$ is the true standard error of the mean. Similarly, the zone $\bar{X} \pm 2.58\ \sigma_{\bar{x}}$ is the 99 percent confidence interval. The zone for any other confidence coefficient may be found in Appendix D. The selection of the appropriate confidence coefficient is discussed on page 238.

For example, we wish to estimate the mean diameter of the population of ball bearings in Table 9–2, which is assumed to be unknown. We take sample No. 1 (column 3) and proceed as above (assume sampling without replacement; all units are in thousandths of an inch):

$$\bar{X} = +.14$$

$$s_{\bar{x}} = \frac{s}{\sqrt{n}} \sqrt{1 - \frac{n}{N}} = \frac{1.81}{\sqrt{50}} \sqrt{1 - \frac{50}{565}} = \frac{1.81}{7.07} (.955) = .244$$

Use this value as an estimate of the true standard error of the mean $\sigma_{\bar{x}}$. The error involved is a minor one for larger samples.

Compute $\bar{X} \pm 1.96\ s_{\bar{x}}$ as the 95 percent confidence interval for the population mean:

$$\bar{X} + 1.96 s_{\bar{x}} = .14 + 1.96(.244) = .14 + .48 = +.62$$
$$\bar{X} - 1.96 s_{\bar{x}} = .14 - 1.96(.244) = .14 - .48 = -.34$$

Our best point estimate of the population mean is therefore the sample mean, $+.14$, but this estimate is subject to a margin of error defined by the 95 percent confidence limits of $+.62$ and $-.34$. This probability statement needs some interpretation. For any particular sample, the confidence interval either includes the population mean or it does not—we do not know. The *objective* probability is either 100 percent or zero (in this case it does, since we know the population mean is zero). Strictly speaking, the statement means that if a very large number of samples of size n are drawn, and the confidence interval is computed for each, 95 percent of these intervals will include the population mean.

On the other hand, using a subjective interpretation of probability, we can make the more straightforward statement that there is a 95 percent chance that the population mean lies within the confidence interval. In other words, one should be willing to bet, with odds of 19 to 1, that the population mean lies in the interval $+.62$ to $-.34$, based only on the sample information.

Chart 9–3 shows the means and confidence limits for this sample

Chart 9–3

95 PERCENT CONFIDENCE LIMITS FOR THE POPULATION MEAN
OBTAINED FROM 6 SAMPLE MEANS OF
BALL-BEARING DIAMETERS ($n = 50$)

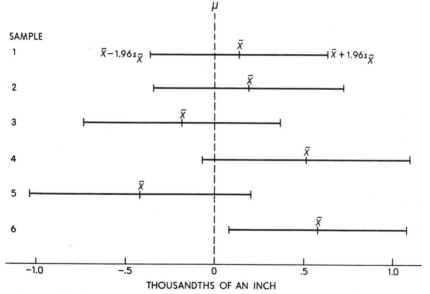

THOUSANDTHS OF AN INCH

Source: Table 9–2 (except sample 6).

and for the other four samples of 50 ball bearings listed in Table 9–2.

The means and intervals all vary, but the latter all include the population mean μ, shown as a dashed line. The confidence interval for a sixth sample, however (not shown in Table 9–2), fails to include the true mean. Of all such possible confidence intervals, then, 95 percent include the population mean.

The confidence interval around a sample mean might be likened to a quoit aimed at a peg—the population mean. Then 95 percent of the quoits will ring the peg. If a bigger quoit is used—say the wider 99 percent confidence interval of $\bar{X} \pm 2.58s_{\bar{x}}$—then 99 percent of the quoits will be ringers.

A 99 percent confidence interval can be computed as $\bar{X} \pm 2.58\ s_{\bar{x}}$ and similarly for any other confidence coefficient, using the table of areas under the normal curve. The 99 percent interval for ball-bearing sample No. 1 is:

$$\bar{X} \pm 2.58s_{\bar{x}} = +.14 \pm 2.58(.244) = +.14 \pm .63$$

Hence, we can say, in subjective terms, that there is a 99 percent chance that the population mean lies between the confidence limits of $-.49$ and $+.77$.

Which Confidence Coefficient Should Be Selected?

Raising the confidence coefficient from 95 to 99 percent increases our degree of assurance that the confidence interval contains the population value, but it also makes our estimate less precise, since the confidence interval itself has been widened by 32 percent (i.e., from 1.96 to 2.58 standard errors). In deciding which confidence level to use, we must understand that the primary purpose of the confidence interval is to report or communicate to others the results of the sample. The confidence interval is a convenient way of expressing the sampling error by giving an interval that is likely to include the population mean. The confidence level chosen therefore is sometimes rather arbitrary. In particular, the 95 percent level is often used in the social sciences, and the 99 percent level in the natural sciences where precision is higher. Other levels should be chosen, however, when we can balance the value of a precise estimate against the cost of missing the true value.

Any economic or business report that cites the mean (or other statistic) of a probability sample should give the reliability of this value in terms of a confidence interval or some other use of $\sigma_{\bar{x}}$ as a measure of the sampling error. For example, a Census Bureau's *Monthly Report on the Labor Force* says, "The chances are about 19 out of 20 that the difference between the estimate and the figure which would have been obtained from a complete census is less than the sampling variability indicated below" (followed by a table showing various sample sizes and the corresponding 95 percent confidence intervals). A statistic having a large sampling error may be useless; at any rate, the error should be stated. The report should also point out that this reliability measure does not include the effect of bias due to nonsampling errors in sample design, incomplete coverage of sample, bias of respondent, etc. These errors should be discussed in qualitative terms.

Errors in Confidence Intervals

The confidence intervals just described may be inaccurate because (1) the standard error of the mean estimated from a single sample is not equal to the true standard error and (2) the sample means may not be quite normally distributed. These errors are appreciable in small samples, but they become insignificant in larger samples. Thus, in the example cited above, increasing the sample size from 50 to 250 reduced the discrepancy in the standard error of the mean from 8 to 2 percent.

PROPORTIONS

The foregoing discussion of statistical inference has been applied to the arithmetic mean. This is an important measure of any variable. It should be noted, however, that many different statistical measures can be submitted to a similar type of statistical inference—medians, standard deviations, and so on. The three essential tools in such analysis are (1) the designated measure as found within the sample, (2) the standard error of the measure involved, and (3) the sampling distribution of the measure.

In this section we apply the principles of statistical inference to the *proportion.* As noted earlier, a proportion represents an *attribute* of a population rather than the average value of a *variable.* This might be the proportion of defective pieces in a lot of bolts produced, the proportion of consumers who plan to buy a color television set, and so on.

It was pointed out in Chapter 3 that a proportion may be considered a special case of the arithmetic mean in which all the values are ones or zeroes. Our discussion about the sampling distribution of means thus applies for the most part to proportions also. In particular, the sample proportion is an unbiased estimate of the population proportion. That is, if all possible random samples of a given size were drawn from a population, the mean of the sample proportions, or the expected value, would equal the population proportion. We will use the symbols p_s and p to denote the proportion of items in the sample and population, respectively, that have a given characteristic. Similarly, q_s and q denote the proportion of items that do not have that characteristic. Hence, $q_s = 1 - p_s$ and $q = 1 - p.$

The Binomial versus the Normal Distribution

The sampling distribution of a proportion (like that of the mean) is the distribution of its values that could be obtained from all possible random samples of size n taken from a population. Sample proportions follow the binomial distribution,[10] though for larger samples (say, when np and nq are above 5) the normal approximation can be used instead, as described in Chapter 6.

We can set up confidence intervals by use of a binomial table, such as Appendix F or G for samples up to 25 in size. For example, suppose we wish to test a sales letter by sending it to 20 households selected at random from a mailing list. We receive five replies, a proportion of .25. What proportion of returns may be expected from the whole list with 95 percent confidence limits? The sample result may produce 0, 1, 2, . . . successes or the equivalent proportions of 0, .05, .10, ... In Appendix F, with $n = 20$, $p = .25$, the values $r = 2$ through 9 include a probability of .962 [with $P(r \leq 1) = .024$ and $P(r \geq 10) = .014$] which is the closest we can get to .95. The confidence limits then are roughly 2 to 9 returns out of 20 letters sent, or 10 to 45 percent for returns to be expected from the complete mailing.

However, statistical inference based on the binomial distribution involves complex technical difficulties, such as those arising from the discreteness of the distribution and the asymmetry of confidence intervals. Further, it is difficult to make a valid inference based on a small sample alone (when the normal approximation cannot be used), without also considering prior information. We will show how to combine prior information and binomially distributed sample data for decision making in Chapter 13. In the present chapter, therefore, we will restrict the discussion to large samples (where np and nq are over 5), so that a nearly normal distribution can be assumed. The analysis is thereby simplified, and the concepts developed for the mean can be carried over and applied directly to the proportion.

The Standard Error of a Proportion

The standard error of a sample proportion is the standard deviation of the p_s's in all samples of the same size that might be drawn from a population. As in the case of the mean, the standard error of a proportion equals the standard deviation of the population divided by the square root of the sample size. In the case of the proportion, however,

---

[10] This is true assuming a very large population, or sampling with replacement. The reader is advised to review Chapter 6 on the binomial distribution and its normal approximation before proceeding.

the standard deviation of the population is $\sigma = \sqrt{pq}$. Hence the standard error of a sample proportion is:

$$\sigma_{p_s} = \sqrt{\frac{pq}{n}}$$

For example, if $n = 100$ and $p = .20$:

$$\sigma_{p_s} = \sqrt{\frac{.20 \times .80}{100}} = \frac{.40}{10} = .04 \qquad \text{or } 4\%$$

Finite Population Correction.   As in the case of the mean, the standard error of a proportion depends more on the absolute size of the sample n than on its relation to the size of population n/N. If the sample makes up a large part of the population, however, the same finite population correction applies as in the case of the mean. The formula is then:

$$\sigma_{p_s} = \sqrt{\frac{pq}{n}} \sqrt{1 - \frac{n}{N}}$$

Thus, if the whole lot or population had a size of only $N = 500$ in the above example, we would have

$$\sigma_{p_s} = \sqrt{\frac{.20 \times .80}{100}} \sqrt{1 - \frac{100}{500}}$$
$$= .04 \times .9 = .036 \qquad \text{or } 3.6\%$$

The Confidence Interval for a Proportion

Suppose that the management of a large grocery chain is interested in estimating what proportion of its customers would prefer a self-service display of prepackaged meat to a meat counter serviced by a butcher. The market research department is assigned to make a study leading to such an estimate.

A random sample of 400 customers is taken, and it turns out that 220, or 55 percent, are in favor of the self-service display. It is extremely unlikely that the population constituting *all* customers would divide in preference exactly in this proportion. How, then, do we estimate the interval in which the true proportion falls with, say, a 95 percent degree of confidence? The analytical principles are the same as those used in constructing confidence intervals for the arithmetic mean. Only the measures are altered to fit the present case.

The standard error of a proportion, as we saw a moment ago, ideally requires the population value of p for its calculation. This we do not know, or we would not be faced with the problem of estimating the interval within which it falls. The common practice is to assume that p has the value of p_s found in the sample and to make the substitution accordingly. Hence, the estimated standard error for the sample proportion is:[11]

$$s_{p_s} = \sqrt{\frac{p_s q_s}{n}}$$

$$= \sqrt{\frac{.55 \times .45}{400}}$$

$$= .0249 \text{ (rounded to .025)}$$

Using the normal distribution (since np_s is well over 5), the 95 percent confidence interval is $p_s \pm 1.96 s_p$, or about two standard errors on each side of .55. Therefore, we are 95 percent confident that the true proportion of customers favoring self-service meat counters lies somewhere between 50 and 60 percent.

As in the case of the arithmetic mean, and for the same general reasons, we could construct intervals of varying degrees of confidence, based upon appropriate multiples of the standard error of the proportion laid off around the value for p_s observed in the sample.

HOW BIG SHOULD A SAMPLE BE?

In planning a sample survey, is it necessary to sample 100 items? 1,000? Or all we can afford? The answer depends mainly on two factors: (1) the economic value of the information contained in the sample and (2) the cost of sampling. The value of sample information and the cost of the sample both increase as sample size increases. The optimum sample size is that which balances the cost and value of the sample. Determination of optimum sample size is discussed in Chapter 14. In this section we will discuss two related questions: (1) How large a sample is needed to obtain a given degree of precision in the sample estimate? and (2) How do we balance sample precision against the cost of sampling?

The Mean

The relation between precision of the sample mean and size of sample is:

---

[11] The formula shown is the one almost universally used, although it is biased. An unbiased estimator would have $n - 1$ in the denominator instead of n. However, for large samples, the difference is trivial. See W. Cochran, *Sampling Techniques* (2d ed.; New York: John Wiley, 1963), p. 33.

$$\sigma_{\bar{x}} = \frac{\sigma}{\sqrt{n}}$$

ignoring the finite population correction for simplicity.

To estimate how big n should be, there are three steps:

1. Determine how small the standard error of the mean $\sigma_{\bar{x}}$ must be in order to obtain the necessary precision. The precision depends on how the results are to be used.
2. Take a random sample of any convenient size and compute the sample standard deviation s as an estimate of σ, the population standard deviation.
3. Substitute the desired value of $\sigma_{\bar{x}}$ and the estimated σ in the above equation and solve for n. This size of sample will give the necessary precision. If a larger sample is then taken, its standard deviation can be used to provide a revised estimate of σ and hence $\sigma_{\bar{x}}$.

The size of the population is usually a negligible factor, as pointed out earlier. However, if the sample makes up more than about 5 percent of the population, the finite population correction should be applied to the above equation.

As an example, suppose it is desired to estimate the population mean of ball-bearing diameters within .3 thousandths of an inch at the 99 percent confidence level (i.e., 2.58 $\sigma_{\bar{x}} = .3$ thousandths). Take a sample of convenient size and compute s as an estimate of σ, for example, sample No. 1 in Table 9–2, where $n = 50$ and $s = 1.81$.

First, determine the desired $\sigma_{\bar{x}}$:

$$2.58\sigma_{\bar{x}} = .3$$

$$\sigma_{\bar{x}} = \frac{.3}{2.58} = .116$$

Now, substitute these values in the equation $\sigma_{\bar{x}} = \sigma/\sqrt{n}$ and solve for n:

$$.116 = \frac{1.81}{\sqrt{n}}$$

Transposing,

$$\sqrt{n} = \frac{1.81}{.116} = 15.6$$

Squaring both sides,

$$n = 244$$

Therefore, a sample of 244 ball bearings (including the original 50) should be taken. Actually, in this example somewhat less than 244 would suffice, since 244 is a significant part of the total population of 565 ball bearings and the finite population correction should be applied. In general, however, when we are sampling from large populations the finite population correction can be ignored.

The cost of a survey includes a constant factor—for setting up the project, overhead, etc.—and a variable factor—so much per item sampled. Suppose it costs $300 to set up the ball bearing inspection and $1 per measurement. Then the total cost $C(n)$ in dollars is:

$$C(n) = 300 + 1n$$

The executive can then compare the cost with the precision of the sample result for various possible sizes of sample, in order to choose among them. Thus, for the ball-bearing example:

n	$s_{\bar{X}}$*	Cost
50	.256	$350
250	.127	550

* In thousandths of an inch.

Since the cost increases directly with the size of sample, and reliability increases only with the square root of sample size, there are diminishing returns, and at some point the slight increase in reliability will not justify the added cost of sampling.

The reliability and cost of a survey depend not only on the size of sample but also on the sampling plan itself. The principal plans are discussed in Chapter 12. For example, instead of a simple random sample, the reliability of a given-sized sample can be increased by stratification, or the unit cost can be reduced by cluster sampling.

Proportions

The size of a simple random sample needed to reduce the standard error to any desired level can be computed for a proportion in the same way as with the mean. Suppose we wish to determine the proportion of customers preferring self-service in the grocery example with a sample standard error of only .02, or two percentage points. This corresponds to 95 percent confidence limits of $p_s \pm 1.96(.02)$ or $p_s \pm .04$. From the trial survey cited above, p is tentatively .55. Then we solve for n in the equation $s_{p_s} = \sqrt{(p_s q_s)/n}$, as follows:

$$.02 = \sqrt{\frac{.55 \times .45}{n}}$$

Transposing,

$$\sqrt{n} = \frac{\sqrt{.55 \times .45}}{.02} = \frac{.4975}{.02} = 24.9$$

Squaring,

$$n = 620$$

It is necessary to sample about 620 customers (or 220 in addition to those already sampled), therefore, in order to obtain a value of p_s that has a standard error of only .02. The increase in precision with larger samples can be balanced against the increased cost, as in the case of the mean above.

Taking Several Samples

Instead of setting the size of a single sample in advance, we can take several smaller samples. Thus, if a shopper inspects a basket of apples in a grocery store, she may find all visible apples perfect and take the basket, or note several rotten ones and reject it out of hand. But if only one or two seem doubtful, she may probe further before deciding. This commonsense notion is supported by sampling theory. That is, if a small sample shows very good or very bad results, we can make a decision forthwith, and only in the borderline case is more sampling necessary.

As an example, *acceptance sampling* is a procedure for sampling a lot of goods in order to determine whether to accept it as conforming to standards or to reject it. A purchaser may wish to sample the quality of a shipment of goods received, or a manufacturer may submit his own output to acceptance sampling at various stages of production. Acceptance sampling plans include single sampling, double sampling, and sequential sampling.

The single-sampling plan specifies the sample size and the number of defective units in the sample that will cause the entire lot to be rejected. This procedure has been described above.

In a double-sampling plan, a smaller sample can be taken to begin with. If it contains a specified number c_1 or fewer defective units, the lot is immediately accepted; if it contains more than c_2, a larger number, the lot is rejected. In the intermediate case, however, a second larger sample is taken. Then, if the combined number of defectives in the two samples is c_2 or less, the lot is accepted; otherwise, it is rejected. Double sampling is preferable to single sampling in reducing the total amount of inspection on very good or very poor lots that can be judged promptly on the first sample with a known probability of making a mistake. It also has the psychological advantage of giving a tentatively rejected lot a second chance. When many second samples are required, however, double sampling may be more complicated and expensive than single sampling.

In sequential sampling, the size of sample is not determined in advance. Instead, a decision is made after each observation or group of observations to (1) accept, (2) reject, or (3) suspend judgment and continue sampling until a decision is ultimately reached. Sequential methods permit reaching a decision on the basis of even fewer observations than other plans in the case of very good or very bad lots, but the procedure may be complex in operation.

Sequential sampling is also used in statistical quality control, in which samples of only four or five items are taken in sequence during a manufacturing process and the arithmetic means of some measurement are plotted on a chart. Thus, if a mean falls outside the tolerance limits, the machine can immediately be stopped and corrected before it turns out more defectives. This subject is illustrated in Chapter 10.

SUMMARY

Statistical inference is the process of making a generalization or prediction about a population value, or parameter, based on a sample value, or statistic. This may be a single-valued point estimate, or a range of values designated as an interval estimate. The process is first described for the mean of a simple random sample.

If all possible means of large samples are drawn from a population, the sampling distribution tends to follow a normal curve. The proportion of items that fall within a given area under the normal curve may be determined from Appendix D. This proportion represents relative frequencies, or the probability that a single item (e.g., a sample mean) will fall within the segment.

An experiment is presented to show how sample means cluster about the population mean—the cluster being closer and hence the precision greater for larger samples. The sampling distribution of the mean must be clearly distinguished from the distribution of individual values in the population or the somewhat similar distribution of individual values in the sample itself (Chart 9–1). The tendency of the sampling distribution of the mean to form a normal curve as n increases in size, whatever the type of population, is called the central limit theorem.

The sample mean is said to be an unbiased estimator of the population mean because its expected value equals the population value. The expected value is the mean of a distribution of all possible means for a given size of sample. The sample mean is also said to be efficient because its sampling distribution usually clusters more closely about the population value than does, say, the median.

The standard error of the mean (i.e., the standard deviation of all possible sample means) measures the precision of the sample estimate. It is related to the population standard deviation and the sample size as follows: $\sigma_{\bar{x}} = \sigma/\sqrt{n}$. However, since σ is usually unknown, the standard error of the mean can be estimated from the standard deviation of a single large sample by the formula $s_{\bar{x}} = s/\sqrt{n}$. This expression should be multiplied by $\sqrt{1 - n/N}$, the finite population correction, if the sample size n is more than about 5 percent of the population size N.

Since sample means are normally distributed, the probability is 68 percent that a single sample mean will fall within the interval $\mu \pm \sigma_{\bar{x}}$. The probability for any other intervals can be found in Appendix D.

We can estimate that the population mean falls within a certain confidence interval, based on the sample mean and standard deviation, with a predetermined probability—say, 95 or 99 percent—of being correct. Thus, $\bar{X} \pm 1.96 \, \sigma_{\bar{x}}$ is the 95 percent confidence interval for the mean; that is, if we state that the population mean falls within this zone, we will have a 95 percent chance of being correct. We can increase the confidence coefficient—say to 99 percent—but only at the cost of making the estimate less precise by widening the confidence interval. The choice depends on the problem. In any case, the confidence interval and coefficient should be stated in reporting the results of sample surveys.

Inferences may be made about sample proportions in much the same way as with means. In fact, a proportion may be considered a special case of a mean in which the attributes, such as defectives and nondefectives, are valued 1 and 0, respectively, and averaged to find the percentage defective.

The standard error of a proportion is $\sigma_{p_s} = \sqrt{(pq)/n}$, where p is the population proportion and $q = 1 - p$. This is estimated as $s_{p_s} = \sqrt{(p_s q_s)/n}$ when sample values are used.

The sampling distribution of p_s follows a binomial distribution, but for large samples (say, when np and nq are greater than 5) the distribution is approximately normal, so we assume normality here both because it is valid for most practical problems and because it is simpler than using the binomial distribution.

A 95 percent confidence interval may then be laid out around the sample proportion (i.e., $p_s \pm 1.96s_{p_s}$) to include p, the population proportion, with a 95 percent chance of being correct. Other degrees of confidence are handled similarly.

The size of a sample can be determined by solving the equation $\sigma_{\bar{x}} = \sigma/\sqrt{n}$ for n in estimating the mean, where $\sigma_{\bar{x}}$ measures the required precision, and σ is estimated from a trial sample. Since precision increases with $\sqrt{n}$ and the cost of sampling increases with n, the precision and cost should be contrasted for several sizes of samples, as an aid in determining sample size. Similarly, for a proportion, the size of sample needed to reduce the standard error s_{p_s} to any desired value can be obtained by solving for n in the formula $s_{p_s} = \sqrt{(pq)/n}$, using an estimated value of p. The question of optimal sample size is discussed further in Chapter 14.

Instead of a single sample, two or more smaller samples can be used, as in acceptance sampling. This procedure has the advantage of promptly signaling a decision whether to accept or reject a shipment or stock of material in case the quality is very good or bad. Further sampling is required only in the intermediate case.

PROBLEMS

1. Explain the following concepts:
   a) Bias versus sampling error.
   b) Sampling distribution of the mean.
   c) Central limit theorem.
   d) Standard error of the mean.
   e) Confidence interval for the mean.

2. Explain:
   a) How to minimize bias in sampling.
   b) The concept of the proportion as a special case of the mean.
   c) The relation between the distribution of proportions and the normal distribution.
   d) A 90 percent confidence interval for a proportion.

3. You are employed by a concern that has just received a shipment of 80 sheets of ⅛-inch insulating board for use in the manufacture of power transformers. You are directed to check a random sample of these boards for thickness, using a zero- to one-inch micrometer. Thickness is a major characteristic affecting the quality of the board, and consequently, the quality of the transformer. The actual measurements of all 80 sheets are shown below. The sheets are numbered from 01 to 80, reading down the columns as indicated by the column headings.

THICKNESS OF 80 SHEETS OF ⅛-INCH
INSULATING BOARD
(In Thousandths of an Inch)

01–10	11–20	21–30	31–40	41–50	51–60	61–70	71–80
123	125	128	125	125	124	126	124
122	123	127	121	125	125	125	123
125	125	125	122	125	124	127	123
122	128	125	123	125	123	125	124
127	124	125	124	124	125	127	125
123	123	124	121	125	126	119	124
127	124	124	123	127	122	125	128
121	123	123	121	119	127	125	125
125	124	128	119	125	125	124	127
122	123	128	124	118	127	125	123

a) Take a random sample of five sheets, using the two-digit random numbers in Table 9–1 and following the procedure described in the text. For example, if the first number selected at random is 43 (row 6, column 8), select sheet 43 with thickness 125; then proceed in any direction in Table 9–1, discarding duplicates and numbers over 80.

b) What is the mean thickness of your sample? This is an unbiased estimate of the mean of the whole shipment. Since this proved to be 124.24 thousandths of an inch, what is the sample error?

c) Find the standard error of the sample mean.

d) You can probably get a closer estimate of the mean thickness of the whole shipment by sampling more sheets. Continue your sampling, therefore, until you have added 5 sheets, for a total of 10 in all, using the same method as before.

e) What is the mean thickness of your larger sample of 10 sheets? What is its error?

f) On the average, how much would you expect to reduce the error of the sample mean by taking a sample of 10 rather than of 5 sheets? (Ignore finite population correction.)

4. As publication manager of a weekly magazine, you wish to develop guaranteed circulation figures to use in soliciting advertising. A study of the copies sold each week during the last few years fails to reveal any marked trend or seasonal movement, the circulation figures tending to fall into a fairly normal distribution pattern. During this period, the mean circulation was 556,000 copies, and the standard deviation was 8,000 copies.

a) If you guaranteed the sale of at least 552,000 copies of next week's issue, what chance would you take of failing to reach this figure?

b) How many copies of your next issue would you print if you wanted to avoid more than one chance in a thousand of being caught short?

c) What average (mean) circulation figure would you be 99 percent safe in guaranteeing for a prospective advertiser over the next year (52 weeks)? Compare this result with that in (*a*) above.

5. *a*) A machine, when in adjustment, produces parts that are normally distributed and have a mean diameter of .300 inches with a standard deviation of .040 inches. If the machine is in adjustment, what is the probability that the mean value of a random sample of four parts will fall between .290 and .304 inches?

 b) What would happen to the standard error of the mean if we increased the sample size from 4 to 16?

6. A population is known to have a mean $\mu = 85$ and a standard deviation $\sigma = 15$.

 a) What is the probability that the mean of a sample of size 36 will fall in the interval 83 to 87?

 b) What is the probability that the mean of a sample of size 81 will fall in the interval 83 to 87?

 c) How large a sample is needed to be 95 percent sure that the sample mean will fall in the 83 to 87 interval?

7. A mill produces cedar shingles that average 4.0 millimeters in thickness at the tapered end. The standard deviation of the process is .2 millimeters. The thickness of a sample of four shingles is measured every hour. The sample mean is used to determine whether the manufacturing process is operating satisfactorily, as follows: If the mean thickness of the four shingles is 3.7 millimeters or less, or 4.3 millimeters or more, the machine is stopped and readjusted. If the mean is between 3.7 and 4.3 millimeters, the process is continued.

 a) What is the probability that the machine will be readjusted after any given sample test, if the process average remains at 4.0 millimeters?

 b) What is the probability that the machine will be readjusted if the process average were to shift to 4.2 millimeters? To 3.9 millimeters?

 c) What is the probability of continuing the process after any given sample if the process average shifts to 4.3 millimeters? To 3.4 millimeters? To 4.2 millimeters?

8. "A sample of 40 from a population of 400,000 will give nearly as precise an estimate of the population mean as a sample of 40 from a population of 4,000, provided the standard deviations of the populations are the same." Is this statement reasonable? Give figures to support your answer.

9. A random sample of 64 is drawn from the records of daily output of a large group of employees in order to estimate the population mean. The sample shows a mean of 136 units and a standard deviation of 24 units. Calculate a 98 percent confidence interval for the mean output of all employees.

10. A random sample of 400 accounts receivable is selected from the 2,000 accounts due a firm. The sample mean is found to be $165.50, with standard deviation of $26.00. Set up a 95 percent confidence interval as an estimate of the population mean. Interpret the meaning of this interval.

11. A certain company employs 400 executives. A sample of 36 is taken in order to estimate the average age of all executives. The results of the

sample are $\overline{X} = 51.0$ and $s = 4.0$ years. Calculate a 99 percent confidence interval for the mean age of all executives.

12. A random sample of 324 sales made during the year in a department store has a mean size of $10.50 and a standard deviation of $2.70. Total number of sales is not known.

    a) Construct a 95 percent confidence interval for the average size of all sales made during the year.

    b) Construct an 80 percent confidence interval for the above.

13. A random sample of 225 orders from a batch received by a certain firm has an average size of $12.74 and a standard deviation of $2.45. Construct a 95 percent confidence interval for the average size of all orders (625) received in this batch.

14. How large a sample would be needed to estimate the mean life of a new type of incandescent lamp within 24 hours, with no greater risk than 1 chance in 20 of being wrong. The standard deviation of burning life is estimated at 200 hours.

15. *a*) The planning commission in a city wished to estimate the mean number of inhabitants per dwelling unit in the city. It selected a simple random sample of 500 dwelling units and obtained the following results: $n = 500$, $\Sigma X = 2,200$, $\Sigma X^2 = 11,680$, where X is the number of inhabitants in a dwelling unit. Calculate a 95 percent confidence interval for the mean number of inhabitants per dwelling unit in the city.

    b) Suppose that there were 10,000 dwelling units in the city. Set up a 95 percent confidence interval for the total population of the city. (Hint: A population total can be estimated as $N\overline{X}$ and the standard error of this estimate as $Ns_{\overline{x}}$.)

16. A random sample of 81 out of the 225 graduating seniors of a college received an average starting salary of $900 a month, with a standard deviation of $117. Give a 90 percent confidence interval for the mean starting salary for all 225 graduating seniors.

17. *a*) What is the standard error of a proportion for samples of size 100, if $p = .1$? If $p = .8$? If $p = .5$? For what value of p is the standard error the greatest? The least?

    b) What is the standard error of the proportion, if $p = .2$ and $n = 100$? If $n = 400$? Just how does the size of sample affect the standard error?

    c) Can we use the normal distribution for making inferences about proportions if $p = .04$ and $n = 25$? If $p = .15$ and $n = 60$? Explain.

    d) Compute the standard error of the proportion if $p = .1$, $n = 100$, and the population size $N = 400$. Is this standard error larger or smaller than if N is infinitely large?

18. A survey of consumer buying plans reveals that 10 percent of a sample of 2,500 families plan to buy a new refrigerator during the next year. Assume that an unbiased simple random sample was used. Set up a 99 percent confidence interval to estimate total refrigerator sales for the whole population of 50 million families. Interpret this forecast.

19. If, in a sample of 600 economics students drawn from schools throughout the country, 360 are sons of businessmen, what is the 90 percent confidence interval for the proportion of all economics students who are sons of businessmen?

20. You wish to make a market survey to estimate the proportion of housewives who prefer your new product to competitors' products. You would like the error in estimating the proportion to be no greater than four percentage points, with a confidence coefficient of 95.45 percent. The sales department offers a preliminary guess that about 20 percent of housewives might prefer your product. If the survey costs $500 to set up, and $5 an interview, about how much should the whole survey cost?

21. In a work sampling study, a machine operator is observed at 100 moments of time selected at random during the workweek. He is found to be doing productive work in 80 of these observations.
    a) Construct a 95 percent confidence interval for the proportion of time that the operator is doing productive work. Interpret this result.
    b) How many observations are needed to determine the true proportion of productive time during the workweek within five percentage points at the 99 percent confidence level?
    c) If the operator had been productive 70 percent of the time, would the sample size in question (b) have to be larger or smaller? Why?

22. A survey is planned to determine the average annual family expenditures for medical expenses of employees in a given company within $50, at the 90 percent confidence level. A pilot study provides an estimate of $334 as the standard deviation of medical expenditures. How large a random sample is needed to yield an estimate with the necessary precision?

23. The controller of a department store takes a sample of 64 monthly statements to be mailed to credit-card holders and finds that the average amount owed is $28, with standard deviation of $12. How many accounts should he sample, in total, if he wishes to estimate the mean amount owed within $1, with only 1 chance in 20 of being outside that range?

24. A manufacturer wishes to estimate the mean tensile strength of angle irons received from a supplier. The standard deviation for samples tested in past shipments has been about 10 pounds. How many should be tested so that there will be no more than one half of 1 percent chance that the error will exceed 2 pounds?

25. Past experience indicates that the standard deviation of the amount of gasoline consumed per year by motorists in a certain area is 50 gallons. How
large a sample must be taken for the estimate of the true mean consumption to have a .99 probability of being within 10 gallons of the actual true
mean?

26. A survey is planned to measure the amount of time that children listen to
television. A preliminary check indicates a mean listening time per week
of about 15 hours with a standard deviation of 5 hours. It is desired to
estimate the mean listening time per week within a half hour, at the 99
percent confidence level.
   a) If the overhead cost of the survey is $500, plus $10 per interview, what
   total cost should be budgeted for the survey?
   b) After the survey is completed, the mean is found to be 18 hours and
   the standard deviation 6 hours. What additional cost, if any (excluding
   overhead), should be budgeted to get a revised estimate of mean listening time, in the light of this new information?

27. The market research department of a certain company was allocated $40,000
to make a survey on the potential sales of a new product. A sample of stores
through which the company distributed its product was to be selected. The
new product was to be introduced in this sample of stores and the sales
noted over a period of three months. The average sales per store per month
would then be used to estimate the total sales potential of the new product.
   Suppose that it costs $10,000, plus $300 per store, to conduct the sample. From past experience with similar products it is estimated that the
standard deviation of sales per store per month is 68 packages of the product.
   a) How large a sample can be taken for the amount allocated? What
   sampling error in the estimate of average sales per store per month can
   be expected?
   b) Suppose that actually a sample of 80 stores was selected. In these stores,
   the average sales per store per month was 84 packages and the standard
   deviation of the monthly sales for the stores was 52 packages. Using
   these estimates, make an estimate of the total *annual* sales of this product
   if it were to be distributed through 80,000 stores. Calculate a 95 percent confidence interval about this estimate. (See hint given in Problem
   15 above.)
   c) What probability would you assign to the possibility that estimated
   total annual sales was off by more than eight million packages? By
   more than five million packages?

28. A large appliance manufacturer needs a current estimate of the retail sales
of his appliances as an aid in production planning. Accordingly, he plans
to take a random sample of retail outlets and obtain sales on a monthly
basis. To aid in planning the survey, a preliminary sample of 60 retail outlets is selected. The results are $n = 60$, $\Sigma X = 1,104$, $\Sigma X^2 = 22,034$, where
X is the appliance sales (units) by store in the past month.

    a) The manufacturer desires that the survey estimate of the mean sales per store be accurate within ± 1 appliance at the 95 percent level. How large must the total sample size be to achieve this precision?

    b) The cost of the survey is estimated at $2,000 plus $40 per store sampled. What is the cost of the survey designed in part (*a*)?

    c) Assume that the manufacturer distributes through 28,000 retail outlets. What will be the sampling error associated with the estimate of total monthly sales of appliances? (See hint given in Problem 15 above.)

29. The consumer research division of an automobile manufacturing firm has a budget of $3,000 for a survey to determine the proportion of consumers who prefer a new design for the radiator grill. The estimate should be correct to within five percentage points, with a 95 percent confidence coefficient. Assume a simple random sample. Cost of the survey is $1,000 for overhead plus $5 an interview.

    Can this proportion be estimated with the required precision for $3,000, assuming $p = .50$? Explain.

30. A television distributor finds that about 22 percent of the potential customers who enter his store buy a television set. Moving to another city, he wishes to estimate this percentage for the new location within ± 4 percent, at the 90 percent confidence level. How many observations should he take?

31. As a ginger ale producer entering a new territory, you need an estimate of the consumers who prefer to buy ginger ale in cans. A consulting firm agrees to make a survey of ginger ale buyers for $2,000, plus $4 an interview. Let $p = .50$, as a guess, and assume a simple random sample.

    a) How much will the survey cost if the error in estimating the proportion is to be no greater than five percentage points at the 90 percent confidence level?

    b) How much will the survey cost if the error is not to exceed five percentage points at the 98 percent confidence level?

SELECTED READINGS

Selected readings for this chapter are included in the list that appears on page 290.

10. TESTS OF HYPOTHESES

WE CAN MAKE a statistical inference either by estimating that the population mean (or other parameter) lies within a certain *confidence interval* or by *testing a hypothesis*. The sampling error $\sigma_{\bar{x}}$ is used in either case. Confidence intervals were considered in Chapter 9. In testing a hypothesis we first set up a hypothesis concerning the true population value of the mean μ, or some other parameter. Then we decide on the basis of a sample whether to accept or reject this hypothesis. If the sample value is close to the hypothetical value, we accept the hypothesis; otherwise we reject it.

In the "classical" theory of statistical inference described in this chapter, one makes a decision either to accept or reject a hypothesis on the evidence of sample information alone. In Chapters 13 and 14 we will extend the analysis to include the judgment of the decision maker and the economic payoffs involved, using the Bayesian approach to arrive at the optimal decision.

The test of hypothesis approach is also useful in business and the social sciences for *reporting* purposes. In this sense, it serves to describe the sampling error associated with a given sample and to describe how likely it is that the sample result could have occurred by chance alone.

An Example

Consider a specific example. In the manufacture of safety razor blades the width is obviously important. Some variation in dimension must be expected due to a large number of small causes affecting the production process. But even so, the average width should meet a certain specification. Suppose that the production process for a particular brand of razor blade has been geared to produce a mean width of .700 inches. Production has been underway for some time since the cutting and honing machines were last set, and the production manager wishes to know whether the mean width turned out is still .700 inches, as intended.

This may be treated as a problem in statistical inference. It would be possible, of course, actually to measure all of the hundreds of thousands of blades turned out and to ascertain the mean width directly. But this would be expensive and very time-consuming. A better alternative would be to reason in terms of a sample. The statistical population of blade widths covers *all* the blades coming from the production line in the future under given technical controls. Since the production process was initially set up to give a mean width of .700 inches, the statistical hypothesis is posed that the true mean of this population is .700 inches. But the process could have gotten a little out of line, and management wishes to know whether .700 inches is still the mean width of all blades.

Accepting the Hypothesis. We have posed the hypothesis that the mean width of razor blades is .700 inches; that is, $\mu_h = .700$, where μ_h is the hypothesized mean. The hypothesis seems reasonable since the machine was adjusted to this width. Suppose we draw a simple random sample of 100 blades from the production line. We measure each of these carefully and find the mean width of the sample to be .7005 inches. The standard deviation in the sample turns out to be .010 inches. That is,

$$n = 100$$
$$\bar{X} = .7005 \text{ inches}$$
$$s = .010 \text{ inches}$$

For the hypothesis $\mu_h = .700$ to be true, the sample mean $\bar{X} = .7005$ inches would have to be drawn from the sampling distribution of all possible sample means whose overall mean is .700 inches.

Now, the important question arises: If the true mean of the population *really* were .700 inches, how likely is it that we would draw a random sample of 100 blades and find their mean width to be as far away as .7005 inches or farther? In other words, what is the probability that a value could differ by .0005 inches or more from the population mean *by chance alone?* If this is a high probability, we can accept the hypothesis that the true mean is .700 inches. If the probability is low, however, the truth of the hypothesis becomes questionable.

To get at this question, compute the standard error of the mean from the sample:

$$s_{\bar{x}} = \frac{s}{\sqrt{n}} = \frac{.010}{\sqrt{100}} = .001 \text{ inches}$$

Since the difference between the hypothetical mean and the observed sample mean is .0005 inches, and the standard error of the mean is .001 inches, the difference is equal to .5 standard errors. By consulting Appendix D, we find that the area within this interval around the mean of a normal curve is .19 × 2 = 38 percent, so that 100 − 38 = 62 percent of the total area falls *outside* this interval (see dashed lines in Chart 10–1). If .700 inches were the true mean, therefore, we

Chart 10–1

SAMPLING DISTRIBUTION OF MEANS OF
RAZOR-WIDTH SAMPLES OF SIZE 100
(Hypothetical Mean = .700 Inches)

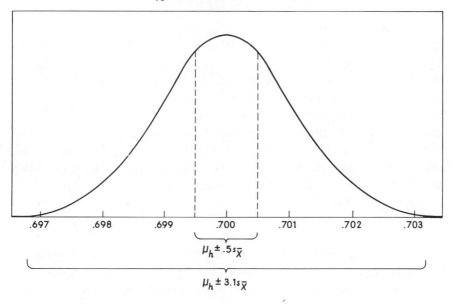

should nevertheless expect to find that about 62 percent of all such possible sample means would, *by chance alone,* fall as far away as $.5s_{\bar{x}}$ or farther. Therefore, the *probability* is 62 percent that our particular sample mean could fall at least this far away.

Remembering that we had substantial reason to accept the hypothesis in the first place—the process having been adjusted to yield a population mean of .700 inches—we should continue to hold to the hypothesis and attribute to mere chance the appearance of a .7005-inch mean in a single random sample of 100 blades.

Rejecting the Hypothesis. Later, after production has gone on for some time, the query again arises: Is it reasonable to believe that the true mean width of blades produced remains .700 inches? Since the process was adjusted to yield that figure, the hypothesis still seems reasonable. We could then test it by taking another random sample of 100 blades. This time the standard deviation is still .010 inches, so the standard error of the mean is still .001 inches, but the mean is now .7031 inches.

In order to test the hypothesis that the true mean of the population is .700 inches, we again go through the same line of reasoning. If the true population mean really were .700 inches, how likely is it that we should draw a random sample of 100 blades and find their sample mean to be as far away as .7031 inches?

Since the difference between the hypothetical mean of .700 inches and the actual sample mean of .7031 inches is .0031 inches, and the standard error of the mean is .001 inches, the difference is equal to 3.1 standard errors of the mean (i.e., .0031/.001 = 3.1). This is the standard normal deviate z introduced in Chapter 6.

Now, if .700 inches really were the population mean, we know from Appendix D that 99.8 percent of all possible sample means, for random samples of 100, would fall within 3.1 standard errors around .700 inches (see wide bracket in Chart 10–1). Hence, the probability is only one fifth of 1 percent that we would get a sample mean falling as far away as ours does.

We have two choices:

1. We may continue to accept the hypothesis (i.e., leave the production process alone), and attribute the deviation of the sample mean to chance. But this is an exceedingly slim chance—one fifth of 1 percent.
2. We may reject the hypothesis as being inconsistent with the evidence found in the sample (hence, correct the production process).

In this case, if we decide on the sample information alone, we would probably make choice (2) and conclude that the mean width of blades from that production line was not really .700 inches. We would reject the hypothesis as being inconsistent with the evidence found in the sample. We would then be wrong only when the hypothesis was actually true and by chance alone a sample mean fell as far away as 3.1

standard errors. But on the average this would occur only twice in 1,000 times.

The Choice between Accepting and Rejecting the Hypothesis. Ultimately, in our example, the choice between letting the production process alone and the alternative of stopping the process to make adjustments depends upon other factors in addition to the sample evidence. The cost of incorrectly stopping the process and the cost of allowing a faulty process to continue are certainly relevant. In addition, the past history of this manufacturing process also influences the choice. If the process rarely goes out of adjustment, we would be more inclined to attribute a far-out sample mean to chance than we would if the process frequently went out of adjustment. The problems of incorporating prior judgment and economic losses are discussed in Chapter 13.

The hypothesis testing analysis, however, is itself helpful. It deals with the evaluation of the sample and the conclusions that may be drawn from that evidence alone. Accepting the hypothesis means that the sample evidence is not in disagreement with the hypothesis. Rejecting the hypothesis means that the evidence is strongly against the hypothesis.

A legal analogy may help in understanding the reasoning involved. In a sense, the hypothesis is on trial and is considered innocent until proved guilty. The evidence is found in the random sample. Before the hypothesis is condemned, the evidence must prove it guilty—not with absolute certainty, but beyond reasonable doubt. The particular form which the evidence takes is the probability that a value as different as the sample mean could have been drawn if the hypothesis were true. If this probability is high, we can accept the hypothesis. On the other hand, if this probability is low, the hypothesis is doubtful. The lower the probability, the progressively greater is the doubt that the hypothesis could be correct. Finally, if the probability is so low that it appears unacceptable to believe that a value as different as the sample mean could have arisen solely by chance, the hypothesis is rejected. It is judged guilty beyond reasonable doubt.

In the first example just considered, the probability was quite high (62 percent) that a discrepancy of .0005 inches could be attributed to mere chance. Therefore, we accepted the hypothesis, particularly since we had pretty good reason to believe in it before the sample was drawn. We could easily view the hypothetical mean of .700 inches as compatible with the findings of the sample and the operations of chance. But in the second example given ($\bar{X} = .7031$ inches), the probability was so low (one fifth of 1 percent) that such a large difference could arise by chance, that the hypothesis ($\mu_h = .700$ inches) was rejected as being untrue.

It is important to note that while rejection of a hypothesis implies that the hypothesis is false, *acceptance of a hypothesis does not necessarily prove that the hypothesis is true.* It may be that the hypothesis is in fact false (i.e., that the true mean μ differs slightly from μ_h) but the sample does not have sufficient precision (i.e., the sampling error is too large) to be able to detect the difference. We shall examine this possibility in more detail shortly.

TYPE I AND TYPE II ERRORS

Understandably, the question can be raised: What critical value should we select for the probability of getting the observed difference $[z = (\bar{X} - \mu_h)/\sigma_{\bar{x}}]$ by chance, above which we should accept the hypothesis and below which we should reject it? This value is called the *critical probability* or *level of significance,* and is denoted α (alpha). The answer to this question is not simple, but to explore it will throw further light on the nature and logic of statistical inference.

Only four possible things can happen when we test a hypothesis. We may be wrong because we:

1. Reject a hypothesis that is really true (a Type I error), or
2. Accept a hypothesis that is false (a Type II error).

Or, we may be right because we:

3. Accept a true hypothesis, or
4. Reject a false hypothesis.

The types of errors noted as possibilities 1 and 2, respectively, are known either as Type I and Type II errors or as errors of the first kind and errors of the second kind.

Type I Errors

In a long run of cases in which the hypothesis is in fact *true* (although we do not know it is true, for otherwise there would be no need to test it), we will necessarily either be wrong as in 1 or right as in 3. That is to say, if we make an error it will have to be Type I. Suppose we should adopt 5 percent as the critical probability, accepting the hypothesis when the probability of getting the observed difference by chance exceeds 5 percent and rejecting the hypothesis when this probability proves to be less than 5 percent. This amounts to the decision to accept the hypothesis when the discrepancy of the sample mean is less than

1.96 standard errors (i.e., $z < z_a$) and to reject the hypothesis when the discrepancy is more than 1.96 standard errors. Using this value as the critical probability, we would expect to make a Type I error 5 percent of the time. This is because even when the hypothesis is true, 5 percent of all possible sample means still lie farther away than 1.96 standard errors. And whenever by chance we get one of these, and the hypothesis is true, we would make the mistake of rejecting a true hypothesis.

Or, we might choose 1 percent as the critical probability, which would correspond to a discrepancy between hypothesis and sample mean equal to 2.58 standard errors. When the hypothesis is in fact true, only 1 percent of all possible sample means would lie farther away than 2.58 standard errors. We would make a Type I error only when by chance alone we happened to draw one of these. Which is to say, we would now make an error of the first kind only 1 percent of the time.

Clearly, then, the proportion of cases in which we would make an error of the first kind, that of rejecting a true hypothesis, can be made as small as we wish simply by reducing the value for the critical probability. In fact, the percentage of cases in which we would expect to make an error of the first kind is *precisely equal* to the critical probability adopted.

Just Significant Probability Level. In many studies, the critical probability is used to describe the statistical significance of a sample result. For example, an economist collects some data on, say, interest rates and the demand for money. He hypothesizes some relationship and wishes to see if the data support his thesis. He tests the hypothesis to rule out the alternative that the observed relationship occurred by pure chance. He then reports his sample result as "significant at the 1 percent level." Such a statement is a report to the reader that has the following meaning: (1) if we were to set up a statistical hypothesis (and the particular hypothesis is either stated or is obvious from the context of the problem), and (2) if we were to test this hypothesis using a critical probability (or significance level) of 1 percent, then (3) we would reject the hypothesis and rule out a chance relationship.

Significance levels (critical probabilities) of 10, 5, 1, and .1 percent are often used in reporting sample data. The smallest of these probability values is chosen at which the hypothesis can be rejected. In other words, the *just significant probability level* is reported.

To make this clear, suppose that the analyst in the razor blade example were reporting the results of a sample of 100 razor blades to a superior. With a sample mean $\bar{X} = .7031$ and a standard error $s_{\bar{X}} =$

.001, the sample mean is 3.1 standard errors away from the hypothesized mean. The analyst might therefore describe the sample mean as "significantly different from .700 inches at the 1 percent level of probability." The use of a 1 percent critical probability would reject any sample mean outside $\mu \pm 2.58\ s_{\bar{x}}$. Note that the sample result could *not* be described as significant at the .1 percent level, which would require a deviation of 3.28 standard errors. This use of the hypothesis testing procedure, therefore, is a reporting or communication technique. It is used in the same manner as a confidence interval to describe the sampling error associated with a given sample.

Type II Errors

So far we have concerned ourselves only with the first kind of error. But there is also the second kind—the possible error of accepting a false hypothesis. The lower the value we set for the critical probability, in general the fewer the hypotheses we will reject. But the chances are then increased of accepting more hypotheses which are false. We can buy safety in one direction only at the expense of danger in the other.

Unfortunately, it is impossible to predict in general the percentage of times we should expect to commit an error of the second kind on the basis of any particular value adopted for the critical probability. The reason for this is that the chance of accepting a false hypothesis depends also upon how far away from the true value the particular hypothesis happens to be. Remember that sample means tend to cluster around the true means of the populations from which they are drawn. If the hypothetical mean is far away from the true mean, it is unlikely that a sample mean will be drawn which appears consistent with the hypothesis. If the hypothetical mean is false but not far from the mark, an error of the second kind is much more likely to be made.

In a long run of instances in which hypotheses are actually false, some will be farther from the true mean than others. Therefore, it is impossible to predict in general the probability of accepting false hypotheses. We can appreciate, however, that the chances of accepting false hypotheses are increased as fewer hypotheses are rejected due to the use of a lower value for the critical probability. The problem of balancing Type I against Type II errors is discussed below.

Operating Characteristic Curves. The exact probability of making a Type II error depends upon how far the true mean μ of the population is away from μ_h, the hypothetical mean. This can best be illustrated by an operating characteristic curve or OC curve, as shown in Chart 10–2.

The vertical scale of Chart 10–2 shows the probability of committing a Type II error (i.e., accepting the hypothesis when it is false). The horizontal scale shows all possible values for the true mean of the population relative to the hypothetical mean μ_h. Thus, if the true mean were one standard error less than μ_h, it would be at the point $-1\sigma_{\bar{x}}$ on the horizontal axis. Panel A represents the use of a critical probability of .05, and panel B a critical probability of .01. In either case, the probability of a Type II error can be found for any possible value of the true mean. Thus, in Chart 10–2A, if the true mean were three standard errors below the hypothetical mean $(-3\sigma_{\bar{x}})$, the probability of a Type II error would be .15, as shown by a dashed line. Similarly, if the true mean were two standard errors below the hypothetical mean $(-2\sigma_{\bar{x}})$, the probability of a Type II error would be .48.

When the true mean is exactly at the hypothetical mean $(\mu = \mu_h)$, a Type II error is impossible. Then the distance from the top of the curve to 1.00 represents the probability of a Type I error. Thus, since .95 is the probability of accepting the hypothesis when $\mu = \mu_h$, then .05 is the probability of rejecting it (when it is true), that is, of committing a Type I error. Except at this exact point, then, the probability of a Type II error decreases from nearly .95 toward zero as the distance between μ and μ_h increases.

Balancing Type I against Type II Errors

In testing hypotheses, we face two dangers: that of rejecting a true hypothesis and that of accepting a false hypothesis. The danger of committing a Type I error can be made as low as we please by reducing the value chosen for the critical probability; but this can be done only at the expense of increasing the danger of committing a Type II error. This can be seen by comparing the two curves in Chart 10–2. The probabilities of a Type II error on Chart 10–2B (with the more stringent critical probability of .01) are higher at every point than on Chart 10–2A.

The classical approach to statistical inference would leave the balancing of these risks and the determination of the critical probability to the judgment of the analyst. In the razor-blade example, a Type I error would mean falsely condemning the accuracy of a production process which was in fact operating as intended. A Type II error would mean continued production of a product which in fact was not meeting specifications. The economic penalty of the Type I error might be an expensive shutdown to look for a nonexistent trouble. The economic consequences of the Type II error might be the loss of consumer good-

Chart 10–2

PROBABILITY OF ACCEPTING THE HYPOTHESIS
FOR ALL POSSIBLE MEANS

(Operating Characteristic Curves)

A

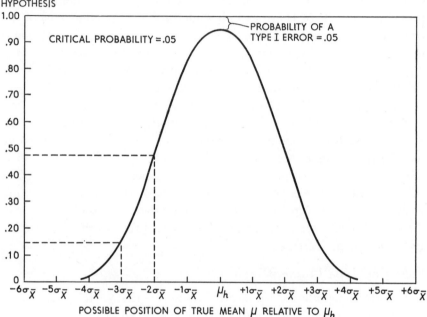

PROBABILITY OF
A TYPE II ERROR:
ACCEPTING THE
HYPOTHESIS

POSSIBLE POSITION OF TRUE MEAN μ RELATIVE TO μ_h

will as the customers later found the product unsatisfactory. (They might get razor burn with undue frequency, or find that the average blade did not fit into the razor.) With these potential economic consequences in mind, it would be up to management to set the value for the critical probability where, in its judgment, the best compromise is reached between risks of incurring the two types of errors.

In the Bayesian approach to statistical inference, the economic risks as well as the judgment of the decision maker are included in a formal decision-making procedure. This approach is the subject of Chapters 13 and 14.

Effect of Sample Size on Probability of Errors

So far the discussion of hypothesis testing has been in terms of some particular size of sample. So long as a given sample size is assumed, the

Chart 10–2 (Continued)

B

PROBABILITY OF
A TYPE II ERROR:
ACCEPTING THE
HYPOTHESIS

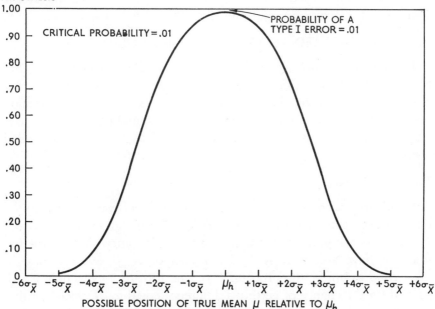

POSSIBLE POSITION OF TRUE MEAN μ RELATIVE TO μ_h

risk of a Type I error can only be reduced at the expense of increasing the risk of a Type II error. There is, however, a way of reducing the chance of accepting a false hypothesis without at the same time increasing the chance of rejecting a true hypothesis. By taking a larger sample the *combined* chance of committing either error can be reduced.

As the size of the sample drawn is increased, $\bar{X}$ will tend to fall closer to the actual value for μ, since $s_{\bar{X}}$ is decreased. With any particular value for the critical probability, Type I errors will be made with the same relative frequency, whatever the sample size. But as $\bar{X}$ is pulled in closer to μ (as is the tendency in taking a larger sample), $\bar{X}$ will in fewer instances appear consistent with a value other than μ, that is, with a false hypothesis regarding μ.

Thus, by taking a larger sample, the chance of a Type II error (accepting a false hypothesis) is reduced, while the chance of rejecting a true hypothesis can be held constant by using the same value for the critical probability. The combined chance of error will be smaller if we can reduce one component while we hold the other chance component constant. Just as we might expect, fewer overall mistakes of statistical inference will be made the larger the size of sample used.

In summary, therefore, the probability of a Type II error decreases with increases in (1) the critical probability α, (2) the size of sample (for a given value of α), and (3) the value of $\mu - \mu_h$.

TWO-TAILED TESTS VERSUS ONE-TAILED TESTS

In the form of testing hypotheses so far discussed, the probability has been calculated of getting a discrepancy as large as or larger than that observed by adding together the two "tails" of the sampling distribution beyond the number of standard errors corresponding to $(\bar{X} - \mu_h)$. This is referred to as "testing in both directions" or as a "two-tailed test."

Two-Tailed Tests

In the first example the probability of 62 percent was attached to the likelihood of getting a discrepancy as large as or larger than that observed $(.5 s_{\bar{x}})$, regardless of the sign of the discrepancy, that is, whether it might have arisen by $\bar{X} \geq .7005$ inch or $\bar{X} \leq .6995$ inches. In the second example, the probability of one fifth of 1 percent was calculated for the chance of getting a difference equal to or exceeding that observed $(3.1 s_{\bar{x}})$, whether that difference be above or below .700 inches.

There are three related reasons for testing in both directions when testing a single numerical value (such as .700 inches) as being the true mean of the population:

1. The hypothesis is in theory formed before the sample is drawn; hence, we don't know in advance whether the observed discrepancy between μ_h and $\bar{X}$ will have a positive or a negative sign.
2. An observed discrepancy of any particular size would be equally harmful to the hypothesis, whether it had a positive or a negative sign.
3. A hypothesis must not be rephrased to incorporate any of the information found in the very same sample which is used to test it.

The last point requires a bit of expansion. The hypothesis that the mean width of blade is .700 inches is a *single-valued hypothesis;* it says not greater than that, not less than that. If, on finding $\bar{X}$ equal to 0.7031 inches we had calculated only the probability of getting by chance a sample mean as large as or larger than .7031 inches, we would have subtly shifted our initial hypothesis to the hypothesis that the population mean is *not greater than* .700 inches. Implicitly, we would have wound up testing a different hypothesis than the one intended, and simply because of the sign of the discrepancy which was found after the sample had been drawn.

In the razor-blade case it seemed quite appropriate to test the single-valued hypothesis of .700 inches, that is, to test in both directions, since presumably we would be just as concerned about blades being too wide as being too narrow.

One-Tailed Tests

In other cases, however, it might be appropriate to test in one direction only; that is, to test what can be called a *multivalued hypothesis.*

If we were concerned with the strength of parachute cords, we would not be worried about their being too strong; we would worry only about their being too weak. If for safety's sake they were designed, let us say, to have a mean breaking point of 1,000 pounds, we would be interested in the hypothesis that the true population mean was 1,000 *or more* pounds.

Should a sample mean greater than 1,000 pounds be found in a random sample drawn, it would immediately be accepted as consistent with the hypothesis. Only if $\bar{X}$ should be less than 1,000 pounds would a question arise concerning the validity of the hypothesis. It would then be appropriate to ask the question: If the mean of the population truly were 1,000 pounds or more, what is the probability of getting by chance a sample mean which falls below 1,000 pounds by as much as the one observed? That is to say, the particular sign of the observed difference now would have a bearing on the truth or falsity of the hypothesis as stated. It is appropriate in this case to test in but one direction, that is, in terms of the probability of getting by chance a sample mean which lies below 1,000 pounds by an amount equal to or greater than that observed.

One important change is made when applying a one-tailed test instead of a two-tailed test, namely, the multiple of the standard error which corresponds to any given critical probability. In a two-tailed test, $1.96\sigma_{\bar{x}}$ corresponds to a 5 percent critical probability, whereas $1.65\sigma_{\bar{x}}$ is the multiple of the standard error associated with 5 percent in a one-tailed test. When testing in both directions, $2.58\sigma_{\bar{x}}$ goes with 1 percent as the critical probability. But for testing in a single direction, the similar combination is $2.33\sigma_{\bar{x}}$ and 1 percent. These can be read from Appendix D for various areas under the normal curve.

For a 5 percent critical probability under a two-tailed test and one-tailed test, respectively, see Chart 10–3.

Chart 10–3

AREAS OF REJECTION—5 PERCENT CRITICAL PROBABILITY

A. Two-Tailed Test   B. One-Tailed Test

REJECTION AREA   REJECTION AREA   REJECTION AREA

$-1.96\sigma_{\overline{X}}$ μ_h $1.96\sigma_{\overline{X}}$ $1.65\sigma_{\overline{X}}$ μ_h

AREA ← 2.5% → | ← 95% → | ← 2.5% →   ∞ ← 5% → | ← 95% → ∞

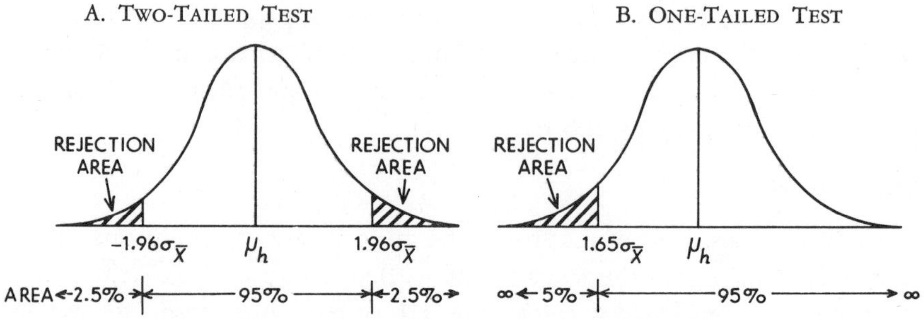

TESTS OF DIFFERENCES BETWEEN ARITHMETIC MEANS

We now consider another important aspect of statistical inference, namely, tests of the significance of differences between sample means. This phase is concerned with the following problem: Given an observed difference between the means of two random samples, each drawn from a different population, is this difference to be taken as signifying a real difference between the true means of the populations involved?

To handle this problem it is necessary to introduce the concept of a new sampling distribution, the sampling distribution of *differences* between means. We can think of this distribution as being formed in the following manner.

On the basis of random sampling from two separate populations, the sampling distributions of the arithmetic means $\overline{X}_1$ and $\overline{X}_2$ would be formed. Each of these sampling distributions is of the same type we have been discussing.

Now imagine that from each of these sampling distributions a sample mean is drawn at random and that the difference between this pair of sample means is noted. Then a second pair of sample means is selected at random, each from its own sampling distribution. The difference between this second pair almost certainly would be different from that found between the first pair, due to chance alone. We can imagine the process carried on repeatedly. Then we would have an indefinitely large number of values representing the differences between all possible pairs of sample means which could be drawn at random from their respective populations. These differences would form a theoretical distribution known as the sampling distribution of the difference between two means.

We know the following things about this new distribution:

1. According to the central limit theorem, the sampling distribution of differences tends to be normal; which is to say that differences between pairs of sample means will be normally distributed, provided that the sample size is large.
2. The mean of the distribution of differences will be the true difference between the population means ($\mu_1 - \mu_2$). That is, the sample difference ($\bar{X}_1 - \bar{X}_2$) is an unbiased estimator of the population difference. This follows from the proposition that the mean of the differences between any two series of values is equal to the difference between their respective means.
3. The standard deviation of the distribution of differences may be estimated by the formula

$$s_{\bar{X}_1 - \bar{X}_2} = \sqrt{s_{\bar{X}_1}^2 + s_{\bar{X}_2}^2}$$

In this formula $s_{\bar{X}_1}$ is the standard error of the mean for the sampling distribution of $\bar{X}_1$ and $s_{\bar{X}_2}$ is the similar measure for the sampling distribution of $\bar{X}_2$. The value $s_{\bar{X}_1 - \bar{X}_2}$ is known as the *standard error of the difference between two means.*[1]

With this important new sampling distribution in mind, we can carry forward our discussion of the present phase of statistical inference in terms of specific examples.

Suppose a trucking firm is testing two brands of truck tires for their wearing ability in order to decide if one brand has greater average mileage than the other. One hundred tires of brand No. 1 are put on the firm's trucks and the mileage records are kept until the tires are worn out; similarly, 144 tires of brand No. 2 are put on trucks and the mileage is recorded. Both brands of tires are placed at random on the firm's trucks to guard against any systematic bias because of characteristics or usage of certain trucks.[2] (A difference in sample size is used in this example merely to emphasize that the two samples need not be

---

[1] In this discussion s represents the standard error estimated from a sample; if the true population value were known, the symbol σ would be used, with appropriate subscript.

The variance (s^2) of the difference is the sum of the variances of the individual means. As a graphic check, the standard error of each mean can be laid off as a side of a right triangle; then the standard error of the difference can be read off as the hypotenuse (Pythagorean theorem).

[2] A better statistical design, perhaps, would call for putting both brands on the same truck to reduce differences due to truck characteristics and usage. For more on this technique of pairing observations, see W. J. Dixon and F. J. Massey, *Introduction to Statistical Analysis* (3d ed.; New York: McGraw-Hill, 1969), pp. 119–23.

equal in size for this method to be applicable.) The following means and standard deviations result (the subscripts referring to the brand number):

Tire Brand No. 1	Tire Brand No. 2
$n_1 = 100$	$n_2 = 144$
$\bar{X}_1 = 37.4$ thousands of miles	$\bar{X}_2 = 36.8$ thousands of miles
$s_1 = 5.1$ thousands of miles	$s_2 = 4.8$ thousands of miles

The test gives tire brand No. 1 an advantage of $\bar{X}_1 - \bar{X}_2 = .6$ thousand miles in average mileage. Nevertheless, because we are quite aware of chance variations that may occur in random sampling, we do not immediately jump to the conclusion that brand No. 1 is longer wearing than brand No. 2. We are led to wonder if the difference in mean mileage observed in the samples arose by chance or whether there is in fact a difference in average mileage between all tires of brand No. 1 and all tires of brand No. 2. That is to say, we wish to know if the observed difference between the sample means indicates a real difference between the means of the two populations.

The Null Hypothesis

Our manner of solving this problem is to set up and test the so-called null hypothesis. This means that we pose the hypothesis that there is *no* difference in average mileage between brand No. 1 and brand No. 2, and then proceed to test that hypothesis against the evidence found in the samples.

The null hypothesis states that the mean of the sampling distribution of differences is equal to zero. This is because the mean of the sampling distribution of differences is known to be $(\mu_1 - \mu_2)$, and the hypothesis is that there is no difference between these population means.

The observed difference of .6 thousand miles between the two random sample means is, in effect, one observation drawn at random from the sampling distribution of all possible differences between pairs of random sample means. We can therefore ask the question: If the mean of the sampling distribution of differences really were zero, what is the probability that we would get a difference between two sample means at least as large as .6?

Since the sampling distribution from which .6 came tends to be normal, we can answer this question as soon as we know the value for the standard error of the difference between means. This is computed as follows from the basic formula $s_{\bar{x}} = s/\sqrt{n}$:

$$s_{\bar{X}_1} = \frac{5.1}{\sqrt{100}} = .51 \qquad s_{\bar{X}_2} = \frac{4.8}{\sqrt{144}} = .40$$

$$s_{\bar{X}_1 - \bar{X}_2} = \sqrt{s_{\bar{X}_1}^2 + s_{\bar{X}_2}^2}$$

$$= \sqrt{(.51)^2 + (.40)^2}$$

$$= \sqrt{.4201}$$

$$s_{\bar{X}_1 - \bar{X}_2} = .65$$

Accepting the Null Hypothesis. Thus, it turns out that the observed difference between the sample means is less than one standard error of the difference ($.6/.65 = .92$ standard errors, to be exact). If the true difference between the population means really was zero, the probability is nevertheless 36 percent that a difference at least as large as .6 thousand miles would appear by chance. It would appear that there is no compelling evidence to be found in the samples that a real difference exists in average mileage between the two brands. In this case it is said that the difference between the sample means is too small to be significant—that is, too small to signify an indisputable difference between the population means.

Rejecting the Null Hypothesis. Let us take the same case again, but assume $\bar{X}_1$ had come out 38.6 instead of 37.4 thousand miles. Now the observed difference between the sample means is $38.6 - 36.8 = 1.8$ thousand miles. This in turn is equal to 2.8 standard errors of such differences (i.e., $1.8/.65 = 2.8$). Since 2.8 is greater than the 2.58 standard errors associated with a .01 probability level, the observed sample difference is significant at the .01 level.

Actually, if there really were no difference between μ_1 and μ_2, the probability of getting an observed difference equal to or greater than 2.8 standard errors in either direction would be only one half of 1 percent. It appears highly unlikely, therefore, that the difference between the means of the samples could have appeared solely by chance in this case. The null hypothesis may very well be rejected.

The Choice between Acceptance and Rejection. In the first instance above, a difference in the sample means of .6 thousand miles or more could occur by chance 36 percent of the time. Most observers, on the basis of the sample alone, would accept the hypothesis. Such an acceptance would imply either (1) that there was no difference in mean wearing ability of the two brands of tires and the observed sample difference was due to chance, or (2) that there was a difference but the samples were too small to detect the difference. On the other hand, a difference in sample means of 1.8 thousand miles is significant at the .01 level and strongly indicates a real difference in mean wearing ability.

What would be the conclusion if, for example, the difference in the sample means were one thousand miles or 1.5 standard errors $(1/.65 = 1.5)$? The probability of a difference in sample means this large or larger is 13 percent. In such a case, we conclude that the sample gives some evidence that one tire is longer wearing than the other on the average, but the possibility that the sample result is due to chance cannot be ruled out. In other words, on the basis of the sample alone, the results are inconclusive.

If some action must be taken—e.g., deciding which tire to purchase —evidence other than the sample would be included in the decision analysis. The past reputation of the tire manufacturers, the prices of the two brands, and the savings associated with longer tire wear should be considered. In the classical statistical approach, these factors should be incorporated in the determination of the appropriate Type I and II error probabilities. In the Bayesian approach, these factors are explicitly included in the decision-making procedure (see Chapters 13 and 14).

Confidence Intervals for the Difference between Sample Means

Rather than testing the hypothesis that there is no difference in population means, we may wish to estimate the actual difference between the means. The procedure, in principle, is identical with that employed earlier in estimating the mean of a population on the basis of the mean of a random sample drawn from that population. The only difference is that the sampling distribution of differences (and its associated measures) is employed in forming the appropriate confidence intervals in the present case.

We wish to estimate $(\mu_1 - \mu_2)$, which is known to be the mean of the sampling distribution of differences. From this sampling distribution we have one observation $(\bar{X}_1 - \bar{X}_2)$, based upon random sampling. Then 68 percent of such observations would be expected to lie within $s_{\bar{X}_1 - \bar{X}_2}$ of the mean difference; 95 percent would be expected to lie within $1.96 s_{\bar{X}_1 - \bar{X}_2}$ of $(\mu_1 - \mu_2)$ etc. Consequently, we should have a 68 percent degree of confidence that an interval specified as $(\bar{X}_1 - \bar{X}_2) \pm s_{\bar{X}_1 - \bar{X}_2}$ would include the value $(\mu_1 - \mu_2)$ and a 95 percent degree of confidence that the interval $(\bar{X}_1 - \bar{X}_2) \pm 1.96 s_{\bar{X}_1 - \bar{X}_2}$ would include the true difference between the population means.

In the second example above, the observed difference is 1.8 thousand miles, with a standard error of .65 thousand miles. We may estimate, therefore, that the true difference between the population means lies within the interval 1.8 thousand miles $\pm$ 1.3 thousand miles (i.e., 1.96 times the standard error) and hold a 95 percent degree of confidence that our estimate is correct. The 95 percent confidence limits are then .5 thousand miles and 3.1 thousand miles for the superiority of tire No. 1 over tire No. 2 as regards average mileage.

If the confidence interval based upon $\pm 3 s_{\bar{x}_1 - \bar{x}_2}$ is computed to give a degree of confidence of 99.7 percent that the true difference is located within its boundaries, the confidence limits work out to be minus .15 thousand miles to 3.75 thousand miles for the difference between brands No. 1 and 2 in average mileage. This result—the appearance of the negative sign for the lower limit of the confidence interval—might puzzle the student, but it really need not. All it means is that for us to be 99.7 percent confident that we have located the real difference in average mileage between the two brands we should have to grant that superiority *might* lie to a small extent with brand No. 2.

TESTS OF HYPOTHESES FOR A PROPORTION

Let us suppose that the problem of the self-service meat counter in Chapter 9 has come up in a somewhat different way—and for purposes of exposition assume that we know nothing of the calculations made heretofore.

Assume that a nationwide survey by a grocery trade association had suggested that customers of chain stores were equally divided in their preference between self-service meat counters and counters serviced by butchers. The management of a regional chain is somewhat impressed by this finding, but it recognizes that regional differences can exist. Management has decided that it will replace butcher-serviced counters if it can get compelling evidence that its particular group of customers favors self-service in a proportion greater than one half.

Now, in this case the nationwide survey has suggested the hypothesis that the true proportion is .50, and only if this is refuted by regional evidence will management decide otherwise. Further, management is interested only in the alternative hypothesis that the true proportion is *greater than* .50; therefore a one-tailed test is the appropriate one.

Let us assume that a random sample of 400 customers is drawn. From the hypothesis that the true population proportion is .50 (i.e., $p_h = .50$), we proceed to calculate the standard error of a sample proportion which would correspond to that hypothesis, namely,

$$\sigma_{p_s} = \sqrt{\frac{p_h q_h}{n}}$$

$$= \sqrt{\frac{.50 \times .50}{400}}$$

$$= .025 \text{ or } 2.5\%$$

Suppose that the proportion of customers favoring self-service in the sample turns out to be .55; then the difference between the sample proportion (p_s) and the hypothetical proportion (p_h) is .05. In terms of multiples of the standard error, this is

$$z = \frac{p_s - p_h}{\sigma_{p_s}} = \frac{.55 - .50}{.025} = \frac{.05}{.025} = 2 \text{ standard errors}$$

Only 2.3 percent of the area under a normal curve falls *above* 50 percent by more than two standard errors in that one-tailed direction (see Appendix D). Hence, the probability is only 2.3 percent that such a large proportion could occur by chance if the true proportion were no greater than .50. We should have to make our decision on the grounds discussed earlier. But the probability of 2.3 percent that chance alone could have created this evidence is surely a low probability. Therefore a conclusion that the regional chain's true population proportion is greater than .50 is strongly indicated.

The Test of a Difference between Two Proportions

Suppose that a manufacturer of farm implements is interested in whether farmers in state No. 1 differ significantly from farmers in state No. 2 with respect to the proportion preferring the make of tractor which he sells. He takes separately a random sample of 100 farmers in each state and finds that the proportion preferring his make is .40 in state No. 1 and .30 in state No. 2. Should this difference in sample proportions be taken as signifying a difference in the true proportions?

The line of statistical reasoning by which this question is answered is already familiar from earlier discussions. Only the new, appropriate measures need to be introduced. The sampling distribution of ($p_{s_1} - p_{s_2}$) may be taken to be fairly normal in large samples because of the central limit theorem.

The standard error of a difference between two independent sample proportions p_{s_1} and p_{s_2} is:

$$\sigma_{p_{s_1} - p_{s_2}} = \sqrt{\sigma_{p_{s_1}}^2 + \sigma_{p_{s_2}}^2}$$

Since the symbolism is going to be a little complicated, it will be more convenient to write this in squared form, which is known as the *sampling variance* of the difference between two proportions. Hence,

$$\sigma^2_{p_{s_1}-p_{s_1}} = \sigma^2_{p_{s_1}} + \sigma^2_{p_{s_2}}$$

That is, the sampling variance of the difference between two independent proportions is the sum of their sampling variances.[3]

Since $\sigma^2_{p_s} = pq/n$ in each case, the above formula may be written

$$\sigma^2_{p_{s_1}-p_{s_2}} = \frac{p_1 q_1}{n_1} + \frac{p_2 q_2}{n_2}$$

where the subscripts 1 and 2 refer to the two states, respectively.

Now, in the present case, we would set up and test the *null hypothesis* that there is *no* difference in the true population proportions involved. Our hypothesis states that $p_1 = p_2$; hence, the observed difference between the sample proportions p_{s_1} and p_{s_2} is caused by sampling errors.

Since we do not know p_1 and p_2, the best estimate of their common value is the weighted mean of the sample proportions (using the sample sizes as weights). This is most easily accomplished by adding the *number* of farmers preferring the tractor in both samples and dividing this total by the total number of farmers. There are 70 farmers preferring the tractor (40 from state No. 1 and 30 from state No. 2) out of 200 farmers sampled, and so the weighted mean proportion is $\bar{p} = 70/200 = .35$.

The sample variance then is:

$$\sigma^2_{p_{s_1}-p_{s_2}} = \frac{\overline{pq}}{n_1} + \frac{\overline{pq}}{n_2}$$

$$= \frac{.35 \times .65}{100} + \frac{.35 \times .65}{100}$$

$$= .00455$$

To find the standard error of the difference we extract the square root:

$$\sigma_{p_{s_1}-p_{s_2}} = .0675 \quad \text{or} \quad 6.75\%$$

In the way now familiar, we express the observed difference of the sample results from the null hypothesis as a ratio to the standard error of such differences. Since the null hypothesis assumes the true difference to be zero, the calculation which we want amounts to:

---

[3] As a graphic solution or check, lay off $\sigma_{p_{s_1}}$ and $\sigma_{p_{s_2}}$ as the sides of a right triangle; then $\sigma_{p_{s_1}-p_{s_2}}$ is the hypotenuse. This is the Pythagorean theorem.

$$z = \frac{p_{s_1} - p_{s_2}}{\sigma_{p_{s_1} - p_{s_2}}} = \frac{.40 - .30}{.0675}$$
$$= 1.48$$

so that the observed difference deviates from the null hypothesis by 1.48 standard errors.

Consultation of Appendix D shows that deviations of this size, regardless of sign, from a true value of zero, are expected to occur by chance alone in 14 percent of all possible samples. In other words, the probability is about 14 percent that this big a spread could occur by chance alone, were the null hypothesis true. This is not significant at the 5 or 10 percent level. Therefore, based on the available evidence, we would probably accept the null hypothesis and attribute the sample results to mere chance. We do not have sufficient evidence to reject the null hypothesis, that is, to conclude that there is a real difference between the two states sampled. This does not prove that $p_1 = p_2$; the evidence is inconclusive. The manufacturer should consider increasing the size of the samples, so that for any given critical probability chosen the overall likelihood of committing an error of inference would be reduced.

APPLICATIONS IN STATISTICAL QUALITY CONTROL

Hypothesis testing is widely used in the field of statistical quality control, as a systematic method of detecting or predicting trouble in manufacturing processes. This technique also helps to reduce waste and improve product quality and design. The principal types of control charts are for variables, or measurable characteristics, and for attributes, or traits that are either present or absent (e.g., a "go not-go" gauge test) or nonmeasurable (e.g., color).

Statistical quality control permits the partitioning of the total variation of a product into two components: (1) *Chance variation* is that which results from many minor causes that behave in a random manner. This type of variation is permissible, and indeed inevitable, in manufacturing. (2) *Assignable variation* is a relatively large variation that can be attributed to special nonrandom causes. It may be excessive in amount so as to require correction. Tool wear, a change in the raw material, a new operator, improper machine setting—all can produce assignable variations. The value of quality control lies in its power to detect the assignable variations in a process quickly; in fact, these variations are often discovered before the product becomes defective.

In a control chart, the hypothesis is posed that the average "level" of a manufacturing process is unchanged. Now, the means of even small samples tend to be normally distributed about the population mean, provided the population is normal. So, assuming that only chance variation is present, we can predict that 99.73 percent of the sample means will fall within the interval $\mu \pm 3\sigma_{\bar{x}}$, the limits arbitrarily used in American industry. The same is true for sample proportions (e.g., the proportion of spark plugs found defective), which tend to be normally distributed when n is large. If a sample mean or proportion falls within these limits, the hypothesis is accepted and the process is allowed to continue. But if a sample value falls outside the 3σ limits, assignable variation is suspected and the hypothesis is rejected. The operation is then stopped and corrected.

In an $\bar{X}$ chart, or control chart for averages, the horizontal scale is designated by subgroup number. Subgroups are samples taken in a certain order. The vertical scale is labeled $\bar{X}$. At the point $\bar{\bar{X}}$ (the mean of several samples) on the vertical scale, a horizontal central line is drawn. Above and below this line at a distance of $3\sigma_{\bar{x}}$, parallel dashed lines are drawn. These are the upper and lower control limits (UCL and LCL). The means of subgroups are then plotted at equal intervals from left to right.

It is customary to use small samples—say, four or five items each—in $\bar{X}$ charts in order to signal a prompt alert if a process goes wrong. For simplicity, the standard deviation is usually estimated from the average of the ranges of several samples, by means of a table.[4]

As an actual case, take a ceramic disk used in the capacitor of a television set. The diameter of the disk must be controlled after it is fired in a kiln, a process subject to numerous sources of variation. The fired diameter of the disk is specified as 500 thousandths of an inch. An inspector takes 20 subgroups of five each and records the readings in thousandths of an inch as deviations from .500 inches.

The results appear in Chart 10–4. All points are within the control limits except for subgroup 12, which signals trouble at that point, perhaps calling for shutdown and repairs. In this case, however, it was found that the process was free of assignable variation, since the lot from which subgroup 12 was drawn had been previously rejected because it did not meet density standards, but had been processed through error. Therefore, a revised mean and control limits were computed from the remaining 19 subgroups, and drawn on the right side of Chart 10–4, for further control of the process.

---

[4] For this table, and a further discussion of quality control, see the first edition of this book, Chapter 25.

Chart 10–4

X̄ CONTROL CHART FOR FIRED DIAMETER OF CERAMIC DISCS

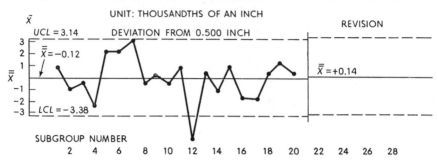

The fact that sample averages follow the normal distribution when assignable variation is absent can be used to detect trouble in a process even though no points may have gone beyond the control limits. With trouble absent, the sample averages should be distributed at random about the central line, with more points near the line than far from it. Then, if an excessively long run—say, seven points or more—occurs on one side of the central line, the evidence is that assignable variation has entered the process, causing a shift in process level, even though no points may have fallen beyond the control limits.

Furthermore, if an upward or downward trend is noted in the points on the control chart, the evidence also indicates that assignable variation is present. This is frequently the result of gradual tool wear. Thus it is evident that in many cases a control chart, if properly interpreted,

Chart 10–5

p CHART FOR SPARK PLUG INSPECTION

(24 Lots of 200 Spark Plugs Each)

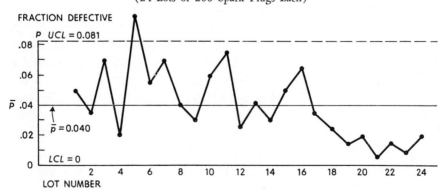

can give an indication of impending trouble even though no points have actually exceeded limits. Corrective action can then be taken to avoid production of any unsatisfactory items.

Control charts are also used for attributes, such as a p chart for the proportion of units that are defective in some way. This chart generally gives best results when the sample size is large—say, at least 50. The central line is placed at $\bar{p}$, the average fraction defective, where $\bar{p}$ is the number of defectives divided by the total number inspected. The control limits are $3\sigma_p$ from the central line.

As an example, Chart 10–5 shows the results of a visual inspection of 24 lots of spark plugs each comprising 200 plugs. It is seen that lot 5 exceeded the upper control, but that the last eight lots are all below the central line, indicating that the process may have improved as a result of the warning alarm given by lot 5.

SUMMARY

We can make a statistical inference either by constructing a confidence interval (as described in Chapter 9) or by testing a hypothesis. In the latter case we set up a hypothesis regarding the value of the parameter—say, the mean. If the sample mean is close to the hypothetical mean, we accept the hypothesis; otherwise we reject it.

In the case of the razor-blade machine that was set to produce blades of average width .700 inches, a sample of 100 blades was tested, with $\bar{X} = .7005$ inches and $s = .010$ inches, so $s_{\bar{X}} = s/\sqrt{n} = .001$ inches. Since the sample mean was only .5 standard errors away from the hypothetical mean, the probability was 62 percent of getting such a discrepancy by chance, so the hypothesis was accepted. In a second trial, however, with $\bar{X} = .7031$ inches, the hypothesis ($\mu_h = .700$ inches) was rejected since it was quite unlikely that such a discrepancy could occur by chance alone. A reasonable hypothesis is usually accepted unless the probability is quite low (say, under 5 percent or even 1 percent) that the discrepancy of the sample value could be attributed to chance. The problem is where to get this critical probability below which we will reject the hypothesis. Rejection of a hypothesis indicates a belief that the hypothesis is false. Acceptance of a hypothesis, however, does not necessarily prove that the hypothesis is true. It may be that the sample is too small to detect a significant difference.

We can make two types of errors in testing hypotheses:

1. Type I: rejecting a true hypothesis.
2. Type II: accepting a false hypothesis.

We can easily control the chance of making a Type I error, since this equals the critical probability that is set in advance. Unfortunately, for a given size of sample, we can reduce the chance of making a Type I error only at the cost of increasing the risk of making a Type II error. The chance of making the latter error is unknown, since it depends on how far the hypothetical mean is away from the true mean.

By taking a larger sample, the combined chance of making either error can be reduced. In particular, if the critical probability is held constant, the chance of a Type I error also remains constant in a larger sample, but the chance of a Type II error is reduced.

An operating characteristic or OC curve shows the probability of making a Type II error (that is, accepting the hypothesis when it is false) for a given critical probability, depending on how far the true mean is from the hypothetical mean. The farther these means are apart, the smaller is the probability of a Type II error.

The critical probability used in hypothesis testing is determined, in the classical approach to statistical inference, by balancing the consequenecs of Type I and Type II errors. If a Type I error would be serious relative to a Type II error, the critical probability should be set relatively low. When the relative costs cannot be determined, critical probabilities are often set at the arbitrary values of 5 or 1 percent.

In the Bayesian approach to statistical inference (Chapters 13 and 14) the economic consequences as well as the prior judgment of the decision maker are included with the sample in making a decision.

Business and economic studies often report a sample result as, for example, "significant at the 1 percent level." Such a statement describes the sampling error associated with a sample and indicates that an implied hypothesis would be rejected if a 1 percent critical probability were used. Significance levels of 10, 5, 1, and .1 percent are commonly used, and the smallest probability at which the hypothesis will be rejected is reported.

In testing hypotheses, we may make either a two-tailed or a one-tailed test. The two-tailed test takes into account the areas under both tails of the normal curve (Chart 10–3). It is appropriate in many practical situations because we are concerned with discrepancies either above or below the hypothetical mean. In case we are concerned only with discrepancies in one direction from the hypothetical mean, however, it is appropriate to use the one-tailed test, which takes into account only the area under one tail of the normal curve. The decision rule is then to reject the hypothesis if $(\overline{X} - \mu_h)/s_{\overline{X}}$ exceeds the following values:

Critical Probability Chosen	Two-Tailed Test	One-Tailed Test
5 percent.............	1.96	1.65
1 percent.............	2.58	2.33

We can also test whether the difference between two sample means signifies a real difference between the population means or whether the observed difference is merely due to chance. To do this, we find the standard error of the difference (theoretically, the standard deviation of a distribution of differences between many pairs of sample means). This is computed from the standard errors of the individual means. Then we can test the null hypothesis (that there is no difference between the population means) by expressing the difference between the sample means as a ratio of their standard error. If this ratio is small, we accept the null hypothesis; otherwise we reject it, depending on the probability that the difference could be due to chance (from Appendix D), and balancing the consequences of Type I and II errors as before. We can also set up a confidence interval around the difference between the sample means, based on its standard error, as was done earlier.

Tests of hypotheses may be applied to proportions by computing the standard error, based on the hypothesized proportion p_h. Then the deviation of the sample proportion from this value ($p_s - p_h$) is divided by the standard error to determine whether it is large enough to be significant. Thus, if the standardized deviation is 1.96 or more (in a two-tailed test), it is significant at the 5 percent critical probability level, and so on (Appendix D).

We can also test whether the difference between two proportions ($p_{s_1} - p_{s_2}$) is significant by dividing the difference by its standard error, where $s^2_{p_{s_1} - p_{s_2}} = s^2_{p_{s_1}} + s^2_{p_{s_2}}$. If this standardized difference is 1.96 or more, it is significant at the 5 percent level, etc., just as above. When we test the null hypothesis that there is no difference between p_1 and p_2, we use the average value of the sample proportions, weighted by the size of the two samples, to compute the standard error of the difference.

Statistical quality control is an application of hypothesis testing in manufacturing. Control charts are used to separate the normal chance variation from assignable variation (attributable to nonrandom causes) so that the latter can be promptly recognized and remedied.

The $\overline{X}$ chart for variables is used to control the average value or "level" of a characteristic. On an $\overline{X}$ chart, horizontal lines are drawn at the estimated population mean on the vertical scale and at $3\sigma_{\overline{X}}$ control limits above and below it. Subgroup averages are plotted at equal intervals along the horizontal axis.

Nearly all of the points should fall within the control limits of an $\bar{X}$ chart if chance variation alone is present. If a point falls outside the limits or if about seven or more consecutive points fall on one side of the central line or if they show an upward or downward trend, assignable variation is probably present. This should be corrected promptly.

The example of the ceramic disk illustrates how to interpret a control chart and revise the limits if necessary.

The control of attributes may be achieved through the use of p charts for proportion of units that are defective. These charts are constructed and interpreted in much the same way as the control charts for variables.

PROBLEMS

1. Distinguish between:
    a) Confidence intervals and tests of hypotheses.
    b) Type I and Type II errors.
    c) How to find the probability of Type I and Type II errors from an operating characteristic curve.
    d) One-tailed and two-tailed tests.
    e) Use of hypothesis testing for decision making and for reporting.

2. Explain:
    a) How to test a hypothesis that a sample proportion .45 is significantly less than .50.
    b) The null hypothesis for the difference between two sample proportions.
    c) How to determine whether or not a process is capable of meeting specifications in quality control.

3. Distinguish between:
    a) Chance variation and assignable variation in quality control.
    b) $\bar{X}$ charts for variables and p charts for attributes.
    c) Two situations in which the pattern of points on a control chart would indicate trouble even if no points actually fall outside control limits.

4. A random sample of 144 building bricks has a mean weight of 6.9 pounds and a standard deviation of .3 pounds. Is it likely that this sample comes from a brickyard that produces bricks with a mean weight of 7.0 pounds?

5. A grocery chain store adopts a policy of issuing trading stamps on all purchases. Prior sales had averaged $16.00 per customer over the past year, with a standard deviation of $4.80. At the end of a trial period with the new stamps, a random check of 400 customers shows average sales of $16.80. Have the stamps increased the average sales?

6. A machine, when in adjustment, produces parts that have a mean diameter of .300 inches with a standard deviation of .012 inches. A random sample of 36 parts yields a mean diameter of .303 inches. Is the machine probably still in adjustment or not? Give reasons.

7. If we change the critical probability from 5 to .1 percent, what is the effect on:

   a) The probability of rejecting a true hypothesis?
   b) The probability of accepting a false hypothesis?

8. The engineers of a machine tool manufacturer design a new machining operation with the expectation that it will require an average time of 24 minutes, with a standard deviation of 4 minutes.

   a) If 64 operations are sampled from this process, what is the probability that the sample mean $\overline{X}$ will overstate the process average μ by one minute or more? What is the probability of an error of one minute or more in either direction?
   b) After the process is set up, 64 operations are actually sampled, with $\overline{X} = 25.3$ minutes and $s = 4$ minutes. Test the null hypothesis that the process average is still 24 minutes.

9. A machine produces baling wire with average breaking strength of 70 pounds, when properly adjusted. After the machine has been in operation for some time, a sample of 36 pieces is tested. They have a mean strength of 68 pounds and a standard deviation of 3 pounds.

   a) Does the machine need readjustments? (I.e., test the hypothesis that $\mu = 70$.) Explain the significance of computations.
   b) Construct a 99 percent confidence interval around the sample mean and explain its meaning.

10. a) Suppose the null hypothesis is $\mu_h = 14.0$, $n = 25$, $\sigma = 2.0$, and the critical probability is .05. Using Chart 10–2, what would be the probability of a Type II error if the actual μ of the population were 15.0? If the actual μ were 14.5?

   b) What would be the probability of a Type II error if the sample size were increased to 36 and the actual μ were 15.0? If the actual μ were 14.5?
   c) What would be the probability of a Type II error for $n = 25$, if a .01 critical probability were to be used and the actual μ were 15.0? If the actual μ were 14.5?

11. The standard time for a certain assembly operation is 2.4 minutes. Jones has been observed and timed in this operation 32 times over the past two weeks with the following results: $X =$ observed time in minutes for Jones to complete the assembly operation; $n = 32$, number of observations of Jones; $\overline{X} = 2.8$ minutes; $\Sigma X = 89.6$; $\Sigma X^2 = 320.63$.

   If the evidence is sufficiently strong that Jones is not meeting the standard on the average, then he is to be retrained. What conclusion can you draw from the sample result? What action should be taken?

12. A certain pneumatic tool is designed so that it should operate on a pressure of no more than 20 pounds per square inch. Management was receiving complaints from purchasers that the pressure necessary to operate the tools was in excess of the 20 pounds psi standard. To check this, 40 tools were selected from current production, and the operating pressure was checked on each under controlled conditions. The results were X = pressure in pounds per square inch to operate a given tool; $n = 40$; $\Sigma X = 740$; $\Sigma X^2 = 14{,}041$.

 a) Is a one- or two-tailed test appropriate in this situation?

 b) What can you conclude from the statistical test of the hypothesis?

 c) Does your answer to (b) reply to the objection raised by the customers? Why or why not?

13. A manufacturer of incandescent lamps is testing to see if the average life of the lamps he is manufacturing is above or below the standard of 2,000 hours. To check, the manufacturer proposes to take a sample of 200 lamps and to determine the life of each. He plans on using a 1 percent critical probability (two-tailed). From past experience, the standard deviation of the burning life of this type of lamp is known to be about 1,000 hours.

 a) What is the hypothesis?

 b) What is the meaning of a Type I error in this situation? What is the probability of a Type I error?

 c) Suppose that the true mean life deviates by 100 hours from the standard. What is the probability that the sample will be able to detect the difference?

 d) Suppose that the true mean life deviates by 200 hours from the standard. What is the probability that the sample will be able to detect this difference?

 e) Suppose that the true mean life differs from the standard by 150 hours. How large a sample would be necessary to detect this difference with only 1 chance in 10 of making a Type II error?

14. A large distributor of cosmetics has kept his outstanding accounts receivable to a mean age of 18 days, over the past year. This average is considered a standard by which to measure the efficiency of the credit and collections department. During the current month, however, a random check of 100 accounts yields an average of 20 days, with a standard deviation of 9 days.

 a) Is this result significantly different from the standard at the 5 percent level of significance? At the 2 percent level? Give reasons.

 b) If management has reason to believe that the collection of accounts is becoming slower, and is interested only in the possibility that the average age has increased, is the sample result significantly greater than the standard at the 5 percent level? At the 2 percent level?

15. The credit manager for an oil company claims that the average balance on statements mailed to credit-card holders is at least $32. To check this claim, an auditor takes a sample of 64 statements and finds that the average amount owed is $30, with a standard deviation of $12. On the basis of the sample evidence, what can we say about the credit manager's claim?

16. An auditor for another oil company takes a sample of 36 credit-card statements. He obtains a mean balance of $34 and a standard deviation of $10.
Is there a significant difference in the mean balance of credit-card statements between this company and that of Problem 15 above?

17. As purchasing agent for an electrical equipment manufacturer, you wish
to compare the average thickness of two shipments of ⅛-inch insulating
board, consisting of 200 sheets just received from manufacturer A and 200
sheets from manufacturer B. This board is used in the manufacture of power
transformers. Thickness is a major characteristic affecting the quality of the
board and consequently the quality of the transformer. All sheets are
measured for thickness using a zero- to one-inch micrometer, with the following results, in thousandths of an inch:

	Mfr. A	Mfr. B
Mean	124.76	125.36
Standard deviation	2.63	3.10
Sample size	200	200

Regarding these lots as random samples of each manufacturer's output, does
this test indicate that manufacturer B's product is superior in thickness, or
is the difference probably due to sampling error? Show computations and
explain your answer.

18. Observations are made on the time required to check out customers in a
supermarket. For a sample of 36 customers, it takes Mary an average of 6
minutes with a standard deviation of 3 minutes. It takes Joan an average of
8 minutes per customer with a standard deviation of 5 minutes. Is the difference in average time between the girls significant at the 5 percent level?
(Use a two-tailed test.)

19. A coffee company was testing two new types of jars for its brand of instant
coffee. To conduct the test 200 stores were selected, and each type of jar
was introduced to one half of the stores. Sales records were kept for each
store. The sales of the new jars were expressed as a percentage of previous
monthly sales. For jar A, the average sales increase was 3 percentage points
with a standard deviation of 20 percentage points. For jar B, the average
sales increase was 8 percentage points with a standard deviation of 24 percentage points.
   a) Is there significant evidence that the average sales increase for jar A is
      greater than 0 percent?
   b) It there significant evidence that the average sales increase for jar B is
      greater than 0 percent?
   c) Is there a significant difference between the sample means?

20. Suppose two brands of cigarettes are tested for burning time with the purpose of deciding whether one brand is longer burning than the other. One
hundred cigarettes of brand No. 1 are burned under test conditions and
the length of burning time is noted, and 144 cigarettes of brand No. 2 are
similarly tested. The following means and standard deviations result (the
subscripts referring to the brand number):

	No. 1	No. 2
	$n_1 = 100$	$n_2 = 144$
	$\bar{X}_1 = 9.36$ minutes	$\bar{X}_2 = 9.00$ minutes
	$s_1 = 0.83$ minutes	$s_2 = 1.20$ minutes

Estimate the difference in the mean burning time between the two brands and determine a 95 percent confidence interval for this difference.

21. The loan department of a certain bank specializes in loans to small businesses. For these loans, it is important to have an accurate evaluation of the financial standing of the business. To make this evaluation, a credit officer reviews the financial statements and application forms, and even interviews the applicant if desired and forms an opinion of the applicant's credit rating. This is expressed as an integer between 0 and 9, 9 being an excellent rating and 0 being the rating of a very poor credit risk.

The management of the bank wished to be sure that the two credit officers, Green and Gray, were using the same standards in giving credit ratings. Accordingly, 30 applicants were selected at random, and Green and Gray were asked to make an independent evaluation. The results are shown below:

Application Number	Green Evaluation X_1	Gray Evaluation X_2	Difference d
1	8	7	1
2	5	3	2
3	6	7	-1
4	9	9	0
5	1	2	-1
6	4	2	2
7	5	5	0
8	8	6	2
9	7	4	3
10	5	6	-1
11	2	1	1
12	2	2	0
13	1	0	1
14	6	7	-1
15	5	4	1
16	3	3	0
17	6	6	0
18	6	5	1
19	4	5	-1
20	3	1	2
21	6	6	0
22	5	4	1
23	4	4	0
24	5	5	0
25	4	3	1
26	3	5	-2
27	4	3	1
28	8	9	-1
29	8	5	3
30	4	3	1
Total	147	132	+15
Mean	4.90	4.40	0.5
Sum of Squares	849	726	53

Management realized that there would be differences in the evaluation of individual applicants but wanted the credit officers to give the same average evaluation.

a) Using the evaluations of the 30 applicants by Green and Gray as separate samples, test the hypothesis that there is no difference in their evaluations, on the average. Is the observed difference significant?

b) The fourth column in the above table shows the difference *d* between the evaluation of Green and Gray. Using this set of 30 observations as one sample, test the hypothesis that the mean of the difference *d* is equal to zero. Is the observed difference significant?

c) Compare the two methods of (*a*) and (*b*) for evaluating the differences between means. Why is the second more efficient than the first?

22. Refer to Problem 8 in Chapter 4. Is the observed difference in the averages of the two types of lamps significant?

23. A production supervisor wished to estimate the percentage of time a certain machine was idle because of breakdowns, delays, etc. Since it would be difficult to keep accurate records, a sampling procedure was instituted. Accordingly, the status of the machine was checked by the supervisor over a period of four weeks at random times (i.e., the times were selected in advance, using a table of random numbers). This procedure is known as *work sampling*. A total of 300 checks were made on the machine, and in 24 instances the machine was idle.

a) Estimate the percentage of idle time on the machine and calculate a 90 percent confidence interval about the estimate.

b) Determine if the percentage of idle time is significantly less than 10 percent.

24. In a brand-preference survey of 1,600 consumers in a given area, 760 expressed a preference for Brand A and 840 for all other brands combined.

a) Construct a 95 percent confidence interval for the proportion favoring brand A.

b) Is the proportion of customers who prefer brand A significantly less than one half?

c) Is the proportion favoring brand A in this city significantly different from that in another city, where 600 out of 1,200 favored brand A?

d) Construct a 99 percent confidence interval for the difference tested in part (*c*).

25. The median life of a certain electronic tube is claimed by the manufacturer to be 600 hours. You draw a random sample of 100 from a shipment of these tubes and find that only 23 last over 600 hours. Do you believe the manufacturers' claim? Why? (Hint: 50 percent of the values exceed the median.)

26. After finding that 23 out of 100 electronic tubes from manufacturer No. 1 outlast 600 hours, you order a shipment of similar tubes from manufacturer No. 2 and find that 52 out of a random sample of 200 outlast 600 hours. Is

there a significant difference in the durability of the two manufacturers' tubes? Explain.

27. As research director of a flour-milling company, you are asked to measure consumer reaction to a newly developed cake mix by having consumers compare it to brand B, which will be its principal competitor.

   a) Assuming that you intend to run a taste test for a random sample of consumers in Chicago and that you want, if possible, to determine consumer preference within a margin of 4 percent (at the 2σ confidence level), how large a sample should you plan on taking? Assume *a priori* that half the consumers favor your product.

   b) If the results of the first 400 interviews indicate that your new product is favored by 57 percent of the consumers interviewed, are you safe in assuming that there is a real preference for your product?

   c) In a second set of 300 interviews made in Miami, 60 percent favored your new product. Assuming that both samples were taken at random and that the interviewing was done without bias, is there a significant difference in the preference for your product between the two cities?

28. The following are data obtained by the management of a department store in a study of delinquent time payment accounts. In a sample of 600 time-payment accounts opened by individuals who had resided in the community for more than five years, 58 had become delinquent at one time or another. In a sample of 400 time-payment accounts for individuals who had resided in the community for less than five years, 26 had become delinquent.

   a) Is the difference between the two significant at the 5 percent level?

   b) What is a possible fallacy in interpreting this difference, whether significant or not?

29. The market research department of the Bodhauser Beer Company conducted a taste test to determine if consumers could distinguish Bodhauser Beer from its chief competitor, Schultz. Accordingly, 200 beer drinkers were selected, given unmarked samples of both beers, and told to state a preference.

   Because it was feared that the order in which the different beers were presented to the test group might affect their preference, the group was broken into two parts; half (Group 1) were given Bodhauser before Schultz, and the other half (Group 2) were given Schultz before Bodhauser. The results are shown in the table below:

	Group 1	Group 2
Number in group	100	100
Number preferring Bodhauser	54	58

   a) Ignoring the order in which the beer was presented (i.e., lumping both groups together), was there significant evidence that either beer was preferred over the other?

   b) Were the initial fears that the order might affect the preference substantiated? That is, is there evidence from the experimental data that the two sampled groups differed?

30. One of the critical component parts of a product manufactured by your company is a size $\frac{5}{16}$-inch carbon steel bolt. In order to meet product specifications this bolt must have a hardness rating between 77.5 and 89.5 on the Rockwell "B" Hardness Scale. Following a heat treatment designed to produce the desired hardness, a sample of four bolts is drawn at random from each lot, and each bolt is tested for hardness. Ten of these samples, taken in consecutive order, are tested on the Rockwell "B" scale and show the following means ($3\sigma_{\bar{x}} = 4.26$):

Sample	$\bar{X}$
1.....................	85.375
2.....................	81.875
3.....................	86.125
4.....................	83.250
5.....................	84.125
6.....................	84.125
7.....................	85.625
8.....................	86.375
9.....................	86.625
10....................	87.625
Total....................	851.125

(Ten samples are used here to minimize computations. In practice, however, at least 20 or 25 samples are needed for reliable results.)

a) Set up an $\bar{X}$ chart to control the hardness of these bolts and plot central line, control limits, and subgroup means.

b) Does the heat-treating process appear to be in statistical control? If so, what is your best estimate of the average hardness rating of all bolts produced by this process?

c) If any points are out of control, revise the limits accordingly and plot the results on the chart.

31. A test of 2,000 transistors, in 20 lots each containing 100 transistors, shows 10 percent defective on the average. What is the maximum percentage defective the inspector should allow on the next lot for it to be within $3\sigma_p$ control limits? (Note: $\sigma_p = \sqrt{pq/n}$.)

32. A quality control engineer is about to set up a control chart for a production process. The process, when in control, produces items with a mean of 40 and a standard deviation of 5. For simplicity, we assume that there are two states in which the process is out of control, one with a process mean of 48 and the other with a process mean of 36. Both have a process standard deviation of 5 (there is never any change in the variability of the process). The costs (economic losses) for these various events are shown in the table.

Possible Events: Process Average Is	Action: Accept the Process	Action: Reject the Process
36.........................	$ 800	0
40.........................	0	$1,200
48.........................	1,000	0

The quality control engineer wants to use an $\bar{X}$ Chart, sample size four, having control limits $40 \pm k\sigma_{\bar{x}}$. He wishes to select an optimal value for k, Accordingly, he constructs the following table:

Process Average Is	Average (Expected) Costs		
	$k = 1$	$k = 2$	$k = 3$
36	A	B	C
40	D	E	F
48	G	H	I

a) Find the values A through I to fill in the table.

b) Explain how you might go about deciding what value of k to use.

SELECTED READINGS

BOWKER, A. H., and LIEBERMAN, G. J. *Engineering Statistics*. 2d ed. Englewood Cliffs, N.J.: Prentice-Hall, 1963.

Contains an authoritative and readable treatment of the application of statistical inference to quality control problems.

DIXON, W. J., and MASSEY, F .J. *Introduction to Statistical Analysis*. 3d ed. New York: McGraw-Hill, 1969.

An excellent reference source on the use of statistical inference in a variety of situations. Chapters 6, 7, and 8 discuss statistical inference, estimation, and tests of hypotheses. Chapter 14 discusses Type II error in detail.

FREUND, J. E., and WILLIAMS, F. J. (rev. by PERLES, B., and SULLIVAN, C.). *Modern Business Statistics*. Englewood Cliffs, N.J.: Prentice-Hall, 1969.

Chapters 8, 9, and 10 are a readable treatment of sampling, estimation, and hypothesis testing. Appendix II is an elementary treatment of quality control applications.

GUENTHER, WILLIAM C. *Concepts of Statistical Inference*. New York: McGraw-Hill, 1965.

An extended treatment of inference at the elementary level.

HAMBURG, M. *Statistical Analysis for Decision Making*. New York: Harcourt, Brace & World, 1970.

Chapters 5 through 8 roughly parallel the material on statistical inference given in this text.

HOEL, PAUL G. *Introduction to Mathematical Statistics*. 4th ed. New York: John Wiley, 1971.

Presents the mathematical foundations of statistical inference at an intermediate level for those with a background in calculus.

RICHMOND, SAMUEL B. *Statistical Analysis*. 2d ed. New York: Ronald Press, 1964.

Chapters 6 to 8 provide a rigorous treatment of statistical inference.

SCHLAIFER, ROBERT. *Introduction to Statistics for Business Decisions.* New York: McGraw-Hill, 1961.

  Chapters 10 and 11 present a treatment of the classical theory of statistical inference from a Bayesian decision-theory approach.

WALLIS, W. ALLEN, and ROBERTS, HARRY V. *Statistics: A New Approach.* New York: The Free Press, 1956.

  Part III treats a wide variety of topics in inference, with many examples.

11. FURTHER TEST PROCEDURES: t, χ^2, and F DISTRIBUTIONS

IN THE PREVIOUS two chapters, the discussion of statistical inference was based upon the assumption that a large sample was taken, resulting in a sampling distribution for the sample mean or proportion that was approximately normal. But many other sampling situations are not covered by this assumption. In particular, we may have a small sample of a variable or attribute; we may have an attribute that is classified into more than two categories, so that the binomial distribution is not applicable; we may have samples from two or more populations to evaluate simultaneously; or, finally, we may have data that are measured in terms of ranks, rather than on an interval scale. The purpose of this chapter is to extend the results of the previous chapters to these special cases. To do this, we shall introduce three new distributions, the t, χ^2 (chi-square), and F distributions, as well as nonparametric statistical tests.

TESTS OF MEANS: SMALL SAMPLES

The assumption of large samples which has been made in Chapters 9–10 was necessary to insure (1) that the sampling distribution of the sample mean was approximately normal and (2) that little error was introduced by estimating the population standard deviation σ by the sample standard deviation s. Because of these properties, large sample estimation is quite generally applicable, making possible statistical inferences without any specific assumption about the shape of the distribution from which the sample was drawn. But in certain situations it is not possible or economical to obtain a large sample. Does this mean that statistical probability statements cannot be made in these situations? The answer to this question is a strong no, together with the qualification that additional assumptions or other methods are necessary. One method of dealing with small samples can be used when the population distribution from which the sample is drawn is *normal* or approximately normal. There are two cases, depending on whether σ is known.

Case A: Sampling from a Normal Population, σ Known. The central limit theorem discussed in Chapter 9 states that means of large samples are approximately normally distributed. The same is true for small samples, provided the population from which the sample is drawn is normal (i.e., means of samples, both large and small, from normal populations are normally distributed). And if the standard deviation σ is known, the analysis can proceed exactly as in the previous two chapters. The standard error of the sample mean is, as before, $\sigma_{\bar{x}} = \sigma/\sqrt{n}$ (times the finite population correction $\sqrt{1 - n/N}$ if it applies). Confidence intervals for the population mean, as well as tests of hypothesis, can be formulated in the same way as before.

Case B: Sampling from a Normal Population, σ Unknown. When the population standard deviation σ is not known, it must be estimated from the data in the small sample. To handle the sampling error in both the sample mean $\bar{X}$ and the sample standard deviation s, a new sampling distribution must be introduced.

This symmetric but nonnormal distribution is called the t distribution. The ratio t (like the standard normal deviate z) is defined as the deviation of the sample mean from the population mean expressed in standard error units. That is,

$$t = \frac{\bar{X} - \mu}{s_{\bar{x}}}$$

where $s_{\bar{x}}$, the standard error of the mean, is computed from s, the standard deviation of a sample, by the formula $s_{\bar{x}} = s/\sqrt{n}$ (times $\sqrt{1 - n/N}$ if needed).

The sampling distribution of t differs for each size of sample. There is one t distribution for samples of size 10, another for size 11, and so on. Hence, the values of t corresponding to the 5 and 1 percent probability levels are not 1.96 and 2.58 as in the normal curve, but depend on the sample size, as shown in Table 11–1.

Table 11–1 is abstracted from the more detailed t table in Appendix M. In this table the first column lists the "degrees of freedom" rather than sample size. The concept of *degrees of freedom* (represented by the abbreviation *df* or the symbol *d*) is important and occurs repeatedly

Table 11–1

VALUE OF t AT 5 AND 1 PERCENT
PROBABILITY LEVELS

Degrees of Freedom (d)	.05	.01
10	2.228	3.169
20	2.086	2.845
30	2.042	2.750
∞	1.960	2.576

throughout this chapter. It refers to the number of independent observations used to make a sample estimate. In calculating the sample mean $\bar{X}$, all n observations are independently determined. However, in calculating the sample variance, the sample mean $\bar{X}$ is used in the formula $s^2 = \Sigma(X - \bar{X})^2/(n - 1)$. There are only $n - 1$ independent terms in the summation in the numerator, because once $n - 1$ deviations from the mean have been determined, the last one is fixed (since the sum of all deviations about the mean must equal zero). Hence there are $n - 1$ degrees of freedom in any inference using s^2 as an estimate of σ^2. In general, the number of degrees of freedom is the number of observations minus the number of unknowns or parameters estimated from the data.

The t distribution depends upon the number of degrees of freedom in the sample estimate. In calculating confidence intervals or testing hypotheses about a single sample mean $\bar{X}$, there are $n - 1$ degrees of freedom. Since Appendix M gives values up to 30 degrees of freedom, we can define a small sample, for the purpose of using this table, as one in which n is 31 or less. The t distribution looks more and more like the normal distribution as n increases in size, so the t values approach the corresponding values for the normal distribution. These are listed in the last row of the table. The probabilities in the heading of the table refer to the sum of the two-tailed areas under the curve that lie outside the points $\pm t$. The values of t are listed in the body of the table. For a single-tailed area, divide the probability by two.

As an example, for a sample of size 8, enter the row $n - 1 = d = 7$; then 5 percent of the area under the curve falls in the two tails outside the interval $t = \pm 2.365$. That is, $2\frac{1}{2}$ percent of the area falls in each tail, and 95 percent of the area falls within the interval $t = \pm 2.365$. A t value of 2.365 therefore should be used in setting up a 95 percent confidence interval for the mean when the sample size is 8.

Confidence Intervals

As an example, a manufacturer wishes to estimate the average weight of a large shipment of 20-gauge uncoated steel sheets received from a supplier. The estimate is to be expressed as a 95 percent confidence interval centered on a sample mean. He selects 8 pieces at random, and finds that the sample mean is 148.4 pounds per hundred square feet, while the standard deviation is 2.07 pounds. The standard error of the mean is then

$$s_{\bar{x}} = \frac{s}{\sqrt{n}} = \frac{2.07}{\sqrt{8}} = .73 \text{ pounds}$$

To find the 95 percent confidence interval, he finds $t = 2.365$ in the table as described above. The confidence interval is then

$$\bar{X} \pm t \cdot s_{\bar{x}} = 148.4 \pm 2.365(.73) = 148.4 \pm 1.7 \text{ pounds}$$

He can then state that the average weight of the whole shipment lies between 146.7 and 150.1 pounds, with a 95 percent chance of being correct.

Testing Hypotheses

Alternatively, the manufacturer in the foregoing problem might wish to test whether the mean weight of the sample of steel sheets (148.4 pounds) was significantly less than the specification of 150 pounds called for in his purchase order. That is, we test the null hypothesis that $\mu \geq 150$ pounds. Since the manufacturer is concerned only with deviation below the specification, we use a one-tail test. Suppose we select a 5 percent level of significance. We now compute the deviation of the sample mean from this hypothetical mean in units of the estimated standard error (.73 pounds) as follows:

$$t = \frac{\bar{X} - \mu_h}{s_{\bar{x}}}$$

$$= \frac{148.4 - 150}{.73} = -2.19$$

In order to find a one-tailed 5 percent probability point, we look up the 10 percent (two-tailed) point in Appendix M for $d = 7$ degrees of freedom. This value is 1.895. Since the absolute value of t, -2.19, is greater than 1.895 we can reject the hypothesis that $\mu \geq 150$ pounds at the 5 percent level. The sample mean is significantly less than the specification of 150 pounds at the 5 percent level of significance.

Test of the Difference between Population Means

In Chapter 10, we tested the difference between two population means involving large samples. If the samples from each population are small, the t distribution may be used in similar fashion to test for the differences in population means. Additional assumptions are required: (1) The two sampled populations are normally distributed, and (2) the standard deviations of the two populations are equal.[1] Consider a sample of size n_1 from the first population and size n_2 from the second population. Then:

$\bar{X}_1$ and $\bar{X}_2$ are the sample means from the two populations

s_1 and s_2 are the sample standard deviations from the two populations

The first step is to obtain a common or *pooled* estimate for the standard deviation for both populations. Denoting this pooled estimate by s_{po} we have:

$$s_{po} = \sqrt{\frac{(n_1 - 1)\, s_1^2 + (n_2 - 1)\, s_2^2}{n_1 + n_2 - 2}}$$

Since the standard deviations of the two populations are assumed the same, this is the best estimate of the standard deviation in each population. We can then calculate the standard error for each sample mean as:

$$s_{\bar{X}_1} = \frac{s_{po}}{\sqrt{n_1}} \quad \text{and} \quad s_{\bar{X}_2} = \frac{s_{po}}{\sqrt{n_2}}$$

Finally, the sampling error of the distribution of differences in sample means is (as in the previous chapter):

$$s_{\bar{X}_1 - \bar{X}_2} = \sqrt{s_{\bar{X}_1}^2 + s_{\bar{X}_2}^2}$$

By substituting the values of $s_{\bar{X}_1}$ and $s_{\bar{X}_2}$ above, this can be simplified to:

$$s_{\bar{X}_1 - \bar{X}_2} = s_{po} \sqrt{\frac{1}{n_1} + \frac{1}{n_2}}$$

Finally, the ratio:

$$t = \frac{(\bar{X}_1 - \bar{X}_2)}{s_{\bar{X}_1 - \bar{X}_2}}$$

---

[1] Tests are also available when the standard deviations are not assumed equal. See W. J. Dixon and F. J. Massey, *Introduction to Statistical Analysis* (3d ed.; New York: McGraw-Hill, 1969), p. 119.

is distributed as a t distribution with $(n_1 + n_2 - 2)$ degrees of freedom. This may be used to calculate confidence intervals or test hypotheses about the population means μ_1 and μ_2.

Example. A company is interested in knowing if there is a difference in the average salary received by foremen in its two divisions. Accordingly, samples of 12 foremen in the first division and 10 foremen in the second devision are selected at random. Based upon past experience, foremen's salaries are known to be approximately normally distributed, and the standard deviations of salaries in the two divisions are about the same. The sample results are:

	First Division	*Second Division*
Sample size.....................	$n_1 = 12$	$n_2 = 10$
Average monthly salary of foremen in sample.................	$\bar{X}_1 = \$1050$	$\bar{X}_2 = \$980$
Standard deviation of salaries in sample.......................	$s_1 = \$68$	$s_2 = \$74$

The null hypothesis is: $\mu_1 - \mu_2 = 0$. The alternative hypothesis (two-tailed) is: $\mu_1 - \mu_2 \neq 0$.

We first calculate the pooled estimate of the common standard deviation:

$$s_{po} = \sqrt{\frac{(n_1 - 1)\, s_1^2 + (n_2 - 1)\, s_2^2}{n_1 + n_2 - 2}}$$

$$= \sqrt{\frac{11\,(68)^2 + 9\,(74)^2}{12 + 10 - 2}} = 70.8$$

Then,

$$s_{\bar{X}_1 - \bar{X}_2} = s_{po}\sqrt{\frac{1}{n_1} + \frac{1}{n_2}} = 70.8\sqrt{\frac{1}{12} + \frac{1}{10}}$$

$$= 30.3$$

And,

$$t = \frac{(\bar{X}_1 - \bar{X}_2)}{s_{\bar{X}_1 - \bar{X}_2}} = \frac{(1050 - 980)}{30.3} = 2.31$$

The 5 percent t value from Appendix M with $(n_1 + n_2 - 2) = (12 + 10 - 2) = 20$ degrees of freedom[2] is 2.086. Since 2.31 is greater than this, we can reject the null hypothesis at the 5 percent level

---

[2] Note that the combined sample size $(n_1 + n_2)$ was reduced by two to obtain the degrees of freedom. This is because two sample estimates—those for $\bar{X}_1$ and $\bar{X}_2$—were used in the formula for $s_{\bar{X}_1 - \bar{X}_2}$.

of significance. There is a significant difference in the foremen's mean salaries in the two divisions.

CHI-SQUARE TESTS

Business and economic data are often classified by attributes into two or more categories. Problems of statistical inference involving two categories (defective, not defective; pass, fail) were represented by proportions in Chapters 9 and 10. We shall now extend the analysis to several categories of classification and to problems involving relationships between attributes.

In previous chapters, the emphasis has been on the sample proportion or percentage in a given category. In this chapter, the emphasis will shift to the frequency or numerical count of items in a category.

The Chi-Square Distribution

Before proceeding, we introduce a new theoretical concept called the chi-square (χ^2) distribution. The χ^2 variable is composed of sums of squared normal random variables. That is, if y_i is a variable that has a standardized normal distribution ($\mu = 0$, $\sigma = 1$), and if the y_i's are independent, then the expression

$$\chi^2 = \left(\sum_{i=1}^{d} y_i^2 \right)$$

has a chi-square distribution. The only parameter of this distribution is d, the degrees of freedom, and d represents the number of independent terms in the summation expression above. Since it involves only squared terms, the χ^2 distribution is always positive. The expected value or mean of the χ^2 distribution is also d, that is $E(\chi^2) = d$; and the variance of the χ^2 distribution is $2d$.

The chi-square distribution for several values of d is shown in Chart 11–1. Note that the distribution is skewed to the right. Appendix N, at the end of the book, is a table of the values of χ^2 for selected *right-tail* probabilities. A small section of Appendix N is reproduced in Table 11–2. For example, this table shows that with six degrees of freedom there is a 99 percent chance that a χ^2 value will exceed .872 and only a 1 percent chance that it will exceed 16.812.

Frequencies and the Chi-Square Distribution

The χ^2 distribution is useful in statistical tests involving comparison between observed frequencies and those that would occur under some

Chart 11–1

CHI-SQUARE DISTRIBUTION

FOR 2, 6, AND 12 DEGREES OF FREEDOM

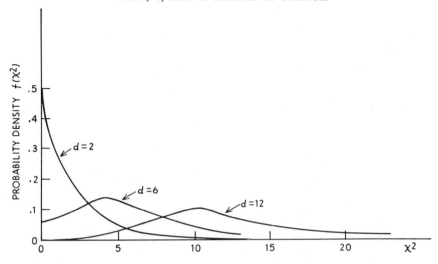

theoretical assumption. To illustrate, suppose a firm is testing a new food product on two groups of women—working women and housewives. The product is compared to one already on the market, and the women express a preference for one or the other. The firm is interested in knowing if the woman prefer the new product to the old, and if there is a difference between the groups. A hypothesis is formulated as follows:

$$p = p_1 = p_2 = .5$$

where p is the true proportion and p_1 and p_2 are the proportions preferring the new product in the sampled groups. The hypothesis states

Table 11–2

VALUES OF CHI-SQUARE FOR SELECTED RIGHT-TAIL PROBABILITIES

Degrees of Freedom d	Right-Tail Probability				
	.99	.95	.50	.05	.01
2	.0201	.103	1.386	5.991	9.210
6	.872	1.635	5.348	12.592	16.812
12	3.571	5.226	11.340	21.026	26.217

SOURCE: Appendix N.

that preferences are equal between groups and between products. Or in other words, the hypothesis implies that the two samples could have come from the same population having $p = .5$.

Let n_1 and n_2 be the sample size and r_1 and r_2 be the number preferring the new product in each group. The values r_1 and r_2 are from a binomial probability distribution. But if the samples are large enough, the normal approximation to the binomial distribution may be used and the expressions

$$\frac{r_1 - n_1 p}{\sigma_{r_1}} \quad \text{and} \quad \frac{r_2 - n_2 p}{\sigma_{r_2}}$$

will each be approximately normally distributed with mean $\mu = 0$ and standard deviation $\sigma = 1$ (that is, they will be standardized normal deviates). Here σ_r is the standard deviation of the binomial distribution (see page 131). Then:

$$\sigma_{r_1} = \sqrt{n_1 p q} \quad \text{and} \quad \sigma_{r_2} = \sqrt{n_2 p q}$$

where $q = 1 - p$. Further, by squaring each term and adding, we obtain the expression:

$$\left[\left(\frac{r_1 - n_1 p}{\sigma_{r_1}} \right)^2 + \left(\frac{r_2 - n_2 p}{\sigma_{r_2}} \right)^2 \right]$$

Recall that the chi-square variable is a sum of squared normal variables. Also note that there are two independent terms in the expression. Hence, the above expression has a χ^2 distribution with two degrees of freedom.

To see how to use this result, we will continue the example and substitute numbers. Suppose the results are as shown in Table 11–3.

Table 11–3

	Group 1 *Working Women*	Group 2 *Housewives*
Sample size..............$n_1 = 100$		$n_2 = 225$
Number preferring new product.................$r_1 = 56$		$r_2 = 130$
Standard deviation (with hypothesis $p = .5$)......$\sigma_{r_1} = \sqrt{100(.5)(.5)}$ $= 5$		$\sigma_{r_2} = \sqrt{225(.5)(.5)}$ $= 7.5$

Then:

$$\chi^2 = \left(\frac{r_1 - n_1 p}{\sigma_{r_1}}\right)^2 + \left(\frac{r_2 - n_2 p}{\sigma_{r_2}}\right)^2 = \left(\frac{56 - 50}{5}\right)^2 + \left(\frac{130 - 112.5}{7.5}\right)^2$$

$$= 2.44 + 5.44 = 7.88$$

Note that if the two groups differ in their responses or if the responses differ from 50 percent, the value of χ^2 will tend to be large. From Table 11–2 (or Appendix N) we can see that a χ^2 value of 5.991 or more could occur by chance with only 5 percent probability. Our value of 7.88 is greater than this, and hence we can reject the hypothesis at the 5 percent level of significance. There is evidence in the data that the two samples do not come from the same population with $p = .5$.

This simple example illustrates the basic idea that the χ^2 distribution can be used in testing hypotheses about frequencies. Rather than proceeding as above and basing our computations on only one of the two possibilities (i.e., only on the number preferring the product), it is easier to count the frequencies for all categories (in this case, both for the number preferring and the number not preferring the product). In general, suppose there are k such categories ($i = 1, 2, \ldots, k$). Let O_i be the observed frequencies in the ith category and let E_i be the expected or theoretical frequency in the ith category. Then the χ^2 statistic is defined by:

$$\chi^2 = \sum_{i=1}^{k} \frac{(O_i - E_i)^2}{E_i} \tag{1}$$

This formula will be illustrated shortly.[3] It is important to note that O_i and E_i in the formula are frequencies (that is, number of occurrences), not relative frequencies (proportion of occurrences).

Hypotheses about Proportions for Several Populations

To illustrate the general χ^2 formula above and to generalize the example in the last section, consider the following problem. A manufacturing facility plates and polishes parts which are later assembled into a final product. A large number of defects have been occurring, and management thinks the polishing operation may be the cause. Four operators polish the parts, using similar machines. To test if there are differences in the defective rate between polishers, 200 finished parts were randomly selected for each operator and the number of defects

---

[3] For the example of this section, it can be shown that formula (1) is algebraically equivalent to the expression above used to illustrate the chi-square test for two samples.

Table 11–4

NUMBER OF PARTS DEFECTIVE AND NOT DEFECTIVE (O_i)

IN SAMPLES FOR FOUR OPERATORS

| | Operator Numbers | | | | | Proportion | Expected Value |
	1	2	3	4	Total	(p)	$E_i = np$
Defective........	21	15	8	16	60	0.075	15
Not defective....	179	185	192	184	740	0.925	185
Total..........	200	200	200	200	800	1.000	200

noted. The results are shown in Table 11–4. The null hypothesis is formulated, stating that there is no difference between the operators or, alternatively, that the four samples could all have come from the same population. This hypothesis is:

$$p = p_1 = p_2 = p_3 = p_4$$

The proportion defective over all four operators was $p_s = 60/800 = .075$. We use this as an estimate of p, the overall proportion defective. If the hypothesis is true and $p = .075$, then the expected number defective for each operator is $np = .075(200) = 15$; and the expected number not defective for each is $200 - 15 = 185$. These are the expected frequencies, the E_i in Formula (1). The observed or O_i values are given in Table 11–4. Note that there are eight categories, the defective and nondefective for each of the four operators. Thus we calculate the χ^2 statistic as:

$$\chi^2 = \sum_{i=1}^{8} \frac{(O_i - E_i)^2}{E_i}$$

$$= \frac{(21 - 15)^2}{15} + \frac{(15 - 15)^2}{15} + \frac{(8 - 15)^2}{15} + \frac{(16 - 15)^2}{15}$$

$$+ \frac{(179 - 185)^2}{185} + \frac{(185 - 185)^2}{185} + \frac{(192 - 185)^2}{185} + \frac{(184 - 185)^2}{185}$$

$$= \frac{86}{15} + \frac{86}{185} = 6.20$$

Degrees of Freedom. Before deciding whether the observed χ^2 value of 6.20 is larger than expected by chance, we need to determine the degrees of freedom (d). Since there are eight terms (one for each category), it might at first seem that $d = 8$. However, all the category values are not independently determined. Once the number of defects for any operator is known, the number of nondefects is immediately known as 200 minus the number of defects. Hence only four of these categories are independently determined. In addition, we estimated p, the overall proportion defective from the data, thus using up one additional degree of freedom.[4] To see this, note that once the overall number of defects is set at 60 (i.e., 7.5 percent), only three of the defect categories can be independently determined—the last being 60 minus the sum of the remaining. Hence there are only 3 degrees of freedom in the χ^2 value obtained above.

Looking in Appendix N for $d = 3$, we see that the .10 value for χ^2 is 6.251. This is greater than our value of 6.20. Hence, we cannot reject the hypothesis at the 10 percent level of significance. There is not sufficient evidence to say that the four operators differ in the number of defects they produce.

Contingency Tables

The example in the previous section tested the hypothesis that the defective rate was independent of the machine operator. This can be extended to test a similar hypothesis of independence of two factors having any number of attribute categories. This analysis is usually presented in terms of a contingency table, which lists the possible categories of one variable along the top of the table and the categories of the other along the side. For example, suppose that an analyst suspects that attendance at movies is related to age. He selects a sample of people and classifies them by age and by the number of movies they have seen in the last month. The data are shown in Table 11–5. Examination of this table seems to indicate that people in the 14 to 19 and 20 to 25 age categories tend to go to the movies relatively more often than either

Table 11–5

SAMPLE OF PERSONS CLASSIFIED BY AGE AND
FREQUENCY OF MOVIE ATTENDANCE (O_i)

Number of Movies Attended in Past Month	Age Group (Years)					Total	Proportion (p_s)
	Under 14	14–19	20–25	26–35	Over 35		
None	85	15	22	120	189	431	.469
1	43	20	22	33	68	186	.202
2	31	31	34	43	75	214	.232
3 or more	3	28	30	18	10	89	.097
Total	162	94	108	214	342	920	1.000

---

[4] Note that in the previous example, the value of p was part of the hypothesis and not estimated from the data. Hence, in that case, no degrees of freedom were utilized in estimating p.

older or younger people. However, there is the possibility that there is no difference in movie attendance by age and that the observed results in the table are due to chance.

To answer this, we formulate the hypothesis that the two factors (age and movie attendance) are statistically independent.[5] This hypothesis implies that attendance at movies in each age category is in the same proportion as the attendance pattern of the whole group (also, that within any movie attendance category, the age distribution is the same as that for the whole population). On the basis of this hypothesis of independence, the theoretical frequency for each cell (i.e., each age-movie attendance category) can be calculated. These frequencies are shown in Table 11–6. The number 75.9 in the first row and first col-

Table 11–6

THEORETICAL OR EXPECTED FREQUENCIES
ON ASSUMPTION OF INDEPENDENCE (E_i)

Number of Movies Attended in Past Month	Age Group (Years)					Total
	Under 14	14–19	20–25	26–35	Over 35	
None..............	75.9	44.0	50.6	100.3	160.2	431
1.................	32.8	19.0	21.8	43.3	69.1	186
2.................	37.6	21.9	25.2	49.7	79.6	214
3 or more..........	15.7	9.1	10.4	20.7	33.1	89
Total.............	162.0	94.0	108.0	214.0	342.0	920

umn is calculated, for example, as follows: first, $431/920 = 46.9$ percent of the 920 people in the whole group saw no movies last month. If the factors are independent, we would expect 46.9 percent of the 162 people in the under-14 age category also to have seen no movies last month; and $.469 \times 162 = 75.9$. The other values in the table are calculated in a similar manner.

The 20 values in Table 11–6 represent the theoretical or expected frequencies—the E_i values of Formula (1). The observed frequencies, the O_i values, are from Table 11–5. And the χ^2 statistic is calculated as:

$$\chi^2 = \sum_{i=1}^{20} \frac{(O_i - E_i)^2}{E_i}$$

$$= \frac{(85 - 75.9)^2}{75.9} + \frac{(15 - 44.0)^2}{44.0} + \cdots + \frac{(10 - 33.1)^2}{33.1}$$

$$= 163.1$$

[5] Statistical independence was described in detail in Chapter 5.

Degrees of Freedom.   There are 20 cells in Tables 11–5 and 11–6 and 20 terms in the summation used to calculate the χ^2 value. However, not all of these are determined independently. Note that the 4 row totals and 5 column totals were used in determining the E_i values. Hence, we should reduce the degrees of freedom to allow for this. The total reduction in degrees of freedom is 8 (4 rows + 5 columns − 1) since once eight row or column totals are determined, the last is also determined. Thus the total degrees of freedom for the χ^2 value calculated above are $20 - 8 = 12$.

In general, for contingency tables:

d = Degrees of freedom
  = Number of cells − number of rows − number of columns + 1
     (rows × columns)

Returning to our example, the χ^2 value for 12 degrees of freedom and .01 level of significance is 26.217 (from Appendix N). Since our observed value of 163.1 is considerably greater than this, we can reject the hypothesis at the .01 level. There is substantial evidence that attendance at movies is related to age.

Contingency tables may be used to test for the existence of a relationship not only between variables that have numerical values[6] (as was the case for both movie attendance and age in our example), but also for factors that have only qualitative characteristics. For example, people could be classified by political party, by sex, and by race; charge accounts can be classified as up-to-date, overdue, and in default.

Cell Size.   The use of the χ^2 distribution in this chapter rests upon an approximation of the normal distribution.[7] Recall that in Chapter 6 in the discussion of the normal approximation to the binomial (page 143), we used a rule of thumb to guarantee an adequate approximation. This rule specified that the sample size had to be large enough so that np and nq would both be greater than 5 (where $q = 1 - p$). The same rule applies to the E_i or expected frequencies in any cell in using the chi-square distribution. All the E_i values should be at least 5. If some E_i values are less than this, it may be necessary to combine categories so that the rule is satisfied. Note that in our example the smallest expected frequency was 9.1 in Table 11–6.

---

[6] For such variables, regression analysis (Chapter 16) may also be used to test for statistical dependence.

[7] This is a normal approximation to the binomial or multinomial distribution. There is also the problem of adjustment for the approximation of a discrete by a continuous distribution. For relatively large samples, this is not a problem. For small samples, corrections may be necessary. See references at the end of the chapter.

Goodness-of-Fit Tests

In general, all chi-square tests used in this chapter represent a test of the "goodness of fit" of observed data to some theoretical or hypothesized frequencies. But the term goodness of fit has generally come to mean testing as to whether some observed data are likely to have come from a hypothesized probability distribution.

In some applications of operations research models to business decision problems, assumptions are made involving certain probability distributions. For example, in queuing or waiting line problems (see Chapter 15), an assumption is often made that customer arrivals follow a Poisson distribution. The goodness-of-fit test can determine whether this assumption is correct. Similarly, the t and F distributions discussed

Table 11–7

FREQUENCY DISTRIBUTION

HOURLY EARNINGS OF 214 APPRENTICE MACHINE TOOL OPERATORS

Hourly Earnings	Midpoint	Number of Operators $f = O_i$
$2.25 and under $2.35.....................	$2.30	2
$2.35 and under $2.45.....................	2.40	23
$2.45 and under $2.55.....................	2.50	49
$2.55 and under $2.65.....................	2.60	63
$2.65 and under $2.75.....................	2.70	45
$2.75 and under $2.85.....................	2.80	25
$2.85 and under $2.95.....................	2.90	3
$2.95 and under $3.05.....................	3.00	4
Total.................................		214

in this chapter require a normal population, and the goodness-of-fit test can be used to check this assumption.

As an example, consider the sample of 214 apprentice machine tool operators discussed in Chapter 2. The frequency distribution of the sample is repeated as Table 11–7.

Suppose we wished to test the hypothesis that this represents a sample from a normal distribution. The sample mean $\bar{X}$ is 2.609 and the sample standard deviation s is .136. We use these as estimates of the corresponding population values for μ and σ. Based upon these estimates and the hypothesis of normality, the probability of a sample value falling in each interval given in Table 11–7 can be calculated. For example, the probability of a sample observation in the interval "$2.45 and under \$2.55" is calculated from the standard normal deviate (z) in Appendix D as follows:

$$z_1 = \frac{2.45 - 2.609}{.136} = -1.169 \text{ and } P(z < -1.169) = .1212$$

$$z_2 = \frac{2.55 - 2.609}{.136} = -.434 \text{ and } P(z < -.433) = .3322$$

Finally, $P(-1.169 < z < -.434) = .3322 - .1212 = .2110$.

Probabilities for the other intervals can be calculated in a similar manner and are shown in Table 11–8. In the last column, the expected

Table 11–8

EXPECTED PROBABILITIES AND FREQUENCIES

NORMAL DISTRIBUTION WITH $\mu = 2.609$, $\sigma = .136$, and $n = 214$

Hourly Earnings	Normal Probability	Expected Frequency (E_i)
Under \$2.35	.0285	6.1
\$2.35 and under \$2.45	.0927	19.8
\$2.45 and under \$2.55	.2110	45.2
\$2.55 and under \$2.65	.2865	61.3
\$2.65 and under \$2.75	.2314	49.5
\$2.75 and under \$2.85	.1117	23.9
\$2.85 and under \$2.95	.0321	6.9
\$2.95 and above	.0061	1.3
	1.0000	214.0

frequencies in each interval are calculated by multiplying the probabilities by the total number of sample items, 214.

The last column of Table 11–8 represent the E_i for Formula (1). Note that the last interval has an expected frequency of only 1.3. Since this is less than 5, it violates our rule of thumb on cell size frequencies. To deal with this, we simply group this interval with the adjacent one and make a new category "\$2.85 and above." This interval has an expected frequency of $6.9 + 1.3 = 8.2$ and an observed frequency of $3 + 4 = 7$.

Finally, the χ^2 statistic can be calculated using Formula (1), obtaining, except as modified above, the observed frequencies (O_i) from Table 11–7 and the expected frequencies (E_i) from Table 11–8. That is,

$$\chi^2 = \sum_{i=1}^{7} \frac{(O_i - E_i)^2}{E_i}$$

$$= \frac{(2 - 6.1)^2}{6.1} + \frac{(23 - 19.8)^2}{19.8} + \cdots + \frac{(7 - 8.2)^2}{8.2}$$

$$= 4.27$$

Degrees of Freedom. After combining the last two intervals, seven intervals remain. However, two degrees of freedom were used in estimating μ and σ from the sample data. A third degree of freedom was used in making the total expected frequencies (214) match the sample number. Hence there are only $7 - 3 = 4$ degrees of freedom remaining in the χ^2 term above.

From Appendix N, the χ^2 value for the .10 level of significance is 7.779 for 4 degrees of freedom. The observed χ^2 of 4.27 is well below this. In fact it is close to the expected value of χ^2 with 4 degrees of freedom, which is 4.0. Hence, we cannot reject the hypothesis of normality. The data could easily have come from a normal population.[8]

The above procedure may be used in exactly the same fashion to test if observed data are consistent with a Poisson, binomial, exponential, or other probability distribution. Note that the number of degrees of freedom in each case is: degrees of freedom = number of intervals, minus number of parameters estimated in fitting the data, minus one (for matching total frequencies).

THE F DISTRIBUTION AND ANALYSIS OF VARIANCE

In this section we will treat two types of statistical tests: (1) those dealing with hypotheses about population variances, and (2) those comparing the means of several populations. Although these may seem like quite different tests, they have in common the use of the F distribution.

The F Distribution

Suppose we have two independent random variables y_1 and y_2, each having a χ^2 probability distribution with d_1 and d_2 degrees of freedom respectively. Then the ratio:

$$F = \frac{y_1/d_1}{y_2/d_2}$$

---

[8] Note that this is the same conclusion that was reached by the graphic method of using normal probability paper (Chart 6–5) in Chapter 6.

has an F distribution. The F distribution has two parameters, d_1 and d_2, the degrees of freedom in the numerator and denominator respectively. To indicate this, the F variable is sometimes written as $F(d_1, d_2)$.

The F variable cannot be negative (since neither of the χ^2 variables can be negative) and has an expected value of approximately 1.0.[9] Since there is a different F distribution for each value of d_1 and d_2, it would require a book to furnish complete tables. Instead, Appendix O shows values of F only for the right-tail probabilities of .05 (in light type) and .01 (in bold type). A portion of Appendix O is shown as Table 11–9. The table indicates, for example, that when $d_1 = 6$ and $d_2 = 10$,

Table 11–9

SELECTED VALUES FOR F DISTRIBUTION
RIGHT-TAIL PROBABILITIES

.05 (Light Type) and .01 (Bold Type)

d_2 (Denominator)	d_1 (Numerator)		
	2	6	10
2	19.00	19.33	19.39
	99.01	**99.33**	**99.40**
6	5.14	4.28	4.06
	10.92	**8.47**	**7.87**
10	4.10	3.22	2.97
	7.56	**5.39**	**4.85**

there is a .05 chance that the F variable will be larger than 3.22 and a .01 chance that it will exceed 5.39.

Test of Equality of Population Variances

One application of the F distribution is in testing the hypothesis that the variances (σ_1^2 and σ_2^2) of two normally distributed populations are equal. Suppose two samples of size n_1 and n_2 are drawn from different populations, with the resulting sample means $\bar{X}_1$ and $\bar{X}_2$ and sample standard deviations s_1 and s_2. Consider the ratio:

$$F[(n_1 - 1), (n_2 - 1)] = \frac{s_1^2/\sigma_1^2}{s_2^2/\sigma_2^2} \tag{2}$$

---

[9] The actual expected value of the F distribution is $d_2/(d_2 - 2)$. Note that for very small sample sizes in the denominator of the F ratio, this can be far from 1.0.

Now:

$$\frac{s^2}{\sigma^2} = \frac{\Sigma(X - \bar{X})^2}{(n - 1)\sigma^2}$$

and the term:

$$\frac{\Sigma(X - \bar{X})^2}{\sigma^2}$$

is a sum of squared normal deviates; thus both the numerator and denominator of Formula (2) have χ^2 distributions divided by $(n - 1)$ degrees of freedom. Hence, from our definition, the ratio in Formula (2) has an F distribution with $(n_1 - 1)$ and $(n_2 - 1)$ degrees of freedom, respectively.

We can now use Formula (2) to test the hypothesis that two population variances are the same. If the hypothesis is true $(\sigma_1^2 = \sigma_2^2)$, then the σ's cancel in Formula (2) and the F ratio is reduced to:

$$F[(n_1 - 1), (n_2 - 1)] = \frac{s_1^2}{s_2^2}$$

If the hypothesis is true, s_1 and s_2 should have nearly the same value, and hence the ratio should be about one. If the ratio is far from one, there is evidence that the hypothesis is not true. The F distribution will determine the limits of chance variation.[10]

Example. Two samples are drawn from two normal populations with $n_1 = 11$, $s_1 = 28.0$, $n_2 = 7$, and $s_2 = 21.4$. Prior to using the t test (described earlier) we may wish to test the assumption that the two populations have equal variances. The F ratio is:

$$F(10,6) = \frac{s_1^2}{s_2^2} = \frac{(28.0)^2}{(21.4)^2} = 1.71$$

Since this is less than the 5 percent F value for 10 and 6 degrees of freedom (from Table 11–9 or Appendix O), which is 4.06, we cannot reject the hypothesis at the 5 percent level. The differences between the sample variances could easily be attributable to chance.

---

[10] This is a two-tailed test, and can be rejected if F is either too small or too large. Note, however, that our tables of the F distribution only give right-tail values. By simply always placing the larger s^2 in the numerator of the ratio, the right-tail value can always be used. (Alternatively, the reciprocal of the right-tail value is the left-tail value.)

Analysis of Variance: Testing the Equality of Several Population Means

Tests of the difference between two sample means were discussed earlier in this chapter and in the previous chapter. It is sometimes useful to test the hypothesis that the means of several sampled populations are equal. For example, an instructor may try different methods of instruction (standard lecture method, programmed instruction, or audiovisual instruction) on different sections of a course. Each of these methods represents a different experimental condition or treatment. The instructor may wish to know if the observed differences on a final examination are a result of the different treatments or could be attributable to chance variation. Or the experiment may involve three groups of students (lower level undergraduates, upper level undergraduates, and graduate students), and the instructor may wish to simultaneously esti-

Table 11–10

SALES IN 12 STORES (CASES PER MONTH)

THREE PROMOTION METHODS

	Point-of-Sale Ads	Newspaper Ads	Use of Demonstrator
	5	10	23
	3	15	18
	10	8	16
	6	7	11
Group average......	$\bar{X}_1 = 6.0$	$\bar{X}_2 = 10.0$	$\bar{X}_3 = 17.0$

mate the separate effects of the instruction method and the grade level by testing the statistical significance of each set of factors. The F test can be used for this purpose, as illustrated below.

Suppose a firm is interested in three methods for promoting a new food product: (1) point-of-sale advertising material, (2) newspaper advertising, and (3) use of a demonstrator in the store. To test the effectiveness of the three methods, each is tried in a sample of 4 stores of about equal size (a total of 12 stores). The sales in cases per month are shown in Table 11–10. Note that the group averages vary from 6 to 17. The overall average, $\bar{\bar{X}}$, is 11.0. The firm wishes to know if the observed differences are significant, or if they could be attributable to chance variation.

Before analyzing the problem, two assumptions are made: (1) Store sales within each group (that is, for each promotion method) are normally distributed. (2) The variances of sales within each group are the same. That is:

$$\sigma_1^2 = \sigma_2^2 = \sigma_3^2 = \sigma^2$$

The null hypothesis is that there is no difference in the population means; that is, $\mu_1 = \mu_2 = \mu_3 = \mu$.

The technique used to test this hypothesis is called *analysis of variance*. It involves breaking the total variation of all observations about the overall mean $\overline{\overline{X}}$ into two parts:

1. The within-group variation—the variation of the individual observations about the group means.
2. The between-group variation—the variation of the group means about the overall mean.

Within-Group (Pooled) Variance. Note that the sample variance for point-of-sale advertising for the four stores is calculated as:

$$
s_1^2 = \frac{\displaystyle\sum_{i=1}^{4}(X_i - \overline{X}_1)^2}{n-1}
$$

$$
= \frac{(5-6)^2 + (3-6)^2 + (10-6)^2 + (6-6)^2}{4-1} = \frac{26}{3} = 8.67
$$

We can alternatively describe this as:

$$
s_1^2 = \frac{\text{Sum of squared deviations}}{\text{Degrees of freedom}} = \frac{SSD_1}{df_1} = \frac{26}{3}
$$

Values of $s_2^2 = SSD_2/df_2 = 38/3$ and $s_3^2 = SSD_3/df_3 = 74/3$ can be calculated similarly. Recall the assumption that the population variances in each group were equal. Hence, each of the s^2's is an estimate of the common variance σ^2. To combine or pool these estimates into a common estimate, we add the SSD for each group and divide by the sum of the degrees of freedom:[11]

$$
\text{Estimated } \sigma^2 = \frac{SSD_1 + SSD_2 + SSD_3}{df_1 + df_2 + df_3} = \frac{SSD_w}{df_w} = \frac{138}{9} = 15.3
$$

where the subscript w refers to "within group."

Between-Group Variance. The calculation of the sum of squared deviations of the sample (group) means about the overall mean is:

---

[11] When there is the same number of sample items in each group, as in our example, this procedure is the same as averaging the sample variances.

$$SSD_b = (\bar{X}_1 - \bar{\bar{X}})^2 + (\bar{X}_2 - \bar{\bar{X}})^2 + (\bar{X}_3 - \bar{\bar{X}})^2$$
$$= (6 - 11)^2 + (10 - 11)^2 + (17 - 11)^2 = 62$$

where the subscript b refers to "between group." Dividing this SSD_b by two degrees of freedom (three groups less one degree of freedom used to estimate $\bar{\bar{X}}$), we have an estimate of the between-group variance $\sigma_{\bar{X}}^2$:

$$\text{Estimated } \sigma_{\bar{X}}^2 = \frac{SSD_b}{df_b} = \frac{62}{2} = 31$$

That is, 31 is an estimate of the variability of means of samples of size $n = 4$ (the number of stores in each group) about the overall population mean.

If the hypothesis $\mu_1 = \mu_2 = \mu_3 = \mu$ is true, then the samples from each group can be considered as samples from the same population with mean μ. Recall from Chapter 9 that in this case the sampling error $\sigma_{\bar{X}}^2 = \sigma^2/n$ is also a measure of the variability of sample means. Rewriting this as $n\sigma_{\bar{X}}^2 = \sigma^2$, we see that if the hypothesis is true,[12] $n \cdot$ (estimated $\sigma_{\bar{X}}^2$) is an estimate of σ^2. If the hypothesis is not true, on the other hand, $n \cdot$ (estimated $\sigma_{\bar{X}}^2$) could be expected to be larger than σ^2.

Total Variance. The total variance of the individual observations about the overall mean can also be calculated as:

$$\text{Total variance} = \frac{\sum_{i=1}^{12} (X_i - \bar{\bar{X}})^2}{12 - 1} = \frac{SSD_t}{df_t}$$
$$= \frac{(5 - 11)^2 + (3 - 11)^2 + \cdots + (11 - 11)^2}{11}$$
$$= \frac{386}{11} = 35.1$$

where the subscript t means "total." The results of all these calculations may be summarized in the analysis of variance table shown as Table 11–11. Note that the sum of the squared deviations and degrees of freedom for the between-groups and within-groups categories add to the totals for the two categories.

The last column in Table 11–11 gives two independent estimates for

---

[12] The formula and calculation shown are valid if the sample size in each group is the same, as in our example. If this is not true, then

$$SSD_b = \sum_{i=1}^{k} n_i(\bar{X}_i - \bar{\bar{X}})^2$$

where there are k groups and n_i is the sample size in the ith group.

Table 11–11

ANALYSIS OF VARIANCE TABLE

Source of Variation	Sum of Squared Deviations		Degrees of Freedom	Estimate of σ^2
Between groups........	$n \cdot SSD_b = 4(62) =$	248	2	124.0
Within groups.........	$SSD_w =$	138	9	15.3
Total.................	$SSD_t =$	386	11	

σ^2, the common variance within each group. If the hypothesis is true, these should differ from each other only by chance variation. Previously, the F distribution was shown to represent the ratio of two sample variances. Hence, the F distribution can be used to test our hypothesis. We calculate the ratio:

$$F = \frac{\text{Estimate of } \sigma^2 \text{ from between groups}}{\text{Estimate of } \sigma^2 \text{ from within groups}} = \frac{124.0}{15.3} = 8.10$$

Looking up the .01 value of F in Appendix O for two and nine degrees of freedom, we find 8.02. Since the observed value of 8.10 is greater than this, we can reject the hypothesis at the 1 percent level of significance. The variation between group means is too large to be attributable to chance.

Two Factors of Classification

Suppose the firm in the previous example was interested in testing three formulas for its product in addition to the three methods of promotion. For example, on a frozen meat pie, the crust might be hard, medium, or soft. To test both these factors, two experiments could be designed. However, it is much more efficient to test for both factors at once. This is called *two-way analysis of variance.* A design such as that shown in Table 11–12 might be utilized. With this design, six stores are used to estimate the effect of each promotion method and, similarly, six stores are used for each crust formula. But the total design requires only 18 stores.

Suppose the design is carried out, the experiment conducted in the 18 stores, and the results obtained are as shown in Table 11–13. There are two sampled stores in each cell, and these are shown separated by a comma.

The Experimental Model. Before proceeding to analyze these data, we must examine the underlying experimental model that is assumed. We define the following terms.

Table 11–12

EXPERIMENTAL DESIGN TO TEST EFFECTIVENESS OF
PROMOTION METHODS AND CRUST FORMULAS

Promotion Method

Differences in Product	column row	Point-of-Sale Ads 1	Newspaper Ads 2	Use of Demonstrator 3	Total
Hard crust........1		2 stores	2 stores	2 stores	6 stores
Medium crust.....2		2 stores	2 stores	2 stores	6 stores
Soft crust........3		2 stores	2 stores	2 stores	6 stores
Total...........		6 stores	6 stores	6 stores	18 stores

Overall Mean. The overall mean is the expected value over all rows and columns, called μ.

Row Effects. Row effects are the effects of the different crust formulas. They are measured as differences from the overall mean. Let R_i be the row effect for the ith row.

Column Effects. Column effects are the effects of the different promotion methods. Again these are measured as deviations from the overall mean and are designated C_j for the column effect of the jth column.

Interaction Effects. The effect in any cell is assumed to be a sum of the row and column effects. However, sometimes there is an interaction effect in which the effect in the cell is greater (or less) than the combined row and column effects. For example, two drugs taken separately may have little effect, but if taken in combination may have substantial

Table 11–13

SALES OF A SAMPLE OF 18 STORES FOR
SELECTED PROMOTION METHODS AND CRUST FORMULAS

(In Cases per Month)

Promotion Method

Differences in Product	column row	Point-of-Sale Ads 1	Newspaper Ads 2	Use of Demonstrator 3	Average
Hard crust........1		10,6	14,8	18,16	12.0
Medium crust.....2		3,8	12,9	19,15	11.0
Soft crust........3		5,4	10,7	23,11	10.0
Average.........		6.0	10.0	17.0	

effects. The results are called interaction effects. In our example, the existence of interaction effects might mean, for instance, that newspaper advertising was effective when used in combination with a soft crust formula, but not with a hard crust. We designate the interaction in the ith row, jth column, by I_{ij}. The experimental model is then:

$$X_{ijk} = \mu + R_i + C_j + I_{ij} + \epsilon_{ijk}$$

where X_{ijk} is the kth observation in the ith row, jth column; μ is the overall mean; R_i, C_j, and I_{ij} are the row, column, and interaction effects; and ϵ_{ijk} is the residual or unexplained variation.

Estimates. Our first task is to estimate these effects.

Overall Mean. Let $\bar{\bar{X}}$ be the overall average of the sample data. In our example, $\bar{\bar{X}}$ is 11.0. This is an estimate of μ.

Row Effects. Let $\bar{X}_{R_1}$ be the average of the sample items in the ith row. For example, $\bar{X}_{R_1} = 12$. Then $(\bar{X}_{R_1} - \bar{\bar{X}})$ is an estimate of the row effect R_i. That is:

$$\bar{X}_{R_1} - \bar{\bar{X}} = 12 - 11 = \quad 1 \quad \text{(estimate of } R_1\text{)}$$
$$\bar{X}_{R_2} - \bar{\bar{X}} = 11 - 11 = \quad 0 \quad \text{(estimate of } R_2\text{)}$$
$$\bar{X}_{R_3} - \bar{\bar{X}} = 10 - 11 = -1 \quad \text{(estimate of } R_3\text{)}$$

Column Effects. Let $\bar{X}_{C_j}$ be the average of the sample items in the jth column. For example, $\bar{X}_{C_1} = 6$. Then $(\bar{X}_{C_j} - \bar{\bar{X}})$ is an estimate of the column effect C_j. That is:

$$\bar{X}_{C_1} - \bar{\bar{X}} = \quad 6 - 11 = -5 \quad \text{(estimate of } C_1\text{)}$$
$$\bar{X}_{C_2} - \bar{\bar{X}} = 10 - 11 = -1 \quad \text{(estimate of } C_2\text{)}$$
$$\bar{X}_{C_3} - \bar{\bar{X}} = 17 - 11 = \quad 6 \quad \text{(estimate of } C_3\text{)}$$

Interaction Effects. Let $\bar{X}_{ij}$ be the average of the sample items in the cell in the ith row and jth column. For example, $\bar{X}_{11} = 8$. Then the interaction effect is:

$$\begin{array}{ccccc} \text{Cell Average} & - & \text{Row Effect} & - & \text{Column Effect} & - & \text{Overall} \\ \bar{X}_{ij} & & i\text{th row} & & j\text{th column} & & \text{mean} \end{array}$$

Using our symbols and simplifying slightly, this becomes[13]:

$$(\bar{X}_{ij} - \bar{X}_{R_i} - \bar{X}_{C_j} + \bar{\bar{X}})$$

which is an estimate of interaction effect I_{ij}. The estimated interaction effects in our example are shown in Table 11–14.

Analysis of Variance. A question can be raised about whether the row, column, and interaction effects represent merely chance variation,

---

[13] Note that the sign of $\bar{\bar{X}}$ is plus in this second formula. This results because $\bar{\bar{X}}$ is subtracted twice (once in $\bar{X}_{R_i}$ and again in $\bar{X}_{C_j}$) and must be added back once to give the correct estimate.

Table 11–14

ESTIMATED INTERACTION EFFECTS

	Column		
Row	1	2	3
1...................	1.0	0	−1.0
2...................	−0.5	0.5	0
3...................	−0.5	−0.5	1.0

or whether they represent significant differences. An analysis of variance similar to that used in the last section can be used to answer this question. The hypothesis is that there are no significant row, column, or interaction effects. Then the variance is broken down into parts as before. The general approach is given in Table 11–15. Here, c repre-

Table 11–15

TWO-FACTOR ANALYSIS OF VARIANCE

Source of Variation	Sum of Squared Deviations (SSD)	Degrees of Freedom (df)	Estimate of σ^2
Rows...............	$c \cdot n \left(\sum_{i=1}^{r} (\bar{X}_{R_i} - \bar{\bar{X}})^2 \right)$	$r - 1$	SSD/df
Columns.............	$r \cdot n \left(\sum_{j=1}^{c} (\bar{X}_{C_j} - \bar{\bar{X}})^2 \right)$	$c - 1$	SSD/df
Interactions..........	$n \left(\sum_{i=1}^{r} \sum_{j=1}^{c} (\bar{X}_{ij} - \bar{X}_{R_i} - \bar{X}_{C_j} + \bar{\bar{X}})^2 \right)$	$(r - 1)(c - 1)$	SSD/df
Within-Group (cells) (also called residual)..	$\sum_{i=1}^{r} \sum_{j=1}^{c} \sum_{k=1}^{n} (X_{ijk} - \bar{X}_{ij})^2$	$(r \cdot c)(n - 1)$	SSD/df
Total...............	$\sum_{i=1}^{r} \sum_{j=1}^{c} \sum_{k=1}^{n} (X_{ijk} - \bar{\bar{X}})^2$	$(n \cdot r \cdot c - 1)$	

sents the number of columns, r the number of rows, and n the number of observations in each cell.[14]

The results for our example are given in Table 11–16. Recall that the hypothesis is that the row, column, and interaction effects are zero. If this is true, the four numbers in column (4) of Table 11–16 are all

---

[14] It is possible to have a design with a different sample size in each cell. However, this case will not be treated here.

Table 11–16

ANALYSIS OF VARIANCE

Source of Variation (1)	Sum of Squared Deviations (2)	Degrees of Freedom (3)	Estimate of σ^2 (4)	F Ratio (5)
Rows..............	$3 \cdot 2(1^2 + 0^2 + (-1)^2) = 12$	$3 - 1 = 2$	6.0	.417
Columns..............	$3 \cdot 2((-5)^2 + (-1)^2 + 6^2)$ $= 372$	$3 - 1 = 2$	186.0	12.910
Interactions..........	$2(1^2 + 0^2 + (-1)^2$ $+ (-.5)^2 + \cdots$ $+ 1^2) = 8$	$(3-1)(3-1) = 4$	2.0	.139
Within-Group (cells) (or residual).......	$(10 - 8)^2 + (6 - 8)^2$ $+ (14 - 10.5)^2 + \cdots$ $+ (11 - 17)^2 = 130$	$(3 \cdot 3)(2 - 1) = 9$	14.4	
Total..............	$(10 - 11)^2 + (6 - 11)^2$ $+ (14 - 11)^2 + \cdots$ $+ (11 - 11)^2 = 522$	$(2 \cdot 3 \cdot 3 - 1) = 17$		

estimates of σ^2, the within-cell variance. The last value, 14.4, is a direct estimate of this, and is not affected if the hypothesis is false. However, this is not so for the other values in column 4; they are expected to be greater than σ^2 if the effects are not zero.

Again the F distribution may be used to decide if the observed values are greater than chance would allow. The ratios of the first three values in column 4 to the fourth value are shown in column 5. These are the F ratios. They are to be compared to the F values in Appendix O for the appropriate degrees of freedom. Note that $F(2,9) = 8.02$ for the .01 level of significance. Since the column's F ratio of 12.91 is greater than this, the column effects are significant at the .01 level. However, the row and interaction F ratios are less than one and are not significant. This means that the different promotion methods varied significantly in sales effectiveness, but the differences in crust formulas did not, nor were interactions between promotion methods and crust formulas significant.

Further Remarks. The above analysis included interactions on the assumption that they might be present. However, if there is *a priori* reason to believe that there is no interaction effect, the analysis may be simplified and may be done in terms of the row and column effects only.[15]

---

[15] In this case, the interactions SSD and *df* are added to the within-group SSD and *df*, and the analysis done as above. Note that if there is only one observation per cell, the interaction effects cannot be estimated and must be assumed equal to zero. Row or column effects may also be combined in a similar fashion if they prove nonsignificant

The example above illustrated a design involving only two factors (promotion method and crust formula). In general, any number of factors may be included in the analysis. Although the calculations are more complicated, the basic concepts are the same as illustrated above.

Note that the use of the F distribution was based upon the assumption of an underlying normal distribution. Further, in the analysis of variance tests, it was assumed that the variances of the sample items within groups (or within cells) were equal. Although these assumptions are necessary for the mathematical derivation of the F tests given, studies have shown that the tests are relatively insensitive to moderate violations of these assumptions. As long as the underlying distributions are not bimodal or very skewed, and the within-group (or cell) variances are roughly equal, the F distribution is a good approximation and the F test results are reasonably valid.

The amount of calculation involved in estimating effects and testing hypotheses using analysis of variance can be quite large, as was shown in the example involving two factors. Fortunately, computer programs are widely available which enable these calculations to be performed quickly and accurately.

NONPARAMETRIC TESTS

The statistical tests described earlier in this chapter generally require assumptions about the underlying distribution from which the sample was taken. In particular, the t tests and F tests require an assumption of normality.[16] There are a whole series of statistical tests, generally called *nonparametric* or *distribution-free* tests, which do not require such assumptions.

In addition, many of the nonparametric tests can be used on *ordinal* scale data. An ordinal scale requires only that items can be ranked in some order. For example, a consumer in a market test may be asked to rank his preferences for a group of products. Data in this form are often encountered in personnel research, marketing research, and studies of organizational behavior. The statistical tests presented earlier are not capable of dealing with such data.

Many of the nonparametric tests require relatively few calculations.

---

and this agrees with *a priori* judgment. It would be possible, in our example, to reduce the analysis back to a one-way analysis of columns by combining the row and interaction *SSD* and *df* with the residual.

[16] Although the chi-square test for contingency tables and goodness of fit relies upon a normal approximation, it does not make any assumption about the distribution from which the sample was taken. Hence, the chi-square is often classified as nonparametric when used for these purposes.

They are, thus, not only short-cut methods, but are also easier to understand for those less versed in statistical procedures.

While there are a great many nonparametric techniques, two are presented in this chapter to give the reader some idea of the use of these tests. Source books of nonparametric tests are listed at the end of the chapter.

Rank-Sum Test for Two Independent Samples

The rank-sum test for two independent samples[17] is an example of a group of nonparametric tests that are based upon the ordering or ranking of items in samples. In this particular case, two samples are randomly drawn from two populations. The test procedure is designed to test the hypothesis that the two populations are the same. The test makes no assumption about the underlying population distribution and requires only that the observations can be ranked by some criterion.

Consider an example. Suppose a firm, in an effort to improve safety in its factory, was experimenting with a two-day safety training program for workers. A group of eight workers was selected at random and put through the training course. A second group of seven workers was also selected at random to be the control or comparison group. These seven did not attend the training course. Two months after the training program, a supervisor who did not know which employees attended the program was asked to rank all 15 workers in terms of their safety behavior (eg., their use of protective equipment). Rankings go from 1 (best safety behavior) to 15 (poorest safety behavior). The resulting ranks for the 15 employees are: 6*, 2*, 7, 12, 9*, 14, 4, 3*, 1*, 5*, 15, 8*, 11*, 13, 10, where the asterisks indicate those who were trained.

If there were no difference between the trained and untrained groups, the rankings should be more or less evenly scattered between the two groups. On the other hand, if the training had any effect, those trained should have somewhat lower (that is, better) rankings. Examination of the data gives some indication that this is so. But the observed results might be due to chance.

To examine this, a statistic T is calculated indicating the sum of the ranks for one group. Here the sum of the asterisk numbers is $T_1 = 45$ for group 1 (those trained). Tables have been calculated for the probability distribution of T. In particular, Appendix P shows the upper (u) and lower (l) critical values of T for .05 and .01 critical probabilities.

---

[17] This test is known in different variations as the Wilcoxon two-sample test and the Mann-Whitney test.

Table 11–17 is a part of Appendix P. In our example, $n_1 = 8$ (number trained) and $n_2 = 7$ (number not trained). The critical values from Table 11–17 (or Appendix P) are 46 and 82 for the .05 level of significance. If the calculated value of T_1 falls within these limits, we cannot reject the hypothesis at the .05 level of significance. If the observed value of T_1 is less than or equal to 46, or greater than or equal to 82, the hypothesis that the groups are equal can be rejected at the .05 level. In our case, since the observed value of $T_1 = 45$ is less than the lower limit of 46, we can reject the hypothesis. There is significant evidence that the training affected safety performance.

Note that the critical probabilities in Table 11–17 and Appendix P

Table 11–17

CRITICAL VALUES OF T_1

[Lower (l) and Upper (u) Limits]
For Selected Values of n_1 and n_2

| | | n_2 | | | |
| | | 7 | | 8 | |
n_1		.05	.01	.05	.01
6	l	27	24	29	25
	u	57	60	61	65
7	l	36	32	38	34
	u	69	73	74	78
8	l	46	42	49	43
	u	82	86	87	93

are two-tailed critical values. That is, the hypothesis can be rejected if T_1 is either too large or too small. For one-tailed tests, the tables can be used with the probabilities reduced by one half (that is, at the .025 and .005 levels) or more detailed tables can be used (see the references at the end of the chapter).

In the example, we used the sum of the ranks for the trained group, T_1. We could alternatively have used the sum of the ranks of the non-trained group ($T_2 = 75$). In general, $T_1 + T_2 = (n_1 + n_2)(n_1 + n_2 + 1)/2$.

Ties. If there are ties in the rankings, each tied element is given the average rank of those tied. For example, if the supervisor had thought that worker 2 (rank 2) and worker 8 (rank 3) were actually equal as regards safety practices, we would give each a rank of 2.5 and proceed as before.[18]

Larger Sample Sizes.   The tables of Appendix P are usable for sample sizes up to 10. For samples larger than this the statistic T_1, under the assumption that the hypothesis is true, is approximately normally distributed, with mean

$$\mu_{T_1} = \frac{n_1(n_1 + n_2 + 1)}{2}$$

and standard deviation

$$\sigma_{T_1} = \sqrt{\frac{n_1 n_2(n_1 + n_2 + 1)}{12}}$$

where, as before, T_1 is the sum of ranks of the first sample and n_1 and n_2 are the sample sizes of the first and second samples, respectively. Then:

$$z = \frac{|T_1 - \mu_{T_1}| - 1/2}{\sigma_{T_1}}$$

is the standardized normal deviate z, tabled in Appendix D. The $\frac{1}{2}$ in the above formula represents an adjustment for continuity, since we are approximating a discrete distribution for T_1 by the continuous normal distribution.

Rank-Sum Test for Several Independent Samples

The rank-sum test can be extended to the case where there are three or more independent samples from different populations.[19] This test is the nonparametric equivalent of the analysis of variance procedure using the F test described earlier in this chapter.

Suppose there are k independent samples. Let n_i be the sample size in the ith group, and $\Sigma n_i = n$. As before, let all n items be ordered in a single ranking. Let T_i be the sum of the ranks for the ith group.

The null hypothesis is that all the populations are the same. Under this hypothesis, the statistic H, where

$$H = \frac{12}{n(n + 1)} \sum_{i=1}^{k} \frac{T_i^2}{n_i} - 3(n + 1)$$

---

[18] If more than one fourth of the elements are ties, this rank-sum procedure should be modified as described in the advanced texts.

[19] This is called the Kruskal-Wallis test.

has approximately a chi-square distribution with $k - 1$ degrees of freedom. The approximation is adequate if the sample size in each group is three or more. As before, ties are given the average rank of the tied items.[20]

Example. To illustrate this technique, we will use the data from Table 11–10. The 12 stores are ranked in terms of sales and the rankings are shown in Table 11–18. Note that two stores have sales of 10

Table 11–18

SALES RANKINGS IN 12 STORES

THREE PROMOTION METHODS

	Group 1 Point-of-Sale Ads	Group 2 Newspaper Ads	Group 3 Use of Demonstrator
	11	6.5	1
	12	4	2
	6.5	8	3
	10	9	5
Total	$T_1 = 39.5$	$T_2 = 27.5$	$T_3 = 11$

cases per month and are tied for the sixth and seventh rankings. They are each assigned a rank of 6.5.

Substituting in the formula:

$$H = \frac{12}{n(n+1)} \sum_{i=1}^{k} \frac{T_i^2}{n_i} - 3(n+1)$$

$$= \frac{12}{12(13)} \left[\frac{(39.5)^2}{4} + \frac{(27.5)^2}{4} + \frac{(11.0)^2}{4} \right] - 3(13)$$

$$= 7.875$$

Referring to Appendix N, we find that the chi-square value for $k - 1 = 3 - 1 = 2$ degrees of freedom is 7.824 for the .02 significance level. Since our observed value of H is greater than this, we can conclude that the groups differ at the .02 level of significance. This agrees with the result obtained earlier, although the significance level is not as small (.01 versus .02). Note, however, that to apply the nonparametric test, we did not have to make the assumption of normality nor the assumption that the variances within groups were equal. In fact, no assumptions were required about the underlying populations.

---

[20] If one-fourth or more of the items are tied, adjustments should be made to correct for this.

Other Nonparametric Tests

There are a large number of other nonparametric tests. For example, there are tests for matched samples, tests for medians, tests for runs, and rank correlation tests, to mention only a few. The references at the end of the chapter describe many of these tests.

Nonparametric tests have become increasingly popular in recent years. Not only are they easy to use and interpret, but they require fewer assumptions than comparable parametric tests. Finally, many of the nonparametric tests are almost as powerful as comparable parametric tests (in having a high probability of rejecting the null hypothesis when it is false), even when the assumptions of the latter are true.

SUMMARY

In this chapter, a variety of advanced procedures for testing hypotheses were introduced.

The first was the use of the t distribution when sampling from a normal population with small samples. Tests of hypotheses about means and differences between two sample means were considered.

The chi-square distribution provides a test of hypotheses about frequencies. Three specific examples involved testing differences in population proportions for several populations, contingency tables, and the goodness-of-fit case.

The F distribution was used to test hypotheses about population variances and extended to testing for differences in means for several populations. The analysis of variance technique was introduced for this, and for the two-factor experiment case.

Finally, two nonparametric tests were considered: a test for the difference between two populations for independent samples, and a test for differences in several populations. Both tests were based upon the ranking of sample items. Nonparametric tests are simple and require no assumptions about the underlying populations.

PROBLEMS

1. Explain:
   a) Why the means of large samples follow the normal distribution while the means of small samples may deviate significantly from normality.
   b) Why, when taking a small sample from a normal population, the normal distribution can be used for statistical inference if σ is known while the t distribution must be employed if σ is not known.

2. Management is interested in the average wait for a customer at a checkout counter during certain peak periods in a supermarket. A sample of 16

customers is taken at random, and their waiting times are noted. The mean waiting time was seven minutes with a standard deviation of three minutes. Can we conclude (with 95 percent confidence) that the mean waiting time was not less than five minutes? (Assume that the population to be sampled is normal.)

3. A random sample of 25 is drawn from the records of daily output of a large group of employees in order to estimate the population mean. The sample shows a mean of 136 units and a standard deviation of 24 units. (Daily output is normally distributed.)

   a) Calculate a 98 percent confidence interval for the mean output of all employees.

   b) Does the mean output of 136 units differ significantly from the standard output of 144 units set by management? Explain.

4. The Alvin Chemical company is contemplating adding some petroleum storage tanks at its distribution center in Chicago. It is common practice in this company to obtain several estimates from its own engineers of the cost of such capital expenditures. The average of these estimates is then used as the expected expenditure figure in capital budget planning.

   For the storage tanks in Chicago, five estimates were obtained:

	Estimate
Estimator	*(Millions of Dollars)*
Pearson......................	.$ 9
Neyman.......................	14
Fisher........................	8
Wald..........................	9
Hotelling.....................	10

   Noting the diversity in the estimates, Mr. Alvin, the president, wonders if it would not be possible to put some outside limits (say with 95 percent confidence) as maximum and minimum estimated expenditures.

   a) Provide Mr. Alvin with such an interval estimate.

   b) What assumptions is it necessary to make to give this estimate? Discuss the validity of these assumptions.

5. The manager of a fleet of cars was investigating differences in maintenance and repair costs for two makes of automobiles in his fleet. He selected a sample of 15 cars of each make and calculated the maintenance and repair cost per mile over the past year for each car. The results are:

$$n_1 = 15 \qquad\qquad n_2 = 15$$
$$\bar{X}_1 = \$.018 \text{ per mile} \qquad \bar{X}_2 = \$.025 \text{ per mile}$$
$$s_1 = .015 \qquad\qquad s_2 = .021$$

   Is there evidence from this data that the two makes differ significantly in average maintenance and repair costs per mile?

6. A credit officer for a bank surmised that whether or not a person had a savings account was one indication of credit-worthiness. Accordingly he

selected a sample of 150 customers from his files and classified them according to payment defaults:

	No Defaults	Defaults
Have savings account	87	3
No savings account	48	12

Based upon this information, is the credit officer's surmise correct?

7. A firm was examining alternative methods of packaging a new product. Two package designs (designated red and blue) were tested on consumers in three cities. In city A, 200 consumers were sampled and 60 percent preferred the red package; in city B, 100 consumers were selected, with 72 percent preferring the red package; in city C, 300 consumers were sampled, with 54 percent preferring the red package. Is there evidence from these data that there are significant differences between cities in the percentage who prefer the red design?

8. The manager of a computer facility has collected data on the number of times that service to users was interrupted (usually because of machine failure) in each week for the past 50 weeks.

Interruptions per Week	Number of Weeks
0	16
1	20
2	9
3	3
4	2
Total	50

Test the hypothesis that service interruptions are a random phenomenon (that is, come from a Poisson distribution).

9. A colleague of yours claims that he has no use for a table of random numbers, since he can generate random digits in his head that "are as random as any in tables." You are skeptical, but decide to test him by having him "generate" 100 of his "random digits." You have classified these hundred digits by the frequency of occurrence.

Digit	Frequency	Digit	Frequency
0	15	5	14
1	8	6	6
2	15	7	17
3	5	8	7
4	7	9	6
		Total	100

When you point out to your friend that certain digits seem to occur more frequently than others, he responds that it is merely chance variation. Do you agree? Explain.

10. A manufacturer of fire equipment was trying to find out what variables influenced the purchase of a fire extinguisher. One variable, home ownership,

was suggested as a possible influence. A sample of 100 households was selected and classified as shown below.

	Have Extinguisher	No Extinguisher
Home owner	20	50
Not home owner	0	30

Is there evidence from the above data of a relationship between home ownership and possession of a fire extinguisher?

11. The following data will be referred to in Problems 11 to 16. A researcher was experimenting to find better methods of assessing subjective probabilities. He had devised three different methods and conducted an experiment to test the effectiveness of each. A group of subjects was trained in each method. All subjects were then given a test on their ability to assess probabilities, and the tests were scored. The results are shown below:

Method 1	Method 2	Method 3
43.2	36.3	48.7
47.2	49.5	55.3
57.2	41.3	50.3
50.7	52.0	55.2
53.0	42.3	50.5
59.0	34.5	45.2
52.7	33.2	48.3
	41.3	49.3
	35.7	47.8
	45.2	
	47.7	
	50.0	
	44.3	
	47.3	
	51.2	

$n_1 = 7$	$n_2 = 15$	$n_3 = 9$
$\bar{X}_1 = 51.86$	$\bar{X}_2 = 43.45$	$\bar{X}_3 = 50.07$
$s_1 = 5.47$	$s_2 = 6.30$	$s_3 = 3.32$

Suppose the researcher is interested in comparing method 1 with method 2. As a first step, he wishes to test the hypothesis that the variances in the two groups are equal. Perform this test. Is there a significant difference in the variances?

12. Is there a significant difference in the mean scores on the test between the method 1 group and the method 2 group? (Use data in Problem 11.)

13. Test the hypothesis that the means of the three groups are the same. (Use data in Problem 11.)

14. Using the rank-sum test, test to see if there is a difference between scores for the method 1 group and the method 3 group. (Use data in Problem 11.)

15. Using the rank-sum test, test to see if there is a difference between scores for the method 2 group and the method 3 group. (Hint: Use the normal approximation with the data in Problem 11.)

16. Using the rank-sum test, test to see if there are differences among the three groups. (Use data in Problem 11.)

17. Refer to the example on page 297. Test the hypothesis that the variances of foremen's salaries in the two divisions are the same.

18. A firm is market testing three new versions of an instant coffee. Mix 1 is a rough mix, mix 2 is fine, and mix 3 has a sparkly additive. Each mix is placed in nine stores and the sales measured over a period of a month. The stores are selected to be equally mixed, with three large, three medium, and three small stores. The sales in cases per month are shown in the table below:

	Mix 1	Mix 2	Mix 3
Large stores	78,62,40	107,76,87	12,60,45
Medium stores	25,58,43	45,39,12	9,17,58
Small stores	7,16,22	48,15, 3	21,15, 6

a) Estimate the row and column effects.
b) Estimate the interaction effects.
c) Perform the analysis of variance to determine if these effects are significant.

19. A study was undertaken to measure the attitudes of students towards large business firms. A questionnaire was designed and administered to a sample of students classified by race and sex. The results of each questionnaire were classified as favorable, neutral, or unfavorable towards large business firms. The data are:

	White		Nonwhite	
	Male	Female	Male	Female
Favorable	36	40	8	10
Neutral	46	34	6	10
Unfavorable	30	24	14	12
Total	112	98	28	32

a) Test the hypothesis that attitude towards large business firms is independent of sex (over both race categories).
b) Test the hypothesis that attitude is independent of race (over both sexes).
c) Using the designations white male, nonwhite male, white female, nonwhite female as a category set, test the hypothesis that attitude is independent of this categorization.

20. Refer to the data shown in Table 2–3, page 26. Test the hypothesis that the data are a sample from a normal distribution. The sample mean is $\overline{X} = .4251$ and the sample standard deviation is $s = .00082$.

21. Refer to Problem 17 in Chapter 2 (continued as Problem 5 in Chapter 3 and Problem 5 in Chapter 4). Test the hypothesis that the data are a sample from a normal distribution.

22. A study was undertaken to determine the factors influencing handling time for metal plates used in a punch press. The weight of the metal piece was thought to be a determining factor. Accordingly, metal plates were classified as light, medium, or heavy weight, and the handling time (in thousands of a minute) were recorded for a sample of metal plates. The handling times are shown below.

Light		Medium		Heavy	
30	32	30	42	70	64
25	35	56	50	88	105
15	25	30	50	70	80
42	52	64	85	85	105

a) Using only the 12 observations in the two top rows, test the hypothesis that weight does not influence handling time.

b) Using all 24 observations, test the hypothesis that weight does not influence handling time.

23. Refer to Problem 22 above. Suppose the observations in the two top rows were recorded for machine operator no. 1 and those in the bottom two rows for operator no. 2. Perform a two-way analysis of variance to determine if handling time varies by operator as well as by weight.

24. A study is being made of the amount of time patients with a certain disease spend in the hospital. Six male and six female patients with this disease are selected at random from each of the three hospitals in a city and the number of days spent in the hospital for each patient is recorded. The data are given below.

	Hospital A	Hospital B	Hospital C
Male Patients	15,19,21	20,26,32	28,32,36
	26,22,17	29,21,24	24,30,32
Female Patients	28,22,24	29,20,26	35,38,30
	16,19,26	27,29,25	29,34,33

a) Is there significant evidence in the above data that the length of hospital stay varies from hospital to hospital?

b) Is there evidence of different treatment for males and females?

c) Is there any interaction between hospital and sex in terms of length of hospital stay?

SELECTED READINGS

Dixon, W. J., and Massey, F. J. *Introduction to Statistical Analysis*. 3d ed. New York: McGraw-Hill, 1969.

Covers a wide variety of statistical test procedures. Chapters 8 and 10 treat tests involving the *t* and *F* distributions and analysis of variance. Chapter 13 treats applications of the chi-square distribution. Chapter 17 deals with nonparametric tests.

HAMBURG, M. *Statistical Analysis for Decision Making.* New York: Harcourt, Brace & World, 1970.
 Chapter 9 is a readable discussion of chi-square and analysis of variance at a moderate level.

KRAFT, C. K., and VAN EDEN, C. *A Nonparametric Introduction to Statistics.* New York: Macmillan, 1968.
 Part II describes a number of nonparametric tests. This book also includes extensive tables of nonparametric statistics.

OWEN, D. B., *Handbook of Statistical Tables.* Reading, Mass.: Addison-Wesley, 1962.
 Contains extensive tables, not only for the t, F, and χ^2, but also for many nonparametric statistics.

PAZER, H. L., and SWANSON, L. A. *Modern Methods for Statistical Analysis.* Scranton, Pa.: Intext Educational Publishers, 1972.
 Chapters 6, 7, and 10 treat the material in this chapter at a comparable level, but in a little more detail.

RICHMOND, S. B. *Statistical Analysis.* 2d ed. New York: Ronald Press, 1964.
 Chapters 11 and 12 present an elementary treatment of chi-square and analysis of variance.

SIEGEL, S. *Nonparametric Statistics.* New York: McGraw-Hill, 1956.
 The basic reference source for nonparametric statistics.

12. SAMPLE SURVEY METHODS

MUCH OF THE MATERIAL that we have studied has been concerned with the interpretation and evaluation of sample information. The emphasis has been primarily on simple random samples. In actual practice, simple random samples are often impossible to obtain or prohibitively expensive. In this chapter we examine some different methods of selecting samples. Some of these methods will be more efficient than simple random sampling; others can be used where simple random sampling would be impossible; others are less costly than simple random sampling.

The first half of the chapter describes the principal sampling methods in common use. A knowledge of these types of samples is essential to an understanding of data collection and the interpretation of results. The second half, on "Measuring the Precision of Sample Statistics," takes up the actual calculation of the standard error and other measures needed in evaluating the mean or proportion in a sample survey. This technical section can be omitted if desired; the formulas are straightforward but a bit complicated.

There are two broad classes of methods of selecting samples: (1) *probability sampling,* including simple random sampling, systematic selection, stratified random sampling, ratio estimation, and cluster sampling, and (2) *nonprobability sampling,* including quota sampling and judgment sampling. These are discussed below.

PROBABILITY SAMPLING

Probability sampling includes all methods of sampling in which the sampled units are selected according to the laws of chance, so that the probability of being included is known (and not zero) for each member of the population. "Selected according to the laws of chance" means using some chance device such as a table of random numbers rather than personal judgment to choose the items sampled. The "probability of being included" may be equal for all units in the population (as in simple random sampling) or it may be, say, "probability proportional to size" (e.g., a company with two million sales having twice the probability of being selected as one with one million sales). In any case, however, the probability must be known, and hence the population itself must be identifiable.

In probability samples one can estimate objectively the precision of the sample results or compare the precision of different types of samples. The precision of probability samples increases (i.e., the sampling error decreases) as the size of the sample increases, whereas errors of judgment persist in larger nonprobability samples. Hence, probability sampling is generally used, wherever feasible, in large-scale surveys.

Simple Random Sampling

Simple random sampling has been used in all our discussions of sampling in Chapters 9 through 11. Simple random sampling means that each possible sample of a given size in the population has an equal chance of being selected.

Systematic Selection

A systematic sample is one in which every kth item (e.g., every 10th item) is selected in a list representing a population or a stratum (a relatively uniform segment) of the population. The number k is called the *sampling interval.* The first number is chosen at random from the first k items, as described below. Systematic selection ensures that the items sampled will be spaced evenly throughout the population.

For example, suppose you wish to take a systematic sample of 6 households from a block of 78 households. First, list and number the households. Then divide 6 into 78; this means that you should select every 13th house. Choose the first household at random from the numbers 1 through 13, using a table of random numbers. Say this is number 6. Now select every 13th house, beginning with number 6—that is, 6, 19, 32, 45, 58, and 71—to complete the sample.

Systematic sampling is often equivalent in its results to random sampling, if the elements in the population occur in a random order. For example, in dealing cards in the game of bridge, each player has a systematic sample (every fourth card). If the cards are shuffled well before the deal, the hand is equivalent to a random sample. Where the elements in the population are considered in random order, the formulas used for simple random sampling apply also to systematic sampling.

Systematic selection has an important advantage over simple random sampling if similar parts of the population tend to be grouped together, that is, if nearby elements resemble each other more than they resemble those at greater distances. For example, residents with similar incomes tend to be located in the same neighborhoods. A systematic selection of a city's blocks, numbered in serpentine fashion as described below, would then include more nearly the same proportion of each income group than a simple random sample.

Systematic selection should not be used, however, if there is some periodic variation in the population corresponding to the sampling interval. For example, in the case of sampling households in a block, if the block were laid out so that every eighth house was a large one on the corner, a systematic sample of every eighth house might include only large corner houses.

Systematic sampling has come into widespread use because it is easy to apply and it usually yields good results. For example, in the 1970 census of population every 20th person was asked several supplementary questions on various subjects. The cost of collecting and compiling information for this 5 percent sample was small compared with that of a complete enumeration or of an independent 5 percent sample survey. At the same time, the reliability of the information was sufficient for almost any purpose.

Stratified Sampling

If a population is made up of fairly uniform parts or strata, the precision of sample results can be improved by *stratification*. That is, the population is first broken down into strata such that the elements within each stratum are more alike than the elements of the population as a whole. Then an assigned part of the sample is drawn from each stratum by random selection (or by one of the other methods to be described later). Stratification is therefore only one step in the complete sampling method; it is always used in conjunction with other procedures.

As indicated above, the strata should be defined so that the significant elements within a stratum are more uniform than they are for the population as a whole. For example, in a study of household incomes a city can be divided into high- and low-income areas so that income varies less within each area that it does in the city as a whole. Here, geographic location provides a useful basis for stratification. In this case, the average income of a stratified random sample generally will be closer to the true average for the whole population than would that of a simple random sample of the same size selected from the city as a whole without stratification. Stratified sampling is thus useful for reducing the sampling error. As an extreme example of how stratification reduces this error, consider the following. A factory has only two categories of workers, each category having only one wage rate. If we were to take a simple random sample of workers in the factory and measure wages, we would have an estimate and some sampling error associated with the estimate. However, if we were able to group the workers by classification into two strata, we could then take a sample of only one worker for each stratum, and we would have no sampling error at all. We would know exactly the wages of all in the factory.

While the above example is artificial, it does illustrate the fact that by taking homogeneous groups and sampling separately from each group, we can gain some accuracy in sampling. A second advantage of stratification is that it gives us separate estimates for parts of the population. This kind of information may be useful for many management purposes.

Stratification should therefore be applied to *heterogeneous* populations, such as humans, since people can be divided into fairly uniform strata—by income, sex, age, or other criteria that affect the variable being studied (e.g., buying habits). Under these circumstances, stratification usually achieves greater precision for a given cost. On the other hand, stratification is unnecessary in *homogeneous* populations, as in measuring the diameter of ball bearings, where there are no discernible strata, such as differences in machine tools or operators, that affect the results.

Example. As an illustration of the use of stratified sampling, let us consider an application in the railroad industry.[1]

The bill for goods shipped (called a waybill) is usually paid to one railroad. However, the goods may have traveled over several different railroads while going from shipper to receiver. Each railroad over which the goods traveled is allocated a portion of the total revenue of the waybill. At one time, this was done by examining all waybills and allocating the revenue on each. A sampling procedure was considered

---

[1] This example is adapted from C. West Churchman, "Applications of Sampling to LCL Revenue Divisions," in *Proceedings: Modern Statistical Methods for Business and Industry,* (Pittsburgh: Graduate School of Industrial Administration, Carnegie Institute of Technology, May 1953).

to reduce the accounting cost of estimating the revenue allocation between railroads.

Table 12–1 shows the distribution of revenues of waybills terminating at a certain junction. Note that this distribution is extremely skewed, with a large number of waybills having small dollar amounts and a few

Table 12–1

FREQUENCY DISTRIBUTION OF WAYBILLS

Waybill Revenue	Number of Waybills	Percent of Waybills	Total Revenue	Percent of Total Revenue
0 to $ 4.99	3,047	56.0	$ 8,868	15.5
$ 5 to $ 9.99	1,074	19.7	7,502	13.1
$10 to $19.99	645	11.8	8,934	15.6
$20 to $39.99	381	7.0	10,695	18.7
$40 and over	298	5.5	21,245	37.1
Total	5,445	100.0	$57,244	100.0

having large amounts. It was decided to stratify the population into five groups—the same as those shown in the table. The waybills were accordingly sorted, and the number of waybills and total freight revenue in each group were ascertained. A systematic sample of each group was selected as shown in Table 12–2. Note how the proportion of each

Table 12–2

STRATIFIED SAMPLE OF WAYBILLS

Group	Revenue	Selected in Sample: All Waybill Nos. Ending In	Approximate Percentage Sample
1	$ 0 to $ 4.99	02, 22, 42, 62, 82	5
2	$ 5 to $ 9.99	2	10
3	$10 to $19.99	2 and 4	20
4	$20 to $39.99	01 through 50	50
5	$40 and over	All	100

stratum sampled varies, from 5 percent of group 1 to 100 percent of group 5. This is an efficient procedure for extremely skewed distributions such as we have here.

Using the percentage of revenue accruing to each railroad in each group (stratum), it is possible to estimate the percentage of total revenue due each railroad.

A few loose ends need to be discussed before we can leave stratified sampling. The first is the question: How many strata and how should they be determined? Oftentimes, the number and the boundaries of strata are determined by administrative convenience. Certain geographic areas, such as counties or states, form natural boundaries. However, there are times when the survey designer can set the number of strata. Then, how many strata should he make? Let us first point out that as long as we can select strata that differ somewhat from each other (with different means or standard deviations for the variable measured) we can continually increase precision. That is, under this circumstance, the larger the number of strata the better. However, in any actual situation, we do not know the content of all possible strata, and some point is reached when we cannot be sure we are breaking the population into strata that differ from each other. At this point, the use of more strata does not increase the precision. And remember that the more strata, the more computations are needed.

Another question is: What size subsample should we take in each stratum? This is discussed on pages 347–50.

Stratification and Nonresponse.   One method of handling nonresponse in a survey is to consider the population as made up of two strata, one being those who respond (e.g., those who reply to a mail questionnaire), and a second stratum of those who do not respond. When a survey is taken, the respondents can be used as one subsample. Then a subsample of the nonrespondents is taken by other means (e.g., by follow-up interviews). This subsample of nonrespondents is then used to provide estimates for the nonresponse stratum.

As an example, suppose that 1,000 mail questionnaires are mailed out and 520 are returned. Thus there are 480 nonrespondents in the sample. Suppose that 1 out of 4 of these are selected at random (120 in all), and interviewers are sent to obtain the desired answers. The total sample size would then be 520 + 120 = 640. However, the values obtained for the 120 nonrespondents would have to be multiplied by 4 to assure the correct weight.[2]

Ratio Estimation

In many business and economic surveys, it is important to estimate not the mean of a population but a *ratio*. As noted elsewhere, the ratio (including the proportion, percentage, fraction, or index number) is

---

[2] For the error formulas and further discussion on this type of sampling, see Leslie Kish, *Survey Sampling* (New York: John Wiley, 1965), pp. 132, 217, 304, 532–62, and other readings listed at the end of this chapter.

the basic summary measure for comparing two attributes, just as the mean is a basic measure for summarizing variables.[3] For example, an accountant may wish to sample a firm's accounts receivable to determine the ratio of balances in overdue accounts to the total balance of all accounts.

A ratio can also be used to estimate a population mean or total. For example, a ratio is often employed to approximate the total number of wild animals in a certain area or the number of fish in a lake. A known number of animals or fish are tagged and released in the area to be surveyed. After allowing sufficient time for them to mix with the group, a number of the animals or fish are caught. The ratio of the number tagged to the total number caught then yields an estimate of the total number of animals or fish. For example, suppose 1,000 fish are tagged and put in a lake, and subsequently 200 fish are caught of which 20 are found to be tagged. That is, there is a ratio of 10 fish for every tagged one in the sample. Since the total number tagged is 1,000, the total number of fish is estimated at 10 times the number tagged, or 10,000 fish.

As another example, the ratio of persons per water meter (say, three to one) is often used to make intercensus estimates of a city's population, since the number of water meters is usually easily obtainable. Similarly, the ratio of number of children in public schools to total population is used to estimate current population, since the count of schoolchildren is readily known.[4]

The use of ratio sampling to estimate a population mean or total depends upon the availability of certain auxiliary data that are related to the variable we are estimating. In the above examples, the number of water meters and the number of schoolchildren were the auxiliary data needed to estimate the total population. If such data are available, then ratio sampling can be quite efficient in reducing sampling error.

Cluster Sampling

Cluster sampling is the procedure by which a population is divided into several groups or clusters. A number of these clusters are then drawn into the sample and a subsample (possibly 100 percent) of ele-

---

[3] Ratios are described in Chapter 2, the binomial distribution in Chapter 6, inferences involving proportions in Chapters 9 and 10, and index numbers in Chapter 18.

[4] The perils in this process are obvious. Trends in the makeup of a city's population may change the ratio over time. Hence, inaccurate estimates will be made if the ratio is not reestimated periodically. At least one large city received a severe shock at the time of the 1970 census, when population estimated as above was quite different from official census figures.

ments is selected from each of the specified clusters. Thus, we are sampling at two stages: the first stage, where a sample of clusters, called *primary sampling units,* is drawn; and a second stage, in which individual elements, called *secondary* or *elementary sampling units,* are taken from the selected clusters.

We shall discuss only two-stage sampling, but there is no reason why three or even more stages could not be employed. For example, in sampling a city we could define the primary unit as the block, the secondary unit as the dwelling unit, and the tertiary unit as the individual. When each cluster is contained in a separate geographic area, cluster sampling is also called *area sampling.* The main advantage of cluster sampling is that it reduces the cost per elementary sampling unit. To understand this, suppose we were taking a sample of business establishments in a certain county. If a simple random sample were selected, the establishments in the sample would be scattered widely over the whole county. It would take interviewers a considerable amount of travel time to obtain the desired results. On the other hand, suppose the county were first broken up into geographic areas (clusters), and a sample of the clusters was taken. Then a subsample of the establishments within the selected areas is determined. With this procedure considerable travel time for the interviewer would be saved, since all of the establishments sampled will be clustered in the areas selected rather than spread randomly over the county.

A second advantage of cluster sampling is that it can be used sometimes where other methods are not applicable. For example, in selecting the sample of business establishments above, a complete list of all the establishments may not be available. However, it would be relatively simple to divide the county into geographic areas and select a number of these clusters as a sample. Business establishments could be listed and sampled within the selected areas without great difficulty. That is, we would have to prepare lists only within the selected areas.

On the other hand, cluster sampling is relatively inefficient. The results of a cluster sample are usually not as precise as those of a random sample of the same size. They can be made equally or more precise only by taking a larger sample. The cost of conducting a survey, however, may still be lower. For example, instead of spending $10,000 to interview a random sample of 1,000 householders at an average cost of $10 each, one might get better results for $9,000 with a cluster sample of 1,500 householders costing only $6 each.

Serpentine Numbering and Systematic Selection. A recommended method of selecting the clusters in area sampling is to number the primary sampling units in a *serpentine* sequence, following a winding path similar to that of a snake (see diagram). For example, in a study of household incomes, the numbering of city blocks should follow a sequence of blocks having about the same average household income. All blocks in such an area should be numbered before proceeding to a lower-income or higher-income area. After the block map has been numbered, the desired number of blocks should be chosen by *systematic* selection (e.g., every 10th block), with a random start, as explained previously.

SERPENTINE NUMBERING
OF CITY BLOCKS

1	2	3	4	5
10	9	8	7	6
11	12	13	14	15

This area sampling design achieves all of the advantages of geographic stratification when blocks in one stratum are numbered before proceeding to another stratum. However, stratification by some other characteristic, such as block size, is sometimes advisable.

Subsampling. After the primary sampling units have been chosen, elementary sampling units are selected from each of these clusters. The selection may be a complete census of the cluster (e.g., all the houses in the block) or a random or systematic sample (e.g., every fifth house).

The cost per interview for a subsample is higher than that for a complete census of the selected clusters. The choice between these alternatives depends in part on the complexity of the interview and the availability of lists. If the questionnaire is simple and no list of elementary sampling units (e.g., households) is available, it is usually cheaper to take a complete census of the selected clusters (e.g., blocks); when a lengthy interview is required, the advantages of subsampling justify the cost of listing and sampling the elementary sampling units.

Note that cluster sampling is used in conjunction with other sample types, such as random or systematic samples, which are needed to select both the primary and secondary sampling units.

We have skirted over some of the major problems associated with cluster sampling, such as: How many clusters? How large should they be? How many units should be in the subsample from the cluster? How do we compare the cost of a cluster sample with other methods? These questions have been left to advanced texts (see "Selected Readings" at the end of this chapter).

Replicated Sampling

Replicated sampling is a technique of selecting independent subsamples of the population (sometimes called "interpenetrating" subsamples). For example, instead of a random sample of 200 elements from some population, one might divide the 200 into 10 subsamples, each consisting of 20 elements. The subsamples are structured exactly the same—that is, they are *replicas* of each other. With replicated sampling, the overall estimate of the mean is the mean of the individual subsample estimates.

One main use of replicated sampling is in determining the sampling error for complicated sample designs, since the calculations are much simpler, as shown on page 359. Also, for systematic sampling, where the sampling error is difficult to estimate unless the elements in the population are in random order, replicated samples may be used to make a simple estimate of the sampling error. Finally, replicated sampling is used to estimate possible measurement error in the survey. Thus, if each subsample is taken from the reports of a separate interviewer, a replicated sample could reveal interviewer bias. The use of replication in nonprobability sampling is described below.

NONPROBABILITY SAMPLING

Nonprobability sampling includes any method of sampling which does not satisfy all requirements of a probability sampling design. This may involve selection of a sample according to personal convenience (to minimize cost) or expert judgment (to increase precision in certain small samples) or under conditions where no complete list is available for objective selection (e.g., a survey of executives who influence corporate buying policy on industrial equipment). Nonprobability sampling methods are important in business and economic research despite the disadvantage that the precision of their results cannot usually be measured objectively. Two principal types of nonprobability sampling are quota sampling and judgment sampling.

Quota Sampling

A quota sample is one in which the interviewer is instructed to collect information from an assigned number, or quota, of individuals in each of several groups—the groups being specified as to age, sex, income, or other characteristic—much like the strata in stratified sampling. Subject to these controls, however, the individuals selected in each group are left to the interviewer's choice rather than being determined by probability methods.

For example, the McGraw-Hill Publishing Company carries out numerous attitude surveys among executives who read industrial magazines, to aid the McGraw-Hill management in the conduct of its own publications. In one such survey, covering chemical industries, the company's interviewers had a complete list of plants but no comprehensive list of individual executives. A stratified, systematic sample of plants was first selected in each area. Given this list, each investigator was instructed to visit several plants and locate and interview a specified number of executives who had some influence on the company's purchasing policy. This quota method was considered by the director of marketing research to be the only feasible way of conducting an industrial survey when the population of respondents could not be identified.

Quota sampling is popular in market surveys and public opinion polls because it is cheaper per sample unit than random sampling and, when carefully controlled, has many of the advantages of stratified random sampling. However, it is subject to two important sources of error: (1) the quotas set for the interviewer represent a rather crude stratification plan for the population, being based on only a few broad criteria, such as age (young, middle-aged, or old) and income (low, middle, or high); (2) since the interviewer is free to select individuals within a quota, he may choose people in convenient locations who may not be typical of the class of the population they have been chosen to represent. For example, in a survey of the number of young children by households, the method of interviewing women who happen to be at home would be apt to yield a sample with too large a proportion of women with young children, because such women are more likely to be at home during the hours in which the interviewing is done than are other women. Therefore, interviewers must be carefully trained to avoid such pitfalls.[5]

Quota sampling has been popular in preelection polls since the 1930s. The pollsters missed Truman's upset of Dewey in 1948, but have done fairly well since, at least at the national level. Thus, Gallup

[5] Sometimes the sample is chosen so that the average age, income, or other pertinent characteristic of the individuals selected is equal to the average for the population. This is sometimes called "controlled" or "purposive" sampling. However, this control does not necessarily mean that the sample will be typical in other respects, such as in buying habits. Furthermore, this method is more difficult to administer than the simpler quota method, so it is used less frequently.

claims an average error of only two percentage points in predicting the winner's percentage of total votes in the six presidential elections of 1952 through 1972.

It is often argued that all large-scale surveys should be based on a probability sampling design because of its greater objectivity. On the other hand, since a much larger quota sample can be taken for the same cost as a smaller probability sample, and because population lists may be unavailable, quota samples are still favored in some circumstances.

Judgment Sampling

A judgment sample is one which is selected according to someone's personal judgment. A judgment sample may be superior to a probability sample (1) in very small-scale surveys, (2) in "pilot studies" which precede major surveys, or (3) in constructing index numbers. Also, they are often less costly than probability samples. Unfortunately, however, judgment samples may be biased, and it is difficult to assess the validity of their results.

Examples of judgment samples in small-scale surveys include the choice of a single plant (i.e., a sample of one) in which to try out a new personnel policy, or the choice of a few typical cities in which to make a market survey. A survey of consumer preferences for shampoo was conducted in San Jose, California, since this city was considered to be typical of the western market for this product. Such a judgment selection was probably superior to choosing a single city at random from a list of all cities in the West. This advantage of judgment selection, however, rapidly diminishes as the size of the sample increases, because there is a steady increase in the precision of a probability sample, while the bias of the investigator persists in judgment sampling.

In pilot studies, which are designed to pretest a questionnaire to be used in any large survey, emphasis is placed on detecting unforeseen difficulties, which can be overcome by revising questions, rearranging the schedule, or training interviewers. For this purpose, respondents in a pilot study are often chosen on a judgment basis in such a way as to overrepresent types of individuals most likely to cause difficulties.

Another type of statistical work in which judgment selection is usually preferred to probability selection is that of index number construction (described in Chapter 18). Consider the problem of choosing the sample of 400 goods and services that make up the *Consumer Price Index* of the U.S. Bureau of Labor Statistics. There should be sample items for each of several broad classes of expenditures made by the typical family. These items should be representative of their classes with respect to price movements, and they should have some importance in themselves. In view of these and similar difficulties, items used in the construction of index numbers are usually chosen according to the judgment of experts in the field. Probability selection in such cases is applied only to classes in which there are a great many items of the same order of importance.

Accordingly, judgment selection is recommended for samples which are too small for the advantages of more objective methods, for pilot studies in which certain types of bias may actually be desirable, and for the selection of components in index numbers. Objective methods of selection, however, are necessary to attain a high degree of reliability in most large samples.

Precision of Nonprobability Samples

The precision and standard errors of probability samples can be measured because the sample statistics follow the laws of chance (e.g., the means of large random samples follow the normal distribution), so that we can set confidence limits or test hypotheses with known probabilities. The standard error of a nonprobability sample, on the other hand, has no such significance, since sampling variation reflects unknown errors of judgment rather than chance.

However, if we take a replicated sample from the items in a nonprobability sample, all of the subsamples reflect about the same judgment factors, since they are replicas in their design. The subsample means, therefore, will vary because of numerous chance factors, and thus may follow a normal distribution. Hence, the standard error of the replicated sample is claimed to have some probability significance.

As an example, the standard error has been computed for a replicated sample of the items priced in the *Consumer Price Index,*[6] using pairs of subsamples for different items (e.g., different models of cars priced) and different stores and different cities to provide a total of 732 city-group relatives. Each of these subsamples is carried forward monthly from a base in December 1963. Then, since many independent factors affect the dispersion of the 732 means, they are believed to be normally distributed, and the standard errors are computed for each month by the formula given below for replicated samples. The validity of these standard errors is controversial. Nevertheless, replicated sampling provides a possible means of making a rough estimate of the precision of nonprobability samples in general.

---

[6] See M. Wilkerson in *Journal of the American Statistical Association,* September 1967, pp. 899–914.

MEASURING THE PRECISION OF SAMPLE STATISTICS

In this section we determine the standard errors of the mean (or total) and proportion in various types of samples, to measure their precision in estimating population values. We will consider random and systematic samples, stratified samples, ratio estimation, cluster samples, and replicated samples in turn.

Random and Systematic Samples

The standard error of a sample mean or proportion is the basic tool in making statistical inferences, such as setting confidence intervals or testing hypotheses. These measures were discussed in Chapters 9 and 10 for a simple random sample. The same applies to a systematic sample drawn from a randomly distributed population. This section is thus offered simply as a review.

The sample mean $\bar{X} = \Sigma X / n$ is an unbiased estimate of μ, the population mean. The sample variance, $s^2 = \Sigma (X - \bar{X})^2 / (n - 1)$ is an unbiased estimate of σ^2, the population variance. And the sample standard deviation, s, is the square root of the variance.

The standard error of the sample mean is estimated as

$$s_{\bar{X}} = \frac{s}{\sqrt{n}} \sqrt{1 - \frac{n}{N}}$$

where n is the sample size and N is the population size. The term $\sqrt{1 - n/N}$ is the finite population correction, used when sampling without replacement from a limited population. It can be ignored if n/N is very small.

The population total and its standard error may be estimated simply by multiplying the sample mean $\bar{X}$ and its standard error $s_{\bar{X}}$ by the number of items in the population N. Thus,

$$\text{Population total} \qquad = T = NX$$
$$\text{Standard error of population total} = s_T = Ns_{\bar{X}}$$

Finally, the sample *proportion* p_s is an unbiased estimate of the population proportion p. Then the estimate of standard error of the sample proportion is

$$s_{p_s} = \sqrt{\frac{p_s q_s}{n}} \sqrt{1 - \frac{n}{N}}$$

where $q_s = 1 - p_s$. Again, the finite population correction on the right can be omitted if n/N is small.

Stratified Sampling

Before introducing the estimation formula for stratified sampling, it is necessary to introduce some notation: Let M_i = the total number of elements (items) in the ith stratum; N = total number of elements in the population = ΣM_i; m_i = the sample size in the ith stratum; $\bar{Y}_i$ = the mean of the sampled elements in the ith stratum; s_i = the sample standard deviation in the ith stratum. Then the estimate of the overall mean is

$$\bar{Y}_s = \Sigma w_i \bar{Y}_i$$

where w_i represents the weight of the ith stratum, computed as

$$w_i = \left(\frac{M_i}{N}\right)$$

The standard error of the overall mean is

$$s_{\bar{Y}_s} = \sqrt{\Sigma w_i^2 \, s_{\bar{Y}_i}^2}$$

where $s_{\bar{Y}_i}$ is the estimated standard error in each stratum. That is,

$$s_{\bar{Y}_i} = \frac{s_i}{\sqrt{m_i}} \sqrt{1 - \frac{m_i}{M_i}}$$

(The last term is the finite population correction—this can be ignored in any stratum in which m_i/M_i is very small.

A few comments will help the understanding of these formulas. Note that the weight w_i is simply the fraction of the population in the ith stratum. The overall mean is simply a weighted average of the means in each stratum, using the relative numbers in each stratum as the weights. The standard error is weighted in a similar fashion (i.e., the variance is weighted by w_i^2).

An example will help to clarify further the meaning of the formulas. Suppose we wish to estimate the mean annual income of a population, which we divide into two strata—a high- and a low-income group. The first stratum is composed of 1,000 members, of which we sample 100. The second stratum contains 2,000 members, of which we sample 500. These numbers are shown, together with the sampling results, in Table 12–3.

Table 12–3

STRATIFIED SAMPLE OF INCOMES

Stratum Number (i)	Items in Stratum (M_i)	Items in Sample (m_i)	Mean Income of Items in Sample $(\bar{Y}_i)$	Standard Deviation of Items in Sample (s_i)
1..............	1,000	100	$10,000	$1,000
2..............	2,000	500	5,000	500
Total.........	3,000 = N	600		

To estimate the average $(\bar{Y}_s)$ for the total population we first have to determine the weights to attach to each stratum. These are:

$$\text{Weight for first stratum} = w_1 = \frac{1,000}{3,000} = \tfrac{1}{3}$$

$$\text{Weight for second stratum} = w_2 = \frac{2,000}{3,000} = \tfrac{2}{3}$$

That is, one third of the population items are in the first stratum and two thirds are in the second stratum. Then the estimate of the population mean is

$$\bar{Y}_s = \Sigma w_i \bar{Y}_i = (\tfrac{1}{3})(\$10,000) + (\tfrac{2}{3})(\$5,000) = \$6,667$$

We next wish to calculate the standard error for this estimate. To do this we must first calculate the standard errors of the mean for each stratum:

$$s_{\bar{Y}_i} = \frac{s_i}{\sqrt{m_i}} \sqrt{1 - \frac{m_i}{M_i}}$$

That is,

$$s_{\bar{Y}_1} = \frac{1,000}{\sqrt{100}} \sqrt{1 - \frac{100}{1,000}} = \sqrt{9,000}$$

$$s_{\bar{Y}_2} = \frac{500}{\sqrt{500}} \sqrt{1 - \frac{500}{2,000}} = \sqrt{375}$$

And the standard error for the population mean is:

$$s_{\bar{Y}_s} = \sqrt{\Sigma w_i^2 s_{\bar{Y}_i}^2} = \sqrt{(\tfrac{1}{3})^2(9,000) + (\tfrac{2}{3})^2(375)}$$
$$= \sqrt{1,167} = \$34$$

It can be demonstrated—though it is not done here—that a simple random sample of 600 items from this same population would have yielded a sampling error of about $100. Hence, stratification was quite efficient in this example.

Allocation of the Sample to Strata: Proportional Allocation. In the example above, we arbitrarily established sample sizes of 100 and 500 in the two strata, respectively. Now, our knowledge of survey sampling procedures is of primary usefulness in designing surveys beforehand rather than *ex post facto.* Hence, the student may wonder at such an allocation of sample items between strata. Would it not have been better to have them more equally distributed? How large a sample should be taken in each stratum?

One simple answer to this is *proportional* allocation, that is, allocate items in the sample to the various strata in the same proportion as the total elements in the population. This is often called a *self-weighting* sample.

As an illustration, suppose that the example given above represented a sample taken a year ago and that we are going to design a new sample. (Assume that the number of elements in each stratum and the standard deviations in each stratum remain the same.) Suppose that our new sample will also be 600 items, but we are free to allocate these items between the two strata as we see fit.

Proportional allocation would mean that since one third of the items of the total population are in the first stratum, one third of the sample items should also be in that stratum. Thus, $m_1 = \frac{1}{3}$ of $600 = 200$. And since two thirds of the items are in the second stratum, it should receive two thirds of the sample. That is, $m_2 = \frac{2}{3}$ of $600 = 400$. Proportional allocation is used if (1) the variability within the strata is approximately constant (i.e., the standard deviations within each of the strata—s_i—are about the same) or (2) little is known about the variability within the strata (hence, we may as well assume that they are about the same).

Proportional allocation has several advantages. It is the intuitively plausible or commonsense method of representing the different parts of the population (like the Supreme Court's proportional representation decree for state legislatures). In addition, it sometimes makes the formulas easier. For example, the estimate of the mean of the population is simply the mean of the sample—no weights are necessary.

Allocation of the Sample to Strata: Optimum Allocation. If there is a considerable amount of variability within the strata, however (i.e., the standard deviations of the items in the strata—the s_i—are of different magnitudes), we can do better than proportional allocation. That is, we can achieve less sampling error by allocating the sample items between strata in an optimum fashion. Note the allocation of sample items in the railroad waybill example on page 335. The fifth stratum (revenue $40 and over) contains $5\frac{1}{2}$ percent of the whole population of waybills and all (100 percent) of this stratum is included in the sample. On the other hand, the first stratum (revenue 0 to $4.99) contains 56 percent of all waybills, but only 5 percent of this group is included in the sample.

Using optimum allocation, we divide the total sample among the strata in such a way that we obtain the smallest sampling error for a given size of sample. The standard error is a function not only of the sample size within each stratum, but also of the variability of these items. To achieve optimum allocation, we assign in proportion to both the size of the stratum and the standard deviation within the stratum. The formula is thus

$$m_i = n \frac{M_i s_i}{\Sigma M_i s_i}$$

where n is the total sample size, M_i refers to the total number of items in the ith stratum, m_i is the sample size in that stratum, and s_i is an estimate of σ_i (the standard deviation of the items in the ith stratum).

To illustrate this, consider the example on page 346. Table 12–4

Table 12–4

STRATIFIED SAMPLE OF INCOMES—
OPTIMUM ALLOCATION

Stratum Number (i)	Items in Stratum (M_i)	Standard Deviation of Items in Stratum (s_i)	Product ($M_i s_i$)
1	1,000	$1,000	1,000,000
2	2,000	500	1,000,000
Total	3,000 = N		2,000,000

shows the number of items (M_i) and the standard deviation (s_i), together with the product $M_i s_i$ and the total $\Sigma M_i s_i$.

Let us take a sample of $n = 600$ items as before. How should they be allocated to minimize sampling error? Using the above formula, the sample size for the first stratum should be

$$m_1 = (600)\frac{1{,}000{,}000}{2{,}000{,}000} = 300$$

and the sample size for the second stratum is also 300.

To review the formulas for sampling error with stratified sampling and to illustrate that optimum allocation does reduce sampling error, let us carry out the calculation of the standard error of the mean with optimum allocation.

When we use these sample sizes and other data from Table 12–4, the standard errors within each of the strata are

$$s_{\bar{Y}_i} = \frac{s_i}{\sqrt{m_i}}\sqrt{1 - \frac{m_i}{M_i}}$$

so that

$$s_{\bar{Y}_1} = \frac{1{,}000}{\sqrt{300}}\sqrt{1 - \frac{300}{1{,}000}} = \sqrt{2{,}333}$$

$$s_{\bar{Y}_2} = \frac{500}{\sqrt{300}}\sqrt{1 - \frac{300}{2{,}000}} = \sqrt{708.3}$$

And the standard error for the population mean is

$$s_{\bar{Y}_s} = \sqrt{\Sigma w_i^2\, s_{\bar{Y}_i}^2} = \sqrt{(\tfrac{1}{3})^2(2{,}333) + (\tfrac{2}{3})^2(708.3)} = \sqrt{574} = \$24$$

This is quite a significant decrease over the previous allocation, which gave a sampling error of $34.

Allocation of the Sample to Strata: Least-Cost Allocation. If there is a difference in cost of obtaining a sample item in the various strata, then we can introduce this cost into the considerations. The formula becomes

$$m_i = n\frac{M_i s_i/\sqrt{c_i}}{\Sigma(M_i s_i/\sqrt{c_i})}$$

where c_i is the cost of sampling one item in stratum i, and m_i, n, M_i, and s_i are as defined above.

To continue our example, suppose that it cost $4 to obtain one item in stratum 1, and $9 to obtain one item from stratum 2. Then for a sample of 600 items, we should allocate to stratum 1:

$$m_1 = (600)\frac{1{,}000{,}000/\sqrt{4}}{(1{,}000{,}000/\sqrt{4}) + (1{,}000{,}000/\sqrt{9})} = 360$$

Similarly, the allocation to the second stratum is $m_2 = 240$.

Thus, because it is cheaper to sample in stratum 1, a larger sample should be taken in that stratum than under the optimum allocation above.[7]

Ratio Estimation

To illustrate the use of ratios in sampling, let us consider an example in detail. A company wishes to estimate its total inventory value at the end of each month. This would require a fairly large sample, since the values of different inventory items are likely to have a large standard deviation—that is, they probably range from a few cents up to hundreds of dollars. We might be able to achieve some improvement by stratification. An easier approach, however, would be to use ratio sampling.

We can take a random (or systematic) sample of items from the inventory, and compare their total current value with their value in the last annual inventory, as in Table 12–5. Then we multiply the percentage change in this sample value by the total annual inventory value, which was taken on a 100 percent basis, to estimate the total current inventory.

This ratio estimate of current inventory has a smaller sampling error than one based on a random sample of the current inventory alone, if the values of an item are related in the two periods. This relationship is shown in Chart 12–1. Here the points showing the relation of annual to current inventory values by item cluster along a diagonal regression line. That is, a major item is likely to have a high value in both periods, while a minor item will have consistently low values. The sampling error of a ratio estimate depends on the standard deviation of the points above and below this line (the standard error of estimate), whereas the sampling error of the mean of a sample of current inventory items depends on the larger standard deviation of the Y values above and below their own mean. We shall carry out this illustration further after introducing notation and formulas.

Notation and Formulas. Let Y denote the unknown variable that we are trying to estimate—the current inventory value per item. Let X denote the variable about which we have complete information—the dollar value per item at the last annual inventory. An inventory item here refers to a particular type of merchandise, such as a certain kind of

---

[7] We could go one step further and ask: Given C dollars, how many items should we take and how should we allocate them? The total sample size is

$$n = C \frac{\Sigma(M_i s_i / \sqrt{c_i})}{\Sigma(M_i s_i \sqrt{c_i})}$$

Then the allocation to strata can be done as above.

Table 12–5

SAMPLE OF 50 ITEMS FROM THE INVENTORY RECORDS OF A COMPANY
VALUES FOR CURRENT AND ANNUAL INVENTORY (IN DOLLARS)

Item Number	Annual Inventory Value (X)	Current Inventory Value (Y)	Item Number	Annual Inventory Value (X)	Current Inventory Value (Y)
1	$ 160	$ 182	26	84	89
2	87	84	27	171	152
3	280	315	28	103	96
4	123	125	29	326	350
5	20	28	30	38	35
6	254	300	31	128	139
7	100	82	32	124	102
8	142	151	33	87	99
9	50	55	34	375	420
10	124	136	35	80	88
11	64	52	36	208	216
12	164	160	37	86	99
13	40	48	38	67	58
14	151	154	39	305	349
15	107	105	40	158	146
16	80	92	41	32	39
17	193	150	42	184	160
18	93	110	43	137	100
19	231	250	44	115	165
20	54	68	45	33	57
21	101	110	46	216	186
22	16	18	47	119	141
23	191	220	48	64	72
24	109	120	49	312	300
25	91	95	50	27	35
			Total	$6,604	$6,903

spark plug or hammer. The value of an item is the number on hand times the cost per unit—not the cost of one unit alone. Thus, in Table 12–5, the value $160 for item 1 might represent 80 hammers at a unit cost of $2.

In our example, we take a sample of 50 items from the inventory and find their total value at each date; that is, ΣX (the annual inventory) and ΣY (the current inventory).[8] Then we calculate the ratio R,

---

[8] There is a slight problem that we have ignored in this simple example. Some items in either the annual inventory or the current inventory might be out of stock. The definition of the population would then have to be a list of all items in stock at both times.

Chart 12–1

RELATIONSHIP BETWEEN ANNUAL INVENTORY AND CURRENT
INVENTORY BY ITEMS, RANDOM SAMPLE OF 50 ITEMS

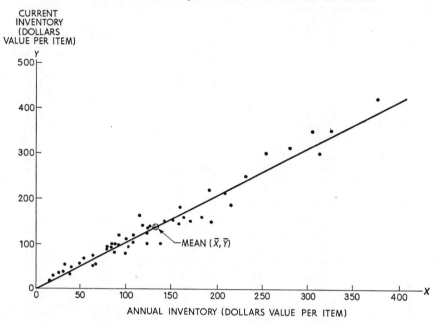

$$R = \frac{\Sigma Y}{\Sigma X}$$

which is an estimate of the unknown true ratio relating the total populations of X and Y. In our example, the ratio compares current inventory with annual inventory. We can use this ratio to estimate the total
of the Y values, as follows: $T_Y = RT_X$, where T_Y is the ratio estimate
of the total of the Y population and T_X is the total of the X population, which is assumed to be known.

The *mean* of the Y values is estimated similarly: $\bar{Y}_R = R\mu_X$, where
$\bar{Y}_R$ is the ratio estimate of the true mean μ_Y of the Y population. This
is to be distinguished from $\bar{Y}$, the mean of the sample items. The value
μ_X is the mean of the X population, which is known. Note that the
sample mean $\bar{X}$ generally will not be exactly the same as μ_X.

The total, of course, is N times the mean. That is, $T_X = N\mu_X$ and
$T_Y = N\bar{Y}_R$, where N is the total number of items.

In our example (Table 12–5), the ratio of current inventory to annual inventory value for the sample of 50 items is:

$$R = \frac{\Sigma Y}{\Sigma X} = \frac{6{,}903}{6{,}604} = 1.0453$$

That is, the inventory, by our estimate, increased 4.53 percent in value from the annual to the current inventory. Suppose the annual inventory totaled \$3,447,519. This is T_X. Then the total current inventory T_Y can be estimated as:

$$T_Y = RT_X = (1.0453)(3{,}447{,}519) = \$3{,}604{,}000$$

Assume there were 24,167 inventory items in the annual inventory (i.e., $N = 24{,}167$), so that the mean value was:

$$\mu_X = \frac{3{,}447{,}519}{24{,}167} = \$142.654 \text{ per item}$$

Then we could estimate the mean value per item of the current inventory as:

$$\bar{Y}_R = R\mu_X = (1.0453)(142.654) = \$149.11$$

Note that this is different from $\bar{Y}$, the mean value of current inventory in the sample, which is \$6093/50 = \$138.06. Thus, our estimate is considerably higher than we would have obtained from a simple random sample.

It may help to ponder this last statement. We are making a higher estimate using ratio sampling than we would have had we considered the sample as a simple random sample. Perhaps this is most easily seen if we consider the estimate of total current inventory. Our estimate from ratio sampling is given above as \$3,604,000. The simple random sample estimate for a total is

$$T_Y = N\bar{Y} = (24{,}167)(138.06) = \$3{,}336{,}000$$

Thus, ratio estimation gives us an estimate that is \$268,000 above that obtained using a simple random sample estimate. Why is this so? The ratio estimate is higher precisely because we realize, from our knowledge of the X variable, that the sample has understated the population total. Note that $\bar{X}$ (the value for the sample) is \$132.08, while the known population value is $\mu_X = \$142.65$. Hence, we adjust the value of $\bar{Y}_R$ upward to correct this understatement. Of course, in some samples it will be necessary to adjust downward, for identical reasons.

It is also important to note that we are dependent upon a close relationship between X and Y for ratio sampling to be efficient. If this were lacking, there would be no sense in making the adjustment as we did above.[9]

Bias and the Ratio Estimate. Unfortunately, the ratio estimate is a biased estimator of the population ratio. That is, the average of the ratios obtained from many samples does not generally equal the true ratio in the population. However, this bias is quite small in large samples, and we can ignore it in this case.

The bias will be negligible even for small samples if the relationship between X and Y can be approximately described by a straight line through the origin. Examination of Chart 12–1 indicates that this is certainly true for our example of estimating current inventory from annual inventory.

The following general rule has been suggested for determining when the bias in a ratio sample is negligible.[10]

The bias in the ratio estimate and the associated standard error are of negligible size if

1. The sample size exceeds 30.

2. Both $\dfrac{s_Y}{\sqrt{n}\,\overline{Y}}$ and $\dfrac{s_X}{\sqrt{n}\,\overline{X}}$ are less than .1.

Standard Error of the Ratio Estimate. The amount of sampling error associated with the ratio R and the ratio estimates $\overline{Y}_R$ and T_Y can be estimated by the following formulas:

$$\text{Standard error of ratio} = s_R = \sqrt{\frac{\Sigma Y^2 + R^2\Sigma X^2 - 2R\Sigma XY}{n(n-1)\overline{X}^2}}\,\sqrt{1 - \frac{n}{N}}$$

ΣXY is the cross-product term and is obtained by multiplying and then summing corresponding values of X and Y. The last term is the finite population correction and may be omitted if the sample is a small percentage of the population.

$$\text{Standard error of mean} = s_{\overline{Y}_R} = s_R\overline{X}$$

$$= \sqrt{\frac{\Sigma Y^2 + R^2\Sigma X^2 - 2R\Sigma XY}{n(n=1)}}\,\sqrt{1 - \frac{n}{N}}$$

$$\text{Standard error of total} = s_{T_Y} = N s_{\overline{Y}_R}$$

---

[9] The ratio estimate is more efficient (i.e., has smaller sampling error for a given size sample) than simple random sampling if the X and Y variables are highly related. A measure of the relationship between X and Y is the correlation coefficient (see Chapter 16) defined as $r = \Sigma xy/\sqrt{\Sigma x^2}\sqrt{\Sigma y^2}$. Generally, the ratio estimate is more efficient than simple random sampling if $r > \frac{1}{2}\,\sigma_x\mu_Y/\sigma_Y\mu_x$.

[10] William G. Cochran, *Sampling Techniques* (2d ed.; New York: John Wiley, 1963), p. 157.

When the true mean μ_X is known, it should be used in place of $\bar{X}$ in the above formulas.

To illustrate, let us continue the example of estimating total current inventory. The standard error for this estimate is, as above,

$$s_{T_Y} = N s_{\bar{Y}_R} = N \sqrt{\frac{\Sigma Y^2 + R^2 \Sigma X^2 - 2R\Sigma XY}{n(n-1)}} \sqrt{1 - \frac{n}{N}}$$

From Table 12–5, we can calculate the following:

$$\Sigma Y^2 = 1,365,701$$
$$\Sigma X^2 = 1,227,238$$
$$\Sigma XY = 1,285,673$$

Recall also that:

$$n = 50$$
$$N = 24,167$$
$$R = 1.0453$$

Since the sample is a very small part of the total population, the finite population correction in the above formula can be ignored. Then:

$$s_{T_Y} = (24,167) \sqrt{\frac{1,365,701 + (1.0453)^2(1,227,238) - 2(1.0453)(1,285,673)}{50(49)}}$$

$$= (24,167) \sqrt{\frac{18,820}{2,450}}$$

$$= 66,980$$

Thus, our estimate of total current inventory is $3,604,000 with a standard error of $67,000. This standard error is only 2 percent of the total, with a sample of 50 items, so ratio estimation is quite efficient in this case. For comparison, the sampling error obtained from simple random sampling is about $314,000.[11]

Before using the standard error to determine confidence limits, we should check the rules given above for determining if bias is negligible. Note that:

---

[11] To see this:

$$s_Y^2 = \frac{\Sigma Y^2 - \bar{Y}\Sigma Y}{n-1} = \frac{1,365,701 - (138.06)(6,903)}{49} = 8,421.9$$

$$s_Y = 91.76$$

$$\text{Estimate of standard error of mean} = s_{\bar{Y}} = \frac{s_Y}{\sqrt{n}} = \frac{91.76}{\sqrt{50}} = 12.977$$

$$\text{Estimate of error in total} = s_{T_Y} = N s_{\bar{Y}} = (24,167)(12.977) = 313,600$$

1. Sample size is greater than 30 ($n = 50$).

2. $\dfrac{s_Y}{\sqrt{n}\bar{Y}} = \dfrac{91.76}{\sqrt{50} \cdot 138.06} = .094$, which is less than .1, and

$\dfrac{s_X}{\sqrt{n}\bar{X}} = \dfrac{85.11}{\sqrt{50} \cdot 132.08} = .091$, which is also less than .1.

Hence, we will not worry about bias in the estimates of T_Y and s_{T_Y}.

Cluster Sampling

Let us consider a single example to illustrate the concepts involved in cluster sampling. Suppose we were interested in estimating the average family income in a certain city. There are 997 blocks in the city, and they are numbered in the serpentine fashion described earlier. Thirty blocks are selected at random. In each selected block, the number of households is determined and a sample of three households is selected. An interviewer is sent to the head of the selected households and total household income determined. The results are shown in Table 12–6.

In this example, the primary sampling unit is the city block and the secondary unit is the household. Note that the number of households in the whole city may not be known. We need to know only the number of households in each of the blocks selected, and this information may be readily obtainable.

Notation and Formulas. Before we can convert the data contained in Table 12–6 into an estimate of the average income in the city, it will be necessary to present the formulas and symbols used in them. Let:

$N =$ the number of primary units (blocks in this example) in the population

$n =$ the number of primary units (blocks) in the sample

$M =$ the total number of secondary units (households) in the whole population

$M_i =$ the number of secondary units in the ith primary unit—the number of households in the ith block

$m_i =$ the number of secondary units sampled in the ith primary unit—number of households sampled in the ith block

$\bar{Y}_i =$ the average of the sampled secondary units in the ith primary unit—average income in the ith block

$T_i = M_i\bar{Y}_i$ the estimate of the total for the ith cluster—total income in the ith block

A simple estimate of the mean of the population (average income per household) for a cluster sample is:

$$\bar{Y}_c = \frac{\Sigma M_i \bar{Y}_i}{\Sigma M_i} = \frac{\Sigma T_i}{\Sigma M_i}$$

Note that this formula does not involve M, the total number of all secondary units (households). Only the M_i, the number of households in the sampled blocks, is required.

Table 12–6

SAMPLE ESTIMATE OF AVERAGE HOUSEHOLD
INCOME IN A CERTAIN CITY

	Block No. (Determined by Random Number)	Number of Households in Block	Average Income of 3 households in Block ($000)	Estimate of Total Income of all Households in Block ($000)
i		M_i	$\bar{Y}_i$	$T_i = M_i\bar{Y}_i$
1	643	45	10.7	480.0
2	346	63	5.7	357.0
3	960	52	7.3	381.3
4	236	54	11.7	630.0
5	730	54	9.6	522.0
6	376	65	5.3	346.7
7	25	71	6.7	473.3
8	203	62	6.3	392.7
9	639	66	5.0	330.0
10	91	55	7.7	421.7
11	505	61	11.7	711.7
12	922	71	9.0	639.0
13	310	57	6.0	342.0
14	459	73	7.7	559.7
15	595	67	11.0	737.0
16	936	67	9.7	647.7
17	879	63	8.3	525.0
18	707	53	8.3	441.7
19	733	66	9.3	616.0
20	166	49	11.7	571.7
21	750	65	7.0	455.0
22	550	59	6.3	373.7
23	425	60	9.7	580.0
24	576	54	10.3	558.0
25	360	57	11.7	665.0
26	721	49	8.3	408.3
27	685	55	10.7	586.7
28	440	56	8.3	466.7
29	297	47	6.3	297.7
30	107	71	7.3	520.7
Total		1,787		15,038.0

The cluster sample estimate $\bar{Y}_c$ is biased, but the bias is small if a fairly large number of primary units (blocks) are sampled.[12]

An estimate of the sampling error of the cluster estimate $\bar{Y}_c$ is

$$s_{\bar{Y}_c} = \sqrt{\left(\frac{N}{M}\right)^2 \frac{\sum M_i^2(\bar{Y}_i - \bar{Y}_c)^2}{n(n-1)} \left(1 - \frac{n}{N}\right) + \frac{(N/n)\sum M_i^2 s_{\bar{Y}_i}^2}{M^2}}$$

where $s_{\bar{Y}_i}$ is the standard error of the estimate of $\bar{Y}_i$ in the ith cluster (the error associated with the estimated average income in a block), and

$$s_{\bar{Y}_i} = \frac{s_i}{\sqrt{m_i}} \sqrt{1 - \frac{m_i}{M_i}}$$

where s_i is the standard deviation of the items sampled in the ith cluster. When M is not known, use the estimate $N\Sigma M_i/n$ instead.

Note that the equation for $s_{\bar{Y}_c}$, the standard error of the cluster estimate, has two parts. The first term is roughly related to the variability *beween* cluster means, and the second term to the variability *within* clusters. The first term generally is the larger. In fact, if the sampled clusters represent a small fraction of the total number (n/N less than about .05), the second term becomes small and can be ignored in calculations.

In our example (Table 12–6) of sampling incomes in a city, the estimate of the mean income per household is

$$\bar{Y}_c = \frac{\Sigma T_i}{\Sigma M_i} = \frac{15,038.0}{1,787} = 8.415 \text{ thousands of dollars}$$

and the estimated sampling error of this mean is

$$s_{\bar{Y}_c} = \sqrt{\left(\frac{N}{M}\right)^2 \frac{\sum M_i^2(\bar{Y}_i - \bar{Y}_c)^2}{n(n-1)}}$$

using only the first term and ignoring the finite population correction ($1 - n/N$) since n is only 3 percent of N. Here $N = 997$, $n = 30$, and M is estimated as:

$$M = \frac{N}{n} \Sigma M_i = \frac{997}{30} (1,787) = 59,388$$

---

[12] An unbiased estimate is also available if M is known. However, the unbiased estimate is generally less efficient than the biased estimate above. See Cochran, *op. cit.*, pp. 300–305, for more details.

Since

$$\sum M_i^2(\bar{Y}_i - \bar{Y}_c)^2 = 437,811 \text{ (calculation not shown)},$$

$$s_{\bar{Y}_c} = \sqrt{\left(\frac{997}{59,388}\right)^2 \left(\frac{437,811}{30(29)}\right)} = \sqrt{.1418}$$

$$= .377 \text{ thousands of dollars}$$

This is a fairly large sampling error—about 4.5 percent of the mean —considering the size of the total sample (90 households). A simple random sample of 90 households would have been more accurate. However, the 90 households in the cluster sample would be considerably cheaper to survey than the equivalent random sample. Furthermore, taking a random sample would have been impossible to do without first compiling a complete list of all households in the city—quite a task!

The method described above is one way in which cluster sampling can be formulated. Other methods are useful for different situations. For example, when the primary units or clusters vary greatly in size, a technique may be used which will make the probability of selecting a cluster proportional to the size of the cluster. In addition, three or even more stages may be used, as noted earlier. These require more complicated formulas, but the basic ideas illustrated above are the same.

Replicated Sampling

We need not illustrate replicated sampling, as the formulas are simple. Suppose that k replicated samples are drawn and for each a mean $\bar{Y}_j$ is calculated. Each $\bar{Y}_j$ is an estimate of the population mean. The overall replicated sampling estimate of the mean is

$$\bar{Y} = \frac{\Sigma \bar{Y}_j}{k}$$

and the estimated sampling error is

$$s_{\bar{Y}} = \sqrt{\frac{\Sigma(\bar{Y}_j - \bar{Y})^2}{k(k-1)}}$$

In words, the standard error $s_{\bar{Y}}$ is determined only from the variance of the sample means $\bar{Y}_j$ themselves,[13] thus avoiding all calculations of variances within and between clusters, within strata, etc.

---

[13] The estimated sampling error $s_{\bar{Y}}$ has $k-1$ degrees of freedom. In determining confidence intervals, therefore, it may be necessary to use the t distribution.

The number of replications k to make depends upon various factors in the design. The value $k = 10$ has been suggested as a good number for a wide variety of applications.[14]

SUMMARY

Information obtained from samples is indispensable in modern business and economic research. It is important, therefore, to plan sample surveys in such a way as to obtain the desired information with maximum precision and minimum cost of time and effort.

Probability sampling includes all methods (such as simple random sampling, stratified random sampling, systematic selection, and cluster sampling) in which there is a known probability of being selected for each individual in the population. Nonprobability sampling includes all other methods, such as quota and judgment sampling. Probability sampling methods have a basic advantage in that the precision of their results can be measured objectively and compared as between different sample designs. This is especially important in very large samples.

A simple random sample of n units is one selected from the population in such a way that each combination of n units has an equal probability of being selected. A table of random numbers is usually used to select items at random.

Systematic sampling is the process of taking observations at equal intervals in a list. When nearby parts of a population are alike, systematic sampling with a random start is superior to simple random sampling in spacing the sampling units more evenly over the population.

A stratified random sample is one in which the population is divided into fairly uniform groups or strata. Then a random sample is drawn from each selected stratum. If the various strata can be made more homogeneous than the population as a whole, a stratified sample will yield more precise results than a simple random sample of the same size.

The total sample must be apportioned to the various strata. Proportional allocation assigns the sample elements to the strata in the same proportions of the total sample as they occur in the population. If the strata differ considerably in variability, then optimal allocation will improve the estimate. Optimal allocation assigns the sample to the strata in proportion to the strata size and standard deviation within the strata. If the cost of sampling varies considerably between strata, then

---

[14] W. Edwards Deming, *Sample Design in Business Research* (New York: John Wiley, 1960), chap. 21. Chapters 6–15 present a thorough treatment of replicated sampling designs.

least cost allocation should be employed to maximize precision relative to cost.

Stratification of a population into respondents and others, and sub-sampling from the nonrespondents, is one method for dealing with nonresponse in surveys.

Ratio estimation focuses on proportions rather than on means. A ratio estimate may also be used to estimate the mean (or total) of one population, using the ratio between the variable to be estimated and an auxiliary variable that is related to the first and about which complete information is available.

The efficiency of the ratio estimate depends upon the correlation between the two variables used in the estimate. If the two variables are closely related, the ratio estimate can have a much smaller sampling error than a simple random sample. The ratio estimate is biased (the average of many ratio estimates would not give exactly the population value), but the bias is negligible if the sample size is large.

Cluster sampling involves (1) selecting groups or clusters as primary sampling units and (2) taking a census or sample of "elementary sampling units" or secondary units within these groups. Cluster sampling is called area sampling when the cluster falls in some geographic division, such as a city block. A cluster sample yields less precise results than a simple random sample of the same size, but the cost may be much less. The clusters are often chosen by systematic selection from a map on which areas are numbered by serpentine order.

There are several methods of cluster sampling. One is to sample the primary units with equal probability and subsample secondary units. Formulas and an illustration of this technique are presented. If the primary units vary greatly in size, they may be selected with probability proportionate to size. Other methods are also available.

The technique of replicated sampling involves drawing several independent subsamples from the population, all using the same sample design. The use of replicated samples makes the estimation of sampling error relatively easy.

Nonprobability sampling (including quota sampling and judgment selection) is the selection of a sample according to personal choice, expert judgment, or under conditions where lack of data prevents a probability selection. It is sometimes recommended when probability sampling is not feasible.

In quota sampling the investigator may choose the respondents from a quota or assigned number of individuals in each designated class. A quota sample is cheaper per unit than stratified random sampling and is popular in market surveys and public opinion polls, despite the serious pitfalls inherent in this method.

Judgment sampling is the selection of a sample based on expert judgment. It is recommended for surveys in which the sample is very small, for pilot studies preceding larger surveys, and for most economic index numbers.

The standard error of a nonprobability sample may possibly be estimated by replication, as in the case of the Consumer Price Index.

The standard error must be calculated for a sample statistic in order to determine its precision as an estimate of the population value. The computation of means, totals, proportions, and their standard errors is illustrated for various types of samples in the last half of the chapter.

PROBLEMS

1. Comment on the following statements:
   a) Sampling errors are due to improper methods of selecting a sample.
   b) Survey results may be made as accurate as necessary by increasing the size of the sample.
   c) A complete census is always preferable to a sample, if time and money permit.
   d) Probability sampling should be used in all large-scale surveys to obtain valid results.

2. Distinguish between:
   a) Probability sampling and nonprobability sampling.
   b) Probability sampling and simple random sampling.
   c) Stratified sampling and quota sampling.
   d) Proportional and nonproportional sampling in stratified samples.
   e) Primary and elementary sampling units in cluster sampling.

3. You wish to conduct a survey of students in a university to determine which facilities they prefer (e.g., swimming pool, bowling alley, cafeteria) in a new student union building that is being planned. Compare the advantages of each of the three pairs of sampling methods in Problems 2(a), 2(c), and 2(d) above for this purpose.

4. Time Inc. made a survey of college graduates to determine their success and satisfaction in life as related to their education record and various other characteristics that would aid *Time Magazine* in analyzing its readership. Using lists supplied by colleges, *Time Magazine* sent questionnaires to all 15,700 graduates whose names began with "Fa" (Farley, Farmer, etc.). Over 9,500 replies were received.
   a) What method of sample selection is this?
   b) What sources of error might distort the results?
   c) Suggest another method of selecting a sample of this size that seems

preferable to you, and show why this method should reduce the errors of response without greatly increasing the cost of the survey.

5. Each student is to select a sample of 25 values of a quantitative variable and compute the average by adding the values and dividing the sum by 25. To insure comparability of results obtained by the various members, the class should agree on the choice of variable and the method of selection to be used. Problems to be considered include:
   a) Are the data readily available?
   b) If values are recorded on cards, might the cards be shuffled to arrange them in random order?
   c) Are the values listed and numbered in order so as to facilitate selection by means of a table of random numbers?
   d) Would systematic selection be effective?
   e) What strata might be constructed for stratified sampling?

6. As a distributor of major household appliances, you wish to survey the potential market for new appliances in your town by interviewing a sample of householders. Plan a cluster sample of the area as follows:
   a) Secure an up-to-date map of the town or one district of a larger city.
   b) Number the blocks, or equivalent areas, in serpentine fashion so as to follow a sequence of blocks having about the same household incomes.
   c) Choose a systematic sample, with random start, of 20 blocks on this map.
   d) Visit the 10th block selected (as an example) and list all house or apartment numbers around the block.
   e) Select a random sample of six houses or apartments from this block, using a table of random numbers.
   f) Comment briefly on the validity of this procedure for the problem at hand.

7. A population is divided into two strata, and a sample is taken from each stratum as follows:

	Stratum 1	Stratum 2
Number of elements in stratum, M_i	1,000	4,000
Number in sample, m_i	100	225
Stratum sample mean, $\bar{Y}_i$	85	75
Σy_i^2 in stratum, where $y_i = (Y_i - \bar{Y})$	9,900	89,600

   a) Estimate the mean for the whole population.
   b) Estimate the standard error of the mean of the whole population.

8. An election is being held in a certain plant to determine if the workers should be represented by a union. To estimate beforehand the preference of the workers, management hires a consulting firm to take a sample of workers. The results are shown below in the table.

Department	No. of Workers in Department	No. of Workers in Sample	No. of Workers in Sample Voting for Unionization
1........................	5,000	100	60
2........................	5,000	50	20
Totals..................	10,000	150	80

a) What estimate should management make of the proportion of workers in the whole plant voting for unionization?

b) What is the sampling error of this estimate?

Hint: The standard error of the proportion in each stratum is

$$s_{p_{s_i}} = \sqrt{\frac{p_i q_i}{m_i}}$$

Use this in the same fashion as the standard error $s_{\bar{y}_i}$.

9. As a dealer in retail hardware you are considering buying out the inventory of a merchant who is going out of business. You have a list of the items that he carried in stock but no exact inventory has been made. There is the added problem of evaluating the worth of these items since many are obsolete or so old and damaged that they are worthless. Accordingly, you decide to take a sample of the items, check the count, and value carefully the sampled items.

The inventory is broken down into three product groups, including a special group for high-valued items. The number of items in each group is shown below. In addition, you make the following rough estimates of the standard deviations of the values of the items for each product group.

Product Category	Items in Product Category	Approximate Standard Deviation
High-value items...........................	100	$120
Paints and paint products...................	400	20
General hardware..........................	500	10
Total......................................	1,000	

Suppose you were considering a total sample of 50 items.

a) How would you allocate the items by proportional allocation? By optimum allocation?

b) Estimate the standard error of the sample mean using proportional allocation and using optimal allocation.

10. A market research firm conducted a survey to estimate the percentage of the population in a certain city that preferred a particular brand of soft drink.

In order to obtain additional information, the city was divided into three areas, corresponding roughly to the high-, medium-, and low-income groups, respectively. A sample was then taken in each area. The results are shown in the table.

Income Area	Approximate Number of Consumers	Number Sampled	Number Preferring Brand X	Percent Preferring Brand X
High.......................	20,000	80	16	20
Medium...................	120,000	150	75	50
Low.......................	60,000	120	72	60
Total.....................	200,000	350	163	

a) Make an estimate of the overall percentage of consumers who prefer brand X.

b) How much sampling error is associated with the above estimate? Compute a 95 percent confidence interval about your estimate above.

Note: Recall that the formula for the sampling error of a proportion is

$$ s_{p_s} = \sqrt{\frac{p_s q_s}{n}} $$

This is equivalent to the $s_{\bar{y}_i}$ in the formula for the estimate of the standard error in stratified samples.

c) If you were to design a survey to be taken for a similar product (i.e., the percentages within the various groups are expected to be the same as above), how would you allocate a proposed sample of 400 among the three income groups? (Let $s_i = \sqrt{p_i q_i}$)

11. The A & B Sporting Goods Company was interested in estimating the annual expenditures for camping gear for the 100,000 family units in the San Jose, California area. In order to obtain information for the designing of a sampling plan, a pilot sample of 100 family units was chosen at random. The estimated annual expenditures for camping gear (U_i) and the annual family income (Z_i) were obtained for each family unit. A summary of these numbers is shown below:

$$
\begin{aligned}
\bar{U} &= \text{average expenditure} = \$26 \\
\Sigma U_i &= 2{,}600 \\
\Sigma U_i^2 &= 130{,}000 \\
s_u &= \$25 \\
\bar{Z} &= \$10 = \text{average income (thousands)} \\
\Sigma Z_i &= 1{,}000 \\
\Sigma Z_i^2 &= 13{,}600 \\
s_z &= \$6 \text{ (thousands)} \\
\Sigma U_i Z_i &= 40{,}000
\end{aligned}
$$

a) Make an estimate of total expenditures for camping gear for the 100,000 family units in San Jose by (i) simple random sampling and (ii) ratio estimation. Assume total annual income for all 100,000 units is known to be $900 million.

b) Compare the two estimates. Why do they differ? Which is more accurate? Why?

c) As an alternative, the San Jose area could have been stratified by geographic area into three economic area groups. Estimates of standard deviations of expenditures for camping gear within each area are provided. How would you allocate your sample of 100 items between these groups? What accuracy would you estimate? Compare this to the simple random and ratio estimates above.

Area	Number of Family Units	Estimated Standard Deviation of Expenditures
High income	30,000	$25
Medium income	40,000	15
Low income	30,000	5
Total	100,000	

12. Mr. Worthy, president of Worthy Products, was considering marketing a new product—an ornamental gadget that could be attached to fenders, bumpers, or hoods of automobiles. The gadget would be sold on a door-to-door basis and some automobile owners might buy two, three, or even more.

There were some 200,000 households and some 250,000 automobiles in the territory which Mr. Worthy intended to canvass. In order to make an estimate of his sales in this territory, Mr. Worthy drew a random sample of 50 households and made sales calls. The results of this survey are shown in the table.

Household Number	Gadgets Sold	Cars in Household	Household Number	Gadgets Sold	Cars in Household
1	0	0	26	0	0
2	0	2	27	0	2
3	2	4	28	2	4
4	0	1	29	0	1
5	0	0	30	0	0
6	0	0	31	0	0
7	0	0	32	0	0
8	0	2	33	0	2
9	0	2	34	0	2
10	1	3	35	1	3
11	0	1	36	0	1
12	0	1	37	0	1
13	0	1	38	0	1
14	0	2	39	0	2
15	0	3	40	0	3
16	0	2	41	0	2
17	0	0	42	0	0
18	0	1	43	0	1
19	0	1	44	0	1
20	0	2	45	0	2
21	1	3	46	1	3
22	2	3	47	2	3
23	1	1	48	0	1
24	0	2	49	1	2
25	0	1	50	0	1
			Total	14	76

a) Treating the sample data as a simple random sample of households, estimate the total sales for all 200,000 households.

b) Using the ratio of sales to number of automobiles in a household, estimate total sales.

c) Compare the two estimates. Why do they differ? Considering possible bias, which estimate do you think is more accurate?

13. A study was undertaken in a certain city to estimate the total number and types of major appliances (refrigerators, stoves, washers, dryers, dishwashers, freezers). The city was first divided into 600 blocks. From aerial photographs and automobile trips about the city, the number of households in each block was estimated. By this process, it was estimated that there were 10,000 households in the city. Next, 30 blocks were selected at random. In each of these blocks all the households were contacted and information about their appliances was obtained. The results are shown in the table.

Block No.	Number of Appliances	Estimated No. of Households
1	64	16
2	48	14
3	42	5
4	94	20
5	70	13
6	40	11
7	31	12
8	21	6
9	49	12
10	73	22
11	85	23
12	47	17
13	39	8
14	60	14
15	66	20
16	32	8
17	53	12
18	64	24
19	110	27
20	95	28
21	137	40
22	49	9
23	63	15
24	54	15
25	59	11
26	80	19
27	64	17
28	110	24
29	73	26
30	103	33
Total	1975	521

a) Estimate the total number of major appliances in the city using the ratio estimate (ratio of number of appliances to number of households in a block).

b) Consider the blocks are clusters, with 100 percent second-stage sampling, and make an estimate of total number of major appliances, using the cluster sampling approach. Does your estimate differ from that in (a)? Explain.

c) How else might you make an estimate of total number of appliances in the city from the data above?

14. An oil company wanted to estimate the average monthly sales for the next month for its approximately 104,000 credit-card customers. The credit-card accounts were filed by account number in 500 drawers, each containing approximately 200 accounts.

It was decided first to draw a random sample of 30 drawers and then a systematic sample of 10 accounts from each drawer selected. The results are shown in the table.

Drawer	Accounts in Drawer	Average Monthly Sales in Sample
1	220	21.67
2	184	19.26
3	200	3.20
4	176	12.17
5	210	5.42
6	208	13.10
7	198	7.15
8	202	10.85
9	206	12.50
10	194	15.47
11	218	17.29
12	217	6.18
13	192	24.53
14	212	8.22
15	202	6.33
16	225	19.13
17	209	7.57
18	208	1.12
19	215	14.71
20	224	6.83
21	216	12.92
22	226	7.21
23	234	34.17
24	196	8.47
25	218	11.16
26	242	9.28
27	200	17.42
28	215	9.64
29	210	22.77
30	204	14.98

a) Estimate the overall average monthly sales for all 104,620 accounts and the sampling error associated with this estimate.

b) What other sampling methods would you suggest that might be more efficient (less sampling error) in this case? How does your method compare with the procedure above in terms of the cost of taking the sample?

15. Consider as a population all the students in your college or department or all the employees in your firm. Determine some variable that you would like to measure for this population, such as the income they expect 10 years after graduation, the average distance they commute to school or work, or the number of hours per week they spend watching television.

a) Design a sampling plan to estimate the information you want. Be sure to define your population exactly. (How do you handle part-time students or employees?) Indicate where you could obtain lists and other information needed for the survey design. Decide upon how accurate you wish the results to be and how large a sample you will need to achieve this accuracy.

b) Prepare a questionnaire to obtain the desired information. Pretest the questionnaire on a group or groups of persons. Is the survey to be done by mail or personal contact? How will you handle nonresponse?

c) Conduct the survey and tabulate your results. Estimate the information you want and determine the sampling error associated with your estimate.

d) Write up this project in a report form indicating: (i) the sampling plan chosen and why it was chosen, (ii) how the survey was conducted, and (iii) the results of the survey.

SELECTED READINGS

COCHRAN, WILLIAM G. *Sampling Techniques.* 2d ed. New York: John Wiley, 1963.

This is a textbook and reference source on sampling theory and technique. It is at a relatively advanced level and would be useful for students who wish to go into more detail and depth.

CYERT, R. M., and DAVIDSON, N. J. *Statistical Sampling for Accounting Information.* Englewood Cliffs, N.J.: Prentice-Hall, 1962.

The early chapters deal with the general theory of sampling. Chapter 7 deals with ratio estimation and Chapter 8 with stratified sampling. The treatment is at a moderate level, and examples of sampling in accounting are included.

DEMING, W. EDWARDS. *Sample Design in Business Research.* New York: John Wiley, 1960.

Contains several examples of sampling in business. Replicated sampling is treated in detail. However, the level is advanced and difficult to follow in many places.

HANSEN, M. H.; HURWITZ, W. N.; and MADOW, W. G. *Sample Survey Methods and Theory.* New York: John Wiley, 1953. 2 vol.

Volume I is an authoritative and thorough treatment of sampling methods and applications.

KISH, LESLIE. *Survey Sampling.* New York: John Wiley, 1965.

A modern, comprehensive treatment, incorporating the experience of the University of Michigan Survey Research Center.

MENDENHALL, W.; OTT, L.; and SCHAEFFER, R. L. *Elementary Survey Sampling.* Belmont, Calif.: Wadsworth, 1971.

A thorough treatment of survey sampling methods at an elementary level.

SLONIM, MORRIS J. *Sampling in a Nutshell.* New York: Simon and Schuster, 1960.

A short, readable treatment of sampling. It covers many important topics, including stratified, cluster, and systematic sampling, with applications.

YAMANE, T. *Elementary Sampling Theory.* Englewood Cliffs, N.J.: Prentice-Hall, 1967.

A good reference source, treating survey sampling at a moderate level.

13. BAYES' THEOREM AND SAMPLING

THIS CHAPTER and the next will investigate the process of making decisions based upon information part of which was obtained from a sample. These chapters bring together the elements of decision making under uncertainty (the subject of Chapters 7 and 8) with the concepts of statistical inference (treated in Chapters 9 and 10). Thus, three factors may contribute to the decision solution: (1) the economic consequences of the various actions; (2) the original probability distribution of the decision maker; and, now, (3) the added information obtained from a sample. Chapters 13 and 14 show how to revise probabilities in the light of sample information and how to evaluate this information in advance to determine whether we should take a sample —and if so, what size of sample—before acting. This chapter treats the case of sampling from binomial distributions. Chapter 14 applies this analysis to the case of normal probability distributions.

In Chapter 8, the concept of the expected value of perfect information (EVPI) was introduced. This represented the economic worth, in a given decision situation, of having a perfect predictor of what event would occur. Chapter 8 also showed how, in general, to evaluate the value of partial information, as long as the reliability of the information could be ascertained. In many decision situations, it is possible to take a sample to obtain partial information.

Any sample estimate has associated with it sampling error and possibly bias, so it is not a perfect predictor. But the sample does give some additional information and should, on the average, improve the decision that is made. Since an improvement in decision making has an economic value, the sample information has a measurable worth to the decision maker; and the larger the sample, the greater the value, since larger samples are more precise. But larger samples cost more money than smaller samples. And so the problem facing the decision maker is to pick an optimum sample size that balances the worth of the sample information with the cost of taking the sample. This sample size might even be zero, meaning that he should act now without sampling. On the other hand, the sample cannot be so large that its cost exceeds EVPI.

A second related question is how the decision maker should act after he has taken a sample. How much weight should he place upon the sample information relative to his prior probabilities? Should he change his decision because of the sample? There are thus two questions facing the decision maker in an uncertain situation: (1) Should he take a sample, and if so, how large? (2) Given that a sample has been taken, what action should be taken on the basis of the sample results? Because this second question—the effect of sampling on decision making—generally is easier to answer than the first, we shall begin with it and return to the first question—on the selection of the sample itself—at the end of the chapter.[1]

PRIOR AND POSTERIOR PROBABILITY DISTRIBUTIONS

In order to introduce the concepts of prior and posterior decision making or "betting" distributions, let us first consider a rather artificial illustration. Suppose there are two large identical opaque jars on the table in front of you. Each of these jars contains 50 Ping-Pong balls. Jar A contains all red-colored balls; jar B contains all white balls. One of the jars is picked by the following random procedure: A fair die is rolled. If a 1 or a 2 turns up, jar A will be picked; if a 3, 4, 5, or 6 turns up, jar B will be picked. You are not allowed to witness the rolling of the die. Now, you are asked to play a game in which you guess which jar is to be selected. It is reasonable to assign a probability of ⅓ to the event "jar A is picked," since the probability of rolling a 1 or 2 out of six faces on the die is ⅓. Similarly, the probability of the event "jar B is picked" is ⅔. Let us call these our *prior* probabilities. These probabilities represent betting odds about which jar is to be selected.

Now, suppose a jar has been selected (which one you do not know), and you are allowed to take a ball from it and look at it before acting —that is, before guessing "A" or "B." The drawing of the ball from the jar is essentially taking a sample of size 1. After the sample, what would be your betting odds (called the *posterior* probability distribu-

---
[1] We consider here taking only a single sample and then acting. This procedure is often desirable, as in making a nationwide business survey involving a large fixed cost. Alternatively, we could take a series of samples and reach a decision whenever the cumulative evidence became convincing one way or the other. These are called sequential sampling plans.

tion) about which jar was selected? It would depend upon the color of the ball that was drawn. Since jar A contains all red balls and jar B contains all white balls, the color of the ball would give us an errorless indicator of which jar was selected. The betting distributions are shown in Table 13–1.

Table 13–1

PRIOR AND POSTERIOR PROBABILITY DISTRIBUTIONS

Event: Jar Selected Is	Prior Probability (Before Draw)	Posterior Probability: Ball Drawn Is	
		Red	White
A.....................	.333	1.0	0.0
B.....................	.667	0.0	1.0
	1.000	1.0	1.0

The important points of this illustration are: (1) we have an initial decision-making probability distribution (column 2)—this is designated as the prior distribution, since it is set up before the sample is taken; (2) this probability distribution is revised after the inclusion of the sample information—this revised distribution is called the posterior probability distribution; and (3) the posterior distribution depends upon the sample outcome. There is a different posterior distribution for each sample result.

Bayes' Theorem

The above example may seem trivial when one jar contained all white balls and the other all red balls. It is not so trivial if we change the problem slightly. Suppose, for example, jar A contains 70 percent red balls and 30 percent white balls, and jar B contains 20 percent red balls and 80 percent white balls. Let us see how to determine the posterior probabilities in this case.

Although we have already introduced Bayes' theorem for dealing with such problems in general (see page 205), it is useful to review it here and apply it in the specific context of sampling. If only one ball is to be drawn, it can be either red or white. We can draw up the joint probabilities in Table 13–2, as was done in Chapters 5 and 8. Recall that a jar (either A or B) was selected at random by rolling the die, and then a ball was selected at random from the designated jar. Hence, we can determine the joint probability of obtaining both a particular jar and a particular color of ball. For example, the joint probability of drawing jar A and then a red ball is $P(A, R)$. From page 101, the joint probability can be written as

$$P(A,R) = P(A) \, P(R|A)$$
$$= (.333)(.70) = .233$$

where $P(A) = .333$, the probability of drawing jar A, and $P(R|A)$ is the conditional probability of a red ball given jar A; it equals .70 since jar A contains 70 percent red balls.

The other joint probabilities in Table 13–2 are computed in a similar

Table 13–2

JOINT PROBABILITY TABLE

Jar	Red Ball Drawn	White Ball Drawn	
A	$P(A, R) = P(A) P(R\|A)$ $= (.333)(.70) = .233$	$P(A, W) = P(A) P(W\|A)$ $= (.333)(.30) = .100$	$P(A) = .333$
B	$P(B, R) = P(B) P(R\|B)$ $= (.667)(.20) = .133$	$P(B, W) = P(B) P(W\|B)$ $= (.667)(.80) = .534$	$P(B) = .667$
	$P(R) \quad = P(A, R) + P(B, R)$ $= .233 + .133$ $= .366$	$P(W) \quad = P(A, W) + P(B, W)$ $= .100 + .534$ $= .634$	1.000

manner. The entries at the bottom of the table are the marginal probabilities of obtaining a given color of ball. That is, one can obtain a red ball either by drawing jar A and then a red ball or by drawing jar B and then a red ball. Thus, the probability of a red ball is the sum of these joint probabilities, that is,

$$P(R) = P(A, R) + P(B, R) = .233 + .133 = .366$$

We are now ready to revise the prior betting distribution. Suppose that we draw a red ball. We ask this question: What is the probability that we have selected jar A, given the draw of a red ball? Symbolically, we want to find the conditional probability $P(A|R)$. From the definition of conditional probability (Chapter 5),

$$P(A|R) = \frac{P(A, R)}{P(R)} \qquad (1)$$

That is, the conditional probability of jar A, given that a red ball was drawn, is equal to the joint probability of jar A and a red ball divided by the marginal probability of a red ball. But a red ball may be drawn either from jar A or B and, hence, the marginal probability may be expressed as the sum of the probabilities of drawing a red ball from jars A and B. That is,

$$P(R) = P(A, R) + P(B, R)$$

But now the probabilites $P(A, R)$ and $P(B, R)$ may be written as in Table 13–2, column 1:

$$P(A, R) = P(A) P(R|A) \quad \text{and} \quad P(B, R) = P(B) P(R|B)$$

We can then rewrite Formula (1) as

$$P(A|R) = \frac{P(A)P(R|A)}{P(A)P(R|A) + P(B)P(R|B)} \tag{2}$$

As cited in Chapter 8, conditional probability expressed in the form of Formula (2) is known as Bayes' theorem. Note that it expresses the posterior probability of jar A given a red ball drawn, $P(A|R)$, in terms of the prior probabilities for jars A and B, $P(A)$ and $P(B)$, and the conditional probabilities of a red ball drawn from jars A and B [$P(R|A)$ and $P(R|B)$].

Substituting the numerical values in Formula (2), we have:

$$P(A|R) = \frac{(.333)(.70)}{(.333)(.70) + (.667)(.20)} = \frac{.233}{.366} = .637$$

The analogous Bayes' theorem formula for $P(B|R)$ is:

$$P(B|R) = \frac{P(B)P(R|B)}{P(A)P(R|A) + P(B)P(R|B)}$$
$$= \frac{(.667)(.20)}{(.333)(.70) + (.667)(.20)} = .363$$

The values $P(A|R) = .637$ and $P(B|R) = .363$ are the revised or posterior probabilities that the jar selected is jar A or jar B, respectively, given that the sample ball was red. If a white ball had been drawn, then the posterior probabilities could be obtained in a similar manner. They are $P(A|W) = .158$ and $P(B|W) = .842$.

These posterior probabilities represent "betting odds" in the same sense that the prior probabilities did. There was a 1/3 chance of jar A before a ball was drawn. After the draw of a red ball, the chance for jar A increased to almost 2/3 (i.e., .637); if a white ball is drawn, the odds for jar A drop to 15.8 chances in 100. These results are generally what we would expect from common sense: the draw of a red ball should increase the chances of jar A, since it contains predominantly red balls; and the draw of a white ball should increase the chances of jar B (and decrease those of A), since it contains predominantly white balls. The use of Bayes' theorem enables us to attach exact numerical values to the changes in the betting or decision-making probabilities.

It will be helpful for further analysis to put the computations of the posterior distribution in table form. The general form of the table and the specific calculations which were performed above are repeated in Table 13–3.

Table 13–3

BAYES' THEOREM: COMPUTATION OF POSTERIOR PROBABILITY

(SAMPLE RESULT: ONE RED BALL)

(1) Event: Jar Selected Is	(2) Prior Probability P(Event)	(3) Conditional Probability P(Sample Result|Event)	(4) Joint Probability P(Sample Result and Event) (Col. 2 × Col. 3)	(5) Posterior Probability P(Event|Sample Result) (Col. 4 ÷ Σ Col. 4)
A......... .333		.7	.233	.233/.366 = .637
B......... .667		.2	.133	.133/.366 = .363
Total.....1.000			.366	1.000

Marginal Proba-
bility = P(Sam-
ple Result)

Column 1 in Table 13–3 lists the possible events; in this case, jar A or B. Column 2 shows the prior (i.e., before sample) probabilities: 1/3 and 2/3 for jars A and B, respectively. Column 3 shows the probability of the sample result, given each of the events. In this case it shows the probability of drawing one red ball from jars A and B, respectively. Column 4 is the joint probability of the event and the sample both occurring. It is obtained by multiplying the values of column 2 by those in column 3.

The sum of the values in column 4 is the marginal probability of the given sample result. In this case, it is the probability of drawing a red ball, obtained by summing the two probabilities—a red ball drawn from jar A and a red ball drawn from jar B.

Column 5 shows the posterior probabilities, obtained by dividing the individual column 4 values by the column 4 total. The total of column

4 is the probability of a red ball, but since the red ball in fact has been drawn, its probability must be "blown up" to 1.0. The other values in column 4, therefore, are "blown up" or increased in the same proportion, giving the column 5 posterior probabilities.

Revision of Probabilities: Binomial Sampling

Let us continue the above illustration for one more step. Suppose that we were to draw a sample of three balls from the unidentified jar that was selected (replacing each after it is drawn). Further suppose that of the three balls, two were red and one was white. How would we obtain the posterior probabilities? First let us ask how we can obtain the conditional probabilities for this sample (2 red, 1 white), that is, $P(\text{sample}|\text{jar A})$ and $P(\text{sample}|\text{jar B})$. Since jar A contains 70 percent red balls, the probability of drawing a sample containing two red balls and one white ball from that jar is simply the binomial probability $P(r = 2|n = 3, p = .7) = .441$ (from Appendix F). Similarly, the probability of the sample given jar B (with 20 percent red balls) is the binomial probability $P(r = 2|n = 3, p = .2) = .096$. With these numbers we can fill in the remainder of Table 13–4 to determine the posterior probabilities.

It is important to understand that both the prior and posterior distributions are betting distributions. Before any sample information, we would bet on jar B with odds of two out of three. After this sample, the odds change considerably in favor of jar A (to .697 probability).

In Table 13–4, the sum of column 4 is .211. This is the probability of obtaining this particular sample (two red, one white) when drawing three balls. Other possible sample results are shown in Table 13–5.

Table 13–4

CALCULATION OF POSTERIOR PROBABILITIES

(SAMPLE OF 2 RED BALLS AND 1 WHITE BALL)

(1) Event: Jar Selected Is	(2) Prior Probability	(3) Conditional Probability $P(r = 2\|$ $n = 3, p)$	(4) Joint Probability (Col. 2 × Col. 3)	(5) Posterior Probability (Col. 4 ÷ Σ Col. 4)
A (with $p = .7$)	.333	.441	.147	.147/.211 = .697
B (with $p = .2$)	.667	.096	.064	.064/.211 = .303
	1.000		.211	1.000

↑
Marginal
Probability of
This Sample

Thus the marginal probability of obtaining a sample with three red balls is .120. And if this sample were to occur, the posterior probabilities would be .958 for jar A and .042 for jar B. The calculations of the results shown in Table 13–5 are not shown, but the numbers can

Table 13–5

POSSIBLE SAMPLES OF SIZE THREE
AND POSTERIOR DISTRIBUTIONS

Sample Result	Marginal Probability	Posterior Probability of	
		Jar A	Jar B
3 red balls..................	.120	.958	.042
2 red, 1 white..............	.211	.697	.303
1 red, 2 white..............	.319	.197	.803
3 white....................	.350	.026	.974
Total.....................	1.000		

be obtained by setting up a table such as Table 13–4 for each possible sample result.

POSTERIOR PROBABILITIES AND DECISION MAKING

The discussion above has concentrated upon the revision of probabilities and neglected the economic information in the decision process. Let us reintroduce the economic payoffs by means of an example. A manufacturer of electronic equipment operates two factories: one that manufactures components and another that assembles the components into complete units. A certain part is shipped from the manufacturing plant to the assembly plant in lots of 5,000 units. It has been very difficult to regulate the quality of this particular part; lots have been received with as few as 1 percent of the parts defective to as many as 20 percent of the parts defective. The fraction defective p (i.e., percent divided by 100) in the last 20 lots received is shown in Table 13–6. Let us suppose that management is willing to use these historical frequencies as a betting distribution about the fraction defective in the next lot.[2]

Economic Analysis before Sampling

When a defective part goes unnoticed and is assembled into the final unit, it affects the performance of the final unit. In such cases, the final

---

[2] Perhaps a more reasonable procedure would call for smoothing this frequency distribution to give some probability to the intermediate values of p. For a procedure to do this, see Chapter 2, pages 36–37.

Table 13–6

FRACTION DEFECTIVE FOR LOTS
OF THE SPECIFIED PART

Fraction Defective (p)	Number of Lots with this Fraction Defective	Relative Frequency
.01	3	.15
.02	5	.25
.05	7	.35
.08	3	.15
.10	1	.05
.20	1	.05
Total	20	1.00

unit has to be torn down and the defective part replaced. The cost of this tearing down and reassembling a final unit is $1.50 each.

An alternative is to inspect the entire incoming lot of parts and to remove all defective parts before assembly. The cost of this 100 percent inspection is 10 cents per part or $500 per lot. A lot of the particular part has just arrived and the manager must decide whether to inspect 100 percent or to use the lot as is. Let us first draw up a payoff table for this decision problem. This is done in Table 13–7.

Columns 1 and 2 come from Table 13–6. Costs in columns 3 and 4

Table 13–7

PAYOFF TABLE FOR ACTIONS "INSPECT 100 PERCENT"
AND "ACCEPT LOT AS IS"

(Lot Size 5,000; Inspection Cost 10¢; Replacement Cost $1.50)

Event: Fraction Defective in the Lot (p) (1)	Probability $P(p)$ (2)	Cost* Inspect 100 Percent (3)	Cost* Accept Lot as Is (4)	Opportunity Loss Inspect 100 Percent (5)	Opportunity Loss Accept Lot as Is (6)
.01	.15	$500	$ 75.00	$425	0
.02	.25	500	150.00	350	0
.05	.35	500	375.00	125	0
.08	.15	500	600.00	0	$ 100.00
.10	.05	500	750.00	0	250.00
.20	.05	500	1,500.00	0	1,000.00
Expected value		$500	$ 382.50	$195	$ 77.50

* Note that we have linear cost equations in this example. Cost of inspection = $500. Cost of accepting as is = ($1.50)(5,000)$p$, where p is the unknown variable (fraction defective). Its expected value $E(p)$ can be calculated to be .051 and, hence, the expected cost can be determined as $E(c) = (\$1.50)(5,000)E(p) = \$7,500(.051) = \$382.50$, as above.

are determined as follows: for 100 percent inspection, cost is 10 cents per unit times 5,000 parts = $500; for accepting the lot as is, the cost is $1.50 per unit replacement cost times the number defective (5,000 × p). For example, when $p = .05$, we expect $.05 \times 5,000 = 250$ defectives, and $250 \times \$1.50 = \$375$. Opportunity losses in columns 5 and 6 are obtained by subtracting the lower of two costs in each row from the higher cost. Expected values are the weighted averages of the figures in each column multiplied by their probabilities and totaled.

As can be seen from this table, the better action is to accept the lot as is, since this action has the lower expected cost, even though this will necessitate some rework at a later time. The EVPI is $77.50 per lot (the expected opportunity loss of the better action). Since this is a fairly substantial amount, the decision maker should investigate ways of obtaining additional information.

Economic Analysis after Sampling

One method of obtaining at least partial information in this situation is by taking a random sample of parts in the lot and inspecting the items in the sample. From the number of defects in the sample we can make some inferences about the fraction defective in the entire lot.

Let us suppose that the manager arbitrarily decided to sample 25 items from the lot and that he found that two of 25 were defective. We now want to investigate what action should be taken on the basis of his prior probabilities and the sample information combined. The decision maker can revise his original or prior betting distribution in the same fashion as in Table 13–4. This is done in Table 13–8.

Compare the posterior probabilities with the prior probabilities. The fraction defective in the sample was $2/25 = .08$. Note that the posterior probabilities for values of p close to .08 have increased (relative to the prior values) and the posterior probabilities for p far from .08 have decreased.

We can now use the posterior probabilities, together with the original costs in Table 13–7 to revise our payoff table, using the same computations as before.[3] (See Table 13–9.) The optimal action remains to accept the lot as it is, since this action has the lower expected cost. However, the expected cost is somewhat more than previously, since the fraction defective in the sample (.08) exceeded the expected fraction

---

[3] We can find the $E(p)$ for the posterior distribution $= .0609$. As an alternate method of finding the expected cost, we have $E(c) = (\$1.50)(5,000)E(p) = \$7,500 \times (.0609) = \$456.75$, as in Table 13–9.

Table 13–8

CALCULATION OF POSTERIOR PROBABILITIES BY BAYES' THEOREM

(SAMPLE OF 25 PARTS WITH 2 DEFECTIVES)

Event: Lot Fraction Defective Is p (1)	Prior Probability $P(p)$ (2)	Conditional Probability* $P(r = 2\|n = 25, p)$ (3)	Joint Probability $P(p)P(r = 2\|n = 25, p)$ (Col. 2 × Col. 3) (4)	Posterior Probability $P(p)P(r = 2\|n = 25, p)$ $\overline{\Sigma P(p)P(r = 2\|n = 25, p)}$ (Col. 4 ÷ Σ Col. 4) (5)
.01 15		.024	.00360	.022
.02 25		.075	.01875	.115
.05 35		.231	.08085	.498
.08 15		.282	.04230	.261
.10 05		.266	.01330	.082
.20 05		.071	.00355	.022
Total 1.00			.16235	1.000

↑
Marginal
Probability of
This Sample

* The values in column 3 were obtained from the binomial tables in Appendix F.

defective (.051) prior to taking the sample (Table 13–7, footnote). Note that the posterior EVPI is still quite large ($68.60 from Table 13–9), indicating that the particular sample result did little to resolve the uncertainty about which action to take. The decision maker could consider taking a second sample before acting.

The sample result "2 defectives out of 25" is only one of many that

Table 13–9

PAYOFF TABLE USING POSTERIOR PROBABILITIES

(SAMPLE OF 25 PARTS WITH 2 DEFECTIVES)

Event: Fraction Defective in the Lot p	Posterior Probability $P(p)$	Cost		Opportunity Loss	
		Inspect 100 Percent	Accept Lot as Is	Inspect 100 Percent	Accept Lot as Is
.01 022		$500	$ 75.00	$425.00	0
.02 115		500	150.00	350.00	0
.05 498		500	375.00	125.00	0
.08 261		500	600.00	0	$ 100.00
.10 082		500	750.00	0	250.00
.20 022		500	1,500.00	0	1,000.00
Expected value		$500	$ 456.75	$111.85	$ 68.60

could have occurred. The other possible results are shown in Table 13–10. The decision action changes if three or more defectives are found in the sample—then 100 percent inspection become the more economical decision. Note that different sample results lead to quite different values of the posterior EOL of the better action, or EVPI. When either very few or very many defectives are found in the sample, the decision to be taken becomes relatively clear (i.e., accept if r is 0

Table 13–10

POSSIBLE RESULTS FOR A SAMPLE OF 25 ITEMS

Sample Result (Number of Defectives) r	Posterior Action	Posterior Expected Cost	Posterior Expected Opportunity Loss
0...............	Accept as is	$212.25	$ 8.05
1...............	Accept as is	333.22	26.95
2...............	Accept as is	456.75	68.60
3...............	Inspect	500.00	63.92
4...............	Inspect	500.00	32.55
5...............	Inspect	500.00	13.00
6...............	Inspect	500.00	4.38
7 or more.........	Inspect	500.00	Very small

or 1; reject if $r > 4$). When a "middle" number of defectives is found (around 2 or 3 out of 25), there remains considerable uncertainty about which is the correct action. This is true of sampling in general. Very good or very bad sample results lead to clear-cut decisions; borderline results are indecisive and may require further sampling.

EXPECTED VALUE OF SAMPLE INFORMATION

In the previous section, we addressed ourselves to the question, "Given that a sample of a certain size has been drawn, what action should be taken on the basis of both prior and sample information?" In this section we examine the question, "Should we take a sample, and if so, how large should it be?" As noted earlier, sampling may be costly, and the larger the sample, the greater the cost. Hence, to take a sample, we must determine that the economic value of the information contained in the sample is worth the cost.

A sample has value because it is expected to reduce uncertainty. After the sample, we are generally more sure than before about which event will occur. Hence, we are less apt to make a costly mistake. To see this, compare the EVPI prior to taking the sample, which is $77.50 (Table 13–7), with the posterior expected opportunity losses (or EVPI's) in Table 13–10. After the sample, the EVPI ranges from near zero (when $r = 7$ or more) to a high of $68.60 (when $r = 2$). All the values are below $77.50, indicating that even the most inconclusive sample result ($r = 2$) often reduces the uncertainty.[4] And the sample result ($r = 0$) has a posterior EVPI of $8.05, a considerable reduction. Thus, a sample result "0 defectives out of 25" makes it almost certain that the correct action is to accept the lot as is. In this case the sample information is quite conclusive.

Another way of determining the value of a given size sample before taking the sample is to compare the expected cost (or profit) before the sample with the expected cost (or profit) if we had taken the sample. The amount by which cost is reduced from the before-sample case to after-sample case gives us the economic value of the sample. The prior expected cost is determined, in our example, as $382.50 from Table 13–7. The posterior expected cost, however, depends upon the particular sample result that might occur. For example, the posterior expected cost would be $456.75 for a sample result of 2 defectives out of 25 (see Table 13–9). Similar expected cost values can be calculated from the posterior distributions associated with other sample results. These calculations are not shown, but the results are displayed in Table 13–10. The lowest posterior expected cost would be $212.25, if zero defectives were observed in the sample. At the other extreme, if three or more defectives were observed, 100 percent inspection is the action chosen, with a certain cost of $500.

How can we compare prior with posterior expected cost if posterior expected cost is represented by several possible values? The answer lies in the use of an average or expectation of the posterior costs. Recall that we can determine the marginal probability of any particular sample result for a given set of prior probabilities. Thus, the probability of exactly 2 defectives out of 25 items is found in Table 13–8 (sum of column 4) to be .162. Similarly, the probability for the sample result "0 defectives out of 25 items" can be found to be .387 (calculations are not shown); the probability for the sample "1 defective out of 25 items" is .286; and so on, as shown in column 2 of Table 13–11.

These probabilities can be used as weights to determine the expecta-

---

[4] It is possible to obtain a specific sample result that actually increases posterior EVPI, although this did not happen in this example. However, the expected posterior EVPI over all sample results will be less than the prior EVPI.

Table 13–11

ESTIMATING POSTERIOR EXPECTED COST, BEFORE SAMPLING

Sample Result (Number of Defectives) r (1)	Probability of Sample Result $P(r)$ (2)	Posterior Expected Cost (3)	Expected Value (Column 2 × Column 3) (4)
0................	.387	$212.25	$ 82.14
1................	.286	333.22	95.30
2................	.162	456.75	73.99
3................	.082	500.00	41.00
4................	.039	500.00	19.50
5................	.020	500.00	10.00
6................	.011	500.00	5.50
7 or more.........	.013	500.00	6.50
	1.000		$333.93

tion or average of the posterior expected costs associated with each possible sample result. This calculation is performed in Table 13–11.

The amount of $333.93 from Table 13–11 is our expectation, before taking the sample, of what the posterior expected cost will be. The value of the sample, called *expected value of sample information* or EVSI, is the difference between the prior expected cost ($382.50) and this value. It is $382.50 − $333.93 = $48.57. This is the amount by which we can expect to reduce cost by taking a sample of 25 items and then acting on the basis of the sample result. If the cost of taking the sample of 25 items is less than $48.57, therefore, the sample should be taken. In our example, inspection cost is only 10 cents a part, or $2.50 for 25 parts, so the sample would be worthwhile.

Note that the expected value of sample information is a value obtained before the sample has been taken—in fact, before the decision has been made about whether a sample ought to be taken at all. It is an *expected* value. Before sampling we do not know how much the sample will save; we do not know even what the sample result will be and, hence, are uncertain what action we will take based upon the sample result. Using the probabilities of the various sample results and computing the expected value, we are determining the "best bet" to make in the decision situation.

Throughout this example we have examined only the possibility of a sample of 25 items. Would not a sample of 20 items or 50 items or 100 items be better? The low inspection cost (10 cents per part versus $1.50 replacement cost) and the initial uncertainty as to fraction defective (as shown by the diffuse probability distribution in Table 13–7) suggest that the optimum sample size should be much larger than 25. On the other hand, it would not pay to take a sample so large that its cost exceeded the expected value of perfect information, which was $77.50. Hence, the sample size should not exceed 775 (since $77.50 ÷ .10 = 775), out of the whole lot of 5,000 parts. We could then take a few sample sizes—say, from 50 to 700—and compute EVSI less sampling cost for each to determine the optimum sample size. These calculations would be tedious and might be more costly to perform than the savings from taking a sample were it not for the availability of computers.[5]

Fortunately, we have techniques for the special case of normal sampling (or the normal approximation to the binomial in this case) that reduce all this computation to a single formula. However, because it is necessary to understand the concept of the expected value of sample information (EVSI) and how it can be obtained in a general case, we have gone through the detailed procedure above. The special case will be the subject of the following chapter.

BAYESIAN VERSUS CLASSICAL APPROACH

There is some controversy in the statistics profession over the validity of the decision-making approach suggested in this chapter. Our approach is in accord with the thinking of the Bayesian school. The more traditional or classical approach to the evaluation of sample information was presented in Chapters 9 and 10. The controversy centers about whether the statistician, as a scientist, should be concerned only with the objective evidence of the sample (classical school) or whether he should also be concerned with the whole decision framework, including any subjective judgment of the decision maker about the probabilities of various events. Bayesian analysis takes into account subjective probabilities and utility values in much the same way as they are intuitively considered by the business executive.

A prior judgment is particularly significant if sample information is meager, as in most small samples. In taking very large samples, where the evidence is overwhelming, the prior judgment well may be discarded. How much additional information is needed for its evidence to "swamp" prior probabilities? Bayes' formula provides an answer in the form of an automatic adjustment: If the sample is small, its results may modify prior probabilities but little; but as the sample increases in size,

---

[5] See R. Schlaifer, *Computer Programs for Elementary Decision Analysis* (Boston: Division of Research, Harvard Graduate School of Business Administration, 1971).

the posterior probabilities approach those shown in the sample, ir-respective of the prior judgment.

Bayesian methods also take into account the economic profits or losses of decisions, as well as the probabilities involved. Thus, in the classical testing of hypotheses discusssed in Chapter 10, we reject a hypothesis if the risk of making a Type I error—rejecting a true hy-pothesis—exceeds some critical probability such as 5 percent. This figure is rather arbitrary, and it does not provide for balancing the rela-tive cost of Type I versus Type 2 errors. It is difficult to balance these errors in classical theory. Bayesian statistics adds the economic dimen-sion to the decision-making process and offers an objective criterion for making decisions: set up a probability distribution and payoff table, then maximize expected profits.

The Bayesian approach thus serves as the completion of the classical theory of statistical inference, through providing the decision maker with a logical framework within which to apply both his judgment and sample evidence, in proper proportions, to the economic consequences of his possible actions.

SUMMARY

The subject of this chapter is the application of Bayes' theorem to decision making under uncertainty. This involves the combination of a prior probability distribution (which may be subjective) with the re-sults of a sample to form a posterior decision-making distribution.

Bayes' theorem is a form of expressing the conditional probability of an event, given a sample outcome, in terms of the prior probability of the event and the conditional probabilities of sample result, given the event.

In the electronic component example, we are given prior probabili-ties for various levels of fraction defective (following the binomial distribution), but if we then take a sample of 25 and find 2 defectives, we can modify the prior probabilities by the sample result, as in Table 13–8, to find the posterior probabilities. These revised probabilities are then used in a payoff table, just as the prior probabilities were, to find the expected cost (or profit) of each possible action. In our ex-ample, the best decision before sampling was to accept the lot as is rather than inspect 100 percent. After taking a sample of 25, however, we arrived at a better decision rule: accept the lot if the sample has 2 or less defectives; otherwise, inspect 100 percent. Each possible sample result has a different posterior distribution and a different posterior expected value.

A sample has economic value because it reduces the uncertainty associated with decision making. The specific value, called the expected value of sample information, is determined by subtracting the expected cost posterior to the sample from the prior expected cost. The expected posterior cost is obtained as an expectation or average of the expected costs associated with the various possible sample results. We can determine if a sample of a given size should be taken at all by comparing the cost of the sample with the expected value of sample information. An optimal sample size can be determined by making this comparison for several sizes of samples, from zero up to the sample size whose cost equals EVPI.

PROBLEMS

1. Explain:
    a) Prior and posterior distributions.
    b) Bayes' theorem.
    c) Conditional and joint probabilities.
    d) Posterior expected cost.
    e) Expected value of sample information.

2. For the example used in the text on pages 373–75, verify the posterior probabilities $P(A|W) = .158$ and $P(B|W) = .842$.

3. Verify the posterior probabilities shown in Table 13–5.

4. Verify the calculations shown in Table 13–10 for the row listed below, as assigned:
    a) The row for 0 defectives.
    b) The row for 1 defective.
    c) The row for 3 defectives.
    d) The row for 4 defectives.

5. In a certain portfolio, 70 percent of the industrial stocks increased in value over the past year while 40 percent of the utility stocks increased. The portfolio contains 80 percent industrial stocks.
    a) If a stock is selected at random, what is the probability that it is one that has increased in value?
    b) Suppose a stock is drawn and noted to be one that has increased. What is the probability that the stock is an industrial stock?

6. Of the firms in a certain industry, the median age of the chief executive officer is 50 years. Of those executives under 50, 65 percent were in marketing before becoming president. Of those over 50, only 45 percent reached the chief executive position through marketing.

   If a chief executive is selected at random in this industry, and if it is noted that he had not reached the top through marketing, what is the probability that he is over 50 years old?

7. The Eastern Motel Association is about to poll its members about whether or not to accept a certain national credit card. The executive secretary of the association feels that he knows pretty well how many (i.e., what percent) of the motels favor accepting the credit card. Suppose he attaches the following probabilities to various percentages in favor:

Percent in Favor of Accepting	Probability of Exactly that Percent
30	.10
40	.30
50	.40
60	.20
	1.00

a) Based on this information, would you guess that a vote for the credit card would win or lose? Hint: For 50 percent in favor, split the .40 probability evenly on both sides of 50 percent.

b) Suppose you drew a random sample of 15 motels and found 8 in favor and 7 opposed. What probabilities would you then assign to "percent in favor of accepting"?

c) After the above sample, what is the probability that a vote will find a majority in favor?

8. An election is being held in a certain plant to determine if the workers should be represented by a union. A few days before the election, management assigns the probabilities below to the events, "Proportion of workers who will vote for unionization":

Event	Probability
.35	.15
.40	.30
.45	.20
.50	.20
.55	.10
.60	.05
	1.00

A sample of 20 workers is chosen at random and the voting intentions of each ascertained, with the following results: 11 will vote for unionization and 9 will vote against unionization.

After the sample, what probabilities should management assign to the events "Proportion of workers voting for union"?

9. From past experience, the fraction of items defective in lots manufactured by a certain process has the following distribution:

Lot Fraction Defective	Relative Frequency
.01	.50
.02	.30
.05	.10
.10	.05
.15	.05
	1.00

A sample of 15 items is taken from a certain lot and no defectives are found. What posterior probabilities would you assign to the events "Lot fraction defective"?

10. The Theta Company manufactures its requirements for part No. 805 in lots of 1,000 units. It has been difficult to control the quality of this product without a complicated readjustment of the manufacturing equipment. The cost of such a readjustment is $300. When such a readjustment has been made, only 2 percent defectives are produced. Without the adjustment, the quality has been quite variable, as shown by the history of the last 20 lots:

Fraction Defective	No. of Lots
.02	5
.05	8
.10	4
.15	2
.20	1
	20

A lot of part No. 805 is about to be manufactured, and management is undecided about whether it should pay for the costly adjustment or take the chance of a large percentage of defectives. Replacement cost of a defective item is $5.

a) Draw up a payoff table and calculate the expected cost of each action using the past frequency data as prior probabilities. Which action is preferable?

b) What is the EVPI?

c) Suppose the manufacturing process was set up and the first 20 items were examined and 2 defectives were found. Should the machine be shut down and an adjustment made at this time or should the manufacturing process be allowed to continue?

11. (Continuation of Problem 10.) Suppose the sample result had been no defectives out of 20 items sampled. What is the expected posterior cost of each action? Which action is preferable? What is the posterior EVPI?

12. (Continuation of Problems 10 and 11.)

a) Find the expected posterior cost for other relevant sample results.

b) What is the expected value of sample information for a sample of 20 items in this decision situation?

c) Suppose it cost $20 plus $2 per item sampled. Should a sample of 20 items be taken?

13. As president of the Alma Mater University Alumni you are planning the annual alumni banquet. There are 1,000 members of the alumni chapter. Based upon the attendance of previous years, you assign the following probabilities to the number attending this years' annual banquet:

Number Attending	Probability
100.....................	.2
200.....................	.2
300.....................	.3
400.....................	.2
500.....................	.1

The banquet is to be held at the Ritz-Oasis, and the banquet manager informs you that you must specify the number you expect to attend within the next few days. He gives you a price of $6 per plate for the exact number specified. Additional dinners (beyond the number specified) may be obtained on the day of the banquet (after registration when exact attendance is known) at a price of $8 each. If fewer dinners are needed than ordered, a partial refund of $2 will be made for each dinner not needed (i.e., $4 will be charged for each dinner ordered that is not needed).

The fee that you will charge the alumni has been set at $10 each for those attending. Because of the short time available it is not possible to use a mail reservation system.

a) Based only on the information given above, how many dinners should you order? What is the EVPI? (Only consider ordering dinners in even hundreds.)

b) Suppose that you select a random sample of 20 alumni and call them on the phone. Eight indicate that they will attend. Using this sample information and that above, what action (number of dinners to order) would you take? What is your EVPI?

14. You produce a product on 10 machines. When these machines are in proper adjustment, they produce with 8 percent defective, and this level of quality is acceptable to your customers. The product is stored in your warehouse in lots of 1,000 units, each lot containing units from the same machine. You have just discovered that one of your machines has been out of adjustment for several days, and that this has gone undetected because an inexperienced tester has been assigned to that machine. The machine, while it was out of adjustment, produced 40 percent defective parts. One tenth (10 percent) of the lots that are stored in your warehouse are 40 percent defective; and since lots are mixed up on being transferred from factory to warehouse, you have no way of knowing, without testing, which are the good lots (8 percent defective) and which are the bad (40 percent defective).

Your customers will balk at receiving a lot which turns out to be bad when they try to use the product, and you value the cost of sending a customer a bad lot (40 percent defective) at $500 per lot (cost of replacing defectives and lost customer goodwill).

You could sell the lots as being of inferior quality, but you would receive $100 per lot less than if they were sold as good quality.

Your assistant has suggested taking a sample of one item from each lot, and deciding upon the basis of the sample whether to sell the lot as a good-quality or inferior-quality one. Sampling involves destruction of the product, and the cost is $15 per item sampled.

Assume three alternatives. You may: (i) sell all the lots as good-quality lots, and incur the cost of $500 on those that turn out bad; (ii) sell all the lots as inferior-quality lots at a discount of $100 from the price of a good lot; or (iii) take a sample of one item from each lot, decide on the basis of the sample whether the lot is good or inferior, and sell it as such.

a) Draw up a payoff table for the first two alternatives. Without sampling, what would be your decision?

b) Assume you sample. If the sampled item is not defective, should the lot be sold as of good or inferior quality? What should be done if the sampled item is defective?

c) What is the expected cost of each of the three alternatives above? Which should be selected?

15. A firm has developed a new machine to harvest plums. The machine offers significant advantages over current methods. It is a small firm, and it knows that if it markets this new machine, and if it is successful, that the large firms in the industry will build similar machines and take most of the market away from it. Hence, the firm wishes to make its decision about introducing the new machine solely upon how many it expects to sell in the first year (that is, before the larger firms have time to react).

The firm decides to concentrate its selling efforts on the owners of large orchards. According to trade association lists, there are about 1,000 orchards large enough to use the machine profitably.

The costs associated with setting up the production line, promotion, etc., will amount to $280,000. The firm plans to sell the machine for $12,000; of this, $8,000 will be the cost of manufacturing and selling.

Prior judgment concerning the number the firm could sell in the first year is represented in the table below:

Number Sold	Percentage of Orchards	Probability of Selling This Number
50.....................	5%	.40
80.....................	8	.30
100.....................	10	.20
120.....................	12	.10

Suppose that management decided to select a sample of orchards from the list of 1,000, contact owners, and attempt to sell the new machine (any sold would be produced as a special order if the decision were made not to introduce the machine to the whole market). A sample of 18 orchards was selected.

a) Suppose that the owner of one of the 18 orchards bought the new machine. What posterior probabilities should be assigned to the percentage of orchard owners who would buy the machine?

b) What decision should the firm make? What is the posterior expected profit?

c) Answer questions (a) and (b) on the assumption that a sample of 10 orchards was contacted and none of the owners bought the new machine

16. Refer to Problem 15 above. Suppose that management had not yet taken a sample but was considering whether or not a sample should be taken. Possible sample sizes were 10, 15, 20, or 25 orchards. Suppose that the sample costs $1,000 plus $100 per orchard. How large a sample should be taken, considering only the four possibilities given above? (Note: This problem requires substantial computation).

SELECTED READINGS

Selected readings for this chapter are included in the list which appears on page 421.

14. DECISION MAKING AND SAMPLING: THE NORMAL DISTRIBUTION

CHAPTERS 7 AND 8 introduced the concepts necessary for decision making under uncertainty and for the evaluation of additional information. Chapter 13 extended these ideas to the case when the additional information came from a binomial sample. In this chapter, we shall consider a special case involving the normal distribution. The chapter treats three topics: (1) decision making when the decision maker's prior judgment can be expressed by a normal distribution, (2) determining the posterior distribution from sample information, and (3) evaluating sample information.

THE NORMAL DISTRIBUTION IN DECISION MAKING

In making decisions under uncertainty, the decision maker can express his subjective feelings about the unknown variable as a probability distribution. In many situations it is reasonable to use the normal distribution for this purpose. When this is done, the decision procedure is simplified. Thus, the expected value of perfect information (EVPI) can be found by a single formula. The choice of the normal distribution as a decision-making or betting distribution implies that the decision maker feels that some value of the unknown variable is the most likely (the mean of the distribution), that the variable is more likely to be close to this guess than far away (the area of the normal distribution is clustered around the mean), and that the unknown variable could as likely be above or below this guess (since the normal distribution is symmetrical).

The normal distribution has two parameters, the mean and the standard deviation. For reasons that will become clear later in the chapter, we will use the symbol M_0 to represent the mean of this

393

normal betting distribution, and S_0 to represent the standard deviation. Also, Y will represent the unknown variable. In finding appropriate values for these parameters to use in his particular situation, the decision maker must phrase some questions for himself. In order to estimate the mean M_0, he must find the middle point of his betting distribution. He should be willing to bet that the unknown variable Y is as likely to fall above as below M_0. Also, since two thirds of the area of the normal curve lies within one standard deviation of the mean, the decision maker should specify a range about M_0 such that there is a two-thirds chance that Y would be in this interval.[1] That is, the decision maker should estimate the value of S_0 such that he would be willing to bet that Y will fall in the interval $M_0 \pm S_0$ with odds of two out of three.

Before using this normal distribution, the decision maker should graph it and check the probabilities implied by the distribution against his judgment.[2] For example, he should judge the odds to be about 95 out of 100 that Y will fall in the interval $M_0 \pm 2S_0$.

Expected Value of Perfect Information

We shall consider now problems involving two alternatives, each of which offers a profit (π) that is a linear function of the unknown variable Y; that is, $\pi = a + bY$. We saw in Chapter 8 that in this case the expected profit for each alternative action is also a linear function of the expected value of Y; i.e., $E(\pi) = a + bE(Y)$ regardless of whether the decision distribution is normal or any other shape. Recall also that in this case the opportunity loss functions can be represented by two connected straight lines as shown in Chart 14–1. In general, the calculation of the expected opportunity loss or EVPI is a tedious matter. However, when the decision distribution is normal, the expected opportunity loss and EVPI can be expressed in a simplified form.

Consider Chart 14–2. Here the normal distribution is superimposed

---

[1] An alternative procedure is to specify a symmetric interval about M_0 (e.g., $M_0 + Q$, the quartile deviation) such that there is an even chance for the unknown variable to be in this interval. Then $Q = 2/3 \, S_0$ or $S_0 = 3/2Q$. This follows from the fact that the normal distribution has about half its area in the interval $M_0 \pm 2/3 \, S_0$. (See Appendix B in Chapter 5 for more details.)

[2] The normal distribution is at best an approximation to one's betting distribution. This distribution is continuous, whereas most decision-making distributions are discrete (e.g., sales are in integer units). Also, the normal distribution has tails that go out in both directions indefinitely, though the probabilities in these tails are quite small. Generally, we would like to truncate our decision distribution at certain points (e.g., sales cannot be negative, so the probabilities of negative sales should be zero). Despite these minor inconsistencies, the normal distribution is quite adequate for many situations.

Chart 14–1

OPPORTUNITY LOSS FUNCTIONS FOR TWO-ACTION
PROBLEM WITH LINEAR PROFIT FUNCTIONS

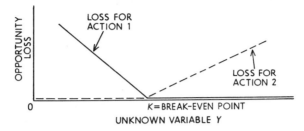

upon the loss function for one of the actions (the one with the higher expected profit). The expected opportunity loss is simply the probability function times the loss function summed over the whole area. The simplified formula for the expected value of perfect information (the EOL of the best action) is:

$$\text{EVPI} = t S_0 L_N(D) \tag{1}$$

where

$$D = \left| \frac{K - M_0}{S_0} \right| \tag{2}$$

In the above formulas, t is the slope of the opportunity loss function; M_0 and S_0 are the parameters of the normal decision distribution; K is the break-even point; and $L_N(D)$ is the *unit normal loss function,* which is found by looking up D in Appendix E. The symbol $| \; |$ means absolute value (ignoring a negative sign).

Chart 14–2

OPPORTUNITY LOSS FUNCTION $L(Y)$
AND NORMAL DISTRIBUTION $P(Y)$

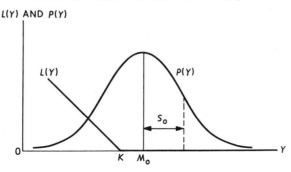

An Example.   A distributor has an opportunity to market his product in a new territory. The fixed cost of this action is $4,000 for advertising, facilities, etc. For each unit sold the distributor will make a profit of $.10. It will thus take sales of 40,000 units to break even ($K = 40,000$).

The distributor is quite uncertain about how many units he will actually sell. However, he is willing to represent his uncertainty about sales by a normal distribution. Suppose that he feels that sales are as likely to be above 50,000 as below 50,000 (that is, $M_0 = 50,000$). Further, suppose he assigns a probability of two thirds to the possibility that actual sales will be in the range of 25,000 to 75,000. Since this range is 50,000 (or M_0) $\pm$ 25,000, the standard deviation $S_0 = 25,000$. When presented with Chart 14–3, the decision maker agrees

Chart 14–3

NORMAL DECISION DISTRIBUTION FOR
POSSIBLE SALES IN NEW TERRITORY

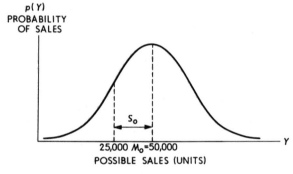

that this adequately represents his betting distribution.

The profit functions are:

$$\text{Open new territory}: \pi = -\$4,000 + (.10)Y$$
$$\text{Do not open new territory}: \pi = 0$$

where Y is the number of units sold. The expected profits are:

$$\text{Open new territory}: E(\pi) = -\$4,000 + (\$.10)(50,000)$$
$$= \$1,000$$
$$\text{Do not open new territory}: E(\pi) = 0$$

And so, with this information, the decision maker should market in the new territory.

The opportunity loss function for this optimal decision is:

$$\text{Opportunity loss} = L(Y) = 0 \qquad\qquad \text{if } Y \geq 40{,}000$$
$$\text{or } L(Y) = (\$.10)(40{,}000 - Y) \qquad \text{if } Y < 40{,}000$$
$$= \$4{,}000 - (.10)Y$$

Using Formulas (1) and (2), we can determine the expected opportunity loss for this decision (which is EVPI, since it is the optimal decision):

$$D = \left|\frac{K - M_0}{S_0}\right| = \left|\frac{40{,}000 - 50{,}000}{25{,}000}\right| = .40$$
$$\text{EVPI} = tS_0L_N(D) = (\$.10)(25{,}000)L_N(.40)$$
$$= (\$.10)(25{,}000)(.2304) = \$576$$

In the above equations, the values of $M_0 = 50{,}000$ and $S_0 = 25{,}000$ represent the decision maker's normal betting distribution. The break-even sales value is $K = 40{,}000$ units. The slope of the loss function is $t = .10$; this is the loss for each unit below the 40,000 break-even level. And, finally, the value of $L_N(D) = L_N(.40)$ is obtained from Appendix E.

Interpretation of EVPI. In the above example the expected value of perfect information is $576. This means that the distributor would pay no more than this amount for information about his exact sales. The information he can get (such as studies of income or market potential) is worth a good deal less than $576, since such information cannot give an exact prediction.

If we reexamine Formulas (1) and (2), we can see what factors influence the value of EVPI.

$$\text{EVPI} = tS_0L_N(D) \tag{1}$$

$$D = \left|\frac{K - M_0}{S_0}\right| \tag{2}$$

Note the following: (a) The symbol t represents the per unit opportunity loss. Hence, the larger t, the larger is EVPI. If t is small, the economic consequences of making the wrong decision are not serious. If t is large, they may be. (b) The larger S_0, the larger is EVPI. The standard deviation S_0 is a measure of the degree of uncertainty in the decision situation. The more the uncertainty, the more valuable the perfect information. (c) The farther the break-even point K is from the expected sales M_0 (in standard deviation units), the larger is

D and the smaller are $L_N(D)$ (see Appendix E) and EVPI. Clearly, if the break-even point is well above or below the expected sales, the decision is relatively certain and additional information is of little value. On the other hand, if $(K - M_0)$ is small, even a little additional information may change the decision, and hence may be valuable.

Another way of looking at EVPI is as the maximum price the decision maker might pay for insurance to guarantee him against a loss.[3] In the distributor example, the decision maker should be willing to pay an insurance premium up to $576. The insurance policy would pay the difference between the revenue from the new territory ($.10 times the number of units sold) and the fee of $4,000 if revenue were less than $4,000.

Another Example. A manufacturer must replace some machinery that has worn out. Two types of machinery can be picked to replace the worn-out equipment. Machinery type A is conventional; it costs $200,000, and has a variable operating cost of $12 per hour (e.g., direct labor, maintenance). Machinery type B is largely automated; it costs $400,000, but has a variable operating cost of only $7 per hour. Both machines produce the same output per hour in quantity and quality.

Because of economic factors, the market for the product is in a state of flux. Hence, the required number of hours of operating time on the machinery is uncertain. Management expressed this uncertainty in terms of a normal distribution with mean $M_0 = 50,000$ and $S_0 = 20,000$ hours.[4]

The cost functions for the two alternatives are

---

[3] Or to guarantee him a profit if he decides not to act, when in fact a profit could have been made. In other words, the insurance would pay the opportunity loss. As a practical example of such a situation, consider the following from a front-page article in *The Wall Street Journal* of December 6, 1966:

"Good Weather, Inc., a Long Island insurance agency that specializes in unusual risks, says that for the past six years a major maker of candy has bought a policy insuring against rain or snow on Valentine's Day. Henry Fox, president of the agency, says, 'Since candy is an impulse-type purchase, the company's retail stores would be left with a large stock if the weather was bad. But people after Valentine's Day won't buy candy in heart-shaped boxes because they're afraid it might be stale. So we insure the manufacturer against the expense of transferring the candy to regular boxes.'

"The policy is for almost $250,000, and the premium is $10,000. It covers various cities in the Northeast and the complex payout formula is based upon the amount of snow or rain and the number of hours that it snows or rains."

[4] Since these hours probably would be spread over several years, discounting procedures are appropriate. Further, tax factors associated with depreciation are relevant to the decision. We have omitted these factors in order to concentrate on the decision analysis. See N. Harlan, C. Christenson, and R. Vancil, *Managerial Economics: Text and Cases* (Homewood, Ill.: Richard D. Irwin, 1962), pp. 239–65, for a discussion of these topics.

Machinery type A: Cost $= C(Y) = \$200{,}000 + \$12Y$
Machinery type B: Cost $= C(Y) = \$400{,}000 + \$\ 7Y$

where Y is the actual number of machine hours used.

The cost functions are graphed in Chart 14–4. Note that by setting

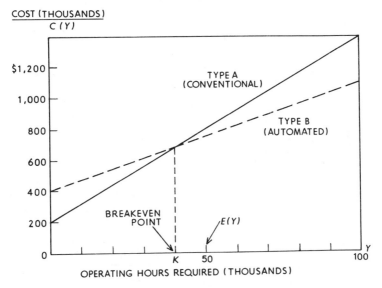

Chart 14–4

COSTS OF TWO MACHINES AS FUNCTIONS
OF OPERATING HOURS

the equations equal to each other, and solving for Y, the break-even point (when the two machinery types have the same cost) occurs at 40,000 hours. If less than 40,000 operating hours are required, the conventional machinery (type A) has less cost. For more than 40,000 hours, the automated machinery (type B) has the cost advantage. And since the expected number of hours $E(Y) = 50{,}000$, the purchase of the type B machinery is the optimal decision.

The same conclusion can be reached by determining the expected cost for the choice of each machine:

Type A: $E(C) = \$200{,}000 + \$12(50{,}000) = \$800{,}000$
Type B: $E(C) = \$400{,}000 + \$\ 7(50{,}000) = \$750{,}000$

Machinery type B has $50,000 less expected cost than its alternative.

The opportunity loss functions are:

Type A: $L(Y) = \$5(Y - 40{,}000) = \$5Y - \$200{,}000$ if $Y > 40{,}000$
  or $L(Y) = 0$ if $Y \leq 40{,}000$

Type B: $L(Y) = 0$ if $Y \geq 40{,}000$
  or $L(Y) = \$5(40{,}000 - Y) = \$200{,}000 - \$5Y$ if $Y < 40{,}000$

They are graphed in Chart 14–5.

Chart 14–5

OPPORTUNITY LOSS FUNCTIONS
FOR TWO MACHINES

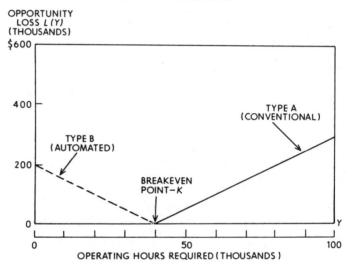

In the above functions, the break-even point K is 40,000 hours. The slope t of the nonzero opportunity loss functions is \$5 (or $-\$5$ for machinery type B). This needs some explanation. The \$5 is the *difference* between the variable operating costs of the two types of machinery ($\$12 - \$7 = \$5$).[5] If machinery type B is purchased and hours required fall below 40,000, the manufacturer incurs costs of \$5 per hour for each hour under 40,000 in excess of what he would have incurred if he had acted optimally.

The expected value of perfect information is:

---

[5] In two-action problems, the slope of the nonzero parts of the loss function is always the difference between the slopes of the profit or cost functions. In the previous examples the slope of one of the profit functions was zero, so that we did not have to make this point.

$$\text{EVPI} = tS_0 L_N(D) \quad \text{where } D = \left| \frac{K - M_0}{S_0} \right|$$

$$D = \left| \frac{40,000 - 50,000}{20,000} \right| = .50$$

$$\text{EVPI} = (\$5)(20,000)L_N(.50) = (\$100,000)(.1978)$$
$$= \$19,780$$

Clearly, the manufacturer should consider obtaining additional information before reaching a decision. Thus, he could make a sample survey of customers, provided its cost was well under $19,780.

DETERMINING THE POSTERIOR DISTRIBUTION

We shall now consider the possibility of obtaining additional information by sampling. Suppose that the unknown variable Y in a decision situation is actually the mean μ of some population. For example, Y might be the average sales per customer for a new product, or the average sales per store in response to an advertising campaign. In this situation, it may be possible to obtain additional information by selecting a sample of items in the population and estimating the unknown population mean μ. In Chapters 9 and 10 we studied the problem of estimating μ from a sample, the sampling error, and the confidence intervals. We now turn to the problem of incorporating these ideas into a decision-making framework.

The Distributions Involved

Since four distributions are involved in the analysis, we shall summarize them below, together with the symbols used. The first two distributions were described in Chapter 9. They represent the behavior of the random variable X and the sample mean $\overline{X}$. The last two distributions represent the uncertainty of the decision maker as to the location of μ, the population mean, both before and after he obtains additional information from a sample. The distributions are listed in Table 14–1 and explained below.

1. *The Population from Which the Sample Is to Be Drawn.* The population is a collection of elements in the real world (e.g., people, houses, accounts) which can be classified by some characteristic (e.g., income, number of rooms, dollars outstanding). By taking a sample of these elements, the decision maker can obtain some information which will help him make his decision. In particular, the sample mean $\overline{X}$ gives an estimate of μ, the unknown mean of the population.

Table 14-1

	Random Variable	Mean	Standard Deviation*
1. Population from which sample is drawn (can be any type of distribution)	X	μ	σ
2. Distribution of the sample mean (normal for large samples)	$\bar{X}$	μ	$\sigma_{\bar{X}}$
3. Prior distribution of the population mean (assumed normal)	μ	M_0	S_0
4. Posterior distribution of the population mean (normal if 2 and 3 are normal)	μ	M_1	S_1

* σ is generally unknown but can be estimated from sample value: $s \approx \sigma$. The $\sigma_{\bar{X}}$ is the standard error of the mean, which can also be estimated from a sample: $s_{\bar{X}} \approx \sigma_{\bar{X}}$.

This population distribution can be of any shape. It will often be skewed to the right in economic phenomena. Like the mean μ, the standard deviation σ is also generally unknown, but for large samples can be estimated from the sample value s with little error.

2. The Distribution of Sample Means. The sample mean, $\bar{X}$, is used to estimate the mean μ of the population from which it is drawn. The sampling distribution of $\bar{X}$ is a theoretical distribution consisting of all possible sample means of a given size drawn from this population.

Assumption (1): The Sampling Distribution of $\bar{X}$ Is Normal. This is not a very restrictive assumption. From the central limit theorem we know that for moderate to large samples the distribution of the sample mean $\bar{X}$ is approximately normal with mean μ (the population mean) and standard deviation $\sigma_{\bar{X}}$, where $\sigma_{\bar{X}} = \sigma/\sqrt{n}$. The value $\sigma_{\bar{X}}$ is a measure of the sampling error of $\bar{X}$. When $\sigma_{\bar{X}}$ is small, the sample contains relatively precise information about μ; when $\sigma_{\bar{X}}$ is large, the sample information gives a more diffuse estimate of μ.

When the standard deviation of the population σ is estimated by the sample standard deviation s, the standard error of the sample mean is calculated as $s_{\bar{X}} = s/\sqrt{n}$.

3. The Prior Distribution. The prior decision-making distribution is a betting distribution representing the decision maker's uncertainty about the unknown value of the mean μ of the population to be sampled. The mean of this prior distribution M_0 is the decision maker's best guess of μ. And the standard deviation S_0 is a measure of his uncertainty about μ. The wider the range of values he believes μ can have, the larger he makes S_0.

Note that the standard deviation of the prior distribution S_0 is not an estimate of the standard deviation σ of the population to be sampled. Such an esimate of σ would often be needed, but it is not at all related to the estimates for the prior distribution. To repeat, S_0 is a measure of the decision maker's uncertain knowledge only about μ, the population mean.

Assumption (2): The Prior Distribution Is Normal. The use of a normal decision-making distribution is quite appropriate in many situations. The normal distribution is symmetric, implying that the decision maker's guess of μ is as likely to be off a given amount in either direction about M_0. The normal distribution has probability clustered around M_0, indicating that the decision maker's guess is more likely to be close to the true μ than to be far away. Finally, using the normal distribution implies betting odds of roughly 2 out of 3 that μ lies in the range $M_0 \pm S_0$ and odds of about 95 out of 100 that μ is in the $M_0 \pm 2S_0$ range.

4. *The Posterior Distribution.* The posterior distribution, like the prior distribution, is a decision-making or betting distribution. It represents the decision maker's uncertainty about the unknown value of μ after taking into account sample evidence. *If the prior distribution and the distribution of sample means are both normal, then the posterior distribution is also normal.*[6] That is, if assumptions (1) and (2) above are satisfied, the posterior distribution is normal. Its mean M_1 and standard deviation S_1 are determined as follows:

$$M_1 = \frac{\dfrac{M_0}{S_0^2} + \dfrac{\bar{X}}{\sigma_{\bar{X}}^2}}{\dfrac{1}{S_0^2} + \dfrac{1}{\sigma_{\bar{X}}^2}} \qquad (3)$$

and

$$\frac{1}{S_1^2} = \frac{1}{S_0^2} + \frac{1}{\sigma_{\bar{X}}^2} \qquad \text{(the denominator in Formula 3)} \qquad (4)$$

Note that:

a) The posterior mean is a weighted average of the prior mean and the sample mean, the weights being the reciprocals of the variances of

---

[6] Actually, the normality of the posterior distribution is rather insensitive to violations in the normality of the prior distribution. According to R. Schlaifer, *Introduction to Statistics for Business Decisions* (New York: McGraw-Hill 1961, p. 309). "If the variance of the decision-maker's true prior distribution is large compared with the sampling variance of $\bar{X}$, he can simplify his calculations with no material loss of accuracy by substituting the mean and variance of his true prior distribution into the formulas which apply to a normal prior distribution."

the two distributions. A smaller variance means a higher precision of the mean and hence a greater weight. Thus, if the prior distribution is relatively narrow (i.e., S_0 is smaller than $\sigma_{\bar{X}}$ and hence $1/S_0{}^2$ is larger than $1/\sigma_{\bar{X}}{}^2$), the prior mean receives greater weight. But if the sample is relatively precise, i.e., $\sigma_{\bar{X}}$ is smaller than S_0 (and hence $1/\sigma_{\bar{X}}{}^2$ is greater than $1/S_0^2$), the sample mean receives greater weight. Also, if there were little prior knowledge, the prior standard deviation S_0 would be very large, and the posterior distribution would reflect almost entirely the sample result.

b) The weight received by the sample mean depends upon n, the size of the sample. Recall that $\sigma_{\bar{X}} = \sigma/\sqrt{n}$. As n increases, $\sigma_{\bar{X}}$ decreases, and the sample becomes more precise. Thus, as sample size increases, the weight received by the sample mean $(1/\sigma_{\bar{X}}{}^2)$ increases, and the posterior distribution is more influenced by the sample result. For very large samples, the prior distribution is swamped and has virtually no effect upon the posterior distribution.

c) The reciprocal of the posterior variance is the sum of the reciprocals of the variance of the prior and the sampling distributions.[7] This implies that the posterior variance (or standard deviation) is smaller than either the prior or sample variance (or standard deviation). In other words, there is less uncertainty in the posterior distribution than in either of the others.

Assumption (*3*): *Two-Action Problem with Linear Profit Functions.* Assumptions (1) and (2) above are enough to guarantee that the posterior distribution is normal. This result may be sufficient to deal with certain decision situations. However, as we did earlier in the chapter, we shall restrict the analysis to problems in which there are only two actions, and the profits (or costs) for each action may be represented by a linear function. This assumption will enable us to reduce the calculation of the expected profit, the expected value of perfect information, and the expected value of sample information, to simple formulas.

An Example

A wholesale merchant had an opportunity to buy a special lot of merchandise for $10,000. The lot contained 100,000 novelty items at a unit cost of 10 cents, which the wholesaler could sell in turn to his customers for 20 cents each. The wholesaler did not think he could sell

---

[7] For further discussion, see *ibid.*, pp. 302 ff.

all 100,000 items but noted that he had only to sell 50,000 to break even. His prior judgment was that he would sell 54,000, but there was some uncertainty about this sales level. The wholesaler expressed his uncertainty about sales in the form of a normal distribution with a mean of 54,000 units and standard deviation of 10,000 units. This meant that the wholesaler would be willing to bet, with even odds, that sales would be above (or below) 54,000, and he would be willing to give 2 to 1 odds that sales would be in the 44,000 to 64,000 range (54,000 $\pm$ 10,000). Such odds reflected his experience with similar merchandise. Let us express these preliminary estimates in terms of sales per customer by dividing the above estimates by 2,000, the number of his customers. Thus, the prior mean is $M_0 = 54,000/2,000 = 27$ and the prior standard deviation is $S_0 = 10,000/2,000 = 5$. In these terms, the decision maker's best guess (M_0) is that he will sell an average of 27 units per customer, and the standard deviation about this guess (S_0) is 5 units per customer. The break-even level of sales (K) is an average of 25 units per customer.

We can express the profit equations as follows:

Profit for action "Buy the lot": $\pi = -10,000 + (.20)(2,000)\mu$
$$= -10,000 + 400\mu \text{ in dollars}$$
Profit for action "Do not buy" $\pi = 0$

In the first equation, μ represents the unknown average sales per customer for the wholesaler's 2,000 customers.

Since the prior mean $M_0 = 27$ is greater than the break-even value $K = 25$, we know that the alternative "Buy the lot" is preferable. The expected profit is

$$E(\pi) = -10,000 + 400M_0 = -10,000 + 400(27)$$
$$= 800 \text{ dollars}$$

Further, we can determine the expected value of perfect information, as we did earlier:

$$\text{EVPI} = tS_0L_N(D) \qquad \text{where } D = \left| \frac{K - M_0}{S_0} \right|$$

Here M_0 is the mean of the betting distribution, S_0 is the standard deviation, t is the slope of the loss function, and $L_N(D)$ is found in Appendix E. Using the prior mean, $M_0 = 27$, and standard deviation, $S_0 = 5$, we have

$$D = \left| \frac{25 - 27}{5} \right| = .4$$

$$L_N(D) = L_N(.4) = .2304 \quad \text{from Appendix E}$$

and

$$\text{EVPI} = 400(5)(.2304) = 461$$

That is, the prior expected value of perfect information is $461.

Suppose that the wholesaler in question decided to obtain additional information in this decision problem by selecting a random sample of 50 customers (from the total of 2,000 customers) and asking each customer how many units he would purchase. Let us suppose that the average of these 50 "purchase orders" is 26 units per customer with a standard deviation of 14.14 units; that is, $\bar{X} = 26$, $s = 14.14$, and $n = 50$ (sample size). The standard error of the sample mean can then be estimated as[8]:

$$\sigma_{\bar{x}} \approx s_{\bar{x}} = \frac{s}{\sqrt{n}}$$

$$= \frac{14.14}{\sqrt{50}} = 2 \text{ units}$$

Since the prior mean (M_0) and the sample mean $(\bar{X})$ are both above the break-even value $(K = 25 \text{ units})$, there would be no reason to reverse the prior decision to buy the lot of merchandise. However, let us determine the posterior distribution anyway.

From Formula (1) we have:

$$M_1 = \frac{\dfrac{M_0}{S_0^2} + \dfrac{\bar{X}}{\sigma_{\bar{x}}^2}}{\dfrac{1}{S_0^2} + \dfrac{1}{\sigma_{\bar{x}}^2}} = \frac{\dfrac{27}{5^2} + \dfrac{26}{2^2}}{\dfrac{1}{5^2} + \dfrac{1}{2^2}} = 26.14$$

From Formula (2):

$$\frac{1}{S_1^2} = \frac{1}{S_0^2} + \frac{1}{\sigma_{\bar{x}}^2} = \frac{1}{5^2} + \frac{1}{2^2} = .29$$

Then:

$$S_1^2 = 1/.29 = 3.45$$

$$S_1 = \sqrt{3.45} = 1.86$$

---

[8] Note that if the sample contains more than 5 percent of the population, the finite population correction factor should be included in estimating $s_{\bar{x}}$. That is, $s_{\bar{x}} = (s/\sqrt{n})(\sqrt{1 - n/N})$, where N is the population size.

The values of $M_1 = 26.14$ and $S_1 = 1.86$ characterize the posterior betting distribution. After the sample, the decision maker's best guess of the value of μ (mean sales per customer) is 26.14 units with a standard deviation of 1.86 units per customer. The posterior distribution is normal, indicating, for example, that the decision maker should be willing to bet, with chances of 2 out of 3, that μ will be within the range 26.14 ± 1.86 or 24.28 to 28.00.

The posterior expected profit is

$$E(\pi) = -10,000 + 400M_1$$
$$= 10,000 + 400(26.14) = \$456$$

And the posterior EVPI is determined as follows:

$$D = \left| \frac{K - M_1}{S_1} \right| = \left| \frac{25 - 26.14}{1.86} \right| = .61$$
$$L_N(D) = .1659 \text{ from Appendix E}$$
$$\text{EVPI} = tS_1L_N(D) = (400)(1.86)(.1659) = \$123$$

Note that the posterior EVPI is considerably reduced from prior EVPI, even though the posterior mean M_1 was moved closer to the break-even point K. This resulted from the large reduction in standard deviation from $S_0 = 5$ to $S_1 = 1.86$, so that there is considerably less chance for a large loss (i.e., for a value of μ considerably below $K = 25$).

It is important to recall that the posterior distribution in the example above was the result of a particular sample ($\bar{X} = 26$, $s = 14.14$, $n = 50$). A different sample result would have led to a different posterior distribution.

EVALUATION OF SAMPLE INFORMATION

In the above section we answered the following question: "Given that a sample has been taken, how should we use the information in the decision process?" We now turn to a different question: "Should we take a sample at all, and if so, how large should the sample be?" We shall answer the above question in two stages: first, we shall calculate the economic worth of a sample of a given size; second (in the next section), we shall determine the optimum sample size—which may be zero, so that no sample is warranted. Additional information, including sample evidence, has value to the decision maker only if there is some chance that the information might change the prior decision. This implies that sample information generally enables us to reduce uncertainty (i.e., posterior expected loss).

Expected Value of Sample Information

Under the assumptions that we have been using in this chapter (two-action problem, linear profit functions, normal prior and sampling distributions), the evaluation of the economic worth of a sample can be accomplished in the six steps below, culminating in Formula (6).

Step 1: Determine the Prior Distribution. The decision maker first finds the mean M_0 and standard deviation S_0 of his prior betting distribution.

Step 2: Determine the Profit Functions. The linear profit (or cost) functions are next determined. This includes the calculation of the break-even value K and the slope t of the opportunity loss functions.

Step 3: Estimate the Precision of the Proposed Sample. Accuracy is measured in terms of the sampling error ($\sigma_{\bar{x}}$) that we expect to obtain with the sample. Since the standard error $\sigma_{\bar{x}}$ is equal to $\sigma/\sqrt{n}$, we must have some estimate of σ, the standard deviation of the population from which the sample is to be taken.[9] This estimate may be obtained from past studies of the population or similar populations, from a pilot sample taken to make such an estimate, or from an educated guess.

Step 4: Estimate the Variance of the Posterior Distribution. This is determined from the prior variance S_0^2 (Step 1) and the sampling error estimate $\sigma_{\bar{x}}$ (Step 3); that is, from Formula (4):

$$\frac{1}{S_1^2} = \frac{1}{S_0^2} + \frac{1}{\sigma_{\bar{x}}^2}$$

Step 5: Determine the Variance Reduction. Designate a quantity S_*^2 which is obtained as follows:

$$S_*^2 = S_0^2 - S_1^2 \tag{5}$$

Note that S_*^2 is a measure of the reduction in the prior variance as a result of taking the sample. Thus, it is a measure of the value of the sample in reducing prior uncertainty.

Step 6: Calculate EVSI. The value of the sample in economic terms is given by the expected value of sample information or EVSI.

$$\text{EVSI} = t S_* L_N (D) \quad \text{where } D = \left| \frac{K - M_0}{S_*} \right| \tag{6}$$

The symbol t represents the slope of the opportunity loss functions, M_0 is the prior mean, K is the break-even point, $L_N(D)$ is tabled in

---

[9] The above formula for sampling error is for simple random sampling. More complicated formulas are necessary for different methods of sampling (e.g., stratification or cluster sampling); see Chapter 12.

Appendix E, and S_* is obtained from Step 5 above. This formula is identical to that for EVPI, with S_* replacing S_1.

The expected value of sample information is a measure of the expected additional profit that will be achieved by acting after the sample has been taken (and using the sample information) rather than acting before sampling. It is an expected value because different sample results will increase posterior profit by differing amounts or may even decrease expected posterior profit.

An Example

Let us continue the example of the wholesaler from pages 404–7. Suppose that the wholsaler had not taken the sample discussed above but was considering the possibility of taking such a sample, say, of 50 items, from his 2,000 customers. He would obtain advance orders from the 50 sample customers. Let us follow through the steps in obtaining EVSI in this illustration.

Step 1. Recall that the wholesaler had a normal prior distribution with mean $M_0 = 27$ items per customer and standard deviation $S_0 = 5$ items.

Step 2. The profit equations were

Action "Buy the lot": $\pi = -10{,}000 + 400\mu$ in dollars
Action "Do not buy": $\pi = 0$

where μ is the unknown average sales per customer. We have previously determined the prior expected profit, $E(\pi) = \$800$, and the prior EVPI $= \$461$. The break-even value K is 25 items per customer, and the slope of the loss function is $t = \$400$.

Step 3. We next need an estimate of σ, the standard deviation of potential orders from the population of 2,000 customers. Let us suppose that from past experience with similar items the wholesaler estimates σ at 25 units per customer. Then we can estimate the sampling error for a sample of size $n = 50$ as:

$$\sigma_{\bar{x}} = \frac{\sigma}{\sqrt{n}} = \frac{25}{\sqrt{50}} = 3.54$$

Step 4. We then estimate the posterior variance as:

$$S_1^2 = \frac{1}{\dfrac{1}{S_0^2} + \dfrac{1}{\sigma_{\bar{x}}^2}} = \frac{1}{\left(\dfrac{1}{5^2}\right) + \left(\dfrac{1}{(3.54)^2}\right)} = 8.33$$

The posterior standard deviation is·

$$S_1 = \sqrt{8.33} = 2.89$$

Step 5. The reduction in the prior variance due to sampling is:

$$S_*^2 = S_0^2 - S_1^2 = 5^2 - (2.89)^2 = 16.67$$
$$S_* = \sqrt{16.67} = 4.08$$

Step 6. The calculation of EVSI follows:

$$D = \left| \frac{K - M_0}{S_*} \right| = \left| \frac{25 - 27}{4.08} \right| = \left| \frac{2}{4.08} \right| = .490$$

$$L_N(D) = L_N(.490) = .2009 \text{ from Appendix E}$$

$$\text{EVSI} = t S_* L_N(D) = (400)(4.08)(.2009) = \$328$$

The value of the sample of 50 items to the decision maker (the wholesaler in this example) is $328. That is, we would expect a sample of this size to reduce uncertainty and to increase posterior expected profit by $328. Recall that the expected value of perfect information is $461. Thus, even such a moderate size sample gives substantial added information (since $328 is about 70 percent of $461).

Factors Influencing EVSI

The size of the expected value of sample information depends on some of the same factors that influence EVPI. In particular, both EVSI and EVPI vary directly with the slope of the loss function (t), the amount of uncertainty shown by the prior standard deviation (S_0), and the closeness of the prior mean to the break-even point $(|K - M_0|)$. In addition, EVSI depends upon the sample size (n) and the dispersion in the sampled population (σ). The larger n, the larger EVSI; but the larger σ, the smaller EVSI, since the sample will have relatively less precision.

OPTIMAL SAMPLE SIZE

In the previous section we assumed a fixed sample size and determined the economic worth of the sample. We now ask the question: "How large should the sample be, including the possibility of $n = 0$, no sample at all?" This is a matter of comparing the value of the sample (EVSI) with the cost of sampling.

Generally, the cost of sampling increases as a linear function of sample size, as shown in Chart 14–6.

The expected value of sample information is also a function of sample size. The larger the sample, the larger EVSI. In Table 14–2 the calculations for EVSI are shown for selected sample sizes for the ex-

Chart 14–6

SAMPLING COSTS

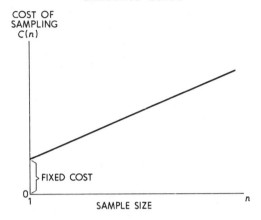

ample above (the wholesaler who is deciding about buying a lot of merchandise).

In Chart 14–7, EVSI is plotted as a function of the sample size n, with a smooth freehand curve drawn connecting the points calculated in Table 14–2, together with the point $n = 0$, for which EVSI $= 0$. Note that EVSI approaches the expected value of perfect information (EVPI) for very large values of n.

Let us suppose that it would cost $100 to set up the sample plus $2 for each item included. This might represent, for example, the cost of organizing the survey ($100) plus the cost of a telephone call and

Table 14–2

CALCULATION OF   EVSI FOR SELECTED VALUES OF n

Wholesaler's Decision to Buy Merchandise

n	$\sigma_{\bar{X}}^2 = \dfrac{\sigma^2}{n}$	$S_1^2 = \dfrac{1}{\dfrac{1}{S_0^2} + \dfrac{1}{\sigma_{\bar{X}}^2}}$	$S_* = \sqrt{S_0^2 - S_1^2}$	$D = \left\lvert\dfrac{K - M_0}{S_*}\right\rvert$	$EVSI = t\,S_*L_N(D)$
20*	31.25	13.89	3.33	.600	$225
50	12.50	8.33	4.08	.490	328
80	7.81	5.95	4.36	.458	369
100	6.25	5.00	4.47	.447	384
200	3.12	2.78	4.71	.424	419

* Actually, for samples as small as $n = 20$, the sampling distribution of $\bar{X}$ may not be normal when sampling from a skewed population. Hence, the calculation of EVSI as shown here is not strictly accurate, since the normality of the sampling distribution of $\bar{X}$ is assumed.

Chart 14–7

EXPECTED VALUE OF SAMPLE INFORMATION
AND COST OF SAMPLING

WHOLESALER'S DECISION TO BUY MERCHANDISE

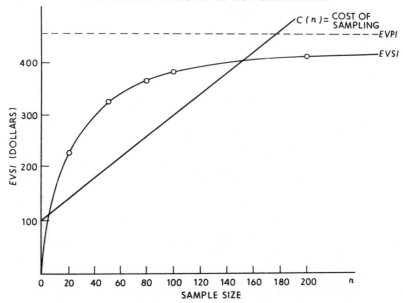

follow-up ($2) to each customer sampled. The sampling cost can be expressed by the equation:

$$C(n) = \$100 + \$2n$$

This equation is also shown in Chart 14–7. From this chart it can be seen that the value of the sample (EVSI) is greater than the cost for values of n between approximately $n = 5$ and $n = 150$. Hence, a sample with size somewhere between 5 and 150 would be preferable to no sample at all.

Expected Net Gain from Sampling

Now let us define ENGS as the expected net gain from sampling, where:

$$\text{ENGS} = \text{EVSI} - C(n) \tag{7}$$

for any given value of n.

ENGS represents the difference between the economic worth of the sample information and the cost of obtaining it. A small sample may

Chart 14–8

EXPECTED NET GAIN FROM SAMPLING

WHOLESALER'S DECISION TO BUY MERCHANDISE

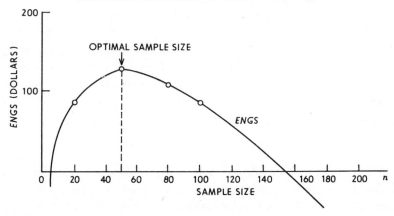

not provide sufficient information to justify its cost. And since the additional value of sample information tends to decline as the sample size increases, a point is reached for large samples where, again, the sample value does not justify its cost. In between, sampling is worthwhile (provided any sample is justified).

The ENGS for our example is plotted in Chart 14–8 as a function of the sample size n. ENGS is maximized at a value of about $n = 50$. This

Chart 14–9

EXPECTED VALUE OF SAMPLE INFORMATION
AND COST OF SAMPLING: SPECIAL CASE

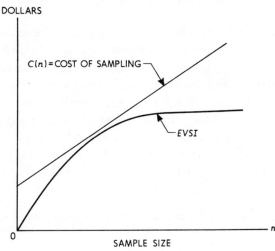

is the *optimal sample size.*[10] The value of the sample exceeds the sample cost by more at this point ($n = 50$) than at any other. Note that ENGS is rather flat in the range $n = 40$ to $n = 80$, indicating that any sample size over this range would be nearly as valuable as the optimum.

It may happen that $C(n)$ is greater than EVSI for all values of n, as illustrated in Chart 14–9. Since the value obtained from sampling (EVSI) never exceeds the sampling cost, no sample should be taken. The decision maker should then act with only his prior information (or find some less expensive means of obtaining information).

SUMMARY

Previous chapters developed the basic framework for combining probabilities, economic information, and sample results to determine optimal decisions. This chapter presents a special case of this general process, which has wide applicability.

First, the use of the normal distribution as a decision-making or betting distribution implies symmetrical, unimodal shaped distribution, with the probability clustered near the center.

Under certain conditions—a two-action problem, linear profit functions, and a normal betting distribution—EVPI can be expressed as a simple formula. In this instance, EVPI depends directly upon the standard deviation of the betting distribution and on the per unit opportunity loss; EVPI depends inversely upon the distance of the break-even point from the mean of the betting distribution.

One way to obtain information in decision situations is to take a sample. In this procedure, four distributions are involved:

1. The population from which the sample is to be drawn can be of any type. The mean of this distribution μ is unknown.
2. The sampling distribution is the distribution of sample means $\bar{X}$ about the true population mean μ. It represents the sampling error associated with estimating μ from the sample mean.
3. The prior distribution represents the decision maker's judgment about the true mean μ of the population to be sampled.
4. The posterior distribution represents the decision maker's judgment about the true mean μ after the information of the sample has been incorporated.

---

[10] The determination of optimum sample size in situations such as the above can be more exactly determined. See Schlaifer, *op. cit.,* chap. 21. The calculated optimum for the above problem is $n = 49$.

The assumptions made in this chapter are:

1. The prior distribution is normal.
2. The sampling distribution of $\bar{X}$ is normal. This assumption is valid for large samples because of the central limit theorem.
3. The decision problem involves a choice between two acts, and the profits (or costs) may be expressed as a linear function of the unknown population mean μ.

If assumptions 1 and 2 above are satisfied, the posterior distribution is normal. Adding assumption 3 enables us to express the expected profit and the expected value of perfect information in simple formulas.

In order to determine whether a sample should be taken and how large it should be, we estimate the expected value of sample information (EVSI). This amount represents the expected economic worth of the sample in improving the decision about to be made. With the assumptions above, the calculation of EVSI for a given sample size n can be reduced to simple formulas.

To determine the optimum sample size, the cost of the sample must be balanced against its value. The expected net gain from sampling (ENGS) is the difference between EVSI and the sampling cost for a given size sample n. If ENGS is plotted on a chart for different values of n, the optimum sample size can be determined at the point where ENGS is largest. If ENGS is always negative, the cost of sampling exceeds its value for all n and no sample should be taken.

Formulas

The formulas in this chapter are summarized below.

Expected value of perfect information for two-action problems with normal betting distribution and linear profit functions:

$$\text{EVPI} = t S_0 L_N(D)$$

where:

$$D = \left| \frac{K - M_0}{S_0} \right|$$

Mean of the posterior distribution for normal prior and normal sampling distributions:

$$M_1 = \frac{\dfrac{M_0}{S_0^2} + \dfrac{\bar{X}}{\sigma_{\bar{x}}^2}}{\dfrac{1}{S_0^2} + \dfrac{1}{\sigma_{\bar{x}}^2}}$$

Reciprocal of posterior variance for normal prior and sampling distributions:

$$\frac{1}{S_1^2} = \frac{1}{S_0^2} + \frac{1}{\sigma_{\bar{X}}^2}$$

Expected value of sample information:

$$\text{EVSI} = tS_* L_N(D)$$

where:

$$D = \left| \frac{K - M_0}{S_*} \right|$$

$$S_*^2 = S_0^2 - S_1^2$$

Expected net gain from sampling:

$$\text{ENGS} = \text{EVSI} - C(n)$$

PROBLEMS

1. Discuss:
   a) The meaning of a normal decision-making distribution.
   b) Why sample information has value.
   c) The distinction between a prior and a posterior distribution.
   d) The effect of sample size on EVSI.

2. In (a) through (d) below, calculate EVPI, using the indicated values of the mean M_0 and standard deviation S_0 of the normal betting distribution, the break-even value K, and the slope of the loss function t.

   a) $M_0 = 100$, $S_0 = 40$, $K = 160$, $t = .5$.
   b) $M_0 = 65$, $S_0 = 15$, $K = 50$, $t = 60$.
   c) $M_0 = 45$, $S_0 = 20$, $K = 50$, $t = .005$.
   d) $M_0 = 120$, $S_0 = 30$, $K = 110$, $t = 1$.

3. Determine the parameters of the posterior distribution in (a) through (d) below. Assume a normal prior with mean M_0 and standard deviation S_0 and a sample of size n with mean $\bar{X}$ and standard deviation s.

   a) $M_0 = 100$, $S_0 = 15$, $\bar{X} = 90$, $s = 25$, $n = 100$.
   b) $M_0 = 42$, $S_0 = 4$, $\bar{X} = 43$, $s = 20$, $n = 35$.
   c) $M_0 = 100$, $S_0 = 5$, $\bar{X} = 90$, $s = 25$, $n = 30$.
   d) $M_0 = 60$, $S_0 = 3$, $\bar{X} = 55$, $s = 10$, $n = 100$.

4. A decision maker has a prior normal distribution with mean $M_0 = 85$ and standard deviation $S_0 = 18$. The standard deviation of the population to be sampled is estimated to be 50. How large a sample must be taken so that the posterior standard deviation S_1 will be 4?

5. The GHK Company was considering a new advertising campaign to increase sales. The advertising would cost $100,000. Management felt that the campaign would "most probably" increase sales by $1 million, but there was considerable uncertainty about this figure. When pressed for more details, management said that there was about one chance in three that the sales increase would be outside the range 0.8 to 1.2 million dollars. GHK was then making a net profit of 12 percent on total sales.

   a) State the profit functions for the two actions: (1) advertise, (2) do not advertise.

   b) State the opportunity loss functions for the two actions. What is the break-even point?

   c) Draw a normal distribution describing GHK sales estimates.

   d) What action should GHK take? What is EVPI?

6. The Flavor Coffee Company was considering the use of a new type of can to package its coffee. The president felt that the new can would have appeal to consumers and would increase sales. In fact, he was willing to bet, with even odds, that sales over the next three years would be increased at least two million pounds. Furthermore, he was willing to bet, again with even odds, that the sales increase would be in the range 1.5 to 2.5 million pounds. Flavor currently makes a net profit (price minus variable cost) of 12 cents on a pound of coffee. The cost of change-over to the new can, however, is large—about $200,000 in new machinery, etc.

   a) Express profits from the new can as a function of sales. What is the break-even point?

   b) State the opportunity loss functions for the actions: (1) use the new can, (2) do not use the new can.

   c) Draw a normal distribution describing Flavor sales estimates.

   d) If Flavor must act now, what actions should be taken? What is EVPI?

7. The ABC Company was trying to decide which machine to purchase to produce a new product:

$$\text{Machine 1: Purchase cost} = \$120,000$$
$$\text{Direct variable cost} = \$2 \text{ per unit}$$
$$\text{Machine 2: Purchase cost} = \$350,000$$
$$\text{Direct variable cost} = \$1 \text{ per unit}$$

Management was undecided about which machine to purchase because of considerable uncertainty about the sales level that would be attained by the new product. Most pessimistic estimates ranged as low as 40,000 units per year while most optimistic estimates were as high as 120,000 units per year. Management felt that 80,000 units per year was perhaps the most probable forecast. Furthermore, they felt that the odds were two out of three that sales would be somewhere between 60,000 and 100,000. The investment was to be considered over a period of three years.

   a) Determine the sales level at which management would be indifferent as to which machine to purchase.

b) Based only on the above information, which machine should be purchased?

c) What is the expected value of perfect information?

8. An election is about to be held in a large plant to see if the workers wish to be represented by a union. Management's expectation that the proportion of workers who will vote for the union is approximately normally distributed. Management feels that there is an equal chance that the proportion voting for unionization will be above or below 40 percent. It also feels that there is an equal chance that the proportion voting for unionization will be within the range 33⅓ to 46⅔ percent as outside this range. A sample of 200 workers is selected at random and their voting intentions are determined; 96 indicate that they will vote for unionization.

a) Describe the probability distribution that management should assign to the event "Proportion of workers voting for unionization" after the sample has been taken. (Use normal approximation to binomial.)

b) Based upon this probability distribution, what is the probability that the union will win the election?

c) What is the probability that the union will win the election, ignoring management's prior judgment and utilizing the sample information only?

9. An employer is concerned with hiring persons who are proficient in a certain manual skill measured on a scale between 0 and 100. The distribution of this skill among applicants for jobs is known to be normal, with mean 50 and standard deviation 10. A test is used for screening purposes as a measure of this manual skill. However, the test is not perfect. The error associated with the test (difference between test score and "true" ability) is normally distributed, with 0 mean and standard deviation 5 points.

An applicant drawn at random scores 60 on the test [Hint: Treat the "true" distribution of skills as the prior distribution and the test error as the sampling distribution. Questions (*a*) and (*b*) apply, then, to the posterior distribution.]

a) What is the probability that the applicant's "true" manual ability is below 50?

b) What is the probability that his "true" ability is above 60?

10. Refer to the example in the text on pages 404–7. Suppose that a sample of 40 customers had been taken with mean $\bar{X} = 24$ and standard deviation $s = 16$.

a) Determine the posterior distribution.

b) Find the optimum action after the sample and the posterior expected profit.

c) What is the posterior EVPI?

11. Refer to the example in the text on pages 409–10.

a) Calculate EVSI for a sample of 40 customers.

b) What is ENGS for $n = 40$?

12. Refer to the same example as in Problem 11. This exercise is a study of the factors influencing EVSI. In each of (*a*) through (*f*) below, calculate EVSI for a sample size $n = 50$ with the indicated change and compare the result with that obtained in the text example. Add a sentence or two to explain the comparison.

   a) Suppose S_0, the prior standard deviation, was 10 rather than 5.
   b) Suppose S_0 was 3 rather than 5.
   c) Suppose the prior mean M_0 was 25 rather than 27.
   d) Suppose M_0 was 32 rather than 27.
   e) Suppose the standard deviation of the population to be sampled was $\sigma = 20$ rather than 25.
   f) Suppose the standard deviation of the population to be sampled was $\sigma = 30$ rather than 25.

13. The Delta Company is considering the introduction of a new product. Delta distributes its products through 8,000 retail outlets. Management expressed its uncertainty about the demand for the new product in terms of a normal probability distribution, with an unknown value of μ being the average sales in units per outlet. The mean of this prior distribution was 50 units per outlet and the standard deviation was 15 units per outlet.

   The new product would involve fixed costs of $100,000 for machinery, promotion, advertising, and working capital. The incremental contribution (price less variable cost) from the sale of each unit was expected to be 22 cents.

   a) Using the above information, what is the best decision—to market or not market the new product? What is EVPI?
   b) Suppose management was considering taking a sample of 100 of the 8,000 retail outlets. The product would be introduced at each of the sampled outlets and the sales would be noted. The average sales per outlet in the sample would then be used as an estimate of the average sales for all 8,000 retail outlets. From past experience, the standard deviation of sales per outlet was estimated at 30 units. What is the EVSI for the sample of 100 outlets?
   c) Suppose the sample of 100 items was actually taken with the following result: $\bar{X} = 59.2$ unit sales per outlet, and $s = 28.7$ unit sales per outlet. What action should be taken posterior to the sample? What is the posterior expected profit? What is the posterior EVPI?

14. Refer to Problem 13 above. Suppose a second sample of 50 outlets was being considered, after the first sample results in part (*c*) had been incorporated in the decision analysis. Should this second sample be taken if the cost of sampling is $200, plus $20 per outlet sampled?

15. As a dealer in retail hardware you are considering buying out the inventory of a merchant who is going out of business. You have a list of the items that he carried in stock but no exact inventory count has been made. There is the added problem of evaluating the worth of these items, since many are obsolete or so old and damaged that they are valueless. Accordingly, you

decide to take a sample of the items, check the count, and carefully value the sampled items.

Before taking the sample you examine the inventory. The owner is asking $225,000 for the lot. You feel, on the basis of your cursory investigation, that it is worth $235,000 to you, but there is much uncertainty about this guess. You feel that there is about one chance in three that your guess could be off as much as $20,000 or more (either high or low).

There are 4,000 different items in the merchant's stock. You estimate that the standard deviation of value by item in the inventory is $50.

Suppose further that the cost of taking a sample of any given size can be described by the equation:

Sampling cost $= \$150 + \$8n$ where n is the sample size

Ignore finite population correction factors throughout to simplify calculations. [Hint: Be sure to express the sampling unit and inventory dollars in the same unit—e.g., since inventory is the total dollars, convert the sample estimate to total dollars (total $= N\bar{X}$); for the error of the sample estimate $(s_{total} = Ns_{\bar{x}})$.]

a) Before consideration of sampling, would you buy the merchandise? What is EVPI? (Assume normality.)

b) What size sample (if any) should be taken? Explain.

16. The Ivanhoe Construction Company has been offered a contract to build a plant for the Zeta Steel and Wire Company. A contract price of $2.8 million has been agreed upon by both parties. Mr. Ivanhoe, the president, has estimated that his cost will be $2.4 million, leaving a profit before taxes of $400,000.

However, Zeta is fearful of losing ground to its competitors and is in a considerable hurry for its new plant. Zeta proposes an incentive contract that would reward Ivanhoe with $50,000 for each month that the project was completed before the scheduled date (20 months from now) and a penalty of $50,000 for each month beyond the target date.

Mr. Ivanhoe is somewhat dubious about agreeing to this provision in the contract. He feels that the contract can be completed in the agreed time (20 months), or even less, if all goes well, but unexpected shortages of materials or other contingencies could considerably delay the project. When questioned further, Mr. Ivanhoe said that 21 months was his "best guess" as to completion time. This would allow for some unplanned delays. He further felt that chances were good (say, two chances out of three) that the completion date would not vary more than three months either way from his guess.

Ivanhoe had an alternative venture that would give a before-tax profit of $300,000. This alternative would have to be forgone if the Zeta project were undertaken.

a) Assume a normal distribution for the time to complete the Zeta project. Based upon this, what should Ivanhoe do and what is the expected profit?

b) Do you think that the assumption of normality is reasonable in this case? Why or why not? If the distribution were not normal, how would it affect your answer to (*a*) above?

Mr. Ivanhoe had been studying the possibility of using some critical path technique (such as PERT or CPM) as an aid in controlling and predicting schedules. Ivanhoe contacted Mr. Wade, of a local consulting firm specializing in critical path methods. After examining Ivanhoe's problem Wade indicated that, using his methods, he could make a reasonably accurate estimate of the time to complete the construction project. This estimate would not be perfectly accurate since all contingencies could not be planned for. Based upon his experience with similar projects, Wade felt that he could estimate completion time within ±1 month with 80 percent probability. Wade's consulting fee for this estimate would be $40,000.

c) Should Mr. Ivanhoe hire Mr. Wade to make an estimate of time to complete the project before Ivanhoe decides to accept or reject the Zeta Steel and Wire contract?

SELECTED READINGS

BIERMAN, H. JR.; BONINI, C. P.; and HAUSMAN, W. H. *Quantitative Analysis for Business Decisions.* 4th ed. Homewood, Ill.: Richard D. Irwin, 1973.

Chapters 7, 8, and 9 deal with revising probabilities and decision making for the normal distribution.

BUZZELL, R. D.; COX, D. F.; and BROWN, R. V. *Marketing Research and Information Systems.* New York: McGraw-Hill, 1969.

Part IV treats the applications of decision theory to marketing problems. Chapter 11 is concerned with revising probabilities, particularly in the binomial case.

DYCKMAN, T. R.; SMIDT, S.; and McADAMS, A. K. *Managerial Decision Making under Uncertainty.* New York: Macmillan, 1969.

Chapters 15 and 17 treat revising probabilities and optimal sample size. Chapter 16 compares the Bayesian and classical methods.

HADLEY, G. *Introduction to Probability and Statistical Decision Theory.* San Francisco: Holden-Day, 1967.

Chapter 9 is an extended treatment of Bayes' theorem and its application to decision problems, with a somewhat different approach.

PRATT, J. W.; RAIFFA, H.; and SCHLAIFER, R. *Introduction to Statistical Decision Theory.* New York: McGraw-Hill, 1965.

An advanced treatment of Bayesian decision theory.

SCHLAIFER, R. *Introduction to Statistics for Business Decisions.* New York: McGraw-Hill, 1961.

Part 2 deals with binomial sampling and Bayes' theorem. Part 4 covers sampling, revision of normal probabilities, and decision making with normal distributions.

WINKLER, R. L. *An Introduction to Bayesian Inference and Decision.* New York: Holt, Rinehart & Winston, 1972.

Chapters 4 and 6 treat the material covered in our last two chapters, but in more depth.

15. MONTE CARLO METHODS IN DECISION MAKING

THE MONTE CARLO method is a means of simulating a real-world situation which involves probability elements. The method is used to determine complex probabilities and estimate expected profits or costs by empirical procedures rather than by theoretical analysis. Many important business decisions involve probabilities that would be difficult to obtain by other methods. Some problems defy direct solution; others would be too costly or time-consuming to solve; and in other cases the experimental conditions cannot be reproduced. Hence, the Monte Carlo method has wide applicability in areas such as inventory problems, scheduling of operations, advertising, allocation of resources, and long-term planning.

This is a simple technique which requires no formulas, only a random number table or a computer. Nevertheless, it brings together the principles of probability distributions, sampling, and decision making (already studied) to provide solutions to complex problems.

Consider, for example, a queuing or waiting line situation, such as an airline check-in counter. Customers arrive in a variable fashion according to some probability process. They may go to each of several check-in counters. The time taken to process a customer is variable (probabilistic) depending on the amount of baggage, the complexity of his ticket, and so on. The manager of this operation may be interested in the probabilities of various numbers of customers in line, or the probability that a customer will have to wait in line more than five minutes. Because of the complexity of such a system, it is impossible— except in very simple cases—to estimate such probabilities by analytic methods.[1] On the other hand, it is possible to simulate such a queuing

---

[1] For a discussion of queuing models, see H. Bierman, Jr., C. P. Bonini, and W. H. Hausman, *Quantitative Analysis for Business Decisions* (4th ed.; Homewood, Ill.: Richard D. Irwin, 1973), Chapter 19.

system and to estimate these probabilities using the Monte Carlo method. An example of this will be given in a few pages.

To understand the Monte Carlo idea, consider a very simple problem such as determining the probabilities of various numbers of heads in five flips of a fair coin. In Chapter 3 we calculated such probabilities using the binomial distribution. In contrast, the Monte Carlo approach might involve a procedure such as the following: (1) obtain a well-balanced coin; (2) flip the coin five times, recording the number of heads that occurred; (3) repeat this step 10,000 times. The relative frequency of occurrence of one head, two heads, etc. in the 10,000 sets of flips should be close estimates of the corresponding probabilities. Thus the Monte Carlo method provides empirical estimates determined by some randomizing procedure such as coin flips or spins of a roulette wheel (which accounts for the name Monte Carlo).

Of course, the Monte Carlo process would not be used on such a trivial problem. Furthermore, the use of randomizing devices such as flipping coins or spinning roulette wheels is not practical, so that tables of random numbers or random number generators on computers are generally used. Since many trials are required, the development of computers has greatly stimulated the use of this simulation method in industry.

MONTE CARLO SAMPLING FROM A DISCRETE DISTRIBUTION

To illustrate the Monte Carlo process and its application to a business decision problem, consider an example in production control. A firm has received a special order for nine units of a given part. The part is manufactured in two stages, the first being a casting operation, and in the second, the part is machined to certain specifications. A batch of units is cast at one time, with a setup cost of $300 plus $100 for each unit cast. Then the machining is done individually at an additional cost of $100 each. At the end of the machining operation for each part, it is inspected to determine if it meets specifications. When nine satisfactory units are produced, no more are machined.

The machining operation is hard to control, and in the past 30 percent of the items have proved defective (that is, not up to specifications). Defects occur at random, so the manager expects the same pattern for this new order.

The problem is to determine how large a batch of parts to cast and to estimate the total cost of filling the order. If too many parts are cast, the $100 spent in casting each is lost, as excess castings have no value. On the other hand, if too few are cast and nine good units are not machined, a new batch must be cast and the $300 setup cost incurred again. Suppose that the manager decides that he will cast 13 items. We shall examine the cost implications of this policy using the Monte Carlo method.

In Table 15–1, a random digit between 0 and 9 is drawn from Appendix L for each item to simulate the machining operation. Since the probability of a defective unit is 30 percent, the digits 1, 2, and 3 are assigned to the defective category and the remainder (4 through 9 and 0) are assigned as good. In trial 1, for example, the first random digit

Table 15–1

MONTE CARLO SIMULATION OF A MANUFACTURING OPERATION
POLICY: CAST A BATCH OF 13 ITEMS

			Costs			
			Casting		Machining,	
Trial	Random Digits*		Setup	Per Unit	per Unit	Total
1	7 ② 7 6 4 5 ① 6 9 4 ② 9 –		$300	$1,300	$1,200	$2,800
2	5 ③ 5 6 6 8 5 ③ 4 0 0 – –		300	1,300	1,100	2,700
3	0 9 ② ③ 7 0 8 4 0 7 ① ③ ③		300	1,300	1,300⎫	3,500
	Second casting: 8 –		300	200	100⎭	
4	9 ③ 6 ② 4 0 8 9 9 6 4 – –		300	1,300	1,100	2,700
5	① ③ ① 7 4 ② 9 5 7 6 ③ 5 ①		300	1,300	1,300⎫	4,000
	Second casting: ② 6 ① 0		300	400	400⎭	
. . .			. . .	. . .	. . .	. . .

Average (5,000 trials) = $3,136

* Circled numbers indicate defective parts; dashes indicate parts cast but not machined.

is 7, indicating a good part machined; the second digit is a 2 indicating a defective (circled), and so on. When the 12th number is drawn, it makes the 9th good one, and thus completes the order. The 13th item, which has been cast, is not machined. The total cost for trial 1 is calculated in the right-hand part of Table 15–1. This cost includes the cost of the casting setup ($300), the cost of casting the 13 units (13 times $100), and the cost of machining the 12 units that were required to get 9 good ones (12 times $100).

Trial 2 repeats this process, but in this case only 11 parts have to be machined before 9 good ones are produced. On trial 3, only 8 of the first batch of 13 are good. Thus the casting operation must be repeated. The manager is now faced with a decision about how many to cast on the second round. Let us suppose he uses a decision rule that he would cast, on the second round, twice the number he is short—in this case, he would cast two, since he is one short of nine good parts. On the second round, the first of the two cast is machined satisfactorily. The total cost on this trial is $3,500. Table 15–1 shows only trials 1 through 5, but the process is repeated on a computer for 5,000 trials. The average cost for these 5,000 trials for this policy (cast 13) is $3,136. This is the Monte Carlo estimate of the expected cost of using this policy.[2]

The Monte Carlo process, in this example, was not used to estimate directly the probabilities of various numbers of defectives, but rather to estimate the expected cost as a function of these probabilities, since it is this cost which interests the manager.

The expected costs for three other policies, similarly estimated from 5,000 trials, are shown in Table 15–2. The costs of casting fewer than

Table 15–2

ESTIMATED COSTS FOR FOUR POLICIES

MONTE CARLO SIMULATION OF A
MANUFACTURING OPERATION

Policy	Estimated Cost
Cast 12 units...............	$3,165
Cast 13 units...............	3,136
Cast 14 units...............	3,132
Cast 15 units...............	3,178

12 or more than 15 units increase progressively. Note that the policy of casting 14 has the lowest estimated cost and therefore provides the best management decision using the Monte Carlo method.[3]

Sampling from a Cumulative Probability Distribution

In the previous example, the Monte Carlo procedure was used to simulate binomial trials (defective versus good). We can go further and apply the same method to any probability distribution. This is done first for discrete distributions, and in the next section, for continuous distributions.

As an example, suppose we know that the daily sales for a product

---

[2] The Monte Carlo process is a sampling process, so the standard error can be calculated as in Chapter 9. The standard deviation (variation from trial to trial) for the 5,000 trials is about $677. Hence the standard error of the estimated mean ($3,136 above) is $\sigma_{\bar{x}} = \sigma/\sqrt{n} = 677/\sqrt{5000} = \$9.60$.

[3] The standard error in footnote 2 indicates that there is not a statistically significant difference between the "cast 13" and "cast 14" policies. The two policies are roughly equivalent. However, casting 14 units is still preferable, since it would involve fewer delays in casting second rounds.

have the discrete probability distribution given in Table 15–3 and that sales are independent from day to day. Our problem is to sample from this distribution to obtain a specific history of sales. The first step is to calculate the cumulative probability distribution as shown in Table 15–3. Next we select a table of random numbers such as Appendix L. Since the probabilities in Table 15–3 have three significant digits, we select a set of random numbers also using three digits (that is, random numbers from 000 to 999). Then the random numbers are assigned to the various events (units sold) to correspond to the cumulative probabilities. Thus the 25 random numbers from 000 to 024 are as-

Table 15–3

PROBABILITY DISTRIBUTION OF SALES

Daily Sales, Units	Probability	Cumulative Probability	Random Number Assignments
50	.025	.025	000 to 024
51	.225	.250	025 to 249
52	.350	.600	250 to 599
53	.250	.850	600 to 849
54	.125	.975	850 to 974
55	.025	1.000	975 to 999
	1.000		

signed to the event "50 units sold"; the next 225 random numbers, from 025 to 249, are assigned to the event "51 units sold"; and so on. We then proceed to draw three-digit random numbers from a table of random numbers. Each random number will determine a daily sales amount, since each three-digit number is assigned to a sales level. The first random number drawn is 504. This falls in the group 250 to 599 that corresponds to sales of 52 units (see Table 15–3). The second random number is 113, which is in the group 025 to 249 and corresponds to sales of 51 units. We continue on with this process of drawing random numbers and generating a history of sales, as shown in Table 15–4.

Note that the probability of drawing, for example, 52 units sold on a given date is exactly equal to the probability shown in Table 15–3, since 350 numbers out of 1,000 were assigned to this event—daily sales of 52. Column 3 in Table 15–4 represents an artificially generated history of sales.

This history of sales could be used in a simulation model to study

Table 15–4

MONTE CARLO SIMULATION OF DAILY SALES

Day	Random Number	Sales
1	504	52
2	113	51
3	360	52
4	559	52
5	149	51
6	837	53
. . .	. . .	. . .

inventory control or the production or purchasing policy for the given product. It might also provide an input for a complex simulation model of the whole firm.

MONTE CARLO SAMPLING FROM A CONTINUOUS DISTRIBUTION

When using the Monte Carlo method to obtain random drawings from a continuous distribution, the process is basically the same. The first step is to determine the cumulative probability distribution of the random variable involved. Suppose, for example, that the manager of the check-in operations of an airline knew that passengers arrived at random to be checked in at an average rate of 18 per hour. If the arrivals were indeed random and independent, then the time between successive arrivals (the *interarrival* time t) follows an exponential distribution (see Chapter 6). With an arrival rate of 18 per hour, the mean time between arrivals is $1/18 = .0556$ hours or 3.33 minutes. Then t can be described by the cumulative exponential distribution shown in Chart 15–1. The chart shows the probability that the time between arrivals will be equal to or less than the indicated number of minutes.

Note that for every value of the cumulative probability there is a corresponding value of t. Also, the cumulative probability ranges from zero to one. By selecting a random number between zero and one, we can find an associated value of t. Thus, if we selected the random number 73 or .73, the associated value of t is 4.3, as shown by the dashed lines in Chart 15–1. By repeatedly drawing random numbers, we can generate a whole series of values for t.

The series of times between arrivals (the third column in Table 15–5) and the associated history of arrivals (last column) represent

Chart 15–1

CUMULATIVE EXPONENTIAL DISTRIBUTION

TIME BETWEEN ARRIVALS

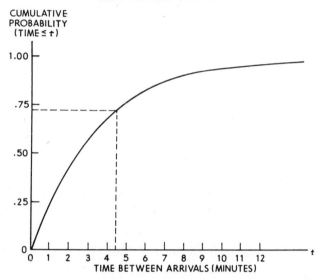

TIME BETWEEN ARRIVALS (MINUTES)

a random sample from the exponential distribution. We shall now see how this history of arrivals can be used to make managerial decisions about the organization of the check-in operations.

Decision Making in a Queuing Situation

Queues, or waiting lines, are common occurrences in many situations where there are random or unscheduled events. Waiting lines are familiar phenomena in barber shops, supermarkets, tool cribs in factories, telephone switchboards, repair shops, and a host of other situations. In all these cases, people, telephone calls, or machines "arrive" in a somewhat random fashion at a "service station" where they await their turn to be "serviced." The time taken to wait on or service an individual may also be a random variable. Queuing theory is the study of the probabilities associated with the length of the waiting line and the time an individual must wait in the queuing system.

As an example, suppose the manager of the check-in counter described in the previous section was trying to decide whether to open a second check-in counter at a given time of day. In the language of queuing theory, each check-in counter is called a channel. Thus he is trying to decide between a one-channel and a two-channel system.

Table 15–5

SIMULATING A HISTORY OF ARRIVALS

USING RANDOM NUMBERS AND A PROBABILITY DISTRIBUTION

Arrival Number	Random Number	Random Time between Arrivals from Chart 15–1	Time of Arrival = Time of Previous Arrival + Time between Arrivals
0			0:00.0
1	.73	4.3	0:04.3
2	.04	0.1	0:04.4
3	.97	11.3	0:15.7
4	.38	1.6	0:17.3
5	.68	3.8	0:21.1
6	.26	1.0	0:22.1
. . .	. . .	. . .	. . .

Suppose he knows that the time between arrivals is distributed as an exponential distribution, as above. Suppose further, that the time taken to wait on a customer (the service time, in queuing terminology), is a constant three minutes per customer. Let us investigate the effects on this system of the Monte Carlo–generated sequence of arrivals from Table 15–5.

This is shown first for the one-channel case in the schematic diagram, Chart 15–2. Time is plotted along a continuous scale running down the length of the diagram. Arrivals are shown at the time they enter the system. They either go directly into service with no wait (for example, arrivals Nos. 1 and 3) or they must wait in the queue until the service channel is free. Arrival No. 2, for example, comes into the system at time 0:04.4. But service started on No. 1 at 0:04.3 and continues until 0:07.3, a three-minute service time. Thus, the service channel becomes free at 0:07.3 and No. 2 can be serviced. The waiting time for No. 2 is thus 2.9 minutes (starting service time 0:07.3 minus arrival time 0:04.4). Note that an arrival may find more than one individual ahead of him. For example, No. 11 finds three individuals ahead of him (plus the one being serviced) when he arrives at time 0:30.0.

Since it is time-consuming to continue the schematic procedure employed in Chart 15–2, let us do the same thing in another form, Table 15–6. In this table the "Time Begun Service," column 3, for the one-channel case is simply either (1) the time of arrival or (2) the "Time Begun Service" for the previous arrival plus three minutes, *whichever*

Chart 15–2

SCHEMATIC DIAGRAM OF THE ONE-CHANNEL QUEUING SITUATION

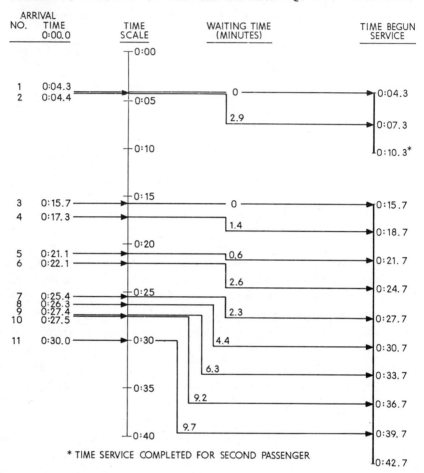

* TIME SERVICE COMPLETED FOR SECOND PASSENGER

is later. This implies that an arrival can go directly into service if the channel is free or must wait until the immediately previous arrival is finished with his service. The waiting time (column 4) is the difference between arrival time and the "Time Begun Service."

For the two-channel case, we use the same history of arrivals. However, the "Time Begun Service" (column 5) for, say, the nth arrival is now determined as (1) the time of arrival or (2) the "Time Begun Service" for the $(n-2)$th arrival (i.e., the arrival before last) plus three minutes, whichever is later.

Because there are two channels, an arrival will have to wait only if

Table 15–6

SIMULATION OF QUEUING SITUATION

Arrival		One-Channel Case		Two-Channel Case	
(1) Arrival Number	(2) Time of Arrival	(3) Time Begun Service	(4) Waiting Time	(5) Time Begun Service	(6) Waiting Time
1	0:04.3	0:04.3	0	0:04.3	0
2	0:04.4	0:07.3	2.9	0:04.4	0
3	0:15.7	0:15.7	0	0:15.7	0
4	0:17.3	0:18.7	1.4	0:17.3	0
5	0:21.1	0:21.7	0.6	0:21.1	0
6	0:22.1	0:24.7	2.6	0:22.1	0
7	0:25.4	0:27.7	2.3	0:25.4	0
8	0:26.3	0:30.7	4.4	0:26.3	0
9	0:27.4	0:33.7	6.3	0:28.4	1.0
10	0:27.5	0:36.7	9.2	0:29.3	1.8
11	0:30.0	0:39.7	9.7	0:31.4	1.4
12	0:35.5	0:42.7	7.2	0:35.5	0
13	0:40.2	0:45.7	5.5	0:40.2	0
14	0:48.2	0:48.7	0.5	0:48.2	0
15	0:48.4	0:51.7	3.3	0:48.4	0
16	0:48.5	0:54.7	6.2	0:51.2	2.7
17	0:49.0	0:57.7	8.7	0:51.4	2.4
18	0:49.1	1:00.7	11.6	0:54.2	5.1
19	0:49.6	1:03.7	14.1	0:54.4	4.8
20	0:50.1	1:06.7	16.6	0:57.2	7.1
21	0:53.6	1:09.7	16.1	0:57.4	3.8
22	1:00.5	1:12.7	12.2	1:00.5	0
23	1:04.0	1:15.7	11.7	1:04.0	0
24	1:06.7	1:18.7	12.0	1:06.7	0
25	1:07.0	1:21.7	14.7	1:07.0	0
26	1:12.0	1:24.7	12.7	1:12.0	0
27	1:12.1	1:27.7	15.6	1:12.1	0
28	1:16.8	1:30.7	13.9	1:16.8	0
29	1:18.0	1:33.7	15.7	1:18.0	0
30	1:24.7	1:36.7	12.0	1:24.7	0
31	1:25.7	1:39.7	14.0	1:25.7	0
32	1:28.2	1:42.7	14.5	1:28.2	0
33	1:31.8	1:45.7	13.9	1:31.8	0
34	1:31.9	1:48.7	16.8	1:31.9	0
35	1:35.4	1:51.7	16.3	1:34.8	0.6
36	1:36.0	1:54.7	18.7	1:36.0	0
37	1:36.1	1:57.7	21.6	1:37.8	1.7
38	1:51.2	2:00.7	9.5	1:51.2	0
39	1:53.1	2:03.7	10.6	1:53.1	0
40	2:05.2	2:06.7	1.5	2:05.2	0
41	2:11.3	2:11.3	0	2:11.3	0
42	2:12.5	2:14.3	1.8	2:12.5	0
43	2:21.5	2:21.5	0	2:21.5	0
44	2:21.9	2:24.5	2.6	2:21.9	0
45	2:26.9	2:27.5	0.6	2:26.9	0
46	2:36.0	2:36.0	0	2:36.0	0
47	2:38.0	2:39.0	1.0	2:38.0	0
48	2:44.2	2:44.2	0	2:44.2	0
49	2:44.7	2:47.2	2.5	2:44.7	0
50	2:45.5	2:50.2	4.7	2:45.5	0

Sum of last 40 items 370.6   29.6

Average wait 9.62   0.74

both channels are being utilized. And if both channels are in use, he must wait until the second arrival before him is finished before he can begin being serviced.

The waiting time (column 6) for the two-channel case is, as before, the difference between the arrival time and the "Time Begun Service" for each arrival.

In Table 15–6 we simulated the waiting times for 50 arrivals covering a period of about 165 minutes. Of course, we could continue the simulation for any number of arrivals. We wish to compare the performance of the one-channel system with the two-channel. We should like to make this comparison when both systems are in equilibrium, that is, when they have been operating long enough to be independent of initial conditions (e.g., starting the queuing process with no waiting line). For this reason we shall exclude the first 10 arrivals from our consideration. Comparing, then, the performance of the two systems for arrivals 11 through 50 we see that the average wait of 9.62 minutes with the one-channel system is reduced to 0.74 minute for the two-channel system. Of course, these estimates are based upon a relatively small sample of arrivals and we should carry out Table 15–6 for many more observations before making a decision about the relative merits of the one- versus two-channel systems.

Note that simulation, in this example, meant the portrayal on paper of a real-world system. The simulation model, as well as other models, can only approximate the elements of the real world, but where actual experience is difficult or impossible to obtain (e.g., why build a second channel to find if one is necessary?), a set of models involving different assumptions can provide an invaluable series of "dry runs."

Monte Carlo Analysis of Decision Trees

In many decision problems, the probabilities of events at a given node in the decision tree may be expressed by a continuous probability distribution (called a *fan*), rather than by a set of discrete events and probabilities. In this case, it is not possible to calculate expected values by the normal process of multiplying probabilities by payoffs and summing. However, the Monte Carlo method can be used to estimate the expected values in these cases.

As an example, consider a manufacturer who must purchase some equipment to produce a new product. Two types of equipment are available. Machine type A is largely manually operated. It costs only $20,000 but has a high variable cost of $4.50 per unit. Further, this

Chart 15–3

CUMULATIVE DEMAND DISTRIBUTION FOR NEW PRODUCT

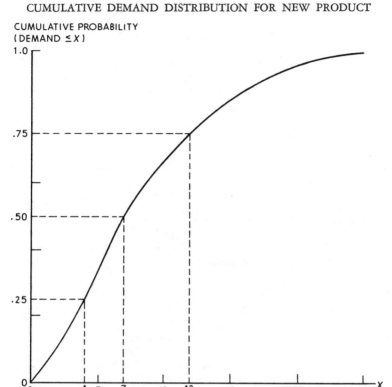

CUMULATIVE PROBABILITY
(DEMAND ≤ X)

DEMAND (THOUSANDS OF UNITS)

machine can produce only 10,000 units annually. An additional 5,000 units may be produced on overtime at a cost of $6.75 per unit.

Machine type B, on the other hand, is more automated but costs $40,000. It can produce up to 15,000 units annually at a cost of $2.50 each, and an additional 7,500 units on overtime at a cost of $3.75 each.

The selling price of the new product has been set at $8.50 per unit, but there is considerable uncertainty about the demand for the product. Management has expressed this uncertainty subjectively in the form of the continuous cumulative probability distribution shown in Chart 15–3. Note that this graph implies that there is a 50 percent chance that demand will be 7,000 units or below, a 25 percent chance that demand will be 4,000 units or below, and a 75 percent chance that demand will be 12,000 units or below. However, there is some chance that demand could be as high as 25,000 units.

Chart 15–4

DECISION TREE FOR MACHINE PURCHASE DECISION

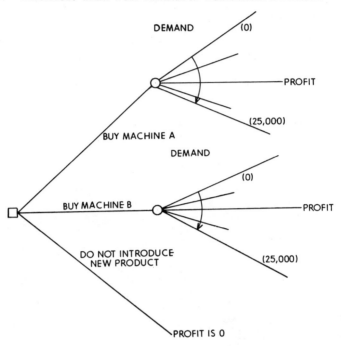

The decision tree for this problem is shown in Chart 15–4. Here the uncertainty about demand is expressed by a fan, indicating that demand can take on any of the values from 0 to 25,000 units.

The amount of profit at the end of the branches in Chart 15–4 depends upon the specific level of demand. The revenue is $8.50 per unit, but the cost depends upon whether or not the unit was produced on overtime. Also, once the overtime limit is reached, no more units can be produced and sales may be lost. In particular, machine A cannot produce over 15,000 units.

The specific profit functions are shown below and are graphed in Chart 15–5. Note that the profit functions are composed of linear segments.[4] The variable X represents the unknown demand in units.

---

[4] If the profit functions were linear over the whole range, we could calculate the expected profit simply by substituting the expected demand in the linear function (see Chapter 8). Because the lines are broken, we cannot do this. Further, we do not know the expected value of this subjective probability distribution.

Chart 15–5

PROFIT FUNCTIONS FOR TWO MACHINES

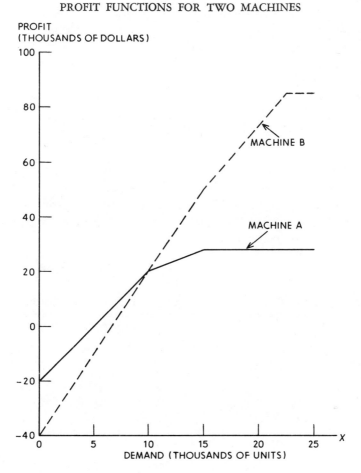

Profit Functions for Machine A

$$\pi = 8.50\ X - 4.50\ X - 20,000$$
$$= -20,000 + 4.00\ X \qquad \text{if } X \leq 10,000$$
$$\pi = 8.50(10,000) - 4.50(10,000)$$
$$\qquad + 8.50(X - 10,000)$$
$$\qquad - 6.75(X - 10,000) - 20,000$$
$$= 20,000 + 1.75(X - 10,000) \qquad \text{if } 10,000 < X \leq 15,000$$
$$\pi = 28,750 \qquad \text{if } X > 15,000$$

Profit Functions for Machine B

$$\pi = 8.50\ X - 2.50\ X - 40,000$$
$$= -40,000 + 6.00\ X \qquad\qquad \text{if } X \leq 15,000$$
$$\pi = 8.50(15,000) - 2.50(15,000)$$
$$+ 8.50(X - 15,000)$$
$$- 3.75(X - 15.000) - 40,000$$
$$= 50,000 + 4.75(X - 15,000) \qquad \text{if } 15,000 < X \leq 22,500$$
$$\pi = 85,625 \qquad\qquad\qquad\qquad \text{if } X > 22,500$$

The Monte Carlo method for this problem involves drawing random numbers; sampling values of demand from Chart 15–3, using the procedure described on page 427; and calculating the profit associated with each level of demand for each machine. A few such Monte Carlo trials are shown in Table 15–7.

Table 15–7

MONTE CARLO ANALYSIS FOR MACHINE PURCHASE DECISION

			Profit	
Trial	Random Number	Demand (Units) from Chart 15–3	Machine A	Machine B
1	.48	6,700	$ 6,800	$ 800
2	.38	5,600	2,400	−6,400
3	.75	12,000	23,500	32,000
4	.93	17,700	28,800	62,800
5	.29	4,600	−1,600	−12,400
. . .	. . .	. . .	. . .	. . .
Average (5,000 trials).		8,290	$ 8,730	$ 9,100

The process has been carried out by computer for 5,000 trials, and the results are shown in the last row of the table. These are estimates of the expected demand and the expected profits for the two machines. Note that machine B has a slightly higher estimated profit than machine A. Therefore, if the manager based his decision upon expected monetary value (without considering risk), he would choose machine B.

This method was illustrated for a simple decision tree (Chart 15–4) but can be used in exactly the same fashion in more complex trees. Furthermore, it is not limited to the piece-wise linear profit functions used in the example, but can be used on functions of any shape. In fact, one important application is in the calculation of expected utility, using curvilinear utility functions such as those illustrated on pages 180 and 184.

RISK ANALYSIS

In the examples thus far, the Monte Carlo procedure has been used to estimate the expected profit or cost for a given decision alternative. This is adequate if the expected monetary value (EMV) decision criterion can be used. However, if the amounts of money are large, and particularly if there is a possibility of negative payoffs, the decision maker would be concerned with the probabilities of the various total payoffs as well as the expected profit.[5] Knowing these probabilities, he can assess the amount of risk involved in a given decision. Risk analysis is a procedure to estimate these payoff probabilities using the Monte Carlo method.

An Example in Capital Investment

One of the most important applications of risk analysis is in the evaluation of major capital investments. Such investments may involve estimates for several unknown factors. For example, the marketing department of a firm may estimate the selling price and market demand for a new product, as well as the market growth, realizable market share, and the life of the product. The accounting and engineering departments may supply estimates of investment cost, and variable and fixed manufacturing costs. Each of these estimates is uncertain, but the uncertainty may be described by a probability distribution for each factor.

The general manager's problem is to evaluate the overall profitability of the project and to assess its risk. That is, he must put together the probability estimates for the several factors in order to estimate the probabilities of various profit levels and the expected profit.

To illustrate this technique, consider an investment that has only two uncertain factors, the per-unit cost and the sales level for a new product. Suppose the estimates for these factors are given as the probability distributions in Tables 15–8 and 15–9. Suppose also that the selling price is a certain $5 per unit and the investment cost is a known $10,000. If we let C represent the unit cost and S the sales (in thousands of units), the profit (in thousands of dollars) is:

$$\text{Profit} = S(5 - C) - 10$$

---

[5] Alternatively, he might wish to use a utility function (Chapter 7), and would need the various probabilities in order to calculate the expected utility.

Let us assume that sales level and unit cost are independent. This is an important assumption, which would not be true if, for example, there were economies of scale (as more units are produced, the unit cost falls). The assumption of independence allows us to sample by the Monte Carlo method independently for sales and cost.

We therefore draw two random numbers, using Tables 15–8 and

Table 15–8

PROBABILITY DISTRIBUTION FOR VARIABLE COST PER UNIT

Cost	Probability	Cumulative Probability	Random Number Assigned
$2.00............	.10	.10	01 to 10
2.50............	.20	.30	11 to 30
3.00............	.40	.70	31 to 70
3.50............	.20	.90	71 to 90
4.00............	.10	1.00	91 to 99
	1.00		(and 00)

15–9, and find the associated values for sales and cost. These are then combined, using the formula above to obtain profit, as shown in Table 15–10. The process is repeated for 25 trials. In actual practice, we should run many more trials, but these 25 will serve to illustrate the procedure.

Table 15–9

PROBABILITY DISTRIBUTION FOR UNIT SALES

Sales	Probability	Cumulative Probability	Random Number Assigned
2,500............	.05	.05	01 to 05
5,000............	.10	.15	06 to 15
7,500............	.25	.40	16 to 40
10,000............	.25	.65	41 to 65
12,500............	.15	.80	66 to 80
15,000............	.06	.86	81 to 86
17,500............	.05	.91	87 to 91
20,000............	.03	.94	92 to 94
22,500............	.02	.96	95 and 96
25,000............	.02	.98	97 and 98
27,500............	.01	.99	99
30,000............	.01	1.00	00
	1.00		

The next step is to classify the 25 profit numbers (last column in Table 15–10) in a frequency distribution. This is done in Table 15–11.

The relative frequencies (last column in Table 15–11) represent estimates of the probabilities for the various profit levels. Thus there is an estimated 12 percent chance for a loss and a 16 percent chance for a profit of $20,000 or more. The decision maker can use this information, together with the estimated expected profit of $12,450 from Table 15–10,[6] in deciding whether or not to make this investment.

Most risk analysis studies in practice involve more than the two factors considered in the example above. In addition, the investment

Table 15–10
MONTE CARLO ANALYSIS OF INVESTMENT DECISION

Trial	First Random Number	Sales S	Second Random Number	Unit Cost C	Profit $S(S-C)$ $-10,000$
1	97	25,000	02	$2.00	$65,000
2	80	12,500	66	3.00	15,000
3	96	22,500	55	3.00	35,000
4	50	10,000	29	2.50	15,000
5	58	10,000	51	3.00	10,000
6	04	2,500	86	3.50	−6,250
7	24	7,500	39	3.00	5,000
8	77	12,500	51	3.00	15,000
9	09	5,000	01	2.00	5,000
10	61	10,000	24	2.50	15,000
11	67	12,500	70	3.00	15,000
12	84	15,000	36	3.00	20,000
13	06	5,000	54	3.00	0
14	69	12,500	54	3.00	15,000
15	44	10,000	59	3.00	10,000
16	77	12,500	28	2.50	21,250
17	75	12,500	61	3.00	15,000
18	46	10,000	71	3.50	5,000
19	24	7,500	96	4.00	−2,500
20	79	12,500	83	3.50	8,750
21	16	7,500	24	2.50	8,750
22	76	12,500	78	3.50	8,750
23	14	5,000	43	3.00	0
24	60	10,000	20	2.50	15,000
25	25	7,500	92	4.00	−2,500
				Average =	$12,450

---

[6] This estimate, based upon only 25 trials, is somewhat high. In this simple case, the expected profit can be calculated to be $11,600. With several hundred trials, the expectation from the Monte Carlo analysis would be closer to the theoretical value.

usually has a life of several years and involves discounting or present value calculations.[7] But the basic idea is the same as in our example, that is, to combine probability estimates for several component factors to obtain probabilities for different levels of profit for the investment as a whole.

An Example in Production Planning

As another example of the same approach, refer back to the example of machines A and B discussed above. Suppose that, from Table 15–7 and its extension of 5,000 trials, we made a frequency distribution for the profit of each machine, rather than merely calculating the expected value. These frequencies for the 5,000 trials are shown in Chart 15–6 in the form of cumulative frequency curves. While the two machines were quite close in expected profit, there is substantial difference in the riskiness.

For example, machine A has a 67 percent chance for some profit, with a 33 percent chance of a loss, while machine B has a 47 percent chance for a loss. Note also that machine B has a 20 percent chance for a loss of more than $20,000, while there is no chance of this large a loss for machine A. On the other hand, machine A is limited to a maximum profit of $28,750 because of its limited capacity (note the abrupt drop in the cumulative curve at this point), while machine B has a 20 percent chance of profits in excess of $40,000. Depending upon his attitude towards risk, the decision maker may prefer the less risky machine A despite its slightly lower expected profit. In fact, if he is sufficiently averse to risk, he may prefer the third alternative of Chart

Table 15–11

FREQUENCY DISTRIBUTION OF PROFIT

Profit (Thousands)	Frequency	Relative Frequency
−$10 and under 0	3	.12
0 and under $10	8	.32
$10 and under $20	10	.40
$20 and under $30	2	.08
$30 and under $40	1	.04
$40 and under $50	0	0
$50 and under $60	0	0
$60 and under $70	1	.04
Total	25	1.00

---

[7] Present value is the present worth of future payments, discounted or adjusted to reflect the time value of money.

Chart 15–6

CUMULATIVE PROBABILITY OF PROFIT

MACHINE PURCHASE DECISION

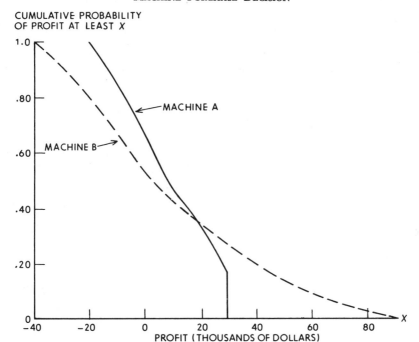

15–4, which is not to introduce the product at all and settle for a zero profit.

SUMMARY

The Monte Carlo method is a simple means of analyzing complex business decisions. This method estimates probabilities and expected profits (or costs) by empirical sampling from probability processes or distributions.

Monte Carlo sampling from a discrete distribution involves the assignment of random numbers to specific outcomes in proportion to their probability of occurrence, drawing a sequence of random numbers, and tabulating the associated outcomes. In this manner, a number of trials or a sequence of outcomes is generated which can be used to estimate expected values or the probabilities of complex events.

For continuous distributions, the Monte Carlo procedure is much the same, utilizing random numbers between zero and one and the cumulative probability distribution.

One application of Monte Carlo analysis is in decision tree problems, where the uncertainty at a given node is represented by a continuous distribution called a fan and the payoff (or utility) function is not linear.

Finally, risk analysis is the application of the Monte Carlo method to assess the risk of a project by combining the probabilities for the several component factors into a probability distribution for different levels of overall profit.

PROBLEMS

1. Select a policy (i.e., a number to cast) and run 20 Monte Carlo trials for the example shown in Table 15–1, on the assumption that the defective rate is 40 percent and the set-up cost is only $200. Estimate from these 20 trials the expected cost of your policy.

2. Refer to the example on pages 431–36. Suppose that the uncertain demand was expressed as a normal distribution with mean of 9,000 units and standard deviation of 4,000 units. Use 25 Monte Carlo trials to estimate the expected cost for each machine.

3. Refer to Problem 2 above. For the 25 trials, calculate a frequency distribution of profit for each machine. Then draw a cumulative frequency curve (such as Chart 15–6) to compare the risk associated with each machine.

4. Refer to the example of pages 437–39. Suppose that the sales level and unit cost were not independent, but instead were related as shown in the table below:

Sales Level (000's of Units)	Probability of a Unit Cost of:						
	$1.50	$2	$2.50	$3	$3.50	$4	$4.50
Under 10...............	0	0	.10	.20	.40	.20	.10
10 and under 20..........	0	.10	.20	.40	.20	.10	0
20 and above.............	.10	.20	.40	.20	.10	0	0

Make 25 Monte Carlo trials for the example using this assumption (that is, produce a table similar to Table 15–10). Determine the frequency distribution for profit (similar to Table 15–11) and calculate the estimated expected profit. How has this new assumption affected the risk of the project?

5. The profit (Y) for a certain decision outcome has a normal distribution with mean of $20,000 and standard deviation of $10,000. Suppose the decision maker's utility function for money can be expressed as follows:

$$u(Y) = .5 \log (Y + 20) \qquad -10 \le Y \le 60$$

where Y is the profit in thousands of dollars. Using 15 Monte Carlo trials, estimate the expected utility for this decision.

6. An investor with $300 is considering the purchase of three stocks, A, B, and C, each selling for $100 a share. He attaches the probabilities shown in the table below to the value (dividends plus market price) of the stocks at the end of one year.

Value at End of Year	Probability		
	A	B	C
$ 90...............		.20	.30
100...............	.50	.20	.10
110...............	.40	.20	.10
120...............	.10	.20	.10
130...............		.20	.40
Total............1.00		1.00	1.00

a) Suppose the investor wishes to buy one share of each stock. Assume that the stocks are independent (i.e., the year-end value of one is not related to the value of any other). Use Monte Carlo analysis to estimate the probability distribution associated with the value of the portfolio of three stocks at year's end. Calculate the mean and variance of this distribution.

b) Compare the mean and variance of the portfolio obtained in (a) above with the mean and variance of each of the alternatives of buying three shares of stock A, or three shares of stock B, or three shares of stock C.

7. Refer to Problem 6. Suppose that a fourth stock, stock D, is available at a price of $100 per share and is unrelated to stocks A and B but is related to stock C as shown by the probabilities in the table.

Value of Stock C at End of Year	Value of Stock D at End of Year					Total Probability
	$90	$100	$110	$120	$130	
$ 90............				.20	.10	.30
100............			.10			.10
110............			.10			.10
120............			.10			.10
130............	.20	.10	.10			.40
Total probability..........	.20	.10	.40	.20	.10	1.00

a) By the use of Monte Carlo analysis, estimate the distribution of year-end value of a portfolio composed of one share each of stocks A, C, and D. Determine the expected value and variance of this distribution.

b) By the use of Monte Carlo analysis, estimate the distribution of year-end value of a portfolio composed of one share each of stocks B, C, and D. Determine the expected value and variance of this distribution.

c) A portfolio of stocks is defined as "efficient" if there is no other portfolio with the same variance having a higher expected value—or, alternatively, if there is no other portfolio with the same expected value having lower variance. Which of the portfolios considered in Problems 6 and 7 are efficient in this sense? Which are inefficient. (Note: Only the portfolios AAA, BBB, CCC, ABC, ACD, and BCD have been considered. There are, of course, others such as AAB—two shares of stock A and one of B, etc. For simplicity, ignore these possibilities.)

8. In the typical "two-bin" inventory situation, an order for replenishment is made when the stock level drops to an amount b. The order is made for an amount q, called the order quantity. It takes a certain number of days, called the "lead time," until the order comes in. During this lead time if sales exceed the order level b, a stock-out condition occurs and sales are lost with cost k. It generally costs a certain amount c_o to place an order and a certain amount c_h to hold one unit of inventory in stock over a period of time (say, a year).

In the usual situation the probability distribution of demand for the product is given as well as the lead time. The constants c_o, c_h, and k are estimated. Then the values for order level b and the order quantity q must be determined to minimize cost over a period of time.

One method of dealing with this problem is to simulate the inventory system for different values of b and q and to use the results of the simulations to determine good values for b and q.

Suppose that the daily demand in units for a certain product is as shown in the table.

Demand (Units)	Probability
0	.10
1	.30
2	.20
3	.10
4	.10
5	.10
6	.05
7	.05
Total	1.00

The lead time is 20 days. Suppose that cost of being out of stock is $k = \$3$ per unit for each stock-out. The cost of placing an order is $c_o = \$10$, and the cost of holding one unit of inventory is 50 cents per month (30 days).

a) Assume that the order quantity q is fixed at 55 units. Simulate 300 days operations for each of three different values of b, the stock level. Estimate the cost for each system. Which value of b is best? Do you think the "best" value of b is greater than or less than the value you obtained?

b) Select three different sets of values for q and b. Simulate 300 days' operations for each set and estimate the cost of the inventory system for each set. Which set gave the lowest cost?

9. The Lakes Ore Company (LOC) wished to expand the number of shipments of iron ore across the lakes. However, the dock facilities at the port were inadequate and new equipment would be needed. During the next season, LOC expected to ship approximately 108 shiploads of ore during the 180 days of peak operations—April 15 to October 12.

LOC had dock space for only one ship and wished to minimize waiting time since a ship's operating cost was $200 per day.

Two different methods of unloading ships were under consideration. One method, A, used considerable manual labor, and required one and a third days (four eight-hour shifts) to unload a ship. This method would cost $500 per ship unloaded. Method B, on the other hand, was considerably more mechanized and cost $700 per ship unloaded. However, ships could be unloaded at a rate of one a day (three shifts).

Suppose that the number of ships arriving during an eight-hour shift followed a Poisson distribution, with mean $m = .20$. Simulate 60 days operations of this system, and estimate the expected cost for each method.

10. Refer to Problem 9 above. Suppose that instead of taking exactly four and three shifts to unload a ship using methods A and B respectively, the unloading times follow the probability distributions below:

No. of Shifts	Probability for No. of Shifts Required	
	Method A	Method B
2........................	0	.20
3........................	.30	.60
4........................	.40	.20
5........................	.30	0
Total....................	1.00	1.00

Simulate 60 days' operations of the system under this assumption and estimate the expected cost for each method. Compare the results to those obtained in Problem 9.

11. The management of a toy company is planning the production schedule for a new toy in advance of the Christmas season. Production must be completed during the summer, before demand for the new toy is known.

There are two models of the toy, the standard version and the deluxe. The deluxe is basically the same product with some frills and minor modifications. The cost (C) of producing the toy is estimated to be:

$$C(X) = \$15,000 + \$2X \quad \text{if} \quad 0 \leq X \leq 10,000 \text{ units}$$
$$C(X) = \$25,000 + \$2X \quad \text{if} \quad 10,000 < X \leq 25,000 \text{ units}$$

where X is the total number of units (standard plus deluxe) produced. Also, there are additional costs of $1 for each deluxe unit produced.

The sales price is $5 per unit for the standard model and $7 for the deluxe. The toy is a fad item and will not be produced again next year. Any unsold units will be disposed of to a large discount store at a price of $2 for the regular and $2.50 for the deluxe.

Management is uncertain about the total demand for the product. This uncertainty is expressed by a subjective continuous probability distribution. Five points on this cumulative distribution are given below. Complete the cumulative distribution by drawing a freehand curve through the points.

Probability of total demand $\leq Y$	1.0	.75	.50	.25	0
Y (units)	20,000	12,000	8,000	6,000	3,000

In addition, management expresses its uncertainty about the percentage share of total demand that the deluxe model would attain by the following five points of a cumulative probability distribution. Again, complete the cumulative distribution with a freehand curve.

Probability that deluxe share $\leq Z$	1.0	.75	.50	.25	0
Z (percent of total)	30%	20%	15%	13%	10%

The share of the standard model is 100 percent minus the deluxe share. Management feels that the share of the deluxe model is independent of total demand.

If demand exceeds production for the deluxe model, such excess demand is lost. However, any excess demand over production for the standard model can be met by substituting the deluxe model (if any are available) at the standard price of $5 per unit.

a) Set up this problem for solution by the Monte Carlo method. Select a production schedule for deluxe and standard units and illustrate your policy by running five trials for the policy selected.

b) Carry out the procedure for an additional 25 trials and estimate the expected profit for your policy.

c) Select another policy and make 30 Monte Carlo trials to estimate the expected profit. Compare this result with that obtained in (*b*) above.

SELECTED READINGS

BIERMAN, H., JR.; BONINI, C. P.; and HAUSMAN, W. H. *Quantitative Analysis for Business Decisions*. 4th ed. Homewood, Ill.: Richard D. Irwin, 1973.

Chapter 20 treats simulation and Monte Carlo analysis, including an example of its use in inventory systems.

HERTZ, D. B. *New Power for Management: Computer Systems and Management Science*. New York: McGraw-Hill, 1969.

Chapters 5 and 6 are a very good treatment of risk analysis from a manager's point of view.

KEMENY, J. G.; SCHLEIFER, A., JR.; SNELL, J. L.; and THOMPSON, G. L. *Finite Mathematics with Business Applications*. 2d ed. Englewood Cliffs, N.J.: Prentice-Hall, 1972.

Applies Monte Carlo analysis to queuing and decision-making situations on pp. 141–56.

MEIER, R. C.; NEWELL, W. T.; and PAZER, H. L. *Simulation in Business and Economics.* Englewood Cliffs, N.J.: Prentice-Hall, 1969.

Chapter 8 treats more advanced Monte Carlo techniques in a nonmathematical way.

NAYLOR, T. H.; BALINTFY, J. L.; BURDICK, D. S.; and CHU, K. *Computer Simulation Techniques.* New York: John Wiley, 1966.

Chapter 4 is an advanced survey of Monte Carlo sampling techniques.

SCHLAIFER, R. *Analysis of Decisions Under Uncertainty.* New York: McGraw-Hill, 1969.

Chapter 13 is a detailed treatment of the application of Monte Carlo analysis to decision problems.

16. SIMPLE REGRESSION AND CORRELATION

RELATIONSHIPS between variables are fundamental in science. The physical sciences have been highly successful in establishing functional relationships or "laws" connecting variables such as temperature and pressure of gas in a closed container, or the distance of an object from the earth and the gravitational pull exerted upon it. The biological and social sciences have had to deal with more complicated situations in which there is less reason to expect exact relationships between variables. The statistical tools of regression and correlation analysis were developed to estimate the closeness with which two or more variables were associated and the average amount of change in one variable that was associated with a unit increase in the value of another variable. The term "regression" refers specifically to the measurement of this relationship. The more general term "correlation" includes regression analysis as well as certain other measures, such as the coefficients of correlation and determination. It is important to explore both the applications and limitations of these powerful tools of analysis in the study of economic relationships. In particular, we shall consider the scatter diagram, curve fitting, the coefficient of determination, estimation of population relationships from sample data, and special problems of time series.

When only two variables are involved, the analysis is described as *simple* regression or correlation. *Multiple* regression or correlation refers to the analysis of three or more variables. This chapter is concerned with simple (two-variable) relationships. The multiple variable case will be considered in Chapter 17.

SCATTER DIAGRAMS

A first step in analyzing relationships between two variables is to plot the data on a dot chart called a scatter diagram. In Chart 16–1A, each dot represents the relation between a family's income and its expenditure for housing. It is evident that families with higher incomes tend to spend more for housing. Thus the two variables are related to, or *correlated* with, each other. The closer the dots cluster around a central *regression line,* as in Chart 16–2A, the higher the correlation. On the other hand, if the dots are scattered at random, in buckshot fashion, we describe the variables as uncorrelated or as having zero correlation.

The correlation between two variables may be described as being *positive,* indicating that high values of one variable tend to be associated with high values of the other variable, and similarly with low values. For example, in Chart 16–1A, the plotted points move upward to the right. When high values of one variable occur with low values of the other, the variables are inversely or *negatively* correlated. Thus, in Chart 16–1B, a larger crop of pigs means a lower price, so the points move downward from left to right.

If the plotted points on a scatter diagram generally follow a straight line, we say that there is a *linear* relationship between the two variables.

Chart 16–1

POSITIVE AND NEGATIVE CORRELATION

A
FAMILY INCOME VERSUS EXPENDI-
TURES FOR HOUSING FOR
SELECTED FAMILIES

B
MILLIONS OF PIGS RAISED
VERSUS PRICE OF HOGS, BY
YEARS

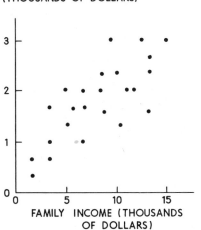

EXPENDITURES FOR HOUSING
(THOUSANDS OF DOLLARS)

FAMILY INCOME (THOUSANDS
OF DOLLARS)

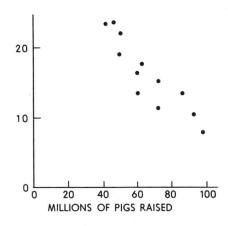

PRICE OF HOGS (DOLLARS)

MILLIONS OF PIGS RAISED

Chart 16–2
LINEAR AND CURVILINEAR CORRELATION

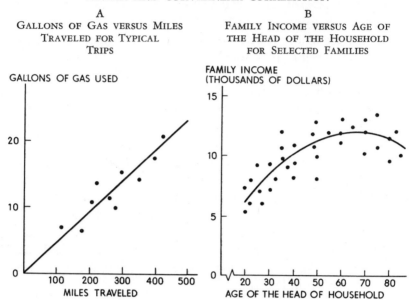

A
GALLONS OF GAS VERSUS MILES
TRAVELED FOR TYPICAL
TRIPS

B
FAMILY INCOME VERSUS AGE OF
THE HEAD OF THE HOUSEHOLD
FOR SELECTED FAMILIES

This is true of Chart 16–2A, where each hundred miles of travel on a trip requires about the same number of gallons of gasoline. Thus the straight line is a good fit to the plotted points. If a curved line gives a better fit, the correlation is said to be *curvilinear* or *nonlinear*. In Chart 16–2B, income at first rises with the age of the head of household, then levels off, and finally falls as retirement age is reached. The curve, as drawn, follows the data more closely than would a straight line.

REGRESSION ANALYSIS

In the previous section, we introduced the scatter diagram as a graphic means of presenting the relationship between two variables. In most business and economic situations, however, we wish to use one of the variables to predict or control the other variable. Hence, we need techniques for prediction and for measuring the error in our predictions. These techniques are called regression analysis.

Fitting a Regression Line

The first step is to express the average relationship between the two variables as a line or mathematical equation. The variable to be predicted is designated as Y, the *dependent* variable. The other variable, X, is the *independent* or predicting variable. The dependent variable is then expressed as some function of the independent variable; i.e., $Y = f(X)$.

The simplest functional form is the straight line. The formula for a straight line is $Y_c = a + bX$, where Y_c is the computed or expected value of Y (i.e., the value on the line for a given value of X and the relationship described by the line.) The constant a is the value of Y_c at the Y axis when $X = 0$, and b is the increase in Y_c for each unit increase in X. The value of b is therefore the slope of the line. When a straight line is used to relate two variables, the regression equation is said to be linear. The slope b is then termed the *regression coefficient.* We will describe how to fit both a linear and a curvilinear regression line below.

An example will serve to introduce the concepts and techniques of linear regression analysis. The personnel manager in an electronic manufacturing company devises a manual dexterity test for job applicants to predict their production rating in the assembly department. In order to do this, he selects a random sample of 20 applicants. They are given the test and later assigned a production rating. It is a common practice to administer an aptitude test to applicants for jobs, especially for types of jobs which require similar skills and for which objective measures of success can be obtained later.

The results are shown in Table 16–1 and Chart 16–3, where each dot represents one employee. The test score is the independent variable. There seems to be a fairly close linear relationship, with the dots clustered along a straight line, and with no extreme deviations.

Our object is to find the values of a and b in the straight line, $Y_c = a + bX$, which will predict production rating (Y_c) for any applicant's test score (X).

Since the points in Chart 16–3 are somewhat scattered, we cannot predict production ratings (Y) exactly. For any given test score, the predicted value Y_c is roughly the average of the production ratings (Y's) with the given test score. Thus, the regression line is often called the *line of average relationship,* indicating that it is a plot of the average values of Y for different values of X. The deviations of the actual ratings from the averages ($Y - Y_c$) are due to various personal differences, flaws in the test as a predictive device, and the omission of other factors that affect test scores.

Graphic versus Mathematical Methods.   Two methods of fitting a regression line are described below: the graphic "freehand" and the method of least squares.

Table 16–1

SCORES ON MANUAL DEXTERITY TEST AND
PRODUCTION RATINGS FOR 20 WORKERS

Worker	Test Score X	Production Rating Y
A	53	45
B	36	43
C	88	89
D	84	79
E	86	84
F	64	66
G	45	49
H	48	48
I	39	43
J	67	76
K	54	59
L	73	77
M	65	56
N	29	28
O	52	51
P	22	27
Q	76	76
R	32	34
S	51	60
T	37	32

Graphic methods in statistical analysis have three advantages over mathematical computations:

1. They save time and labor, unless a computer program is available.
2. Graphic curves are flexible, so they can fit certain curvilinear relations more closely than the more rigid mathematical functions. The graphic analyst can also discount extreme values that distort the least-squares fit.
3. Graphic methods afford a continuing picture of successive steps in analysis. Such a picture aids the observer in planning operations and judging the results. It also provides a visual aid in teaching.

Graphic methods, however, also have three disadvantages:

1. They reflect the subjective errors of the analyst. His personal bias, mistakes in judgment, and optical errors all affect the results. However, mathematical techniques, too, require the analyst to choose the type of equation and data to be used. Mathematical methods are no substitute for personal judgment.

Chart 16–3

GRAPHIC METHOD OF ESTIMATING PRODUCTION RATINGS
FROM TEST SCORES FOR 20 WORKERS

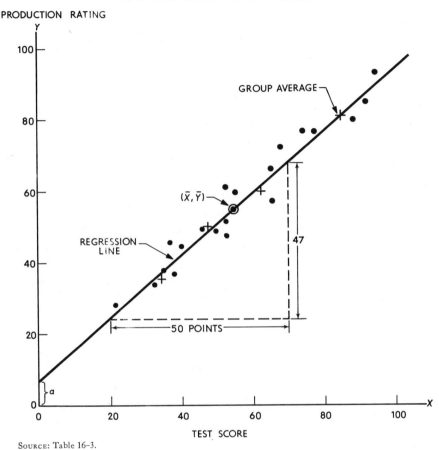

SOURCE: Table 16–3.

2. Because of the subjective element in graphic methods, a skilled analyst is required to draw curves with reasonable accuracy. The amateur may be led astray. Also, high-speed computer programs are available to fit a wide variety of situations.

3. Mathematical curves can be expressed by equations that provide the "best" fit according to some stated criterion. Furthermore, with an equation it is somewhat easier to summarize the relationships, evaluate the results, and predict new observations.

Graphic and mathematical methods may be used in combination to utilize the advantages of each. A graphic regression curve, for example, can be drawn to establish its general location and shape; then an appropriate mathematical equation can be selected for more objective measurement. The graphic curve also serves as a rough check on the accuracy and reasonableness of the mathematical equation. In a research department, the director of research can sketch out a preliminary curve graphically, then set up the program for the proper mathematical computations, and finally check the results against his own sketch.

Graphic Method.   The steps to be followed in the graphic method may be summarized as follows. First, draw the regression line through the plotted points by inspection so that the *vertical* deviations of the dots above and below the line are exactly equal for the series as a whole and are approximately equal for each major segment of the plotted data. These deviations may be marked off accumulatively on the edge of a strip of paper, one above the other, for comparison.

When the dots in the scatter diagram are numerous or widely scattered, the average values of groups of data should be plotted to serve as objective guide points in drawing the regression line or curve. First, divide the data into several groups according to values of X, each group having about the same number of items. Using too many groups will lead to a zigzag pattern in the group averages; using too few groups will make the averages insensitive as guides to the shape of the estimating line.

Second, take the mean of the X and Y values in each group, and plot this group average on the scatter diagram.

Third, draw a smooth line or curve (using a transparent ruler, a flexible spline, or a French curve) between the plotted averages, so that the vertical deviations of the averages above the line exactly equal those below the line over the whole range, and are approximately equal for each of several broad segments along the line. In particular, if the group averages follow a fairly straight line (except for zigzags), plot the overall mean ($\bar{X}, \bar{Y}$) and draw a straight line through this point at such a slope as to equalize approximately the vertical deviations of the group averages on the left of this point and those on the right separately. A curve should be drawn only if the group averages follow an unmistakable curve which is supported by economic logic.

Beginners have a tendency to draw graphic regression curves too steep because they judge goodness of fit by the shortest (or perpendicular) distance from the point to the line rather than by the vertical distance (the direction in which the dependent variable Y is measured) from the point to the line. The use of group averages reduces this error.

In our example of test scores and production ratings, the steps out-lined above have been performed on Chart 16–3. Crosses indicate aver-ages of four groups of points, and the overall average $(\bar{X}, \bar{Y})$ is circled. These averages fall roughly on a straight line, and there is no *a priori* reason why the regression should be curved. A straight line is therefore drawn through the overall average and as close to the group averages as possible. The values of *a* and *b* for the regression line are estimated from the chart. The line crosses the *Y* axis (when $X = 0$) at approxi-mately 4.0. Thus, the intercept *a* is 4.0. Over 50 points of test score, from 20 to 70, the value of Y_c increases from 23 to 70, a difference of 47 units on the production rating scale. Thus, the slope is estimated to be $47/50 = .94$. This is the regression coefficient *b*. The graphic estimate of the regression line can now be written as

$$Y_c = 4.0 + .94X$$

Method of Least Squares. A straight line fitted by least squares has the following characteristics:

1. It gives the best fit to the data in the sense that it makes the sum of the squared deviations from the line, $\Sigma(Y - Y_c)^2$, smaller than they would be from any other straight line. This property accounts for the name "least squares."
2. The deviations above the line equal those below the line, on the average. This means that the total of the positive and negative devia-tions is zero, or $\Sigma(Y - Y_c) = 0$.
3. The straight line goes through the overall mean of the data $(\bar{X}, \bar{Y})$.
4. When the data represent a sample from a larger population, the least squares line is a "best" estimate of the population regression line. This property will be discussed in more detail later.

It is important to stress that the deviations $(Y - Y_c)$ are measured vertically (i.e., along the *Y* axis). The deviations are not perpendicular to the regression line.

For the least squares line, the values of *a* and *b* in the equation $Y_c = a + bX$ are found by solving the two normal equations

$$\Sigma Y = na + b\Sigma X$$
$$\Sigma XY = a\Sigma X + b\Sigma X^2$$

where *n* is the number of pairs of items in a sample.

The computations can be simplified in most problems by measuring both *X* and *Y* as deviations from their means $\bar{X}$ and $\bar{Y}$. These deviations are designated by the small letters x and y, where $x = X - \bar{X}$ and $y = Y - \bar{Y}$. It is not necessary, however, to subtract the mean from each value of X and Y. A simpler procedure is as follows:

1. Compute the product XY, and calculate or look up the squares X^2 and Y^2 in Appendix C for each original pair of observations.
2. Sum these columns. (Steps 1 and 2 can be combined in a single operation on a calculating machine.)
3. Subtract from each sum the *mean times the sum* of the respective variables to get the adjusted sums of the x's and y's expressed as deviations from their means. That is,[1]

Sum	ΣXY	ΣX^2	ΣY^2
Less mean times sum	$-\bar{X}\Sigma Y$	$-\bar{X}\Sigma X$	$-\bar{Y}\Sigma Y$
Equals adjusted sum	$=\Sigma xy$	$=\Sigma x^2$	$=\Sigma y^2$

The sum of the deviations around the means, Σx and Σy, must equal zero, so they drop out of the two normal equations above, which reduce to

$$b = \frac{\Sigma xy}{\Sigma x^2}$$
$$a = \bar{Y} - b\bar{X}$$

where b derives from the second normal equation when $\Sigma x = 0$, and a is obtained by solving the first equation intact to express it in the original units.

For our illustration of test scores and production ratings, the calculations are shown in Table 16–2. We find XY, X^2, and Y^2 for each worker, sum these, and subtract the respective mean times the sum (shown in the box under X and Y) to find Σxy, Σx^2, and Σy^2. Then

$$b = \frac{\Sigma xy}{\Sigma x^2} = \frac{6{,}974}{7{,}395} = .943$$
$$a = \bar{Y} - b\bar{X} = 56.10 - .943(55.05) = 4.2$$

Hence, the regression line is

$$Y_c = 4.2 + .943X$$

---

[1] Note that $\Sigma x^2 = \Sigma(X - \bar{X})^2 = \Sigma(X^2 - 2\bar{X}X + \bar{X}^2) = \Sigma X^2 - 2\bar{X}\Sigma X + n\bar{X}^2$. But since $n\bar{X} = \Sigma X$, we have $\Sigma x^2 = \Sigma X^2 - 2\bar{X}\Sigma X + (n\bar{X})\bar{X} = \Sigma X^2 - \bar{X}\Sigma X$. The formulas for Σy^2 and Σxy can be derived in a similar fashion.

Table 16–2

REGRESSION BETWEEN SCORES ON MANUAL DEXTERITY TEST
AND PRODUCTION RATINGS FOR 20 WORKERS

Worker	Test Score X	Production Rating Y	XY	X²	Y²
A	53	45	2,385	2,809	2,025
B	36	43	1,548	1,296	1,849
C	88	89	7,832	7,744	7,921
D	84	79	6,636	7,056	6,241
E	86	84	7,224	7,396	7,056
F	64	66	4,224	4,096	4,356
G	45	49	2,205	2,025	2,401
H	48	48	2,304	2,304	2,304
I	39	43	1,677	1,521	1,849
J	67	76	5,092	4,489	5,776
K	54	59	3,186	2,916	3,481
L	73	77	5,621	5,329	5,929
M	65	56	3,640	4,225	3,136
N	29	28	812	841	784
O	52	51	2,652	2,704	2,601
P	22	27	594	484	729
Q	76	76	5,776	5,776	5,776
R	32	34	1,088	1,024	1,156
S	51	60	3,060	2,601	3,600
T	37	32	1,184	1,369	1,024
Sum	1,101	1,122	68,740	68,005	69,994
Mean	55.05	56.10			
Less mean times sum............			−61,766	−60,610	−62,944
Equals adjusted sum...........			6,974	7,395	7,050
This is.....................			Σxy	Σx^2	Σy^2

If a job applicant from the same population received a test score of 40, therefore, his production rating could be estimated as

$$Y_c = 4.2 + .943(40) = 42$$

Alternatively, this value might be read graphically from Chart 16–4 (dotted lines).

Chart 16–4

REGRESSION LINE FITTED BY LEAST SQUARES
AND STANDARD ERROR OF ESTIMATE

SCORES AND RATINGS OF 20 WORKERS

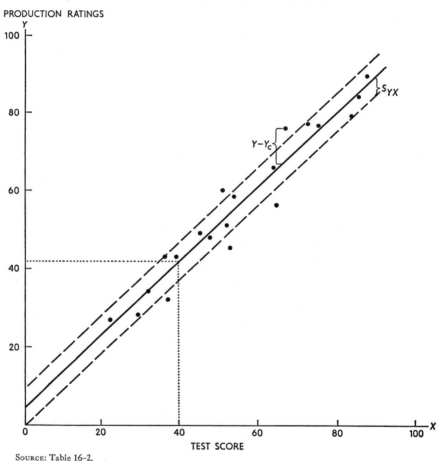

SOURCE: Table 16–2.

Curvilinear Regression

Curvilinear measures of regression should be used whenever (1) the logic of the situation calls for a curved relationship, and (2) the curve actually fits the data better than a straight line. The goodness of fit can be estimated by inspection of the scatter diagram, and determined more precisely from the standard error of estimate, as described in the next section.

We can fit a regression curve by any of three methods: (1) graphic analysis, by drawing a "freehand" curve (perhaps using drawing instruments), (2) fitting a parabola or other polynomial by least squares, or (3) transforming the data into logarithms or other functions so that a linear equation can be appropriately fitted to these functions.

Graphic Analysis. Suppose that a fertilizer manufacturer is conducting an experiment to determine the effects of nitrogen fertilizer upon corn yields. He selects 16 fields and has each planted to corn. Four fields receive no nitrogen, four fields receive 40 pounds each, four fields 80 pounds, and four fields 120 pounds. The results of this experiment are shown in Table 16–3 and Chart 16–5. The average yields for

Table 16–3

NITROGEN FERTILIZER AND CORN YIELDS

Sixteen Fields

	Amount of Nitrogen (Pounds)			
	0	40	80	120
Corn yield	6	40	72	110
(bushels per acre)	12	80	112	122
	18	80	112	130
	36	96	128	142
Total yield	72	296	424	504
Average yield	18	74	106	126

the four groups of fields are listed at the bottom of the table and plotted as circles on the chart. It appears that the four group averages follow a curved line, concave downward. This is logical, since increasing amounts of fertilizer should have successively smaller effects upon corn yield, until some level is reached at which corn yields stabilize or even decline.

A freehand regression curve has been drawn through the four group averages in Chart 16–5 with the aid of a French curve. If there were more points scattered along the X axis, the graphic curve would go close to the group averages, although not necessarily passing through all of them.

If the relationship is really curvilinear, a hand-drawn curve is likely to be a better fit than a straight line fitted by least squares, however impressive the computer printout. The analyst should always plot his data, check for curvilinearity, and consider whether the relationship is

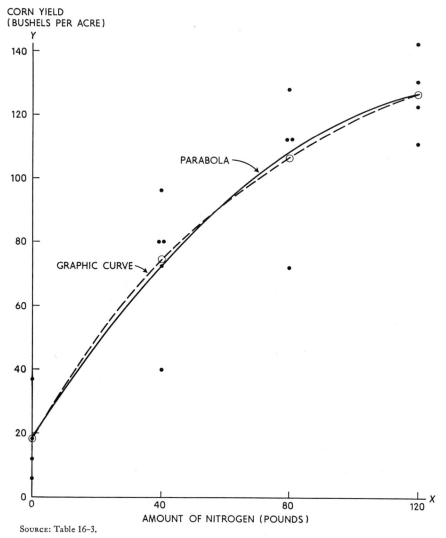

Chart 16–5

NITROGEN FERTILIZER AND CORN YIELDS

SIXTEEN FIELDS

SOURCE: Table 16–3.

logically curvilinear rather than automatically using some straight-line computer program.

Fitting a Parabola.   The degree of success in fitting a mathematical curve depends upon how carefully the functional form of the equation is picked. There are polynomials, logarithmic functions, and many others.

A simple curve is the parabola of the form $Y_c = a + bX + cX^2$. In this equation, a is the height of the curve at the Y axis, b is the slope of the curve at this point, and c determines the direction and degree of curvature. The general shape of a parabola is that of an automobile headlight reflector, pointing either up or down in its usual form. The values of the data will determine automatically what segment of the parabola will be fitted.

A parabola has been fitted to the corn-yield data in Table 16–3 with the following result:[2]

$$Y_c = 18.6 + 1.565X - .005625X^2$$

The parabola is plotted in Chart 16–5. The curve does not pass precisely through the means of the four arrays, though it comes close to doing this. The parabola and graphic curves fit the data about equally well. The parabola is more objective, while the graphic curve is more flexible in being able to approximate types of functions that cannot be represented by simple mathematical formulas.

Use of Logarithms. If the relationship appears curvilinear when plotted on an arithmetic grid, the data can be replotted on semilogarithmic or ratio graph paper (with either variable on the log scale) or on a double-logarithmic graph. Then, if the data follow approximately a straight line on any of these charts, the line can either be drawn graphically with a ruler or fitted by least squares.

In the least-squares method, the logarithms of the appropriate varia-

---

[2] If we use x and y to represent deviations of X and Y from their means, we can solve the following two normal equations to determine the values of b and c in the original equation:

$$\Sigma xy = b\Sigma x^2 + c\Sigma x^3$$
$$\Sigma x^2 y = b\Sigma x^3 + c\Sigma x^4$$

The constant term a can then be calculated from the formula:

$$a = \bar{Y} - b\bar{X} - c\Sigma X^2/n$$

Here, $\bar{X}$, $\bar{Y}$, Σx^2, and Σxy have already been defined and

$$\Sigma x^3 = \Sigma X^3 - \bar{X}\Sigma X^2$$
$$\Sigma x^4 = \Sigma X^4 - (\Sigma X^2)^2/n$$
$$\Sigma x^2 y = \Sigma X^2 Y - \bar{Y}\Sigma X^2$$

This method is not illustrated here, since in practice it is simpler to use multiple regression, as described in Chapter 17. That is, we can treat X^2 as if it were a new variable X_2. Then, if we call the original variable X_1 and change the constants b and c to b_1 and b_2, respectively, the equation for the parabola becomes $Y_c = a + b_1X_1 + b_2X_2$. This is identical with the equation for multiple regression, so we can use the same techniques to find a, b_1 and b_2.

bles are used in place of the original values, and a straight line is fitted just as described above. Thus, if the relationship is linear when plotted on semilogarithmic paper (with Y on the log scale), the equation of the regression line is $\log Y_c = a + bX$. The method of fitting this equation in trend analysis is illustrated in Chapter 19. Conversely, a straight line on semilogarithmic paper with X on the log scale has the form $Y_c = a + b \log X$. Finally, if the relationship is linear when plotted on double-logarithmic paper, the equation is $\log Y_c = a + b \log X$. This equation is a reasonable one to use if Y tends to change by a constant *percentage* for each 1 percent change in X over all X values.

In the nitrogen fertilizer case, plotting corn yields on the log scale of a ratio graph fails to straighten out the group averages. Also, nitrogen cannot be plotted on a log scale because some of its values are zero. Hence, logs won't do. There is a good linear relationship, however, between the logarithms of Sears, Roebuck sales and of U.S. disposable income, illustrated in Chart 16–11 later in this chapter.

Other Transformations. The use of logarithms is a special case of the more general technique of *transformation* of variables to achieve straight-line relations. If the logarithmic relationship is not linear, we can transform a variable into another function, such as the square, the square root, the reciprocal, or combinations of these. Many computer programs incorporate such transformations automatically in the computation of the regression equations.[3] The question of which transformation to use in a specific situation is one of judgment and experience. The analyst should select functions that make sense logically and he should then try several until he finds one that produces a satisfactory linear fit.

Standard Error of Estimate

The usefulness of the regression line for purposes of prediction and control depends on the extent of the scatter of the observations about it. If the observed values of Y vary widely about the line, estimates of Y based on this line will not be very accurate. On the other hand, if the observed values of Y lie quite close to the line, the estimates based on this line may be very good. The measure of the scatter of the actual observations about the regression line is called the standard error of estimate. The standard error of estimate for the population may be estimated from a sample in regression analysis as follows:

---

[3] See *BMD Biomedical Computer Programs*, pp. 15 to 21, for a list of more than 20 transformations or "transgenerations" available in those programs. (Health Services Computing Facility, University of California, Los Angeles, 1968.)

$$S_{YX} = \sqrt{\frac{\Sigma(Y - Y_c)^2}{n - k}}$$

where n is the size of the sample and k is the number of constants in the regression equation.[4] For a straight line, $k = 2$; for a parabola, $k = 3$. If a graphic curve is used, k is estimated as the number of constants that would occur in a mathematical curve of the same general shape.

The value $\Sigma(Y - Y_c)^2$ can be obtained graphically by reading off the vertical (not perpendicular) deviation of each point (Y) from the regression line (Y_c) on the Y scale, squaring each deviation, and summing these squares. The value Y_c can also be computed from the regression equation for each given value of X, to find $\Sigma(Y - Y_c)^2$.

When a straight-line regression has been fitted by least squares, however, it is usually simpler to compute the standard error of estimate by the following formula:

$$S_{YX} = \sqrt{\frac{\Sigma y^2 - b\Sigma xy}{n - 2}}$$

Thus, in our example of test scores and production ratings (Table 16–2):

$$S_{YX} = \sqrt{\frac{\Sigma y^2 - b\Sigma xy}{n - 2}}$$

$$= \sqrt{\frac{7{,}050 - .943(6{,}974)}{20 - 2}}$$

$$= 5.13$$

The standard error of estimate has been laid off in Chart 16–4 above and below the regression line (see dashed lines). If the points are scattered at random about the regression line (i.e., if epsilon $\epsilon = Y - Y_c$ follows a nearly normal distribution), then approximately two thirds

---

[4] The standard error of estimate for the sample itself is $\sqrt{\Sigma(Y - Y_c)^2/n}$. The use of $n - k$ adjusts for sample bias. This number represents the degrees of freedom around the regression line, just as $n - 1$ was used as the number of degrees of freedom around the mean in computing the standard deviation. Whereas the selection of the sample mean as a point from which to measure $Y - \overline{Y}$ uses up only one degree of freedom, the selection of a straight regression line as a base from which to measure the scatter uses up two degrees of freedom: one in requiring that the line pass through the points of means ($\overline{X}, \overline{Y}$) and the other in determining the slope of the regression line.

of the points should lie within this band. Hence, management could predict that an applicant who scored 40 on the test would achieve a production rating of 42 ± 5, or between 37 and 47, with two chances out of three of being correct. This standard error can also be compared with the standard error of estimate based on the use of alternative aptitude tests as predictors, such as mechanical or mathematical ability. (The above confidence interval will be widened slightly if we take into account the sampling error of the regression line itself.)

The standard error of estimate is also useful in determining which of two curves is a better fit. Thus, in the corn yield experiment (Table 16–3 and Chart 16–5), the standard error of estimate around the parabola is

$$S_{YX} = \sqrt{\frac{\Sigma(Y - Y_c)^2}{n - k}} = \sqrt{\frac{4,521}{16 - 3}} = 18.6 \text{ bushels per acre}$$

A straight line (not shown) was also fitted by least squares to the same 16 observations. Its equation is $Y_c = 27.6 + .89X$, and its standard error is

$$S_{YX} = \sqrt{\frac{\Sigma(Y - Y_c)^2}{n - k}} = \sqrt{\frac{5,817}{16 - 2}} = 20.4 \text{ bushels per acre}$$

It appears that the parabola does give more accurate estimates than the straight line, since the average scatter is smaller for the curve even after allowing for the increase in k, the number of constants in the equation.

In other situations the same percentage increase in Y may logically follow a 1 percent increase in X, as noted above. Here, it is rational to fit a straight line to the logarithms of the data. In comparing the goodness of fit for curves fitted to Y vs. log Y, however, we would have to compare a natural value of S_{YX} with a logarithm. Here it is easier to use the coefficients of determination (below), since they are relatives and thus directly comparable.

COEFFICIENT OF DETERMINATION

The coefficient of determination (r^2) is a relative measure of the relationship between two variables. It varies from zero (no correlation) to one (perfect correlation). It may be defined as a measure of the extent to which the independent variable accounts for the variability

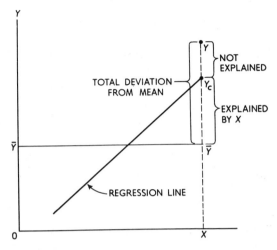

Chart 16–6

COMPONENTS OF THE COEFFICIENT
OF DETERMINATION

in the dependent variable. This concept is illustrated in Chart 16–6. Note that the total deviation of the dependent variable Y from its mean $\bar{Y}$ can be broken into two parts: the deviation of the value on the line from the mean $(Y_c - \bar{Y})$, which is *explained* by the given value of X, and the deviation of Y from the regression line $(Y - Y_c)$, which is *not* explained by X. That is, $(Y - \bar{Y}) = (Y_c - \bar{Y}) + (Y - Y_c)$.

Since the two parts are independent, the total variance of Y may be expressed as the sum of the variances of the two parts:

$$s_Y^2 = s_{Y_c - \bar{Y}}^2 + S_{YX}^2$$

The standard error of estimate (S_{YX}) measures the deviations of the points about the line. Its square represents the variance in Y that remains (i.e., the *unexplained* variance) after the regression line has been fitted to the data. The term $s_{Y_c - \bar{Y}}^2$ is the variance of points on the regression line around the mean value $\bar{Y}$ (or the variance *explained* by the regression line).

By expressing the explained variance as a ratio of the total variance of Y, we obtain the coefficient of determination:

$$r^2 = \frac{s_{Y_c - \bar{Y}}^2}{s_Y^2} = \frac{\text{explained variance}}{\text{total variance}}$$

The coefficient of determination is thus defined as the proportion of the total variance in the dependent variable which is explained by the independent variable.

The *coefficient of correlation* (r) is the square root of the coefficient of determination. It varies from zero (no correlation) to ± 1 (perfect correlation). The sign of r is the same as that of b in the regression equation. Thus, if $r = -1$, all dots are on the regression line sloping down to the right. The coefficient of determination is preferred to the coefficient of correlation for most applications in business and economics because it is a more clear-cut way of stating the proportion of the variance in Y which is associated with X. The coefficient of correlation may suggest a higher degree of correlation than really exists. Thus, if 50 percent of the variance in Y is explained by X (and the other 50 percent is not explained), $r^2 = .50$, but $r = \sqrt{.50} = .71$, a far higher value than .50!

The coefficient of determination may also be expressed as one minus the proportion of total variance which is *not* explained. That is,

$$r^2 = 1 - \frac{S_{YX}^2}{s_Y^2} = 1 - \frac{\text{unexplained variance}}{\text{total variance}}$$

This formula is more convenient for computation than the first one, since the unexplained variance is the square of the standard error of estimate (S_{YX}), which we have already computed in regression analysis.

Thus, in the production rating case, unexplained variance is:

$$S_{YX}^2 = (5.13)^2 = 26.3 \qquad \text{(page 463)}$$

Total variance is:

$$s_Y^2 = \frac{\Sigma y^2}{n-1} = \frac{7,050}{19} = 371 \qquad \text{(Table 16--2)}$$

Then:

$$r^2 = 1 - \frac{26.3}{371} = .929$$

That is, 92.9 percent of the variance in production ratings is explained, or accounted for, by the variance in test scores; only 7.1 percent of the variance is not so explained. The correlation coefficient is

$$r = \sqrt{.929} = .964$$

The coefficients of determination and correlation for a sample may also be defined by the following formula:

$$r_s{}^2 = \frac{(\Sigma xy)^2}{\Sigma x^2 \Sigma y^2} \quad \text{or} \quad r_s = \frac{\Sigma xy}{\sqrt{\Sigma x^2 \Sigma y^2}}$$

The term Σxy measures the degree to which X and Y vary with each other, and the terms Σx^2 and Σy^2 measure the individual variances in X and Y, respectively. The coefficients of determination and correlation are thus measures of the covariance of X and Y relative to the variance of X and Y themselves.

In certain preliminary studies, and particularly in the application of psychology to business problems, a relative measure of degree of relationship between X and Y may be all that is needed. For example, an industrial psychologist may be interested in finding which factors are related to the morale of a group of employees. He may not be interested in explicitly predicting employee morale from the other factors. Thus, he may not wish to use regression analysis, but may still use the correlation coefficient to measure the degree of the relationship between morale and each of the other factors.

Note that the above formula also provides a short-cut method for calculating the coefficient of determination and the coefficient of correlation.

In the production rating case (Table 16–2):

$$r_s^2 = \frac{(6,974)^2}{7,395 \times 7,050} = .933$$

This sample value, however, is biased as an estimate of the true population value of r^2. The best estimate of the latter is, in this example:

$$r^2 = 1 - (1 - r_s^2)\left(\frac{n-1}{n-2}\right)$$

$$r^2 = 1 - (1 - .933)\left(\frac{19}{18}\right) = .929$$

This is the same result as in the formula[5]:

$$r^2 = 1 - S_{YX}^2/s_Y^2$$

---

[5] In this formula, we adjusted for sample bias by using $n - 2$ and $n - 1$ instead of n in computing S_{YX} and S_Y, respectively, to compensate for the loss of degrees of freedom in measuring deviations from the regression line and $\overline{Y}$.

To summarize the chapter thus far, there are three basic measures that describe different aspects of a relationship between X and its dependent variable Y:

1. The regression line provides an estimate of Y for any value of X. The regression coefficient b itself gives the average change in Y for a unit change in X.
2. The standard error of estimate (S_{YX}) indicates the average error in estimating Y from X.
3. The coefficient of determination (r^2) shows what proportion of the variance in Y is explained by the variance in X.

Thus, in comparing the merits of three different aptitude tests in predicting production ratings of workers, one test might yield the greatest gain in production per unit of test score (i.e., the largest b value, assuming equal dispersion of scores); the second might predict production ratings most accurately (smallest S_{YX}); while the third might account for the most variation in ratings (highest r^2). The measure to use depends on the purpose of the inquiry.

DRAWING INFERENCES FROM SAMPLES

Up to this point we have considered regression and correlation measures merely as *descriptions* of the relationship between two variables. However, we are not usually interested in these results solely as they pertain to a particular sample. Almost without exception we are looking for a relationship that will enable us to control or predict new values of the dependent variable from the original set of data.

Thus, regression analysis of business and economic statistics must be approached from the standpoint of (statistical) inference from a particular sample to a "parent population" which includes the given sample and also such future or additional observations as we wish to consider. Both the given sample which we analyze and the actual future values we attempt to control or predict represent only a fraction of all of the possible values that might conceivably be drawn from the population in question. The application of statistical inference to regression analysis leads to the discovery of *general* relationships between variables. This is one of the most challenging and basic problems of scientific research.

The regression line for a sample is only one of a family of regression lines for different samples that might be drawn from the same population. That is, regression measures are subject to sampling error. Nevertheless, we can estimate within what limits the "true" regression line in the population is likely to fall. The theory of estimating population parameters from sample statistics was introduced in Chapters 9 and 10. We can now apply this theory in making statistical inferences about the true values of regression and correlation parameters.[6]

Basic Assumptions

In order to make valid inferences from sample data about population relationships, certain assumptions must be satisfied.

Assumption 1. When we fit a straight line to sample data to estimate the true or population relationship, the latter must also be linear. A similar assumption applies to the curvilinear case. This underlying relationship may be expressed in the form

$$Y = A + BX + \epsilon$$

where A and B are the true (but unknown) parameters of the regression line, and ϵ (epsilon) is the deviation of an actual value of Y from the true regression line. That is, $\epsilon = Y - Y_c$. (The average or expected value of ϵ is zero.) This is the assumption of linearity.

Assumption 2. The standard deviation of the ϵ's is the same for all values of X. This means that there is a uniform scatter or dispersion of points about the regression line. This property is called *homoscedasticity*. Examples illustrating when this assumption is valid and when it is invalid are shown in Chart 16–7. When the scatter is not uniform, a transformation of the data may serve to produce a more even dispersion. For example, if the scatter about the regression line tends to be a

Chart 16–7

SCATTER OF POINTS ABOUT REGRESSION LINE

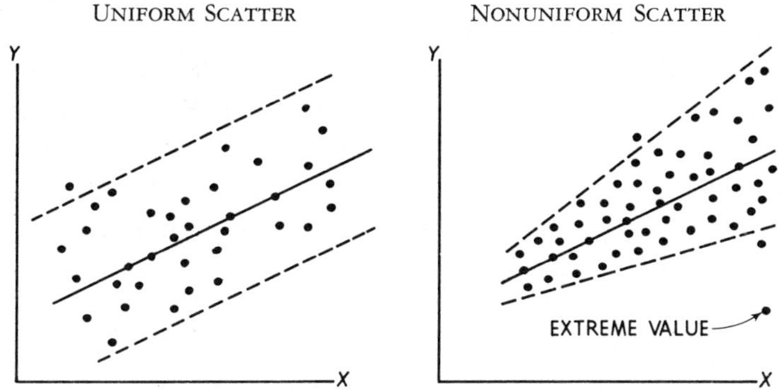

UNIFORM SCATTER   NONUNIFORM SCATTER

EXTREME VALUE

[6] See M. Ezekiel and K. A. Fox, *Methods of Correlation and Regression Analysis* (3d ed.; New York: John Wiley, 1959), chaps. 17 and 19, for a more complete discussion of this topic.

constant *percentage* of the independent variable X, then the use of log Y will make the deviations about the line more uniform.

Assumption 3. The ϵ's are *independent* of each other. That is, the deviation of one point about the line is not related to the deviation of any other point. This assumption of independence is not valid for most time series data. Time series move in cycles rather than randomly about the trend, so that adjoining values (e.g., in two boom years) are closely related. Independent and dependent ϵ's are illustrated in Chart 16–8.

Chart 16–8

INDEPENDENCE OF OBSERVATIONS

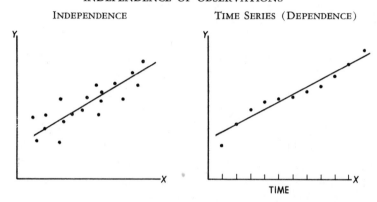

Assumption 4. The distribution of the points above and below the regression line follows a roughly normal curve. This means that the ϵ values are normally distributed.[7]

When these four assumptions are satisfied, the linear regression coefficients and standard error of estimate computed from a sample are efficient, linear, unbiased estimators of the true population values.

In addition to these general assumptions, it is important to distinguish between two cases called the correlation model and the regression model.

Correlation Model. In the correlation model, both X and Y are considered to be random samples drawn from a normal population.[8]

---

[7] The assumption of normality is not necessary if one wishes only to estimate the values of a and b in the regression line. The assumption *is* necessary to make inferences about a and b, using the standard errors s_b and s_{Y_c} considered below, in small samples but not in large samples (because of the central limit theorem). The normality assumption is also necessary (whatever the sample size) in order to make probability statements using the standard error of estimate S_{YX} and the standard error of forecast S_f (below). See A. M. Mood and F. A. Graybill, *Introduction to the Theory of Statistics* (2d ed., New York: McGraw-Hill, 1963), chap. 13, for more detail on the properties of these estimators.

[8] More specifically, the data pairs (X, Y) should represent a random sample from a population that is normal with respect to both variables.

The sample values are thus independent of each other and are normally distributed about their respective means. If this condition is met, together with the four general assumptions listed above, all correlation and regression measures in this chapter may be considered valid.

Regression Model. In the regression model, Y is a random variable, but X is fixed or predetermined at specific values. This is often true of controlled experiments. For example, in measuring the effects of various amounts of fertilizer upon corn yields, the X values in Table 16–3 were determined as 0, 40, 80, and 120 pounds of nitrogen, respectively, in four groups of plots. In this case, regression analysis is valid only for other samples or a population in which the X values are selected in exactly the same manner as in the original sample, for example, for plots of 0, 40, 80 and 120 pounds of fertilizer drawn with the same relative frequency as in this sample. The coefficients of determination and correlation are generally not valid in the regression model.

We now turn to the problem of measuring the sampling error associated with the estimates a and b and the statistical inferences that can be drawn based upon these estimates.

Standard Error of the Regression Coefficient

An inference about a regression coefficient can be made either as a test of significance or as a confidence interval, just as in the case of the mean or a proportion. Either type of inference depends on the standard error of the regression coefficient, as described below.

Testing the Significance of a Relationship. In the first place, it might be useful to know if there is *any* significant relationship between the variables X and Y. Some particular sample may indicate a relationship, even when none exists, by pure chance. If there is no relationship, the true value of B (the regression coefficient of the population) is assumed to be zero. So we set up the hypothesis that $B = 0$. If the sample value b is significantly different from zero, we reject the hypothesis and assert that there is a definite relationship between the variables. To do all this, we compute the standard error of the regression coefficient. This is:

$$s_b = \frac{S_{YX}}{\sqrt{\Sigma x^2}}$$

Here, S_{YX} is the sample standard error of estimate; $x = X - \bar{X}$, and Σx^2 describes the dispersion of X values around their mean. The value s_b is a measure of the amount of sampling error in b, just as $s_{\bar{X}}$ was a measure of the sampling error in the mean $\bar{X}$.

In the production rating example (Table 16–2):

$$s_b = \frac{5.13}{\sqrt{7,395}} = .060$$

The procedure for deciding whether a positive relationship exists between production ratings and test scores may be set forth as follows:

Null hypothesis: $B \leq 0$ (No relationship or a negative one)
Alternative hypothesis: $B > 0$ (Production rating increases as test score increases)

The value of b is .943. If the null hypothesis is true, $B \leq 0$ and b is $+.943$ units from B. In terms of its standard error, this is $.943/s_b = .943/.060 = 16$. Thus b is 16 standard errors from $B = 0$.

If this analysis were based upon a large sample, the one-tailed probability associated with any given deviation could be found from the table of areas under the normal curve in Appendix D. For small samples such as this one (with $n \leq 30$), the t distribution in Appendix M must be used with $n - 2$ degrees of freedom. In either case, a deviation of more than three standard errors is highly significant (except for very small samples). The chance is negligible, therefore, that a deviation as large as 16 standard errors could occur if $B \leq 0$. Hence, we reject the null hypothesis and accept the alternative hypothesis that there is a significant positive relationship between the variables.

The factor b/s_b is sometimes called the *critical ratio*. Thus, in our one-tailed test, if $b/s_b > 1.73$ (for $20 - 2 = 18$ degress of freedom in Appendix M), then b is said to be significant at the 5 percent level.

Confidence Intervals.   A confidence interval for the regression coefficient (b) of a sample is a zone around the sample value which we believe includes the true regression coefficient (B) of the population, with a specified probability—say 95 percent—of being correct. The 95 percent confidence interval for the regression coefficient in a large sample is

$$b \pm 1.96 s_b \quad (\text{Appendix D})$$

In the production rating example, however, with $n = 20$, we look up Appendix M with $n - 2 = 18$ degress of freedom and $P = .05$ to find the confidence interval

$$b \pm 2.10 s_b$$
$$\text{This is } .943 \pm 2.10(.060)$$
$$= .943 \pm .126$$

The manufacturer therefore could make the statement that B is between .817 and 1.069, with a probability of .95 that this statement is correct. Of course, any other degree of confidence could be chosen instead, by reference to Appendix D or M.

Standard Error of a Forecast

It is often important to find within what limits a new observation may be expected to lie. For example, the regression line in Chart 16–4 was used to forecast the production rating for a new applicant who received a test score of 40. The estimated rating was 42 ± 5, where 5 was the standard error of estimate. This measure describes the scatter of the production ratings above and below the regression line fitted to this sample of 20 workers, but it does not take into account the sampling error in the regression line itself. This would vary both in average level and in slope as different groups of workers were tested.

The *standard error of forecast* (S_f) is a measure of the *total* sampling error for any new observation. It is obtained by combining the standard error of estimate and the standard error of the regression line. Standard errors, like standard deviations, may be summed by adding their squares. The resulting formula for the standard error of forecast is:

$$S_f = S_{YX} \sqrt{1 + \frac{1}{n} + \frac{x^2}{\Sigma x^2}} \text{ for each value of } x = X - \bar{X}$$

Here, the 1 under the radical is the standard error of estimate itself, and the other two terms represent the standard error of a point on the regression line.[9]

---

[9] We can express the regression equation in the form $Y_c = \bar{Y} + bx$. The standard error of Y_c for any value of x (the deviation from the mean) will then include the standard errors of both $\bar{Y}$, the average height of the line, and $b(x)$, the slope times the distance of a point X from $\bar{X}$. The standard error of Y_c for any value of x can then be derived by adding the squares of the standard errors:

$$s_{Y_c}^2 = s_{\bar{Y}}^2 + (s_b x)^2$$
$$= \frac{S_{YX}^2}{n} = \frac{S_{YX}^2 x^2}{\Sigma x^2}$$

The standard error of a point on the regression line is therefore

$$s_{Y_c} = S_{YX} \sqrt{\frac{1}{n} + \frac{x^2}{\Sigma x^2}} \text{ for each value of } x = X - \bar{X}$$

This measure provides a confidence interval suitable for estimating the *average* value of Y_c (i.e., the regression line itself) for a group of new observations rather than an individual value of Y. Thus, it might be used in predicting the average test scores for another group of workers, rather than the score for a particular worker.

In the production rating example, $S_{YX} = 5.13$, $n = 20$, and $\Sigma x^2 = 7{,}395$ (Table 16–2). Therefore, the standard error of forecast is:

$$S_f = 5.13 \sqrt{1 + \frac{1}{20} + \frac{x^2}{7{,}395}}$$

The forecast errors for five selected test scores (X) are given in Table 16–4, column 5.

Table 16–4

STANDARD ERROR OF AN INDIVIDUAL FORECAST

TEST SCORES AND PRODUCTION RATINGS OF 20 WORKERS

			Standard Error of	
Selected Value of X (1)	Deviation from Mean, x (2)	x^2 ——— 7,395 (3)	Estimate S_{XY} (4)	Forecast S_f (5)
15	−40	.2164	5.13	5.77
35	−20	.0541	5.13	5.39
55	0	0	5.13	5.26
75	20	.0541	5.13	5.39
95	40	.2164	5.13	5.77

Note: For 95 percent confidence intervals multiply columns 4 and 5 by 2.10.
SOURCE: Table 16–2.

If the calculations for the forecast error are based upon a large sample, and if the values are approximately normally distributed about the regression line, then the chances are about 95 percent that a new observation drawn from the same population will be within 1.96 forecast errors on either side of Y_c. That is to say, the 95 percent confidence interval for a new observation (Y) is $Y_c \pm 1.96\ S_f$.

In the present example, however, with sample size only 20, the 95 percent confidence interval for a new observation is $Y_c \pm 2.10\ S_f$. This interval is shown as the wide band in Chart 16–9. The chances are 95 out of 100, therefore, that a new applicant will achieve a production rating within these limits.

Certain characteristics of Chart 16–9 should be carefully observed. The boundaries of the confidence intervals are curved. The further the X values get from their arithmetic mean, the greater the width of the confidence intervals. This fact points up the danger of extrapolating for values of X that are a considerable distance from $\bar{X}$. When n is small, too, the standard error of forecast substantially exceeds the standard

Chart 16–9

CONFIDENCE INTERVALS FOR AN INDIVIDUAL FORECAST

SCORES AND RATINGS OF 20 WORKERS

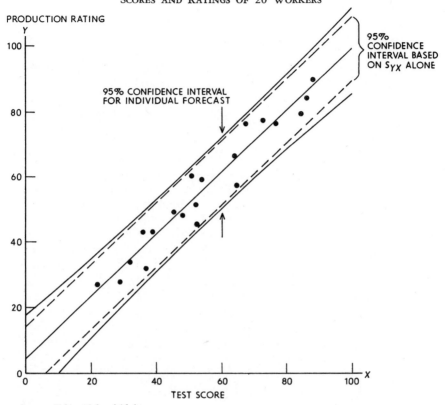

SOURCE: Tables 16–2 and 16–4.

error of estimate, and better reflects the errors of making predictions from small samples.

The forecast error is useful not only for *prediction* but also for *control*. If an observation falls outside the confidence limits, this indicates that it is very likely "out of control" and should be investigated. As a control chart, Chart 16–9 serves much the same purpose as the statistical quality control charts described in Chapter 10. In the present example, management can not only predict that an applicant with test score of 40 will achieve a production rating between 31 and 53 (with probability 95 percent), but they can use these points as control limits. If the applicant's actual production rating falls outside these limits, the

Chart 16–10

MINIMUM CORRELATION IN POPULATION, FOR VARYING OBSERVED CORRELATION (*r*) AND SIZE OF SAMPLE

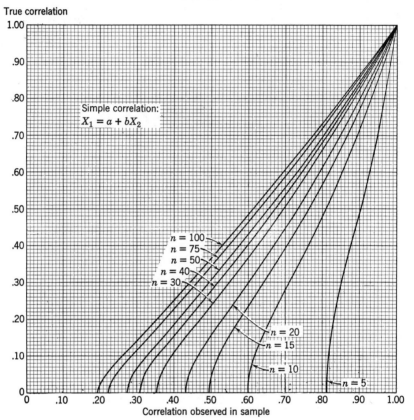

True correlation

Simple correlation: $X_1 = a + bX_2$

$n = 100$
$n = 75$
$n = 50$
$n = 40$
$n = 30$
$n = 20$
$n = 15$
$n = 10$
$n = 5$

Correlation observed in sample

Under conditions of random sampling, one sample out of 20, on the average, will show a correlation coefficient with a ± value as high as that "observed in sample," when drawn from a population with the stated true correlation.
SOURCE: Reprinted with permission from M. Ezekiel and K. A. Fox, *Methods of Correlation and Regression Analysis* (3d ed.; New York: John Wiley, 1959), p. 294.

chart warns the supervisor to investigate. If the employee's production is below 31, it may be possible to identify and remedy the cause of this deficiency; if it is above 53, the factors accounting for this superior performance should also be identified, either as a basis of rewarding the employee or improving work practices generally.

Standard Error of the Coefficient of Determination or Correlation

We will not take up the standard error of the coefficient of determination (r^2) or correlation (r) directly, since this concept involves difficulties that are disproportionate to its rather limited usefulness in business and economics.[10]

The sampling variability of correlation coefficients may be illustrated graphically, however, in Chart 16–10. This chart shows the *minimum* value of the true correlation coefficient for any sample value of *r* at the 95 percent confidence level.

For example, in the production rating case, the coefficient of correlation for the group of 20 workers is $\sqrt{.929}$, or .964. With this value on the *X* axis, use the $n = 20$ curve to find .93 on the *Y* axis. We can say, therefore, that the true correlation for the population is at least .93, with a 95 percent chance of being correct.

If the sample *r* were .60, however, with $n = 10$, we could only say that the true value is at least zero with the same degree of confidence. That is, even if there is *no* correlation in the population itself, 5 percent of all possible samples of size 10 would still yield a correlation coefficient of ±.60 or higher. This chart demonstrates the danger of making inferences about the degree of correlation when *r* or *n* is small.

REGRESSION OF TIME SERIES

The regression of monthly or yearly data may be carried out in the same way as described above. However, time series are not probability samples, but are subject to trends, cycles, and major irregular forces as well as to purely random movements. Hence, problems of interpretation do arise, and there are booby traps to avoid.

Most measures of regression and correlation are theoretically correct only if the residuals $(Y - Y_c)$ are randomly distributed, with uniform dispersion, around each section of the regression line (as described under "Basic Assumptions" above). This is not true of time series. First, the presence of an extreme high or low value (occasioned, say, by a war scare or strike) influences the regression line and the various standard errors in proportion to the square of its deviation and may distort the results.

Second, the absolute residuals tend to get bigger as the industry grows over the years. The use of logarithms to discount this tendency is illustrated below.

---

[10] The standard error of the correlation coefficient can be estimated as $s_r = (1 - r^2) \div \sqrt{n - 1}$. This formula is only applicable to large samples, and even then the distribution of the sample *r*'s is quite skewed when the true value of *r* is far from zero. The value *r*, however, can be transformed into a quantity called Fisher's *z*, whose sampling distribution is nearly normal. For a treatment of confidence intervals and tests of hypotheses using *z*, see W. A. Spurr, L. S. Kellogg, and J. Smith, *Business and Economic Statistics* (Homewood, Ill.: Richard D. Irwin, 1954), pp. 492–93 and Appendix I.

Third, since most time series move in cycles rather than in purely random fashion, there are likely to be runs of several successive positive or negative residuals in a row. This also occurs if a straight line is fitted to a curved relationship. That is, each year's value is related to that of the adjoining year rather than being independent of it. This is called *autocorrelation.* If the residuals are autocorrelated, the standard error of estimate will understate the amount of error likely to be encountered in making forecasts. We will cite a test for appraising the extent of autocorrelation in the example below. If the degree of autocorrelation is greater than could be attributable to chance, the usual standard error formulas are inapplicable.

The regression line itself is often a valid and useful tool for control and forecasting, despite its limitations. But in projecting this line into the future, a careful study is needed to determine whether past relationships are likely to persist. Extrapolation is dangerous but necessary in forecasting. In any case, the two series must be logically related. Otherwise the correlation will be spurious, owing to chance, similar trends, or the common influence of outside factors.

The standard error of estimate, on the other hand, may be of dubious value if its probability significance is impaired by either the autocorrelation or the erratic distribution of residuals ($Y - Y_c$) in time series; hence, a range of one standard error of estimate about the regression line need not include about 68 percent of the items. The same is true of the standard errors of the regression coefficient, the standard error of forecast, and confidence intervals based on these values. It will be shown, however, that these measures may be valid in correlating percentage changes from year to year.

Finally, if two series are growing over the years, the coefficient of determination may grossly overstate the degree of relationship, simply because both series have smaller values in early years and larger ones in later years.

An Example: Forecasting Sales

Suppose we are engaged in forward planning for Sears, Roebuck & Company and wish to establish a quantitative basis for projecting the company's future sales. Since the company distributes a wide variety of consumer products on a nationwide scale, its sales should move pretty closely with United States disposable personal income. Authoritative forecasts of the latter are available. We will therefore correlate the sales and income figures for the post–Korean War period 1953–71, shown in Table 16–5, and use this regression to predict Sears sales for 1972–75.

Table 16–5

SEARS, ROEBUCK NET SALES AND U.S. DISPOSABLE INCOME
1953–71, WITH PROJECTIONS TO 1972–75

Year	Disposable Income (Billions of Dollars) X	Sears Sales* (Billions of Dollars) Y	Percent Change from Previous Year	
			X	Y
1953	252.6	2.982	. . .	. . .
1954	257.4	2.965	1.9	–.6
1955	275.3	3.307	7.0	11.5
1956	293.2	3.556	6.5	7.5
1957	308.5	3.601	5.2	1.3
1958	318.8	3.721	3.3	3.3
1959	337.3	4.036	5.8	8.5
1960	350.0	4.134	3.8	2.4
1961	364.4	4.268	4.1	3.2
1962	385.3	4.578	5.7	7.3
1963	404.6	5.093	5.0	11.2
1964	438.1	5.716	8.3	12.2
1965	473.2	6.357	8.0	11.2
1966	511.9	6.769	8.2	6.5
1967	546.3	7.296	6.7	7.8
1968	591.0	8.178	8.2	12.1
1969	634.4	8.844	7.3	8.1
1970	689.5	9.251	8.7	4.6
1971	744.4	10.006	8.0	8.2
		Projections		
1972	795.	11.219	6.8	7.8
1973	870.	12.477	9.4	11.4
1975	965.	14.100	10.9†	13.4†

* Years beginning February 1.
† Two-year change.
Sources: *Survey of Current Business*, Sears, Roebuck *Annual Reports*. Income projections from *Predicasts*, July 28, 1972; Sears projections from regression equation.

We first plot the data on both an arithmetic scale (not shown) and a double logarithmic scale (Chart 16–11). The relationship appears linear in both cases, so we fit regression lines by least squares to both the natural values and the logarithms. The fits are good, and the coefficients of determination are the same (.994). We have chosen the logarithmic line ($\log Y_c = -2.3681 + 1.1785 \log X$) since the percentage (logarithmic) deviations along the regression line tend to be more uniform (as required by the theory of least squares) than the absolute deviations, which tend to increase with the growth of sales over the years.

Chart 16–11

SEARS, ROEBUCK SALES AND U.S. DISPOSABLE INCOME

1953–71 WITH PROJECTIONS TO 1972–75
(Double Logarithmic Scale)

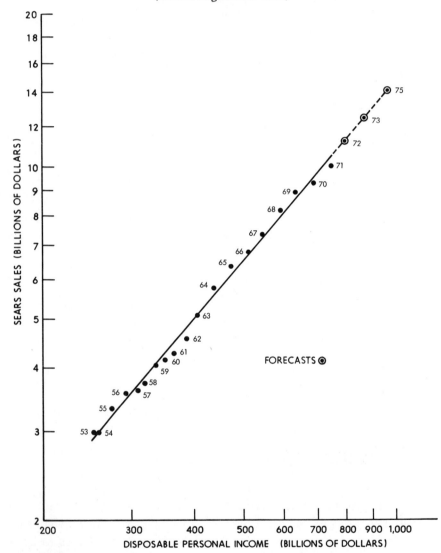

SOURCE: Table 16–5.

Chart 16–11 serves either as a control chart or as a forecasting tool. As a control on 1971 results, note that the recession that year depressed Sears sales 3.8 percent below their "normal" relation with income, but this is not a particularly bad showing since it is only 1.2 times the standard error of estimate of 3.1 percent.

To forecast Sears sales for 1972–75, we can extend the regression line and use the concensus of projections of disposable income by leading economists published in *Predicasts* for July 28, 1972 (Table 16–5, column 3). Substituting these values in the regression equation, we get the estimates of Sears sales shown in Table 16–5, column 2, and Chart 16–11. (The arithmetic straight line was more depressed by the 1970–71 recession and yielded slightly lower forecasts).

Note that since 1971 sales are 3.8 percent below the regression line, the forecast implies that Sears' customers will have to increase their purchases by a corresponding percentage, in relation to their income, to achieve our future estimates. These estimates should be adjusted, therefore, after an appraisal of the business outlook.

Now, how valid are the standard error of estimate and related measures in judging the accuracy of these forecasts? The validity depends in large part on the degree of autocorrelation. Unfortunately, Chart 16–11 clearly shows waves of autocorrelation: the 1957–62 and 1970–71 points are below the regression line and the 1964–69 points are above it.

We can further test the extent of autocorrelation in the residuals around a regression line by computing the Durbin-Watson statistic (d). If this value is 2, there is no autocorrelation; if it is near zero there is a high degree of positive autocorrelation. In the Sears case, $d = .85$, which indicates significant autocorrelation, so the usual standard error formulas are inapplicable.[11] Hence, we cannot estimate the forecast errors in probability terms, even assuming we knew future disposable income for certain. And the income projection itself has an unknown error, which could either increase or decrease the error in the sales forecast.

Another way of estimating the error of forecasts is to make several assumptions and compare the spread of forecasts based on each. Thus, the U.S. Census Bureau makes four different "illustrative projections" of future population based on various assumptions regarding the birth rate, and the reader is left to choose among them as best he can.

---

[11] For details of this test, see Charles R. Frank, Jr., *Statistics and Econometrics* (New York: Holt, Rinehart & Winston, 1971), pp. 276–81, and Appendix E, which shows whether the autocorrelation is significant for various values of d.

Use of Percentage Changes to Provide Valid Measures of Forecast Errors. We can often reduce the autocorrelation of time series, and thus provide a more valid standard error of estimate and related measures, by fitting a regression to the percentage changes from year to year, rather than to the actual data. The results are useful in short-term forecasting.

A regression line has therefore been fitted by least squares to the yearly changes in Sears, Roebuck sales and in disposable income shown in Table 16–5, columns 4 and 5. The plotted residuals (not shown) are more randomly distributed than those in Chart 16–11, and the Durbin-Watson statistic of 1.70 indicates no significant autocorrelation. The various standard errors computed for these percentage changes (e.g. S_{YX}, S_b, and S_f), therefore, are more valid than those computed for the original values. This does not mean, of course, that the forecast itself is necessarily more accurate than one based on original data.

The forecast for 1972 obtained by correlating percentage changes is a 7.8 percent increase in Sears sales over 1971 (Table 16–5), plus or minus a standard error of estimate of 2.9 percentage points. The actual increase for 1972 proved to be 9.8 percent—well within this range. The coefficient of determination, .447, is also more valid than the spuriously high figure of .994 obtained in correlating the original series, which both have rising trends. Unfortunately, however, it means that disposable income explains only 44.7 percent of the variance in yearly percentage changes in Sears sales.

Alternatively, we could correlate the absolute amounts of change each year, but the residuals $(Y - Y_c)$ tend to increase with sales (Y) over the years. The use of absolute changes thus violates least-squares theory, and tends to exaggerate the influence of the later figures.

Finally, we could correlate percentage of the secular trend curve (Chapter 19). These values are shown in Table 19–3, column 8, and Chart 19–7 for Sears, Roebuck sales. Similar deviations could be determined for disposable income. The results bring out the cyclical and other short-term relationships between the two series. The trend line is a more stable base for computing percentages that is the previous year's level, so the scatter of percentages tends to be less erratic. However, in the long run the projections obtained in correlating percentages of trend are increasingly sensitive to errors in extrapolating the trend curve itself.

A more complete analysis would utilize multiple regression (Chapter 17) to relate Sears sales simultaneously to various factors that affect sales (e.g., disposable income, number of stores, and time). We could also project the future trend of Sears' sales over time (Chapter 19). Finally, a detailed study of management policy, consumer preferences, and the general economic outlook is needed to modify the statistical projections. If possible, the analysis should also be carried out for individual lines of merchandise, for different territories, and for the department store and mail-order branches of the business separately, to pinpoint the components of growth.

A CAUTION: CORRELATION DOES NOT MEAN CAUSATION

Before concluding this chapter, a pitfall in logic should be emphasized. The fact that two variables are correlated does not imply in any way that either is a cause of the other. In particular, it is a *non sequitur* to infer that because one event precedes another in time, it is therefore the cause of the other. A student writes a correspondence institute: "I am well pleased with the law course. A month after enrolling, my salary was increased in the amount of 20 percent." *Non sequitur.*

Again, beer consumption and church attendance show a close correlation over the years. This does not mean that beer drinkers seek solace in religion, however, nor that piety induces thirst. Both variables have simply increased with population.

Many business cycle theorists in the past have found that some particular economic factor has correlated with general business activity and hence they have assumed that this factor is "the cause" of business cycles. Unfortunately, economic and business affairs represent a complex of interacting forces. The search for simple cause-and-effect relationships is naïve and unrealistic.

Similarly, large-scale studies have established a correlation between smoking and lung cancer. However, it is a matter of bitter dispute whether smoking *causes* lung cancer, since so many other correlated factors (urban living, smog, tensions, etc.) may also affect cancer.

In general, if factors A and B are correlated, it may be that (1) A causes B, to be sure, but it might also be that (2) B causes A, (3) A and B influence each other continuously or intermittently, (4) A and B are both influenced by C, or (5) the correlation is due to chance.

SUMMARY

Simple regression and correlation analysis is concerned with the study of two logically related variables and how they change together from observation to observation. In most such studies, interest is concentrated on estimating the dependent variable Y from the independent variable

X. These are plotted on a scatter diagram, which shows whether the relationship is close or not, whether it is positive or negative, and whether it is linear or curvilinear.

The basic measures of relationship are the regression line or curve, which describes the average relationship between X and Y; the standard error of estimate, which is the standard deviation of the residuals $(Y - Y_c)$ around this line; and the coefficient of determination, a relative measure of relationship which varies from 0 to 1.

Regression analysis is used in business and economics principally for the purposes of prediction and control. Thus, in correlating the earnings per share (X) with price per share (Y) for a number of stocks, we can predict the price of a stock from the regression line, based on estimated future earnings, or we can use the standard error of estimate to construct a confidence interval around this line and consider the stock unduly high or low in price if it is outside these control limits.

Regression lines or curves can be fitted either graphically or mathematically. In graphic analysis, arrays are constructed by grouping observations for which values of X are approximately equal; a point of means for each array is plotted and a smooth curve is drawn to fit the points of means. If the regression is linear, the line is drawn through $(\bar{X}, \bar{Y})$, the point of means of all observations. The two constants of the linear regression line are its Y intercept a and its slope b, the regression coefficient.

The method of least squares is a means of computing the constants of the regression line so as to minimize the sum of squares of residuals from the line. Thus, in fitting a straight line, $\Sigma(Y - Y_c)^2$ is less than for any other straight line. A straight line fitted by least squares also goes through the overall means of the data and reduces the sum of the plus and minus deviations to zero: $\Sigma(Y - Y_c) = 0$. The computations can be simplified by using the deviations of the variables from their means (i.e., using x and y instead of X and Y).

Curvilinear relations may be expressed by a graphic curve, a parabola, a logarithmic straight line, or some other mathematical function.

A parabola is a curve of the form $Y_c = a + bX + cX^2$. It may best be fitted by treating the X^2 term as a new variable X_2 and then solving the normal equations for multiple regression, using the redefined variables as described in Chapter 17.

To fit a logarithmic straight line, the data may be plotted on semilog or double-log graph paper and a straight line drawn graphically. Alternatively, logarithms may be used in place of either or both of the variables in the calculations of the least squares regression line.

The use of logarithms in regression equations is an example of the transformation of variables. Other transformations, such as the use of square roots or reciprocals, may also be used in regression analysis to permit a linear fit.

Curvilinear methods of regression should be used whenever (1) the logic of the relationship supports a particular type of curve and (2) the standard error of estimate is smaller for this curve than for a straight line.

The standard error of estimate measures the average error of the regression line in providing estimates of Y from given values of X. It may be computed as the standard deviation of the residuals $(Y - Y_c)$ around the regression line or by means of a short-cut formula.

The coefficient of determination (r^2) is a relative measure of relationship. It is the ratio of explained variance to total variance, or 1 minus the ratio of unexplained to total variance. Its square root (r) is the coefficient of correlation.

Total variance is the standard deviation (squared) of the Y values around their mean $(Y - \bar{Y})$. Explained variance is the standard deviation (squared) of the Y_c values around the mean $(Y_c - \bar{Y})$, since this part of the variation in Y can be explained by corresponding changes in X. Unexplained variance is the standard deviation (squared) of Y values around the regression line $(Y - Y_c)$—the variation in Y not explained by X. This is the standard error of estimate, squared. The coefficient of determination is a more direct and unequivocal measure of the proportion of variance in Y explained by X than is the higher-valued coefficient of correlation.

In summary, the regression line, the standard error of estimate S_{YX}, and the coefficient of determination r^2 each measures a different aspect of a given relationship. For many problems of control and prediction, the first two measures will suffice. The coefficient of determination is needed only if the problem calls for a measure of proportionate importance.

When the data used for regression analysis can be considered as a probability sample from a population, we can make statistical inferences based upon the sample data. The assumptions in linear regression analysis are (1) linear relationship between X and Y in the population; (2) uniform scatter about the regression line; (3) the independence of the deviations about the regression line; and (4) a roughly normal distribution of points about the regression line. When these assumptions are satisfied, the sample values a and b are "best" estimates of the population values A and B.

We should also distinguish between the correlation model and the regression model. In the correlation model, both X and Y are assumed to be normally distributed and all correlation and regression statistics are valid estimators. In the regression model, the Y values are normally distributed, but the X values may be arbitrarily limited, as in a controlled experiment. In this case, regression results are valid only for these same X values, and the coefficient of determination is not generally valid.

We can apply tests of significance and confidence intervals to regression results from probability samples in order to make statistical inferences about the parent population. Thus, we can determine whether there is any significant relationship between X and Y by testing the null hypothesis that the population regression coefficient B is zero. If the sample value b divided by its standard error is sufficiently large, according to a table of the normal or t distribution, the relationship is deemed to be significant. We can also compute confidence intervals for b.

By further combining the standard error of the regression line with the standard error of estimate, we obtain the standard error of forecast, which provides confidence limits within which any new observation may be expected to fall. These confidence bands are narrowest at $\bar{X}$; they widen out in either direction. This indicates the danger of estimating Y for values of X that are far from their mean, especially with smaller samples. The forecast error is valuable both in predicting Y and in providing a control chart for Y.

Confidence limits for r are shown in Chart 16–10. The chart illustrates the dangers of making inferences when r or n is small.

Time series present special problems in regression. To illustrate, Sears, Roebuck sales are correlated with disposable personal income for 1953–71 and the regression is used to forecast 1972–75 sales. Plotting the original data on a double-logarithmic scale (Chart 16–11), we find a close linear relationship. However, the residuals around the line prove to be autocorrelated (i.e., successive years' values are too much alike), so the standard error formulas are inapplicable.

In order to reduce autocorrelation, we use the year-to-year percentage changes instead, and find that the various standard error formulas (and r^2) are more valid than in correlating original data.

To determine whether regression relationships will apply in the future, one must make a careful study of management policy, consumer preferences, and general economic trends. Extrapolation of regression curves is dangerous, but it is nevertheless necessary in forward planning.

A pitfall in the use of regression analysis should be noted: correlation between two variables does not, of itself, imply that there is any causal relationship between them.

PROBLEMS

1. Distinguish between:
    a) Linear and curvilinear regression.
    b) The standard error of estimate and the standard deviation of the dependent variable.
    c) The use of regression analysis for prediction and for control.
    d) The coefficient of regression and the coefficient of correlation.

2. Explain:
    a) The method of least squares as applied to regression analysis.
    b) How to test whether there is any significant relationship between two variables.
    c) How to obtain a 99 percent confidence interval for the regression coefficient in a large sample.
    d) How the standard error of forecast is derived from the standard error of estimate.
    e) The coefficient of determination in terms of explained variance, unexplained variance, and total variance.

3. Wheat yields in Kansas in bushels per acre (squared) have a total variance of 25 over many years, of which a variance of 16 can be explained by variations in seasonal rainfall. This year's yield is estimated at 26 bushels an acre based on the season's rainfall of 18 inches.

    Within what range would you predict the yield to be this season on a given farm, with about 95 chances out of 100 of being correct? (Ignore sampling error of regression line itself.)

4. Assume that we conduct an experiment with eight fields planted to corn: four fields having no nitrogen fertilizer and four fields having 80 pounds of nitrogen fertilizer. The resulting corn yields are shown in the table, in bushels per acre.

Field	Nitrogen (Pounds)	Corn Yield Bushels/Acre
1	0	12
2	0	36
3	0	6
4	0	18
5	80	128
6	80	112
7	80	112
8	80	72
Total	320	496

Note: This sample is too small to provide really valid inferences, but it serves to illustrate the methods involved with a minimum of computations.

a) Plot the data as a scatter diagram on an arithmetic chart, and draw a regression line by the graphic method, using group averages as guides.

b) Compute a linear regression equation by least squares. How does this compare with the graphic line when plotted on the chart? Explain the meaning of the regression equation in terms of fertilizer and corn yields.

c) Compute the standard error of estimate. Interpret this value as an aid in predicting corn yields.

d) Predict corn yield for a field treated with 60 pounds of fertilizer, and give the 95 percent confidence limits for this prediction. (Assume a linear relationship and ignore sampling errors in the regression line itself.)

e) Compute the estimated coefficient of determination as 1 minus the unexplained variance over the total variance. What does this figure tell you about the relationship of nitrogen fertilizer and corn yields in general?

5. Refer to the data in Problem 4.

a) Is there any significant relationship between nitrogen fertilizer and corn yields? That is, test the null hypothesis $B \leq 0$ against the alternative hypothesis $B > 0$ at a critical probability of, say, 5 percent.

b) Give the 95 percent confidence interval for the regression coefficient.

c) How is your interpretation of the results in (a) and (b) affected by the fact that the basic data represent a controlled experiment rather than a survey in which both X and Y are normally distributed? (Ignore the small size of the sample.)

6. In the same corn-yield experiment (Problems 4 and 5 above):

a) Compute the standard error of the regression line and its 95 percent confidence limits for fertilizer applications of 0, 40, and 80 pounds, respectively. (See Footnote 9).

b) Compute the standard error of forecast and the 95 percent confidence limits for individual forecasts of corn yield, assuming fertilizer applications of 0, 40, and 80 pounds, respectively.

c) How is your interpretation of the results in (a) and (b) affected by the fact that the basic data represent a controlled experiment rather than a survey in which both X and Y are normally distributed? (Ignore the small size of the sample.)

7. a) If the sample value of r is .60, with $n = 20$, what is the minimum value of the true correlation coefficient of the population at the 95 percent confidence level (Chart 16–10)?

b) If the true correlation coefficient were zero, what sample value would be exceeded by 5 percent of all random samples of size 20?

8. Refer to Table 17–3, page 499. Consider the simple regression between the area of a lot (X) and its price (Y).

a) Verify that the least squares regression equation is $Y_c = 1.453 + .2194 X$. (Refer to Table 17–5, page 504.)

b) Is the relationship between area and price statistically significant?

c) Calculate the correlation coefficient between area and price.
d) A given lot has 18,000 square feet. Estimate the price at which it should have sold. Give a 95 percent confidence interval about this estimate.

9. Refer to Tables 17–3 and 17–5, pages 499 and 504.
   a) Estimate the simple regression line between the elevation of a lot and its price.
   b) Calculate the standard error of estimate.
   c) Is the relationship between elevation and price significant?
   d) Calculate the correlation coefficient between elevation and price.

10. An analyst for a certain company was studying the relationship between travel expenses in dollars (Y) for 102 sales trips and the duration in days (X) of these trips. He has plotted the data, and the relationship is approximately linear. The data are summarized in the table.

	X	Y	X²	XY	Y²
Total...................	510	7,140	4,150	54,900	740,200
Mean...................	5	70			
Adjustment...............			−2,550	−35,700	−499,800
Adjusted total			1,600	19,200	240,000
Adjusted total is..........			Σx^2	Σxy	Σy^2

a) Estimate the regression equation from the above data.
b) What is the practical significance of the value of a (the intercept) in this equation?
c) A given trip is to take seven days. How much money should a salesman allow so that there is only 1 chance in 10 that he will run short?

11. The Certified Foods Company operated a chain of retail food stores. As a means of measuring the efficiency of the various stores, a study was made of the relationship between the number of employees (X) and the average monthly sales volume (Y) for all the stores over the past year. When the data were plotted, the relationship was approximately linear, with the points having a uniform scatter about the line. The data can be summarized as follows: X = the number of employees in each store, Y = the average monthly sales during 1973 for each store in thousands of dollars, n = 100 = the number of stores in the chain, ΣX = 600, ΣY = 1,600, ΣX^2 = 5,200, ΣY^2 = 37,700, ΣXY = 13,600.
   a) Find the line of average relationship (i.e., the regression line). Give a verbal meaning of this equation.
   b) Calculate the coefficient of determination.
   c) Store No. 64 employs 10 persons and has monthly sales of $20,000. Is the performance of this store out of line with the performance of the other stores? How do you know?

12. As the Alma Mater University Alumni secretary in your city, you are responsible for making reservations for the semimonthly alumni luncheons.

Before each meeting you send out letters with return postcards. Each alumnus is asked to return this card if he plans to attend. You find that only a portion of the cards are returned by the time it is necessary to make the reservation, and you are forced to guess about the actual number of lunches that will be necessary.

You have analyzed the data over the past two years (48 luncheons) and have found that there is approximately a linear relationship between the number of reservations received by four days before the luncheon and the actual number present at the luncheon. Therefore, you fit a regression line to the data and find: $Y_c = 20 + 1.50 X$, where Y_c is the estimate of the actual attendance and X is the number of reservations received by four days before the luncheon. You also have $S_{YX} = 5.0$, $n = 48$, $\overline{X} = 20.0$, $\Sigma x^2 = 4,700$, $\overline{Y} = 50.0$, $\Sigma y^2 = 10,575$, $\Sigma xy = 7,050$.

a) Explain the meaning of the regression equation above.

b) Suppose 38 reservations are received for a given luncheon. Calculate a forecast interval at the 95 percent confidence level. (Assume that the deviations about the regression line are normally distributed.)

13. a) Refer to the data in Table 12–5, page 351. Calculate the correlation coefficient between current inventory and annual inventory on an item basis.

b) What is the minimum correlation in the whole population at the 95 percent level of confidence? (Use Chart 16–10, page 476.)

14. A survey of used car dealers in the Salt Lake City area was made by the Newspaper Agency Corporation to determine the relationship between the amount of classified advertising of used cars, and car sales. The table below shows the hundreds of lines of classified ads and the number of cars sold for each of six dealers who used no other advertising medium. (A sample of six items is actually too small to provide valid inferences, but this case serves to illustrate the method of least squares with a minimum of arithmetic.)

Dealer	Hundred Lines of Advertising	Used Cars Sold
A	74	139
B	45	108
C	48	98
D	36	76
E	27	62
F	16	57

a) Plot the data as a scatter diagram on an arithmetic chart, selecting the proper independent variable.

b) Compute a linear regression equation by least squares, and plot it on the chart. Explain the meaning of this equation in terms of advertising and car sales.

c) Find the standard error of estimate.

d) Compute the coefficient of determination. Explain the meaning of r^2 as applied to used car sales.

e) Use the regression to predict used car sales for dealer F on the assumption that he increases his advertising to 50 hundred lines. He does this, but then sells only 70 used cars. How could you use this analysis as a control device for him to judge this result?

f) Could the apparent relationship between classified advertising and used car sales have come about by chance? Or is there a statistically significant relationship between classified advertising and used car sales? To answer this question, test the hypothesis that B, the population regression coefficient, equals zero.

15. A certain mail-order firm used the weight of the incoming mail to estimate the number of orders that would need to be processed. Over a 25-day period the following data were collected:

Day	Pounds of Mail (Hundreds)	Thousands of Orders	Day	Pounds of Mail (Hundreds)	Thousands of Orders
1	1.8	6.4	14	4.1	13.8
2	2.0	8.0	15	4.2	12.8
3	2.0	7.2	16	4.2	16.5
4	2.1	7.5	17	4.2	17.1
5	2.3	6.9	18	4.3	15.4
6	2.6	10.9	19	4.6	16.2
7	2.6	10.3	20	5.0	15.8
8	2.8	9.5	21	5.4	19.0
9	3.1	9.7	22	5.8	19.4
10	3.2	10.6	23	6.0	19.1
11	3.2	12.5	24	6.4	18.5
12	4.0	12.9	25	6.5	20.0
13	4.1	14.0			

a) Calculate the linear regression equation relating the number of orders to the weight of the mail.

b) What is the sampling error associated with the estimated slope b? Are you sure that the true value B is greater than 2.5?

c) Estimate the number of orders for a mail delivery that weighs 500 pounds.

d) Assuming that the points are approximately normally distributed about the regression line, place 95 percent forecast limits on the estimate calculated in (c) above.

16. a) How could you determine whether the regression between test scores and production ratings in Table 16–1 is significantly curvilinear?

b) Since the formula for a straight line is merely a special case of that of a parabola in which $c = 0$, the parabola would seem to fit almost any set of data better than the less flexible straight line. Can you infer, then, that nearly all regressions are significantly curvilinear? Explain.

17. a) Plot Sears, Roebuck sales and disposable income for 1953–71 (Table 16–5, page 479) on an arithmetic chart, with the independent variable on the X-axis.

b) Since the relationship appears reasonably linear, fit a straight line by the graphic method or by least squares, as assigned, to these figures. Give the equation of this arithmetic straight line.

c) Forecast Sears sales for 1972, 1973, or 1975 (whichever is the latest year for which actual sales are available) using the income projections in Table 16–5. Find actual sales and give the percentage error of this forecast, compared with that of the logarithmic straight line in Table 16–5.

d) Compute the standard error of estimate. To appraise its validity, does your chart reveal significant autocorrelation? More than in the case of the logarithmic line (Chart 16–11, page 480)? What other factors might make this value understate the true error of a future projection?

18. As an analyst with Kraftco Corporation, you wish to project the demand for food products as a function of population, using the Federal Reserve index of production in food manufactures and the Census Bureau figures on U.S. population for 1957–71, with projections to 1980, shown in the following table. (The figures are from *Predicasts,* June 25, 1972, and *Business Statistics,* 1971.)

Year	Food Production (1967 = 100) Y	Population (Millions) X
1957	70.9	172.0
1958	72.7	174.9
1959	76.3	177.8
1960	78.4	180.7
1961	80.6	183.7
1962	83.2	186.5
1963	86.2	189.2
1964	89.7	191.8
1965	92.0	194.2
1966	96.7	196.5
1967	100.0	198.6
1968	103.9	200.6
1969	108.3	202.6
1970	111.7	204.9
1971	114.5	207.0
Projections		
1972		208.9
1973		210.8
1975		215.0
1980		228.5

a) Plot these figures on arithmetic, semilog (ratio), or double-log scales to get a linear relation, and fit a regression line by least squares to the appropriate natural values or logs.

b) Forecast the latest year listed in the table for which the actual food index is available. What is the error of the forecast as a percentage of the true value? What part of this error is due to the error in population projection, and what part is due to the regression residual using true population?

c) How could you improve the accuracy of this forecast?

d) Give the standard error of estimate. Does this value appear to be in-validated by significant autocorrelation?

19. An analyst for the Northern Gas Company, you are studying the following data in order to predict daily gas usage as a function of daily average tem-perature. The figures show the 24-hour average temperature and gas usage (in thousands of therms) for 42 weekdays during December and January, excluding holidays.

Day	Temperature X	Gas Used Y	Day	Temperature X	Gas Used Y
1	30°	1,108	23	44°	989
2	29	1,091	24	32	1,114
3	34	1,046	25	35	1,110
4	35	1,029	26	32	1,138
5	39	963	27	30	1,155
6	15	1,297	28	31	1,091
7	16	1,280	29	29	1,194
8	24	1,206	30	19	1,249
9	22	1,202	31	26	1,203
10	11	1,296	32	33	1,105
11	0	1,532	33	32	1,102
12	16	1,375	34	6	1,441
13	6	1,400	35	21	1,307
14	8	1,403	36	33	1,149
15	10	1,350	37	23	1,202
16	28	1,101	38	17	1,273
17	19	1,219	39	31	1,132
18	23	1,177	40	36	1,073
19	34	1,061	41	25	1,233
20	14	1,165	42	17	1,345
21	29	1,188	Total	1,030	50,203
22	36	1,109	Mean	24.52°	1,195.3

a) In order to examine the nature of the relationship, plot a scatter dia-gram on a large-scale arithmetic graph, selecting the proper series as the independent variable. Compute four or five group averages (array means) and mark them with small crosses on the chart. Mark the overall mean $(\overline{X}, \overline{Y})$ with a circle.

b) Draw a straight regression line through $(\overline{X}, \overline{Y})$, using a transparent ruler or string, so as to equalize the vertical deviations from the group averages to the line on each side of $(\overline{X}, \overline{Y})$ separately. Give the equa-tion of this line.

c) Estimate gas usage assuming a forecast of 20 degrees average tempera-ture on a certain day. What gas usage would you expect if the tempera-ture were 40 degrees?

20. The following graphic methods of approximating the standard error of estimate and coefficient of determination may be of interest, though they are not described in the text. Use the data in Problem 19.

a) Estimate the standard error of estimate (S_{YX}) by drawing two lines parallel to the regression line so as to include two thirds of the dots (and hence exclude one sixth on either side). The vertical width of this band, measured on the Y axis, is roughly $2S_{YX}$.

b) Estimate the standard deviation (s_Y) of the gas usages by drawing two horizontal lines to include two thirds of the dots (and hence exclude one sixth top and bottom). The height of this band is roughly $2s_Y$.

c) Then compute the coefficient of determination:

$$r^2 = 1 - \left(\frac{S_{YX}}{s_Y}\right)^2$$

21. You wish to correlate the data in Problem 19 mathematically, as an aid to predicting daily gas usage for your company.

a) Compute the linear regression equation by least squares. Also compare this result with that in Problem 19(b), if it was assigned.

b) What is the estimate of gas usage for a day on which the average temperature is 10 degrees?

c) Compute the standard error of estimate.

d) The average temperature for a certain day in January is expected to be 10 degrees. Establish upper and lower limits (with 95 percent confidence) for gas usage for that day. First compute the standard error of forecast and use it to establish the confidence intervals. How can these limits be used for planning? For control?

SELECTED READINGS

Selected readings for this chapter are included in the list which appears on page 539.

17. MULTIPLE REGRESSION AND CORRELATION

MULTIPLE REGRESSION and correlation analysis enables us to measure the joint effect of any number of independent variables upon a dependent variable. The multiple regression equation describes the average relationship between these variables, and this relationship is used to predict or control the dependent variable. The standard error of estimate is essentially the standard deviation of this variable from its computed values. And, finally, the coefficient of multiple determination measures the proportion of the variance in the dependent variable explained by the other factors. The concepts and techniques in this chapter, therefore, are just extensions of those in simple regression. However, by measuring the simultaneous influence of several factors, we have a more powerful and realistic tool of analysis than in considering only one independent variable, and the use of computer programs facilitates the calculations.

To illustrate the use of several variables, consider the problem of predicting new automobile sales for the coming year. Many factors affect sales, each one explaining a part of the total. Plausible factors include the number of existing motor vehicles registered at the end of the current year, the average age of existing automobiles, the total population 16 years of age or older, the level of disposable income per capita, and the expected retail prices for new automobiles relative to the general price level for consumer goods and services. Here, common sense (and economic theory) should indicate whether each of these variables has a positive or a negative effect upon the sales of new automobiles. It would appear that at least five independent variables would be necessary to explain or forecast variations in car sales.

Multiple regression is often used in connection with forecasting. Such

495

a forecast may be as broad as the general economic outlook for the nation as a whole, or it may be limited to the estimation of the price of a single stock. For example, the Value Line Investment Survey correlates the price of a stock in past years with its earnings per share and dividends (all in logarithms) to determine the estimated future value of the stock. Recommendations for stock purchase are based in part on this "value line" obtained by multiple regression analysis.

MULTIPLE REGRESSION ANALYSIS

The multiple regression equation represents the simultaneous influence of a set of independent variables upon the dependent variable. The linear equation can be written as

$$Y_c = a + b_1X_1 + b_2X_2 + b_3X_3 + \cdots$$

where Y_c is the computed or estimated value of the dependent variable Y and $X_1, X_2, X_3, \ldots$ are the independent variables. The equation is said to be linear since there are no terms such as X_1^2 or X_1X_2 present. The term a is simply the value of Y_c when all the X's are zero. The terms $b_1, b_2, b_3, \ldots$ are the *net regression coefficients*. Each measures the change in Y per unit change in that particular independent variable. However, since we are measuring the simultaneous influence of all variables on Y, the net effect of X_1 (or any other X) must be measured apart from any correlated influence of other variables. This is usually expressed by adding the qualifying statement: "all other variables held constant" or "adjusting for the effect of the other variables." We would say, therefore, that b_1 measures the change in Y per unit change in X_1, *holding the other independent variables constant.*

To illustrate, suppose we wish to predict job performance (Y) of applicants for a given job based on the score of a placement test (X_1) and the interviewer's rating (X_2). The scales are arbitrary. We test a random sample of 18 new employees and later measure their job performance.

In Table 17–1 it can be seen that each successive pair of observations provides a set of values of Y for which X_1 and X_2 are constant. Means of these sets of Y values are presented in Table 17–2. When X_1 increases by 10, the mean of Y increases by 4 (four tenths as much as X_1), and as X_2 increases by 15 or 10, the mean of Y increases by 9 or 6, respectively (six tenths of the change in X_2). Accordingly, the net regression coefficients are $b_1 = .4$ and $b_2 = .6$. In order to determine the intercept value a, note that the regression plane must go through the overall means of the data. Hence:

$$\bar{Y} = a + b_1\bar{X}_1 + b_2\bar{X}_2$$

or:

$$a = \bar{Y} - b_1\bar{X}_1 - b_2\bar{X}_2 = 21 - (.4)20 - (.6)(18.33) = 2$$

So the regression equation is

$$Y_c = a + b_1X_1 + b_2X_2$$
$$= 2 + .4X_1 + .6X_2$$

The net regression coefficient b_1 shows the average effect of a one-unit increase in X_1 (test score) on Y (job performance), holding X_2 constant. That is, b_1 indicates how the test score predicts job performance for men rated alike by the interviewer. The net regression coefficient thus differs from the gross regression coefficient b in simple regression between test scores and job performance in that b shows the

Table 17-1

RELATION OF TEST SCORES AND INTERVIEWER'S RATINGS
TO JOB PERFORMANCE (18 EMPLOYEES)

Employee Number	Job Performance Y	Test Score X_1	Interviewer's Rating X_2
1	5	10	5
2	13	10	5
3	9	20	5
4	17	20	5
5	13	30	5
6	21	30	5
7	14	10	20
8	22	10	20
9	18	20	20
10	26	20	20
11	22	30	20
12	30	30	20
13	20	10	30
14	28	10	30
15	24	20	30
16	32	20	30
17	28	30	30
18	36	30	30
Total	378	360	330
Mean	21	20	18.33

Table 17–2

MEANS OF ARRAYS OF THE DEPENDENT VARIABLE Y'

	$X_2 = 5$	$X_2 = 20$	$X_2 = 30$
$X_1 = 10$	9	18	24
$X_1 = 20$	13	22	28
$X_1 = 30$	17	26	32

Source: Table 17–1.

combined effect of the test score and the intercorrelated effect of the interviewer's rating in predicting job performance.

The regression equation above is the equation of a plane in three-dimensional space, as shown in Chart 17–1. The observed points scatter

Chart 17–1

MULTIPLE REGRESSION PLANE

$$Y_c = 2 + .4X_1 + .6X_2$$

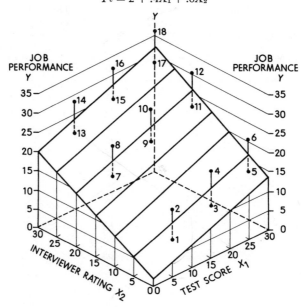

above and below the plane. For linear multiple regression, we assume that such a plane is a good fit to the data. If not, some curvilinear surface may be more appropriate, as described later.

The net regression coefficients may be estimated either by the graphic or the least-squares method. Today, computers provide a variety of fast and accurate programs for least-squares analysis. However, graphic techniques are useful (1) in understanding the basic concepts of multiple regression, (2) to check the assumptions underlying this analysis (e.g., linearity and homoscedasticity), (3) to obtain quick results when no computer is available, and (4) to determine curvilinear relationships when the appropriate equation form is unknown. For these reasons we shall first briefly present the graphic method. This method is feasible if the correlation is fairly high, n is not large, and the independent variables are not too numerous nor correlated with each other.

Graphic Analysis: The Method of Successive Elimination

Let us consider the problem of a certain real estate broker who has purchased a tract of land for subdivision into lots. He wished to know how much the area and the view from these lots contributed to their

Table 17-3
AREA, ELEVATION, AND PRICE FOR 20 RESIDENTIAL LOTS

Lot No.	X_1 Area, Thousands of Square Feet	X_2 Elevation, Feet above Sea Level	Y Price, Thousands of Dollars
1	14.7	155	4.1
2	14.2	155	3.9
3	12.7	158	3.2
4	13.8	158	2.9
5	14.4	155	3.9
6	17.4	157	4.1
7	21.8	172	5.8
8	14.0	170	5.1
9	17.5	175	6.8
10	23.0	185	6.8
11	18.3	185	6.5
12	19.4	205	7.0
13	15.2	215	5.8
14	18.3	195	5.1
15	21.7	178	5.3
16	16.7	160	4.9
17	13.6	205	6.0
18	14.5	190	5.3
19	12.1	203	4.8
20	17.4	125	4.3
Total	330.7	3501	101.6
Mean	16.535	175.05	5.08

Source: Actual data collected by the authors.

value. He also wanted a method for setting a reasonable price on the lots.

In order to obtain some information, the broker selected 20 nearby lots that had been recently sold. He obtained the sale price for each lot and its size (in thousands of square feet). Since he knew the lots at higher altitudes had more value because of the view, he also estimated the elevation of each lot (in feet above sea level). The data are presented in Table 17–3, on the preceding page.

Scatter diagrams showing the relationships between each pair of variables are displayed in Chart 17–2. We see that there is a positive linear correlation between price and area and between price and elevation, but there is no apparent relationship between elevation and area for the 20 lots selected.

The first step in the graphic approach (called the "method of succes-

Chart 17–2

RELATION BETWEEN AREA, ELEVATION, AND PRICE OF 20 LOTS

SCATTER DIAGRAMS

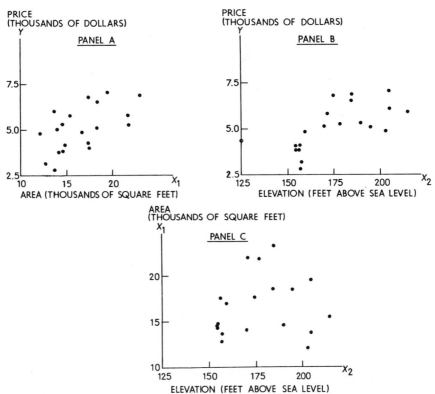

sive elimination") is to determine the simple regression line between the dependent variable Y (price) and the independent variable that is deemed most important. We shall select the area (X_1). This line can be determined by either graphic or least-squares techniques, as described in Chapter 16. (Alternatively, a curved line can be drawn freehand if the relationship is logically curvilinear.) The equation is $Y_c = 1.45 +$ .219 X_1 and is shown in Chart 17–3. The slope of the line indicates

Chart 17–3

REGRESSION LINE BETWEEN PRICE AND AREA

REGRESSION EQUATION: $Y_c = 1.45 + .219X_1$

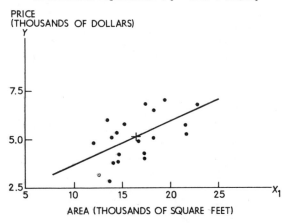

that the price of a lot increases $219, on the average, for every thousand square feet of area. This equation, of course, does not take the elevation of the lot into account.

The next step is to eliminate the effect of area on the price of each lot. This is done by subtracting .219 for each thousand square feet from the price of the lot. This adjustment to a "no area" basis may be done graphically by measuring the residuals or vertical deviations from the regression line in Chart 17–3, or it may be done arithmetically as shown in Table 17–4.

The new price Y' (where $Y' = Y - .219X_1$) represents the price adjusted for differences in the size of the lots. This adjusted price is then plotted against the second independent variable, elevation (X_2), as shown in Chart 17–4.

Note that the adjustment of price for the effect of the size of the lots considerably improved the relationship between price and elevation. (Compare Chart 17–4 with Chart 17–2B.) The regression line be-

Table 17–4

ADJUSTING PRICE OF LOTS FOR EFFECT OF AREA

Lot No.	X_1 Area, Thousands of Square Feet	Adjustment for Area, $.219 \times X_1$	Y Price, Thousands of Dollars	$Y' = Y - .219X_1$ Adjusted Price, Thousands of Dollars
1	14.7	3.22	4.1	0.88
2	14.2	3.11	3.9	0.79
3	12.7	2.78	3.2	0.42
4	13.8	3.02	2.9	−0.12
5	14.4	3.15	3.9	0.75
6	17.4	3.81	4.1	0.29
7	21.8	4.77	5.8	1.03
8	14.0	3.07	5.1	2.03
9	17.5	3.83	6.8	2.97
10	23.0	5.04	6.8	1.76
11	18.3	4.01	6.5	2.49
12	19.4	4.25	7.0	2.75
13	15.2	3.33	5.8	2.47
14	18.3	4.01	5.1	1.09
15	21.7	4.75	5.3	0.55
16	16.7	3.66	4.9	1.24
17	13.6	2.98	6.0	3.02
18	14.5	3.18	5.3	2.12
19	12.1	2.65	4.8	2.15
20	17.4	3.81	4.3	0.49
			Total	29.17
			Average	1.4585

Chart 17–4

REGRESSION LINE BETWEEN ADJUSTED PRICE AND ELEVATION

REGRESSION EQUATION: $Y'_c = -4.09 + .0317X_2$

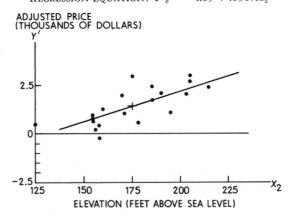

ADJUSTED PRICE
(THOUSANDS OF DOLLARS)
ELEVATION (FEET ABOVE SEA LEVEL)

tween *adjusted* price and elevation is $Y'_c = -4.09 + .0317X_2$. This indicates that the price of a lot increases about \$32 for every foot of elevation—after eliminating the effect of area on price.

We can include the effect of both area and elevation in one equation by taking the term of the first equation that shows the increase in price per unit increase in area and adding it to the second equation, as follows: $Y_c = -4.09 + .219X_1 + .0317X_2$. This is a first approximation to the multiple regression equation.

To refine the estimate, the original price should be adjusted for the effect of elevation (by subtracting .0317 for each foot of elevation). The resulting adjusted price would then be plotted against area (X_1) to obtain a more accurate estimate of the net regression coefficient b_1. After this step, the value of b_2 could be refined, using the improved relationship between Y and X_1. The process could then be repeated until stable values are obtained for b_1 and b_2. Few approximations will be required, however, if the independent variables are not correlated with each other.[1]

Little value can be achieved by following this process further. Our object is merely to describe the graphic method in multiple regression and to clarify the meaning of the net regression coefficient. One can see from this analysis how the value of the net regression coefficient depends upon the other variables in the regression equation.

Finding the Regression Equation by Least Squares

Just as in the case of simple regression analysis, the constants of the linear multiple regression equation are determined by the method of least squares by solving a system of simultaneous linear equations, called the *normal equations,* in which the unknowns are the constants of the regression equation. In order to find the constants in the three-variable linear multiple regression

$$Y_c = a + b_1X_1 + b_2X_2$$

the following three normal equations must be solved:

$$\Sigma Y \quad = na \quad + b_1\Sigma X_1 \quad + b_2\Sigma X_2$$
$$\Sigma X_1Y = a\Sigma X_1 + b_1\Sigma X_1^2 \quad + b_2\Sigma X_1X_2$$
$$\Sigma X_2Y = a\Sigma X_2 + b_1\Sigma X_1X_2 + b_2\Sigma X_2^2$$

---

[1] In this case, the first approximation is very close to the least-squares equation $Y_c = -3.86 + .203X_1 + .0319X_2$, because X_1 and X_2 are uncorrelated. If X_1 and X_2 were highly correlated, a number of successive approximations would be necessary before the graphic fit converged on the least-squares equation. See M. Ezekiel and K. A. Fox, *Methods of Correlation and Regression Analysis* (3d ed.; New York: John Wiley, 1959), chap. 10, also chaps. 14 to 16 for a detailed treatment of multiple curvilinear regression.

These equations can be solved directly, but it is usually simpler to measure each variable as a deviation from its mean, as we did in simple regression. That is, we use small x's and y's, where $x_1 = X_1 - \bar{X}_1$, $x_2 = X_2 - \bar{X}_2$, and $y = Y - \bar{Y}$. This is done most easily by totaling the squares and products of the original X's and Y's, as called for in the above formulas, and then subtracting the *mean times the sum* of the respective variables to get the sums of the small x's and y's as follows:

$$
\begin{array}{cccccc}
\Sigma X_1^2 & \Sigma X_2^2 & \Sigma Y^2 & \Sigma X_1 Y & \Sigma X_2 Y & \Sigma X_1 X_2 \\
-\bar{X}_1 \Sigma X_1 & -\bar{X}_2 \Sigma X_2 & -\bar{Y} \Sigma Y & -\bar{X}_1 \Sigma Y & -\bar{X}_2 \Sigma Y & -\bar{X}_1 \Sigma X_2 \\
\hline
=\Sigma x_1^2 & =\Sigma x_2^2 & =\Sigma y^2 & =\Sigma x_1 y & =\Sigma x_2 y & =\Sigma x_1 x_2
\end{array}
$$

The calculation of the adjusted sums of squares and cross products is shown in Table 17–5 for our example of the price of residential lots.

Table 17–5

MULTIPLE REGRESSION BETWEEN AREA (X_1), ELEVATION (X_2), AND PRICE (Y) OF 20 LOTS

CALCULATIONS OF ADJUSTED SUMS OF SQUARES AND CROSS PRODUCTS

				Symbols					
Sum.........	ΣX_1	ΣX_2	ΣY	ΣX_1^2	ΣX_2^2	ΣY^2	$\Sigma X_1 Y$	$\Sigma X_2 Y$	$\Sigma X_1 X_2$
Mean........	$\bar{X}_1$	$\bar{X}_2$	$\bar{Y}$						
Less*.........				$-\bar{X}_1 \Sigma X_1$	$-\bar{X}_2 \Sigma X_2$	$-\bar{Y} \Sigma Y$	$-\bar{X}_1 \Sigma Y$	$-\bar{X}_2 \Sigma Y$	$-\bar{X}_1 \Sigma X_2$
Gives.........				Σx_1^2	Σx_2^2	Σy^2	$\Sigma x_1 y$	$\Sigma x_2 y$	$\Sigma x_1 x_2$
				Residential Lot Example					
Sum.........	330.7	3501.	101.6	5,657.41	622,729	543.440	1,721.480	18,119.90	57,985.3
Mean........	16.535	175.05	5.08						
Less*.........				−5,468.12	−612.850	−516.128	−1,679.956	−17,785.08	−57,889.0
Gives.........				189.29	9,879	27.312	41.524	334.82	96.3

* Mean times sum.
SOURCE: Table 17–3.

The individual squares and products are not shown because they are usually cumulated in a calculating machine and only the totals need be recorded.[2]

When we express the second and third normal equations in small x's, the terms Σx_1 and Σx_2 equal zero, and the equations become:

---

[2] Since the normal equations for a three-variable problem involve quite a number of sums of squares and products, it is important to choose a system of internal checks when using hand calculators. In this connection a sum variable,

$$X_S = X_1 + X_2 + Y$$

is extremely useful. In addition to the comparatively simple check:

$$\Sigma X_S = \Sigma X_1 + \Sigma X_2 + \Sigma Y$$

the sum of squares of X_S provides the check:

$$\Sigma X_S^2 = \Sigma X_1^2 + \Sigma X_2^2 + \Sigma Y^2 + 2\Sigma X_1 Y + 2\Sigma X_2 Y + 2\Sigma X_1 X_2$$

$$\Sigma x_1 y = b_1 \Sigma x_1^2 + b_2 \Sigma x_1 x_2$$
$$\Sigma x_2 y = b_1 \Sigma x_1 x_2 + b_2 \Sigma x_2^2$$

Substituting the numerical values from Table 17–5, we have:

$$41.524 = 189.29 b_1 + 96.3 b_2$$
$$334.82 = 96.3 b_1 + 9,879 b_2$$

These equations can be solved simultaneously to find b_1 and b_2 as follows: multiply the first equation by $96.3/189.29$, the ratio of the b_1 coefficients. The result is:

$$21.225 = 96.3 b_1 + 48.992 b_2$$

Subtract this from the second normal equation to eliminate b_1. Then,

$$313.695 = 9,830.0 b_2$$

and

$$b_2 = .03191$$

Substitute this value of b_2 in the first normal equation. Solving,

$$b_1 = .2031$$

Finally, substitute both values in the second equation as a check on the arithmetic.

The value of the constant a is

$$a = \bar{Y} - b_1 \bar{X}_1 - b_2 \bar{X}_2$$
$$= 5.080 - (.2031)(16.535) - (.03191)(175.05)$$
$$= -3.864$$

Now, substitute the three constants in the multiple regression equation

$$Y_c = a + b_1 X_1 + b_2 X_2$$
$$= -3.864 + .2031 X_1 + .03191 X_2$$

Thus, for a lot with 15 thousand square feet ($X_1 = 15$) and elevation of 180 feet ($X_2 = 180$), the estimated price would be

$$Y_c = -3.864 + .2031(15) + .03191(180)$$
$$= 4.926 \text{ thousands of dollars, or nearly } \$5,000$$

Curvilinearity. In case any of the variables in multiple regression are believed to have a curvilinear relationship, it is possible to include higher powers of X in the regression equation. The use of such terms is not generally recommended, however, since it is difficult to select the proper algebraic form of the equation from the scatter diagrams of pairs of variables because of the interrelations among the independent variables. Also, more terms might have to be added to the regression equation. It is usually better to transform the data into logarithms, squares, or other functions, if possible, so that linear equations can be fitted by least squares. The use of logarithms is illustrated in the Sears, Roebuck example on pages 478–81. Many computer programs, such as the BMD02R described below, provide a choice of transformations for this purpose.

Dummy Variables. It is possible to include attribute data in multiple regression analysis by the use of dummy variables which take on values of zero and one only. For example, in a study of production orders in a manufacturing plant, one issue might be the effect of designating an order as a "rush" order. This attribute may be included in a regression analysis by adding a dummy variable that has value of one for the rush orders and zero otherwise. The net regression coefficient for this dummy variable would measure the relative effect of the rush order designation.

It is possible to include a factor that has several attribute categories in a similar fashion. For example, suppose a given attribute can be either A, B, or C. To handle this we define two dummy variables (always one less than the number of categories) that take on zero and one values as follows:

Category	First Dummy Variable	Second Dummy Variable
A	0	0
B	1	0
C	0	1

Category A is considered the base of comparison and the regression coefficients of the two variables will then measure the differences between B and A, and between C and A respectively.

Beta Coefficients

In simple regression, the regression line and other calculated values were relatively easy to interpret. In multiple regression, the interpretation is more difficult, since we must sort out the importance of each variable and the interactions between them.

The regression coefficients b_1, b_2, etc. measure the net effect of each variable on the dependent variable Y. But since each of the variables X_1, X_2, etc. may be in different units (in our example X_1 is in thousands of square feet and X_2 is in feet above sea level), it is difficult to ascertain the relative importance of each X in influencing Y. One means of accomplishing this is by using β (beta) coefficients. These are defined to be

$$\beta_1 = b_1\left(\frac{s_{X_1}}{s_Y}\right) = b_1\sqrt{\frac{\Sigma x_1^2}{\Sigma y^2}}$$

$$\beta_2 = b_2\left(\frac{s_{X_2}}{s_Y}\right) = b_2\sqrt{\frac{\Sigma x_2^2}{\Sigma y^2}}$$

The β coefficients are merely the net regression coefficients adjusted by expressing each variable in units of its own standard deviation. This adjustment eliminates the effects of the different size and type of the variables and puts the regression coefficients on a comparable basis. Thus, β shows the change in Y (expressed in standard deviation units) that will be induced by a change of one standard deviation in the independent variable. In our example,

$$\beta_1 = b_1\sqrt{\frac{\Sigma x_1^2}{\Sigma y^2}} = (.2031)\sqrt{\frac{189.29}{27.312}}$$

$$= .535$$

and

$$\beta_2 = b_2\sqrt{\frac{\Sigma x_2^2}{\Sigma y^2}} = (.03191)\sqrt{\frac{9,879}{27.312}}$$

$$= .607$$

That is, for each increase of one standard deviation in X_1 (area) the price increases by .535 standard deviations, while for every increase of one standard deviation in X_2 (elevation) the price increases by .607 standard deviations. The two betas are pure numbers and are comparable. Therefore, elevation is slightly more important than area in explaining the variation in the prices of the lots.

Standard Error of Estimate

Just as in simple regression, the standard error of estimate is in effect the standard deviation of the residuals, $Y - Y_c$. It measures the average scatter of Y values around the regression plane. The standard error of estimate is

$$S_{Y\cdot12} = \sqrt{\frac{\Sigma(Y - Y_c)^2}{n - k}}$$

where n is the number of observations and k is the number of constants in the regression equation. Here, $n = 20$ and $k = 3$. The symbol $S_{Y\cdot12}$

denotes the standard error of estimate of the dependent variable Y regressed against the two independent variables X_1 and X_2.

It is difficult to calculate $\Sigma(Y - Y_c)^2$ directly, so in linear regression we can use the following equivalent formula for computation:

$$S_{Y \cdot 12} = \sqrt{\frac{\Sigma y^2 - b_1 \Sigma x_1 y - b_2 \Sigma x_2 y}{n - k}}$$

In our example,

$$S_{Y \cdot 12} = \sqrt{\frac{27.312 - (.2031)(41.524) - (.03191)(334.82)}{20 - 3}}$$

$$= \sqrt{.4820}$$

$$= .694 \text{ or about } \$700$$

That is, if prices are normally distributed about the regression plane, about two thirds of the prices should fall within $700 of the value estimated from the regression equation (ignoring the minor sampling error in the regression plane itself).

COEFFICIENT OF MULTIPLE DETERMINATION

As in simple correlation, the coefficient of multiple determination is the ratio of explained variance to total variance, or 1 minus the unexplained variance over the total variance. That is,

$$R^2 = 1 - \frac{S^2_{Y \cdot 12}}{s^2_Y}$$

where s_Y^2 is the total variance of the dependent variable Y. In our example, the unexplained variance $(S_{Y \cdot 12}^2)$ was found to be .4820. The estimated total variance (from Table 17–5) is

$$s^2_Y = \frac{\Sigma y^2}{n - 1} = \frac{27.312}{20 - 1} = 1.4375$$

Therefore,

$$R^2 = 1 - \frac{.4820}{1.4375} = .6647$$

About 66 percent of the variance in price, therefore, is explained by the variance in area and elevation of the lots.

The *coefficient of multiple correlation* is the square root of the coefficient of multiple determination. Here,

$$R = \sqrt{.6647} = .815$$

The multiple correlation coefficient is always positive, regardless of the signs of the regression coefficients.

STATISTICAL INFERENCE IN MULTIPLE REGRESSION

When the data used in multiple regression represent a probability sample from some specific population, it is possible to make statistical inferences about the population parameters. In particular, if the population relationship is of the form

$$Y = A + B_1X_1 + B_2X_2 + \epsilon$$

where B_1 and B_2 are the "true" net regression coefficients, A is the true intercept, and ϵ is the residual error or deviation, then the least squares estimates a, b_1, and b_2 are efficient, linear, unbiased estimates of the corresponding population parameters.

The assumptions underlying this estimation procedure are the same as in simple regression, namely,

1. *Linearity:* For fixed values of X_1 and X_2, the mean values of Y lie on a linear plane. This implies $E(\epsilon) = 0$, where $\epsilon = Y - Y_c$.
2. *Independence:* The residuals (ϵ values) are independent of each other.
3. *Uniform Scatter:* The points have a uniform dispersion about the regression plane.
4. *Normality:* The values of ϵ are normally distributed (an assumption not necessary for large samples).

Standard Error of the Regression Coefficient

The regression coefficient b_1 is an estimate of the population parameter B_1. The sampling error associated with this estimate, called the standard error of the regression coefficient, for the case of two independent variables (X_1 and X_2) is

$$s_{b_1} = \frac{S_{Y \cdot 12}}{\sqrt{\Sigma x_1^2(1 - r_{12}^2)}}$$

where r_{12}^2 is the coefficient of determination between X_1 and X_2. Similarly,

$$s_{b_2} = \frac{S_{Y \cdot 12}}{\sqrt{\Sigma x_2^2(1 - r_{12}^2)}}$$

In our example (ignoring the correction for sample bias),

$$r_{12}^2 = \frac{(\Sigma x_1 x_2)^2}{(\Sigma x_1^2)(\Sigma x_2^2)} = \frac{(96.3)^2}{(189.29)(9{,}879)}$$
$$= .0050$$

and the standard errors of the regression coefficients are

$$S_{b_1} = \frac{S_{Y \cdot 12}}{\sqrt{\Sigma x_1^2(1 - r_{12}^2)}} = \frac{.6942}{\sqrt{(189.29)(.995)}}$$
$$= .0506$$

and

$$S_{b_2} = \frac{S_{Y \cdot 12}}{\sqrt{\Sigma x_2^2(1 - r_{12}^2)}} = \frac{.6942}{\sqrt{(9{,}879)(.995)}}$$
$$= .0070$$

We can test the hypothesis that either area or elevation has zero or negative effect (that is, either $B_1 \leq 0$ or $B_2 \leq 0$) by comparing b_1/s_{b_1} or b_2/s_{b_2}. In the case of B_1, the sample value of b_1 is $.2031/.0506 = 4.01$ standard errors away from zero. And the sample value b_2 is $.03191/.0070 = 4.56$ standard errors from a hypothesized $B_2 = 0$. The t value (Appendix M) with $n - k$ degrees of freedom is used to make this test. Here, $n = 20$ and $k = 3$, the total number or constants, so $n - k = 17$. The one-tailed t value at the level of probability is 2.567 for 17 degrees of freedom. Hence, both B_1 and B_2 are significantly greater than zero at the .01 level of significance.

The *standard error of forecast* can be calculated for multiple regression just as in simple regression. See Appendix B at the end of this chapter for the calculations.

USE OF COMPUTER PROGRAMS

In the previous example, the analysis for three variables could be performed easily by hand calculators. With more than three variables, however, the analysis becomes increasingly complicated, since the number of normal equations to be solved for the linear regression equation increases with the number of independent variables. (We cannot visualize a regression plane, as in Chart 17–1, for more than three dimensions, but we can still consider the regression equation as a hyperplane in any number of dimensions.) One solution is to use matrix methods, as described in Appendixes A and B at the end of this chapter. There are also many multiple regression programs available for computers.

We will here describe a typical computer program—specifically the BMD02R multiple regression program[3]—and interpret its printout sheet. This method also illustrates *stepwise* regression, in which the computer carries out the regression by including each independent variable in turn, in order of their importance, so that superfluous variables can be discarded. (In addition, the program permits transformation of variables into logarithms or other functions to achieve linearity, but transformations are not needed here.)

To illustrate this program, we shall expand our illustrative problem.

Table 17–6

CHARACTERISTICS AFFECTING THE PRICE OF 20 LOTS

Lot No.	AREA Thousands of Square Feet X_2	ELEVATION Feet Above Sea Level X_3	SLOPE Degrees X_4	VIEW Scale 1 (Poor) to 9 (Excellent) X_5	PRICE Thousands of Dollars X_1
1	14.7	155	1.5	2	4.1
2	14.2	155	1.8	2	3.9
3	12.7	158	2.9	1	3.2
4	13.8	158	1.0	1	2.9
5	14.4	155	0.5	2	3.9
6	17.4	157	1.0	2	4.1
7	21.8	172	5.7	4	5.8
8	14.0	170	5.4	6	5.1
9	17.5	175	17.5	9	6.8
10	23.0	185	14.5	9	6.8
11	18.3	185	14.4	9	6.5
12	19.4	205	12.2	9	7.0
13	15.2	215	5.0	8	5.8
14	18.3	195	13.1	6	5.1
15	21.7	178	15.2	8	5.3
16	16.7	160	10.1	8	4.9
17	13.6	205	7.4	7	6.0
18	14.5	190	5.8	7	5.3
19	12.1	203	5.1	7	4.8
20	17.4	125	17.3	1	4.3
Total	330.7	3501.	157.4	108.	101.6
Mean	16.535	175.05	7.87	5.40	5.08

---

[3] Described in *BMD Biomedical Computer Programs*, Health Services Computing Facility, School of Medicine, University of California, Los Angeles, January 1, 1964, pp. 233–53. The program output is modified to eliminate some detail and certain statistical measures that are not explained in this text.

Table 17–7

```
BMDC2R - STEPWISE REGRESSION
HEALTH SCIENCES COMPUTING FACILITY, UCLA

PROBLEM CODE                    PRICE
NUMBER OF CASES                  20
NUMBER OF ORIGINAL VARIABLES      5
NUMBER OF VARIABLES ADDED         0
TOTAL NUMBER CF VARIABLES         5
```

VARIABLE		MEAN	STANDARC DEVIATION
PRICE	1	5.08000	1.19895
AREA	2	16.53500	3.15633
ELEVTN	3	175.05000	22.80229
SLOPE	4	7.87000	5.87198
VIEW	5	5.40000	3.13553

CORRELATION MATRIX

VARIABLE NUMBER	1	2	3	4	5
1	1.000	0.578	0.645	0.664	0.879
2		1.000	0.070	0.630	0.396
3			1.000	0.152	0.749
4				1.000	0.608
5					1.000

Suppose that our real estate broker has made estimates of the slope (in degrees) of each lot and has ranked the view on a scale from 1 (poor) to 9 (excellent), in addition to the area, elevation, and price shown in Table 17–3. The results are presented in Table 17–6. We now wish to estimate the importance of each factor in determining the price of a lot.

The BMD program assigns the numbers 1 through 5 to our variables: price, area, elevation, slope, and view. (These numbers differ from the subscripts used above.) The printout in Table 17–7 first shows the means and standard deviations of each variable.[4]

The "correlation matrix" then shows the coefficient of simple correlation between each pair of variables. Note that all the variables are positively related to the dependent variable—price—with correlation coefficients ranging from .578 to .879 (though the net effect of slope (X_1) on price should be negative).

In the stepwise procedure, the program first calculates the simple regression between price and the independent variable that explains the greatest part of the variation in price (the dependent variable). In this

---

[4] The standard deviations, variances, and correlation coefficients in this program are sample values, not adjusted for degrees of freedom. The standard error of estimate, however, *is* adjusted.

Table 17–7—Continued

```
STEP NUMBER     1
VARIABLE ENTERED     5

MULTIPLE R                  0.8787
STD. ERROR OF EST.          0.5881

                    VARIABLES IN EQUATION          VARIABLES NOT IN EQUATION

        VARIABLE    COEFFICIENT  STD. ERROR     VARIABLE      PARTIAL CORR.


        (CONSTANT    3.26574 )
        VIEW    5    0.33597      0.04303       AREA     2      0.52309
                                               ELEVTN   3     -0.04302
                                               SLOPE    4      0.34439




STEP NUMBER     2
VARIABLE ENTERED     2

MULTIPLE R                  0.9135
STD. ERROR OF EST.          0.5158

                    VARIABLES IN EQUATION          VARIABLES NOT IN EQUATION

        VARIABLE    COEFFICIENT  STD. ERROR     VARIABLE      PARTIAL CORR.


        (CONSTANT    1.77976 )
        AREA    2    0.10333      0.04083       ELEVTN   3      0.19185
        VIEW    5    0.29475      0.04110       SLOPE    4      0.09071




STEP NUMBER     3
VARIABLE ENTERED     3

MULTIPLE R                  0.9168
STD. ERROR OF EST.          0.5218

                    VARIABLES IN EQUATION          VARIABLES NOT IN EQUATION

        VARIABLE    COEFFICIENT  STD. ERROR     VARIABLE      PARTIAL CORR.


        (CONSTANT    0.62111 )
        AREA    2    0.11629      0.04451       SLOPE    4      0.21297
        ELEVTN  3    0.00668      0.00854
        VIEW    5    0.25321      0.06746
```

case the variable "view" (number 5) is first included, since $r_{15} = .879$ —the highest value in the top row of the correlation matrix. The next lines show this value, the standard error of estimate (.5881), the co-efficients a and b_5, and the standard error of the latter.

In the next step, a second independent variable is included in the regression. The factor chosen is the one that makes the greatest additional contribution to explained variance. The right-hand column

Table 17–7—Concluded

```
STEP NUMBER      4
VARIABLE ENTERED     4

MULTIPLE R               0.9207
STD. ERROR OF EST.       0.5265
```

VARIABLES IN EQUATION			VARIABLES NOT IN EQUATION	
VARIABLE	COEFFICIENT	STD. ERROR	VARIABLE	PARTIAL CORR.
(CONSTANT	0.24021)			
AREA 2	0.09873	0.04950		
ELEVTN 3	0.01068	0.00983		
SLOPE 4	0.02950	0.03494		
VIEW 5	0.20487	0.08896		

LIST OF RESIDUALS

CASE	RESIDUAL		
1	0.29968	11	0.20937
2	0.14019	12	0.45214
3	-0.27132	13	-0.02269
4	-0.62388	14	-0.64444
5	0.15879	15	-1.07031
6	0.02650	16	-0.63405
7	0.58357	17	0.57611
8	0.27414	18	-0.00541
9	0.60367	19	-0.38660
10	0.04239	20	0.29218

labeled "Partial Correlation" or *partial correlation coefficient* gives an indication at each stage of the relative importance of each of the variables not yet in the regression equation.[5] Thus, the partial correlation coefficient indicates which variable would have the greatest effect (in reducing unexplained variance) if added to the regression. In this step, the variable "area" (number 2) is added. This is helpful, since it reduces the error of estimating lot prices to .5158 thousand dollars.

Variables 3 and 4 (elevation and slope) are then added in turn, but they serve only to increase the standard error of estimate to .5218 and .5265, respectively.[6] (This is reason enough to drop these variables). At the end of step 4, all variables are included in the regression equation.

---

[5] The square of the partial correlation coefficient measures the increase in explained variance from the addition of a given variable, relative to the variance remaining to be explained before the variable was added. See Ezekiel and Fox, *op. cit.,* pp. 192–96, for further explanation.

[6] Paradoxically, the "multiple *R*" also increases, but only because this program fails to correct *R* for loss of degrees of freedom.

Chart 17–5

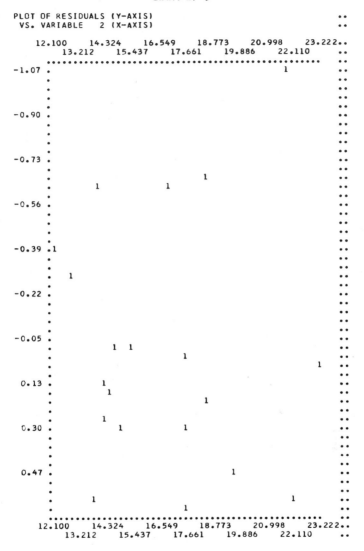

The "List of Residuals" gives the variation in price of each lot not explained by the multiple regression equation. As an optional feature, the computer will plot these residual terms against each of the independent variables. Such a plot is shown in Chart 17–5 for variable 2 (area) and is a useful check on the assumptions of linearity and homoscedasticity (uniform scatter). The scatter seems approximately uniform over the range of the independent variable, and there is no evidence of curvilinearity. (The same is true of the other three plots, not shown.) Hence, we can conclude that the linear and homoscedastic assumptions are satisfied (though the sample size of 20 is too small for us to be certain).

Tests of Significance. The inclusion of the standard errors of the net regression coefficients makes it possible to test for their significance. In particular, we can test whether each coefficient is significantly above zero. The test is performed using the t value (Appendix M) with $(n - k)$ degrees of freedom, where k is the number of constants. For $20 - 5 = 15$ degrees of freedom, the one-tailed t value at the .05 level is 1.753. The variable "view" is significant at this level, since the regression coefficient is 2.30 standard errors ($.20487/.08896 = 2.30$) from zero. And "area" is also significant ($.09873/.04950 = 1.99$). However, neither "elevation" nor "slope" is close to significance at the .05 level (for elevation, $.01068/.00983 = 1.10$; for slope, $.02950/.03494 = .844$. "Elevation" need not be retained on logical grounds, since "view" measures this factor better; and "slope" should have a *negative* effect on price. These are additional reasons to discard these factors and express price as a function of just area and view (Table 17–7):

$$\text{Price} = 1.77976 + .10333 \times \text{area} + .29475 \times \text{view}$$

in thousands of dollars, with standard error of estimate $516.

CAUTIONS IN THE USE OF MULTIPLE REGRESSION

Basic Assumptions

The use of multiple regression formulas in making inferences implies the assumptions that the residuals $\epsilon = Y - Y_c$ are (1) clustered around a linear (not curved) plane, (2) independent of each other, (3) uniform in their scatter, and, for small samples, (4) normally distributed. If these assumptions are not valid, conclusions from multiple regression analysis may be very misleading. Yet they are often overlooked because of the ease in running a computer program and the difficulty of checking the assumptions mathematically. A simple graphic check is to first plot the original variables against each other, as in Chart 17–2, and then, after running the program, to plot the residuals against each independent variable, as in Chart 17–5. The residuals can then be checked visually for these conditions.

The same distinction should be made between the regression model and the correlation model as in simple correlation (see Chapter 16).

A second major source of error in using regression analysis is to extrapolate beyond the range of the data upon which the regression equation was estimated. The equation by itself gives no indication of what lies outside the range of its data—the surface may become curvilinear, for example. Nevertheless, it is sometimes necessary to extrapolate, such as when we make economic forecasts or apply a relationship for one region to another comparable region. For such a projection to be valid, it is essential that the pertinent economic conditions in the extrapolated period or region be essentially similar to those on which the regression analysis was based.

Colinearity

When the independent variables in a multiple regression are highly correlated with each other, the net regression coefficients may be unreliable.[7] This can be seen easily from the formula for the standard error of the regression coefficient in the case of two independent variables:

$$s_{b_1} = \frac{S_{Y \cdot 12}}{\sqrt{\Sigma x_1^2 (1 - r_{12}^2)}}$$

where r_{12}^2 is the coefficient of determination between the independent variables X_1 and X_2.

The standard error is smallest when r_{12}^2 is zero, but as r_{12}^2 approaches one (perfect correlation), the denominator of the equation approaches zero, and the standard error becomes very large, so the regression coefficient itself becomes unreliable. Hence, the standard error is sensitive to the *colinearity* or correlation between X_1 and X_2. This accords with common sense: if X_1 and X_2 move together, it is difficult to distinguish their separate effects on Y. One solution is simply to drop the X that is deemed less important, unless there is a strong logical reason to retain it.[8]

---

[7] For further discussion, see J. Johnston, *Econometric Methods* (2d ed.; New York: McGraw-Hill, 1972), p. 160.

[8] The effects of colinearity may be seen in the computer regression example (Table 17–7). The correlation between elevation (X_3) and view (X_5) is .749 and between slope (X_4) and view (X_5) is .608. Note what happens to the standard error of X_5 as these other two variables are entered in the regression equation. In step 3, the standard error of X_5 increases from .041 to .067 as X_3 is included; it further increases to .089 as X_4 is included in step 4.

While colinearity affects the reliability of *individual* variables in the regression, it may not alter the predictive power of the *total* regression equation. That is, the standard error of estimate may not be increased. The sampling errors of the regression coefficients tend to compensate for each other in the estimate of the dependent variable. Similarly, the sampling error of R^2 is not sensitive to colinearity among the independent variables.

Colinearity may produce some peculiar results in regression analysis besides its effect upon the sampling error of the net regression coefficients. For example, two variables X_1 and X_2 may be highly positively correlated with Y and with each other. But the net effect of X_2, taking X_1 into account, may be negative. This is illustrated in the Sears, Roebuck example below.

Regression of Time Series

Particular caution is needed in dealing with time series. In Chapter 16 we correlated Sears, Roebuck sales with U.S. disposable income for the years 1953–71. In multiple regression, we can also include the number of Sears stores at the beginning of each year and time (years) to see if we can improve the forecast. Logarithms are used, except for time, because (1) the logs of sales and of income have a linear relation (Chapter 16), (2) the log of sales has a linear relation with time itself (Chapter 19), and (3) the logarithmic residuals have a more uniform scatter than the absolute residuals (which increase with the growth of sales), as required in least-squares theory.

Unfortunately the regression coefficient of each of the new variables is negative (and not significantly different from zero). This is illogical, since it implies that for a given income level, the addition of new stores (or the passage of time) would depress sales. Nor do the new variables serve to reduce the standard error of estimate. The villain is colinearity. Hence we drop these variables and revert to the simple regression equation.

This example illustrates three pitfalls of multiple regression, and how to avoid them:

1. Colinearity, avoided by dropping superfluous variables.
2. Curvilinearity, rectified by using logarithms. This transformation also produces a more uniform scatter of residuals (ϵ).
3. Autocorrelation in time series, which invalidates the standard error

of estimate and related measures. This may sometimes be remedied by using yearly percentage changes, as described on page 482.

SUMMARY

Multiple regression measures the simultaneous influence of a number of independent variables upon one dependent variable. A net regression coefficient (e.g., b_1) measures the effect upon the dependent variable of a unit increase in an independent variable, holding the other independent variables constant. The regression equation represents a plane in three-dimensional space or a hyperplane in more than three dimensions.

The multiple regression equation can be estimated either graphically or by least squares. In the graphic method, the dependent variable is first plotted against one of the independent variables and a freehand regression curve drawn; then the vertical residuals from this curve ($\epsilon = Y - Y_c$) are plotted above and below the zero line, with the second independent variable as abscissa. A second curve is drawn, and the residuals from this curve are in turn plotted against a third independent variable (if any) or else are laid off around the first regression curve. The curve is redrawn, and this process is refined by transferring the residuals back and forth until no further improvement occurs in the net regression curves.

The least-squares method can be performed on a hand calculator for three variables, but for more variables it is preferable to use matrix methods (described in the appendixes of the chapter) or a computer program. To calculate the least-squares equation, a set of normal equations must be solved. To make this easier, the sums of the squares and cross products of the variables are adjusted by subtracting the mean times the sum of the appropriate variables to reduce them to deviations from their means.

Curvilinear relations can best be handled by transforming the variables into logarithms or some other function that will make the relationship linear, and then fitting a linear regression equation by least squares.

The standard error of estimate is essentially the standard deviation of the residuals $\epsilon = Y - Y_c$ about the regression plane. And the coefficient of multiple determination is the proportion of the variance of the dependent variable explained by the independent variables. Its square root is the coefficient of multiple correlation. These concepts are equivalent to those in simple correlation.

When the assumptions of linearity, uniform scatter, independence, and normality are satisfied, it is possible to measure the sampling error of the net regression coefficients. These measures can then be used to make statistical inferences about the true regression relationships.

The net regression coefficients can be expressed in common standard-deviation units by multiplying each one by the standard deviation of the appropriate independent variable over the standard deviation of the dependent variable. These β coefficients may be compared for different independent variables, revealing the relative importance of each variable in the regression equation.

Computer programs are widely available for multiple regression analysis; a typical program is described.

Before using multiple regression results, it is important to check the assumptions upon which the analysis is based. Plots of the original variables and the final residuals versus the independent variables provide a graphic check on these assumptions.

Colinearity or correlation between independent variables reduces the reliability of the net regression coefficients, but it may not affect the predictability of the overall regression equation.

Time series involve particular problems. The Sears, Roebuck case illustrates methods of dealing with autocorrelation of residuals, as well as colinearity and curvilinearity.

APPENDIX A: INTRODUCTION TO MATRIX OPERATIONS

Definition of a Matrix

A matrix is a rectangular array of elements (numbers or symbols). An example of a matrix, denoted by the symbol A, is shown below:

$$A = \begin{bmatrix} a_{11} & a_{12} & a_{13} & a_{14} \\ a_{21} & a_{22} & a_{23} & a_{24} \\ a_{31} & a_{32} & a_{33} & a_{34} \end{bmatrix}$$

This matrix is the array of the symbols a_{11} through a_{34}. It has three rows and four columns. Each symbol a_{ij} refers to the element in the ith row and the jth column. A matrix is *rectangular*, indicating that it has the same number of elements in each row and in each column (although the number of rows may not necessarily equal the number of columns).

A matrix with only one row or column is usually called a *vector*. The vector $[a_1, a_2, a_3, \ldots, a_n]$ is an example of a *row* vector (one row), and

$$\begin{bmatrix} a_1 \\ a_2 \\ a_3 \\ \cdot \\ \cdot \\ \cdot \\ a_n \end{bmatrix}$$

is an example of a *column* vector.

The number of rows and columns define the *dimensions* of a matrix. A matrix with 3 rows and 4 columns is said to have dimension 3×4 or, more simply, is a 3×4 matrix. A matrix with the same number of rows and columns is a *square* matrix.

Addition and Subtraction of Matrices

Two matrices may be added (or subtracted) simply by adding (or subtracting) the corresponding elements on an element-by-element basis. That is, the element in the first row and column of A is added to (or subtracted from) the element in the first row and column of B and so on. However, in order to add (or subtract) the matrices, they must be of the same dimensions.

Using an example, if

$$C = \begin{bmatrix} 2 & 4 \\ 3 & 1 \end{bmatrix} \quad \text{and} \quad D = \begin{bmatrix} 1 & 3 \\ 2 & 0 \end{bmatrix}$$

then

$$C - D = \begin{bmatrix} 2 & 4 \\ 3 & 1 \end{bmatrix} - \begin{bmatrix} 1 & 3 \\ 2 & 0 \end{bmatrix} = \begin{bmatrix} 1 & 1 \\ 1 & 1 \end{bmatrix}$$

The Transpose of a Matrix

The transpose of a matrix A (the transpose is designated A') is obtained by interchanging the rows and columns. Thus, for

$$A = \begin{bmatrix} a_{11} & a_{12} \\ a_{21} & a_{22} \\ a_{31} & a_{32} \end{bmatrix} \quad (3 \times 2 \text{ matrix})$$

the transpose

$$A' = \begin{bmatrix} a_{11} & a_{21} & a_{31} \\ a_{12} & a_{22} & a_{32} \end{bmatrix} \quad (2 \times 3 \text{ matrix})$$

Using a numerical example, if

$$B = \begin{bmatrix} 2 & 1 \\ 3 & 4 \end{bmatrix} \quad \text{then} \quad B' = \begin{bmatrix} 2 & 3 \\ 1 & 4 \end{bmatrix}$$

The use of the transpose operation converts a row vector into a column vector and vice versa.

Matrix Multiplication

Matrices may also be multiplied. The rules for matrix multiplication, however, are more complicated than matrix addition. Consider the matrices

$$A = \begin{bmatrix} a_{11} & a_{12} & a_{13} \\ a_{21} & a_{22} & a_{23} \end{bmatrix} \qquad B = \begin{bmatrix} b_{11} & b_{12} \\ b_{21} & b_{22} \\ b_{31} & b_{32} \end{bmatrix}$$

The product $A \times B$ is

$$A \times B = \begin{bmatrix} a_{11} & a_{12} & a_{13} \\ a_{21} & a_{22} & a_{23} \end{bmatrix} \times \begin{bmatrix} b_{11} & b_{12} \\ b_{21} & b_{22} \\ b_{31} & b_{32} \end{bmatrix}$$

$$= \begin{bmatrix} (a_{11}b_{11} + a_{12}b_{21} + a_{13}b_{31}) & (a_{11}b_{12} + a_{12}b_{22} + a_{13}b_{32}) \\ (a_{21}b_{11} + a_{22}b_{21} + a_{23}b_{31}) & (a_{21}b_{12} + a_{22}b_{22} + a_{23}b_{32}) \end{bmatrix}$$

That is, the element in the first row, first column, of the product matrix $(A \times B)$ is obtained by multiplying and then summing the elements of the first row in A and the first column in B; the element in the first row, second column, of the product matrix $(A \times B)$ is obtained by multiplying and then summing the elements of the first row of A and the second column of B; the element in the second row, first column, of $(A \times B)$ is obtained by multiplying and then summing the elements of the second row of A and the first column of B; and so on.

A numerical example will help to illustrate matrix multiplication:

$$C = \begin{bmatrix} 5 & 3 \\ 2 & -1 \\ 1 & 0 \end{bmatrix} \qquad D = \begin{bmatrix} 2 & 1 & 2 \\ 5 & 4 & 6 \end{bmatrix}$$

$$C \times D = \begin{bmatrix} (5 \cdot 2 + & 3 \cdot 5 = & 25) & (5 \cdot 1 + & 3 \cdot 4 = & 17) \\ (2 \cdot 2 + (-1) \cdot 5 = & -1) & (2 \cdot 1 + (-1) \cdot 4 = & -2) \\ (1 \cdot 2 + & 0 \cdot 5 = & 2) & (1 \cdot 1 + & 0 \cdot 4 = & 1) \end{bmatrix}$$

$$\begin{bmatrix} (5 \cdot 2 + & 3 \cdot 6 = & 28) \\ (2 \cdot 2 + (-1) \cdot 6 = & -2) \\ (1 \cdot 2 + & 0 \cdot 6 = & 2) \end{bmatrix}$$

$$= \begin{bmatrix} 25 & 17 & 28 \\ -1 & -2 & -2 \\ 2 & 1 & 2 \end{bmatrix}$$

Dimensions. In order to multiply two matrices, the number of columns in the first matrix must equal the number of rows in the second. Otherwise, multiplication is not defined. The product matrix has the same number of rows as the first matrix and the same number of columns as the second matrix.

For example, a (2×4) matrix (2 rows, 4 columns) can be multiplied by a (4×3) matrix, resulting in a (2×3) matrix:

$$[\text{that is, } (2 \times 4) \times (4 \times 3) \to (2 \times 3)].$$

Note that a (2×4) matrix cannot be multiplied by another (2×4) matrix.

Order of Multiplication. In ordinary multiplication, the order is not important. That is, 5 times 2 gives the same result as 2 times 5. In matrix multiplication, however, the order in which the matrices are multiplied makes a difference. The matrix multiplication $A \times B$ generally does not give the same result as $B \times A$. For example, if

$$A = \begin{bmatrix} 2 & 1 \\ 0 & 1 \end{bmatrix} \quad \text{and} \quad B = \begin{bmatrix} 3 & 0 \\ 1 & 4 \end{bmatrix}$$

then

$$(A \times B) = \begin{bmatrix} 7 & 4 \\ 1 & 4 \end{bmatrix} \quad \text{but} \quad (B \times A) = \begin{bmatrix} 6 & 3 \\ 2 & 5 \end{bmatrix}$$

Hence, when two matrices are to be multiplied it is important to indicate which matrix is on the left (or is first) and which is on the right (or is second).

The Identity Matrix. The identity matrix is a square matrix containing ones along the diagonal and zeros elsewhere. It is usually designated by the symbol I. When the identity matrix is multiplied (either from the left or right) by another matrix of the same dimensions, the result is the original matrix. For example,

$$A = \begin{bmatrix} 2 & 1 \\ 0 & 1 \end{bmatrix} \quad \text{and} \quad I = \begin{bmatrix} 1 & 0 \\ 0 & 1 \end{bmatrix}$$

$$(A \times I) = (I \times A) = \begin{bmatrix} 2 & 1 \\ 0 & 1 \end{bmatrix} = A$$

Matrix Inversion

The inverse of a square matrix A is defined to be a matrix A^{-1} such that

$$A \times A^{-1} = I$$

That is, the product of a matrix times its inverse is the identity matrix I. The inverse for a given matrix may not always exist.[9] But if it does, the inverse A^{-1} may be multiplied by A from either the left or right and will produce the identity matrix. That is,

$$A \times A^{-1} = A^{-1} \times A = I$$

There are several ways to calculate the inverse of a given matrix. We shall present a simple method here without explaining the rationale. The reader is referred to advanced texts for more detail. In general, the calculation of inverses of large matrices (larger than 3×3) is tedious work and should be left to electronic computers.

We start the calculation of the inverse by setting up the matrix to be inverted side by side with the identity matrix. Suppose we wish to invert

$$A = \begin{bmatrix} 5 & 2 \\ 1 & 3 \end{bmatrix} \qquad \text{We set up} \qquad \begin{bmatrix} 5 & 2 \\ 1 & 3 \end{bmatrix} \begin{bmatrix} 1 & 0 \\ 0 & 1 \end{bmatrix}$$

We can then perform any of the following operations on this set of matrices:

1. Multiply any row by a constant.
2. Add (or subtract) any row from another.
3. Multiply a row by a constant and simultaneously add (or subtract) it from another row (a combination of a and b).

Using the operations 1, 2, and 3, the object is to reduce the set of matrices so that the first is in the form of the identity matrix. The second will then be the desired matrix inverse. That is, we wish to arrive at

$$\begin{bmatrix} 1 & 0 \\ 0 & 1 \end{bmatrix} \begin{bmatrix} c_{11} & c_{12} \\ c_{21} & c_{22} \end{bmatrix}$$

where $\begin{bmatrix} c_{11} & c_{12} \\ c_{21} & c_{22} \end{bmatrix}$ is the inverse matrix of $\begin{bmatrix} 5 & 2 \\ 1 & 3 \end{bmatrix}$, our original matrix.

To accomplish this, we proceed as follows: The original matrices are

$$\begin{bmatrix} 5 & 2 \\ 1 & 3 \end{bmatrix} \begin{bmatrix} 1 & 0 \\ 0 & 1 \end{bmatrix}$$

Step 1: Multiply the first row by $\frac{1}{5}$ (using rule 1). This gives

---

[9] A matrix will not have a unique inverse if, for example, two rows are the same. See D. Teichroew, *Introduction to Science in Management* (New York: John Wiley, 1964), chap. 13.

$$\begin{bmatrix} 1 & \frac{2}{5} \\ 1 & 3 \end{bmatrix}\begin{bmatrix} \frac{1}{5} & 0 \\ 0 & 1 \end{bmatrix}$$

Step 2: Subtract row 1 from row 2 (using rule 2). This gives

$$\begin{bmatrix} 1 & \frac{2}{5} \\ 0 & 2\frac{3}{5} \end{bmatrix}\begin{bmatrix} \frac{1}{5} & 0 \\ -\frac{1}{5} & 1 \end{bmatrix}$$

Step 3: Multiply the second row by $1/(2\frac{3}{5})$ or $\frac{5}{13}$ (rule 1). This gives

$$\begin{bmatrix} 1 & \frac{2}{5} \\ 0 & 1 \end{bmatrix}\begin{bmatrix} \frac{1}{5} & 0 \\ -\frac{1}{13} & \frac{5}{13} \end{bmatrix}$$

Step 4: Simultaneously multiply row 2 by $\frac{2}{5}$ and subtract it from row 1 (rule 3). This gives

$$\begin{bmatrix} 1 & 0 \\ 0 & 1 \end{bmatrix}\begin{bmatrix} (\frac{1}{5} - (-\frac{1}{13})(\frac{2}{5})) & 0 - (\frac{5}{13})(\frac{2}{5}) \\ -\frac{1}{13} & \frac{5}{13} \end{bmatrix}$$

or

$$\begin{bmatrix} 1 & 0 \\ 0 & 1 \end{bmatrix}\begin{bmatrix} \frac{3}{13} & -\frac{2}{13} \\ -\frac{1}{13} & \frac{5}{13} \end{bmatrix}$$

Hence,

$$\begin{bmatrix} \frac{3}{13} & -\frac{2}{13} \\ -\frac{1}{13} & \frac{5}{13} \end{bmatrix} \text{ is the desired inverse of } \begin{bmatrix} 5 & 2 \\ 1 & 3 \end{bmatrix}$$

To verify this result we multiply

$$\begin{bmatrix} 5 & 2 \\ 1 & 3 \end{bmatrix} \times \begin{bmatrix} \frac{3}{13} & -\frac{2}{13} \\ -\frac{1}{13} & \frac{5}{13} \end{bmatrix}$$

which gives $\begin{bmatrix} 1 & 0 \\ 0 & 1 \end{bmatrix}$ and is a check on our calculations.

Solution of Simultaneous Equations Using Matrices

Simultaneous equations may be solved by the use of matrices. For example, suppose we had the following three equations with three unknowns:

$$\begin{aligned} 5x_1 + 2x_2 + x_3 &= 10 \\ 3x_2 + 2x_3 &= 8 \\ 4x_1 \qquad\quad + x_3 &= 5 \end{aligned}$$

This set of equations may be expressed in matrix notation as

$$\begin{bmatrix} 5 & 2 & 1 \\ 0 & 3 & 2 \\ 4 & 0 & 1 \end{bmatrix} \times \begin{bmatrix} x_1 \\ x_2 \\ x_3 \end{bmatrix} = \begin{bmatrix} 10 \\ 8 \\ 5 \end{bmatrix}$$

or, letting

$$A = \begin{bmatrix} 5 & 2 & 1 \\ 0 & 3 & 2 \\ 4 & 0 & 1 \end{bmatrix}, X = \begin{bmatrix} x_1 \\ x_2 \\ x_3 \end{bmatrix} \text{ and } B = \begin{bmatrix} 10 \\ 8 \\ 5 \end{bmatrix}$$

we can write

$$A \times X = B$$

Multiplying both sides of this equation by A^{-1} (A inverse) we have[10]

$$A^{-1} \times A \times X = A^{-1} \times B$$

But since $A^{-1} \times A = I$, and $I \times X = X$, we have $X = A^{-1} \times B$.

This, in matrix form, is the solution of our equation. All that is needed is A^{-1}, the matrix inverse. Here the inverse of

$$A = \begin{bmatrix} 5 & 2 & 1 \\ 0 & 3 & 2 \\ 4 & 0 & 1 \end{bmatrix}$$

is

$$A^{-1} = \begin{bmatrix} \frac{3}{19} & -\frac{2}{19} & \frac{1}{19} \\ \frac{8}{19} & \frac{1}{19} & -\frac{10}{19} \\ -\frac{12}{19} & \frac{8}{19} & \frac{15}{19} \end{bmatrix}$$

and the product

$$A^{-1} \times B \text{ is } \begin{bmatrix} \frac{3}{19} & -\frac{2}{19} & \frac{1}{19} \\ \frac{8}{19} & \frac{1}{19} & -\frac{10}{19} \\ -\frac{12}{19} & \frac{8}{19} & \frac{15}{19} \end{bmatrix} \times \begin{bmatrix} 10 \\ 8 \\ 5 \end{bmatrix} = \begin{bmatrix} 1 \\ 2 \\ 1 \end{bmatrix}$$

Since

$$X = A^{-1} \times B, \begin{bmatrix} x_1 \\ x_2 \\ x_3 \end{bmatrix} = \begin{bmatrix} 1 \\ 2 \\ 1 \end{bmatrix} \text{ or } x_1 = 1; x_2 = 2; x_3 = 3$$

This procedure will be applied to regression analysis in Appendix B.

APPENDIX B: MATRIX SOLUTION TO MULTIPLE REGRESSION ANALYSIS

In multiple regression analysis, we must solve the set of normal equations for the values of net regression coefficients. For the case of two independent variables expressed as deviations from their means, the normal equations are

---

[10] Care must be taken to multiply from the same side in both cases.

$$\Sigma y x_1 = b_1 \Sigma x_1^2 + b_2 \Sigma x_1 x_2$$
$$\Sigma y x_2 = b_1 \Sigma x_1 x_2 + b_2 \Sigma x_2^2$$

This can be written in matrix notation as

$$Y = X \times B \qquad \text{where}$$

Y is the vector $\begin{bmatrix} \Sigma y x_1 \\ \Sigma y x_2 \end{bmatrix}$

B is the vector of unknown coefficients $B = \begin{bmatrix} b_1 \\ b_2 \end{bmatrix}$

X is the matrix of sums of squares and cross products $\begin{bmatrix} \Sigma x_1^2 & \Sigma x_1 x_2 \\ \Sigma x_1 x_2 & \Sigma x_2^2 \end{bmatrix}$

In the general case of m independent variables, the normal equations are

$$\Sigma y x_1 = b_1 \Sigma x_1^2 + b_2 \Sigma x_1 x_2 + b_3 \Sigma x_1 x_3 + \cdots b_m \Sigma x_1 x_m$$
$$\Sigma y x_2 = b_1 \Sigma x_1 x_2 + b_2 \Sigma x_2^2 + b_3 \Sigma x_2 x_3 + \cdots b_m \Sigma x_2 x_m$$
$$\Sigma y x_3 = b_1 \Sigma x_1 x_3 + b_2 \Sigma x_2 x_3 + b_3 \Sigma x_3^2 + \cdots b_m \Sigma x_3 x_m$$
$$\vdots \qquad \vdots \qquad \vdots \qquad \vdots \qquad \vdots$$
$$\Sigma y x_m = b_1 \Sigma x_1 x_m + b_2 \Sigma x_2 x_m + b_3 \Sigma x_3 x_m + \cdots b_m \Sigma x_m^2$$

$$\text{Letting } Y = \begin{bmatrix} \Sigma y x_1 \\ \Sigma y x_2 \\ \Sigma y x_3 \\ \vdots \\ \Sigma y x_m \end{bmatrix}, B = \begin{bmatrix} b_1 \\ b_2 \\ b_3 \\ \vdots \\ b_m \end{bmatrix}$$

and

$$X = \begin{bmatrix} \Sigma x_1^2 & \Sigma x_1 x_2 & \Sigma x_1 x_3 & \cdots \Sigma x_1 x_m \\ \Sigma x_1 x_2 & \Sigma x_2^2 & \Sigma x_2 x_3 & \cdots \Sigma x_2 x_m \\ \Sigma x_1 x_3 & \Sigma x_2 x_3 & \Sigma x_3^2 & \cdots \Sigma x_3 x_m \\ \vdots & \vdots & \vdots & \vdots \\ \Sigma x_1 x_m & \Sigma x_2 x_m & \Sigma x_3 x_m & \cdots \Sigma x_m^2 \end{bmatrix}$$

The normal equations are expressed in matrix form, as before, $Y = X \times B$.

To solve this set of equations we need the inverse of the sums of squares and cross products matrix X. And the solution is

$$B = X^{-1} \times Y$$

where X^{-1} is the required inverse.

Example

Using the illustration from page 504, the matrix of sums of squares and cross products is

$$X = \begin{bmatrix} 189.29 & 96.3 \\ 96.3 & 9879.0 \end{bmatrix}$$

Using the procedures described in Appendix A, we find the inverse matrix to be

$$X^{-1} = \begin{bmatrix} .0053092 & -.000051754 \\ -.000051754 & .00010173 \end{bmatrix}$$

Multiplying this by the Y vector we have

$$B = X^{-1} \times Y = \begin{bmatrix} .0053092 & -.000051754 \\ -.000051754 & .00010173 \end{bmatrix} \times \begin{bmatrix} 41.524 \\ 334.82 \end{bmatrix}$$

$$\text{or } B = \begin{bmatrix} b_1 \\ b_2 \end{bmatrix} = \begin{bmatrix} .2031 \\ .03191 \end{bmatrix}$$

or $b_1 = .2031$ and $b_2 = .03191$ as in this chapter.

Standard Error of Regression Coefficients

We shall first designate the individual elements of the inverse matrix X^{-1} by the symbols c_{ij}. Thus,

$$\begin{bmatrix} c_{11} & c_{12} \\ c_{21} & c_{22} \end{bmatrix}$$

is the representation of the inverse above, where $c_{11} = .0053092$; $c_{12} = c_{21} = -.000051754$; and $c_{22} = .00010173$.

Note that $c_{ij} = c_{ji}$ (here, $c_{12} = c_{21}$). A matrix with this property is called *symmetrical*. Note that both X and X^{-1} are symmetrical.

The standard errors of the net regression coefficients can be estimated as functions of the diagonal elements of the inverse matrix.

In the general case,

$$S_{b_j} = S_{Y \cdot 123 \ldots m} \sqrt{c_{jj}}$$

In our example,

$$S_{b_1} = S_{Y \cdot 12} \sqrt{c_{11}}$$
$$S_{b_2} = S_{Y \cdot 12} \sqrt{c_{22}}$$

or

$$S_{b_1} = .6942 \sqrt{.0053092} = .0506$$

and

$$S_{b_2} = .6942 \sqrt{.00010173} = .0070$$

as in the chapter.

Standard Error of the Regression Plane

The sampling error associated with any point on the regression plane can also be measured. Suppose we are interested in measuring the error of the plane at the point $(X_1, X_2, X_3, \ldots, X_m)$. We first measure the distance of this point from the mean of each variable, $x_1 = X_1 - \bar{X}_1$, $x_2 = X_2 - \bar{X}_2$, $x_3 = X_3 - \bar{X}_3$, etc. The standard error of the regression plane can then be expressed as[11]

$$s_{Y_c} = S_{Y \cdot 123 \cdots m} \sqrt{\frac{1}{n} + \sum_{i=1}^{m} \sum_{j=1}^{m} c_{ij} x_i x_j}$$

where

$$\sum_{i=1}^{m} \sum_{i=1}^{m} c_{ij} x_i x_j = c_{11} x_1^2 + c_{22} x_2^2 + \cdots c_{mm} x_m^2 + 2c_{12} x_1 x_2 + 2c_{13} x_1 x_3$$

$$+ \cdots 2c_{1m} x_1 x_m + 2c_{23} x_2 x_3 + 2c_{24} x_2 x_4 + \cdots 2c_{2m} x_2 x_m + \cdots$$

$$+ 2c_{(m-1)m} x_{(m-1)} x_m$$

For our example, let us compute the sampling error of the plane for a point $X_1 = 15.5$ and $X_2 = 165.0$. Since $\bar{X}_1 = 16.535$ and $\bar{X}_2 = 175.05$, $x_1 = 1.035$ and $x_2 = 10.05$.

$$s_{Y_c} = S_{Y \cdot 12} \sqrt{\frac{1}{n} + c_{11} x_1^2 + c_{22} x_2^2 + 2c_{12} x_1 x_2}$$

$$= .6942 \sqrt{\frac{1}{20} + (.0053092)(1.035)^2 + (.00010173)(10.05)^2}$$

$$+ 2(-.000051754)(1.035)(10.05)$$

$$= .6942 \sqrt{.0658} = .1781$$

Standard Error of Forecast

The standard error of forecast is the amount of error associated with making a forecast of a new observation. It includes the standard error of

---

[11] This can be expressed simply in matrix notation as

$$s_{Y_c} = s_{Y \cdot 123 \cdots m} \sqrt{\frac{1}{n} + z' \times X^{-1} \times z}$$

$$\text{where } z = \begin{bmatrix} x_1 \\ x_2 \\ \vdots \\ x_m \end{bmatrix}$$

and z' is the transpose of z. Note also that $c_{ij} = c_{ji}$ because of the symmetry of both X and X^{-1}.

the regression plane plus the scatter about the plane $(S_{Y \cdot 123 \ldots m})$. It is estimated for specific values of the independent variables X_1, $X_2, \ldots, X_m$.

The standard error of forecast is

$$S_f = \sqrt{S_{Y \cdot 12 \ldots m}^2 + s_{Y_c}^2}$$

where s_{Y_c} is the standard error of the regression plane as above. In our example,

$$S_f = \sqrt{(.6942)^2 + (.1781)^2}$$
$$= \sqrt{.5137}$$
$$= .716 \text{ thousand dollars or } \$716$$

PROBLEMS

1. Suppose we have estimated the least-squares linear regression of Y on X_1 and X_2 to be $Y_c = a + b_1X_1 + b_2X_2$. For each of the statements below, indicate briefly why you agree or disagree with the statement.

   a) If b_1 is 12 times as large as b_2, then we may infer that X_1 is considerably more important than X_2 in accounting for the variation in Y.

   b) The number b_1 is intended to measure the expected change in Y in response to a unit change in X_1 with X_2 held constant.

   For all of the remaining statements, suppose further that R^2 is very high, say $R^2 = .98$.

   c) The coefficients a, b_1, and b_2 are estimated to be significantly different from zero.

   d) The estimated relationship is a very close approximation to the true relationship between Y and X_1, X_2.

   e) The observed Y's do not vary much from the calculated Y's.

   f) Variations in X_1 and X_2 account for a very considerable proportion of the observed variations in Y.

   g) The observed residuals ($\epsilon = Y - Y_c$) show no systematic pattern.

   h) Dropping either X_1 or X_2 and estimating the simple regression of Y on the remaining variable would not reduce R^2 very much.

2. In a study of the demand for automobiles, the following regression model was used: $Y_c = a + b_1X_1 + b_2X_2 + b_3X_3$, where Y is expenditures (in billions of dollars) on new cars during year t (the period covered was 1948–61); X_1 is the price index for all cars, new and used, during period t; X_2 is the estimated value of the total stock of automobiles at the end of year $t - 1$, in billions of dollars; and X_3 is the per capita disposable income during year t (in dollars).

   The following results were obtained from the data:

$$Y_c = .0779 - .0201X_1 - .2310X_2 + .0117X_3$$
$$[.0026] \qquad [.0472] \qquad [.0011]$$
$$R^2 = .858$$

where the numbers in the brackets are the standard errors of the respective regression coefficients.

For each of the statements below, indicate briefly why you agree or disagree with the statement.

a) Price has a more important effect on expenditures for new cars than does per capita disposable income.

b) If price increased one index point in a given year, other things being equal, expenditures for new cars would decline by $.0201 billion, on the average.

c) Price does not have an important influence on expenditure for new cars.

d) About 14 percent of the variance in expenditures for new cars must be explained by variables other than stock of automobiles, price, and per capita disposable personal income.

e) The squares of the simple correlation coefficients between Y and the other variables X_1, X_2, and X_3, respectively, must equal .858, that is, $r^2_{Y \cdot 1} + r^2_{Y \cdot 2} + r^2_{Y \cdot 3} = .858$.

f) The fact that the coefficient of X_2 is approximately 10 times as large as the coefficient of X_1 means that X_2 explains considerably more of the variability in Y than does X_1.

g) The residuals $(Y - Y_c)$ are necessarily independent of each other.

3. Annual sales of Tidewater Industries in millions of dollars (Y) correlate with U.S. disposable personal income in billions (X_1) and company advertising expenditures in millions (X_2), as follows, for 1955–72.

$$Y_c = 210 + 18X_1 \text{ (simple regression)}$$
$$Y_c = 175 + 6X_1 + 11X_2 \text{ (multiple regression)}$$

a) What factors are likely to have caused the change in the coefficient of disposable income (X_1) from 18 in the first equation to 6 in the second?

b) If advertising expenditures were to be the same next year as this year (i.e., X_2 held constant), would you expect sales to increase $18 or $6 million in response to a $1 billion increase in disposable income? Explain.

4. The personnel director of the Apex Products Company wishes to determine whether the selling ability of salesmen can be predicted from their education and age. If so, these criteria would provide a valuable aid in selecting the most promising candidates for employment. As a start, 10 salesmen are selected at random and are rated by their supervisor as to sales ability, education, and age. The rating on sales ability covers a seven-point scale, from "Poor" (0) to "Excellent" (6). The education scale varies from "Did not finish high school" (0) to "Has master's degree" (4). The age

scale extends from "Age 20–29" (0) to "Age 60–69" (4). The results are shown below.

Salesman	Sales Ability Y	Education X_1	Age X_2
A	1	0	3
B	1	1	4
C	1	0	2
D	2	2	4
E	2	1	3
F	3	3	1
G	4	2	0
H	4	4	2
I	6	3	0
J	6	4	1
Sum	30	20	20

a) Compute the multiple linear regression equation by the method of least squares to estimate sales ability from education and age. Show all computations.

b) What is the meaning of the net regression coefficient b_1 in this particular case? How would this value differ in meaning from the regression coefficient in simple correlation between sales ability and education alone?

c) How would the reliability of b_1 be affected if the younger men generally had more education than the older men?

5. *a*) Compute the standard error of estimate in Problem 4, and interpret its meaning as applied to predicting the sales ability of future salesmen.

b) Compute the coefficient of multiple determination and interpret its meaning in describing the relationship between sales ability, education, and age for salesmen of this type.

6. The supervisor at Apex Products Company (Problems 4 and 5) is a close friend of employee K. Is his high rating (6.5) of him apparently attributable to favoritism, or can it be reasonably explained by K's education ($X_1 = 4$) and his youth ($X_2 = 1$)? Explain your answer.

7. Arjay Furniture Co. operates a chain of retail stores. As a means of measuring the efficiency of various stores, the management is studying the relationship between the number of employees, the size of the store, and the average daily sales volume for last year. The data can be summarized as follows:

Y = average daily sales for each store in hundreds of dollars
X_1 = number of employees for each store
X_2 = size of each store in hundreds of square yards
n = 103 = number of Arjay stores

The raw data and the necessary adjustments are summarized in the table.

	Y	X_1	X_2	Y^2	X_1^2	X_2^2	YX_1	YX_2	X_1X_2
Total.............	515	618	824	3,975	5,708	9,092	4,090	5,620	5,944
Mean.............	5.0	6.0	8.0						
Less adjustment......				2,575	3,708	6,592	3,090	4,120	4,944
Adjusted total.......				1,400	2,000	2,500	1,000	1,500	1,000
Which is............				Σy^2	Σx_1^2	Σx_2^2	$\Sigma y x_1$	$\Sigma y x_2$	$\Sigma x_1 x_2$

a) Estimate the linear regression equation $Y_c = a + b_1X_1 + b_2X_2$, which predicts monthly sales as a function of the number of employees and the size of the store.

b) Are you sure that the values obtained for b_1 and b_2 in the above equation are statistically greater than zero?

c) Is the regression equation of much use in predicting sales? (Explain your answer.)

d) One of Arjay's newer and larger stores occupies 16 hundred square yards and employs 10 people. Average daily sales have been $1,500. Is this out of line with the experience of other Arjay stores? Explain.

8. A manual dexterity test (X_1) and a finger dexterity test (X_2) were administered to 25 applicants for jobs as aircraft riveters. After these 25 applicants were hired and trained, their performance was measured by the number of rivets set correctly per minute (Y). A multiple regression analysis is to be performed to evaluate the worth of each test in predicting performance of riveters. We have the following:

	Y	X_1	X_2	Y^2	X_1^2	X_2^2	YX_1	YX_2	X_1X_2
Total..........	200	150	125	2,213	1,000	775	1,400	1,225	800
Mean.........	8	6	5						

a) Estimate the linear regression equation, which predicts performance as a function of the two tests.

b) Test the hypothesis that neither test has any predictive value for performance of riveters.

c) Which test do you consider more important in predicting riveting performance?

d) Calculate the coefficient of multiple determination.

e) A new employee scores 9 on the manual dexterity test and 8 on the finger dexterity test. Predict his riveting performance.

9. A study was undertaken at a John Deere farm machinery plant to determine what variables influenced the time taken to handle a piece of flat metal stock to the bump gauge of a punch press. The length and weight of the metal piece were thought to be significant factors. Accordingly, the han-

dling time, weight, and length of a sample of 25 pieces of metal were recorded and are presented in the table.

Item	Time (.001 Min.)	Weight (.1 Lb.)	Length (.1 In.)
1	30	5	35
2	32	12	46
3	15	15	63
4	30	31	67
5	25	6	70
6	25	8	83
7	42	37	88
8	35	23	104
9	42	30	134
10	30	34	151
11	52	17	153
12	50	53	164
13	45	56	173
14	50	41	191
15	70	84	196
16	64	62	198
17	64	66	204
18	70	66	208
19	80	63	238
20	88	80	295
21	105	154	308
22	85	50	310
23	85	184	319
24	105	186	324
25	84	122	394
Total	1,403	1,485	4,516
Mean	56.12	59.40	180.64

a) Estimate the linear regression between the handling time and the length and weight of the pieces of metal.

b) Are the effects of the length and weight statistically significant?

c) Which factor is more important in determining the handling time?

d) Calculate the standard error of estimate and the coefficient of multiple determination.

e) Plot the residuals to check the assumptions of linearity and homoscedasticity (uniform scatter of residuals).

10. An analyst for a manufacturing firm wished to explain the variations that had occurred from period to period in the manufacturing cost per unit of the firm's product. Accordingly, he collected the data over the last 20 quarters. He knew that raw material prices and labor costs had varied considerably over the period, and he estimated an index of these costs. Also, the production rate had fluctuated widely in response to customer demand

and inventories. The production level for each period was measured as a percentage of rated capacity. The data are shown in the table.

Period	Average Manufacturing Cost per Unit	Production Level as Percent of Rated Capacity	Index of Raw Material and Labor Costs
1	$3.65	85	80
2	4.22	78	93
3	4.29	82	107
4	5.43	64	115
5	6.62	50	130
6	5.71	62	128
7	5.09	70	116
8	3.99	90	92
9	4.08	94	94
10	4.38	100	110
11	4.28	104	115
12	4.42	82	117
13	5.11	75	128
14	4.88	84	134
15	4.99	86	135
16	4.57	90	135
17	4.84	94	139
18	5.16	80	142
19	5.67	72	147
20	6.26	60	150
Mean	$4.882	80.10	120.35

a) Determine the multiple regression equation relating cost per unit to production level and raw material cost.

b) Explain the meaning of the coefficients in the regression equation.

c) How well do these factors explain or predict cost per unit?

d) Plot the residuals $(Y - Y_c)$ against each of the independent variables. Is there any evidence of curvilinearity from these plots?

e) For next quarter, the raw materials and labor cost index is expected to drop to 145 and the production level is expected to rise to 80 percent of capacity. What average manufacturing cost per unit would you expect? Should you qualify your estimate as a result of your answer to (d) above?

11. *Note:* This problem requires the use of the matrix multiple regression method (Appendix B, this chapter) or else a computer program.

a) Fit a function of the form $Y_c = a + bX_1 + cX_1^2 + dX_2$ to the data in Problem 10. (Y is manufacturing cost, X_1 is production level, X_2 is raw material and labor cost.)

b) Plot the residuals against the independent variables. Is there any evidence of curvilinearity remaining?

c) Is the coefficient c statistically significant? (Hint: Find standard error of c.)

d) Compare the results of this problem with those of Problem 10.

12. The Value Line Investment Survey computes a multiple regression equation for each common stock showing the typical relationship between its price (X_1), earnings per share (X_2), and dividends per share (X_3) in past years. The following equation was reported for Boeing Airplane Company:

Log normal average value next 12 months
$$= 1.355 + 0.440 \log (.22 \times \text{earnings} + 1.00 \times \text{dividends})$$

a) Explain the meaning of this equation and its use for an investor.

b) What type of linear transformation does this equation illustrate?

c) What other measures or qualifications would be desirable in this survey to aid the investor in appraising the reliability of the equation?

13. You are an analyst interested in estimating future sales for the PPG Industries (formerly Pittsburgh Plate Glass Company). A substantial portion of the company's business is the manufacture of windshields and windows for new automobiles. In addition, the company makes glass and paint products used in new construction. Accordingly, you collect the data below for 1953–70 (in billions of dollars):

Year	PPG Industries Sales Y	Motor Vehicle Mfrs.' Sales X_1	New Construction X_2
1953	.452	24.9	39.1
1954	.431	21.8	41.4
1955	.582	31.5	46.5
1956	.597	26.1	47.6
1957	.621	28.4	49.1
1958	.514	21.4	50.2
1959	.607	27.5	55.3
1960	.628	30.9	53.9
1961	.603	26.8	55.4
1962	.657	33.7	59.7
1963	.778	37.2	63.4
1964	.828	38.6	66.2
1965	.898	47.7	72.3
1966	.942	47.2	75.1
1967	.943	40.4	76.2
1968	1.044	49.6	84.7
1969	1.147	51.5	90.9
1970	1.094	42.5	94.0

SOURCES: Company reports, *Business Statistics, 1971, Survey of Current Business*, June 1972.

a) Find the relationship between PPG sales and the independent variables by multiple regression analysis.

b) Explain the meaning of the multiple regression equation.

c) Is there a significant relationship between PPG sales and each of the independent variables? Explain.

d) Which has more influence on PPG sales—motor vehicle sales or new construction? Give figures.

e) Is there any evidence of curvilinearity or autocorrelation revealed by the residuals $(Y - Y_c)$?

14. *a*) In Problem 13, forecast PPG sales in 1971 based on actual motor vehicle sales of 49.7 billions and new construction of 109.4 billions.

   b) Give a 95 percent confidence interval for this forecast, based on the standard error of estimate alone. What qualifications would you have to make in reporting this figure to management?

   c) Actual PPG sales in 1971 were 1.238 billions. What was the error of your forecast? Was this within your confidence interval?

15. *a*) In order to take into effect growth trends as well as personal income on Sears, Roebuck sales, estimate the multiple regression between the log of Sears, Roebuck sales and the log of disposable personal income (see Table 16–5 and text discussion) as well as the natural value of time, for the years 1953–71.

   b) How does this equation compare with the simple regression equation, excluding time (see text) in accounting for the changes in Sears sales? Explain.

16. Some of the variability in Sears, Roebuck sales may be attributable to the fact that many new retail stores are being opened. The number of stores at the beginning of each fiscal year (February 1) is shown below:

Year	Stores	Year	Stores	Year	Stores
1953	684	1960	741	1967	801
1954	694	1961	747	1968	809
1955	699	1962	747	1969	818
1956	709	1963	748	1970	826
1957	721	1964	761	1971	827
1958	732	1965	777		
1959	736	1966	786		

a) Compute the multiple regression between Sears, Roebuck sales and the independent variables—disposable personal income and number of stores for the years 1953–71—using the logarithms of all variables listed in Table 16–5 and the table above.

b) How does this equation compare with the simple regression equation excluding number of stores (see text) in accounting for the changes in Sears sales? Explain.

c) If this equation is better than the simple regression excluding stores, forecast 1972 Sears sales using the 795 billion income projection in Table 16–5 and the company's report of 836 stores open at the beginning of fiscal 1972.

17. *a*) In order to project the demand for food products, fit a parabola by least squares to the food production index (Y) and population for 1957–71, as given in Chapter 16, Problem 19. That is, change X to X_1, transform X^2 to a second independent variable X_2, and proceed as in multiple linear regression. (Alternatively, you can solve the normal equations given in page 461n to find the constants in the regression equation.)

b) Find the standard error of estimate.

c) Compare this value with that found in Chapter 16, Problem 19(*d*), if you worked that problem. Which curve is a better fit by this criterion?

18. Gotham City maintained a small fleet of automobiles in a special motor pool. These cars were used by the various agencies when special needs arose for temporary use by personnel who did not have their own assigned automobile.

The manager of the pool was trying to determine what factors contributed to the maintenance and repairs costs of the cars under his care. He surmised that factors such as the mileage driven, the age of the car, and possibly even the make of the car contributed to maintenance and repair costs. Accordingly, he collected data on these factors for fifteen cars selected at random from the cars in the pool. The data are shown below:

Car No.	Maintenance and Repair Cost 1972	Mileage Driven in 1972 (thous. of miles)	Age of the Car-years (0 is new car)	Make (coded)
1	$643	18.2	0	A
2	613	16.4	0	B
3	673	20.1	0	A
4	531	8.4	1	B
5	518	9.6	2	B
6	594	12.1	1	A
7	722	16.9	1	B
8	861	21.0	1	A
9	842	24.6	0	A
10	706	19.1	1	A
11	795	14.3	2	B
12	776	16.5	2	B
13	815	18.2	2	A
14	571	12.7	2	A
15	673	17.5	0	B

As a first step in his analysis of these data, the manager calculated the average maintenance and repair cost for new, one-year old, and two-year old cars. The results were:

Age	Number of cars	Average Maintenance and Repair Costs
0	5	$688.4
1	5	682.8
2	5	695.0

Although he was somewhat surprised by the results, the manager concluded that the age of the car did not influence significantly the repair and maintenance costs.

As a next step, the manager calculated the costs by make of car. The results were:

Make	Number of cars	Average Maintenance and Repair Costs
A	8	$713.1
B	7	661.1

He concluded that the pool should, in the future, give preference to purchasing cars of Make B since he would save $52 each per year in maintenance and repairs.

Do you agree with the manager? How would you suggest that he analyze the data? What are your conclusions?

SELECTED READINGS

BRYANT, EDWARD C. *Statistical Analysis.* Rev. ed. New York: McGraw-Hill, 1966.

Chapters 7 and 10 treat simple and multiple regression concisely. Matrix notation is used in the treatment of multiple regression.

CROXTON, F. E.; COWDEN, D. J.; and BOLCH, B. W. *Practical Business Statistics.* 4th ed. Englewood Cliffs, N.J.: Prentice-Hall, 1969, chaps. 14–16 and 21.

Treats a variety of topics in simple and multiple correlation.

DRAPER, N. R., and SMITH, H. *Applied Regression Analysis.* New York: John Wiley, 1966.

An advanced treatment. Covers many practical problems in economics and natural sciences.

EZEKIEL, MORDECAI, and FOX, KARL A. *Methods of Correlation and Regression Analysis.* 3d ed. New York: John Wiley, 1959.

This is the standard book in the field. In the third edition, its major emphasis has been shifted from correlation to regression. Graphic analysis of curvilinear relationships is stressed.

FOX, KARL A. *Intermediate Economic Statistics.* New York: John Wiley, 1968.

Chapters 4, 6, and 7 cover regression analysis, while chapters 10–13 treat multiequation economic models.

FRANK, C. R. JR. *Statistics and Econometrics.* New York: Holt, Rinehart & Winston, 1971.

Covers both regression and multiequation models.

GOLDBERGER, ARTHUR S. *Econometric Theory.* New York: John Wiley, 1964.

An advanced treatment of linear regression, including matrix algebra and systems of simultaneous linear relationships.

JOHNSTON, J. *Econometric Methods.* 2d ed. New York: McGraw-Hill, 1972.

A comprehensive study of the linear normal regression model, autocorrelation, and simultaneous equation problems.

WILLIAMS, E. J. *Regression Analysis.* New York: John Wiley, 1959.

Provides the practical statistician with a compendium of the classical techniques associated with regression analysis.

18. INDEX NUMBERS

INDEX NUMBERS express the *relative* changes in a variable compared with some base, which is taken as 100.[1] The variable may be a single series, such as electric power production, or an aggregate, such as a group of common stock prices. The index number usually represents a sample of such a group. The changes measured may be those occurring over a period of time or those between one place and another.

Many aspects of modern business are described by the use of index numbers. Both government and private agencies are devoting increasing efforts to the construction of index numbers as aids in management and in the interpretation of changes in general economic life. Many businesses use a variety of index numbers for their own internal administrative purposes. Certain statistical publications, notably the *Survey of Current Business,*[2] *Economic Indicators, Business Conditions Digest, Federal Reserve Bulletin,* and the *Trade and Securities Statistics* bulletin of Standard and Poor's Corporation, contain hundreds of economic time series expressed in index number form.

Statistical ingenuity has developed an almost encyclopedic list of uses of business indicators. The most important of these are (1) measures of the economic well-being of the economy, a geographic area, an industry, or a specific business; (2) comparisons of related series for administrative purposes; (3) the use of price indexes as deflators to express a value series in constant dollars; (4) the use of price indexes as escalators in wage and other contracts; (5) specific guides or "triggers" for the

---

[1] The term "index" is sometimes applied to a business indicator expressed in any unit. Thus, pig-iron production in tons may be referred to as an "index" of business activity. In this chapter, however, the term "index number" or "index" refers specifically to a ratio having some base as 100, or to a series of such ratios.

[2] Summary descriptions of 2,500 series may be found in the footnote references of the biennial *Business Statistics* supplement to the *Survey of Current Business.*

initiation of administrative business or government actions; and (6) the basis or orientation for forecasting.

In general, index numbers have the following important advantages, in contrast with actual data:

1. They provide a simple method of comparing changes from time to time or from place to place. It is easy to compare 89 cents for a pound of ham with 38 cents for a quart of milk, but it is not so easy to compare price changes in the two articles over a period of time. Index numbers of the ham and milk prices would indicate the relative change in each price from some given price and which of the two prices had shown the greater change (see Table 18–4). As the number of items increases, this advantage becomes even more apparent.

2. Index numbers facilitate comparison of changes in series of data expressed in a variety of units—for example, dollars, tons, or gallons. Data pertaining to production, sales, inventories, costs, or other aspects of business may also be put into index number form and then compared.

3. They make possible the construction of composites that represent in a single figure some overall measure of business. This simplifies comparisons with other types of data. In February 1973, the U.S. Bureau of Labor Statistics Index of Wholesale Prices stood at 126.9. This single figure indicates the average relation of prices in February 1973 to prices in 1967, the base period for this index, taken as 100. That is, it took 12.69 to buy the same amount of specified goods as could have been bought for $10 in 1967.

Even series expressed in different types of units sometimes can be combined into a meaningful aggregate, provided the combinations make sense. Many examples of such combinations appear throughout this chapter.

4. They describe the typical seasonal patterns of business. The annual peak in department store sales, for instance, regularly occurs in December, while sales of soft drinks are greater in midsummer. These "indexes of seasonal variation" are described in Chapter 20.

KINDS OF INDEX NUMBERS

An examination of any journal of business statistics will reveal many different index numbers which describe changes in various aspects of business and economics. These index numbers may be classified as (1) price indexes, (2) quantity indexes, and (3) value indexes. Some of the most commonly used indexes of these three types, and their principal sources, are listed in Table 18–1. Most of these, but not all, are expressed in relative form.

Table 18–1

SOURCES OF COMMONLY USED INDEXES*

Name of Index	Prepared by	Frequency of Publication	Published Regularly in
A. PRICE INDEXES			
1. Consumer Price Index	U.S. Bureau of Labor Statistics	M	*SCB, FRB, MLR, Business Week, S & P, Ec. Ind., BCD*
2. Wholesale Price Index	U.S. Bureau of Labor Statistics	M	*SCB, FRB, MLR, BCD S & P, Ec. Ind.*
3. Spot Market Prices of 22 Basic Commodities	U.S. Bureau of Labor Statistics	M	*SCB, S&P*
4. Construction Cost Indexes	Engineering News Record	M	*SCB, S&P*
5. Stock Price Averages	Dow-Jones & Co.	H, D, W, M	*SCB, Barron's, S&P, C&FC*
6. Stock Price Index, 500 Stocks	Standard and Poor's Corp.	H, D, W, M	*SCB, FRB, S&P, Ec. Ind., Barron's, Business Week, BCD*
B. QUANTITY INDEXES			
1. Industrial Production	Federal Reserve Board	M	*SCB, FRB, S&P, BCD Ec. Ind.*
2. Business Week Index	*Business Week*	W	*Business Week*
3. Steel Production	American Iron and Steel Institute	W, M	*SCB, Barron's, C&FC, Ec. Ind.*
4. Help-Wanted Advertising	Conference Board	M	*SCB, CBSB, BCD*
C. VALUE INDEXES			
1. Manufacturing Production-Worker Payrolls	U.S. Bureau of Labor Statistics	M	*FRB, S&P*
2. Construction Contracts Awarded (Value)	F. W. Dodge Corp.	M	*SCB, FRB, Ec. Ind.*

* Abbreviations:
H—hourly or shorter intervals; D, daily; W, weekly; M, monthly
SCB—Survey of Current Business (and weekly supplement)
FRB—Federal Reserve Bulletin
MLR—Monthly Labor Review
C&FC—Commercial and Financial Chronicle
S&P—Standard and Poor's *Trade and Securities Statistics*
Ec. Ind.—President's Council of Economic Advisers, *Economic Indicators*
CBSB—Conference Board Statistical Bulletin
BCD—Business Conditions Digest

Price Indexes

Some of the best known indexes are those dealing with prices. Prices have been of widespread interest for centuries as sensitive barometers of industry and trade.

The necessary data for price index numbers arise from the exchange of commodities (1) at different stages of production—raw materials, semifinished goods, and completely fabricated products; (2) at several levels of distribution—industrial, wholesale, and retail; and (3) for a variety of groups of items—consumers' goods, producers' goods, stocks and bonds, durable and nondurable goods.

A *purchasing power index* is the reciprocal of a price index, when both indexes are expressed as ratios with base 1 rather than 100. Taking the wholesale price index of 126.9 for February 1973 as 1.269, its reciprocal is $1/1.269 = .788$, so the corresponding purchasing power index (with base 100) is 78.8. This means that for every dollar's worth of goods one could buy at 1967 wholesale prices, one could buy 78.8 cents' worth in February 1973. Hence, the dollar then was worth only 78.8 cents in comparison with the 1967 dollar.

Quantity Indexes

Quantity indexes measure the physical volume of production, construction, or employment. They are computed for (1) industry in general, (2) specific industries, or (3) specific operations or stages of production or distribution. The data may represent the country as a whole or local trading areas.

Because of the nature of the data, quantity index numbers are frequently less reliable than those based on dollar figures. Historically, business records were designed to include chiefly those aspects of business which could be expressed in monetary units and, consequently, data in physical units for extended periods of time are difficult to obtain.

Value Indexes

Value indexes show the total dollar volume of income, payrolls, sales, and the like. Value is the result of multiplying quantity by price; index numbers of value therefore reflect changes in both quantity and price. The gross national product estimates of the U.S. Department of Commerce are constructed much like other value indexes, but they are expressed in billions of dollars rather than as percentages of a base to avoid the "aura of normality" attached to a base period.

It will be noted that the Federal Reserve Board and *Business Week* indexes of general business activity measure physical volume changes, such as tons of steel and kilowatts of electricity produced, while many regional indexes measure dollar volume, such as factory payrolls and department store sales. Some regional business barometers even combine quantity and value measures, but these indexes are more difficult to interpret.

BASIC METHODS OF CONSTRUCTING INDEX NUMBERS

Simple Index Numbers

A simple index number is constructed from a single series of data which either extends over a period of time or simultaneously represents several different locations. In constructing such an index number, one particular period or place is selected as the base and the item for this base is taken as 100. The other items in the series are then expressed as percentages of this base. A simple index is frequently called a price *relative,* quantity relative, or value relative.

As an example of a quantity relative, an airline executive may wish to compare the changes in air and automobile travel from 1966 to 1971. Since the volume of intercity automobile passenger-miles traveled is over 10 times that of air travel, the executive's purpose would not be accomplished by comparing the changes in actual passenger-miles. The two series can be more easily compared if they are expressed as percentages of passenger-miles traveled in the same base period—say, 1967.

The construction of these simple indexes or quantity relatives is shown in Table 18–2. The three steps are (1) choose the base period

Table 18–2

SIMPLE INDEX NUMBERS OF AIR TRAVEL
AND INTERCITY AUTOMOBILE TRAVEL
IN THE UNITED STATES, 1966–71

Year	Passenger-Miles (Billions)		Index (1967 = 100)	
	Air Travel	Auto Travel	Air Travel	Auto Travel
1966	60.6	902	80	93
1967	75.5	967	100	100
1968	87.5	1,016	116	105
1969	102.7	1,071	136	111
1970	104.1	1,120	138	116
1971	106.3	1,170*	141	121

* Estimated.
SOURCE: *Air Transport Facts and Figures,* 1972, p. 41.

(1967); (2) divide the travel figure each year by the base figure; and (3) multiply the result by 100 (i.e., move the decimal point two places to the right) to express it as a percentage or index number. An index number is written just as a percentage, except that the percent sign ($\%$) is not used. Thus, the 1971 index for air travel is $106.3 \div 75.5 \times 100 = 141$.

This index means that air travel in 1971 was 141 percent of its 1967 volume, an increase of 41 percent. Hence, while automobile travel had increased more than air travel in passenger-miles during this period (203 billion versus 30.8 billion), its relative increase was only 21 percent, compared with 41 percent for air travel.

The increase of the air travel index from *1966* to 1971 was 61 index points, but this is not 61 percent because the base is 80, not 100. The percentage increase was $61 \div 80 = 76$ percent.

A simple index can be computed for any single series of data, such as the price of General Motors stock or a department store's sales. Statistical source books include many indexes of this type. The Bureau of Labor Statistics, for example, publishes monthly price relatives for each of about 2,500 commodities as an aid in comparing individual price changes, in addition to its composite wholesale price indexes.[3]

Composite Index Numbers

Most index numbers in common use are composites. They are constructed according to the principles just described for simple indexes, but they combine several different sets of data. In the following pages, two basic methods of constructing composite index numbers are described: (1) the average of relatives index and (2) the aggregative index. Formulas for both types of indexes are presented on page 551. but it is not necessary to memorize them to understand the procedure involved.

Necessity of Weights. Whenever prices or other data are combined in an index number, the relative importance of each must be taken into account by assigning proper weights to each item. This is necessary because, in reality, no composite index is unweighted. If a set of weights is not explicitly applied, each element of the index automatically (or implicitly) receives some weight. For example, if unit prices of various foods are being added together in the preparation of a composite consumer price index, a given relative change in a higher priced item such as a pound of ham will influence the total more than will the same relative change in a lower priced item such as a quart of milk. Milk, however, should really be weighted more heavily because people consume more; so a system of weights must be used in order to

---

[3] See U.S. Bureau of Labor Statistics, *Wholesale Prices and Price Indexes,* July 1971.

give milk its proper importance in the index. A composite index is thus a *weighted average*[4] of its components.

Average of Relatives Method. Many methods of constructing index numbers have been tried, but the average of relatives method is now used in most leading indexes, such as the Federal Reserve Board's index of industrial production and the Bureau of Labor Statistics' wholesale price indexes. In this method the individual series of price or quantity data are expressed as simple indexes, which are then multiplied by fixed *dollar-value weights* and totaled to yield the composite index.

To illustrate the construction of a *quantity* index, consider an oil company producing aircraft fuel and automobile gasoline. About two thirds of its sales are typically aircraft fuel and one third is gasoline sold through filling stations. An executive wishes to construct a composite index of air and automobile travel and project it into the future as a measure of the potential market for his products. The method is illustrated in Table 18–3. The steps are as follows:

Table 18–3

CONSTRUCTION OF COMPOSITE INDEX
OF AIR AND AUTOMOBILE TRAVEL
BY AVERAGE OF RELATIVES METHOD
(1967 = 100)

Year (1)	Simple Index (1967 − 100) Air Travel (2)	Auto Travel (3)	Weighted Index Air Travel (Column 2 × ⅔) (4)	Auto Travel (Column 3 × ⅓) (5)	Composite Index Air and Auto Travel (Columns 4 + 5) (6)
1966	80	93	53	31	84
1967	100	100	67	33	100
1968	116	105	77	35	112
1969	136	111	91	37	128
1970	138	116	92	39	131
1971	141	121	94	40	134

SOURCE: Table 18–2.

---

[4] The weighted arithmetic mean is used almost universally in computing index numbers, although the weighted geometric mean is theoretically superior for averaging relatives, particularly since they tend to follow a logarithmic normal distribution, with a zero lower limit and infinite upper limit. The geometric mean also minimizes the influence of extremely large relatives, which may distort the arithmetic mean of a small number of items. Nevertheless, the arithmetic mean is used because it is easier to compute and easier to understand than the geometric mean. Also, an arithmetic price index represents changes in the total cost of a bill of goods more accurately than a geometric index, which reflects the average ratios of change in price. That is, the arithmetic mean makes more sense in this connection.

1. Express each individual series as a simple index or relative, by dividing through by the base value. This step is described above. (Columns 1–3 in Table 18–3 are taken from Table 18–2.)
2. Select a dollar-value weight for each series as a measure of its importance in the base year or some other typical period. Divide these weights by their total to express them as relative weights whose sum equals one. In this case the relative importance of air and auto travel *to the company* is measured by the proportion of its dollar sales that go to each industry—⅔ and ⅓, respectively. As a more general example, the Federal Reserve Board weights its component indexes of manufacturing output by "value added by manufacture," from the Census of Manufactures, expressed as percentages of the total weight.
3. Multiply the simple indexes by the relative weights to obtain the weighted indexes (Table 18–3, columns 4 and 5).
4. Add the weighted indexes to obtain the composite index (column 6). This must equal 100 in the base year, since the simple indexes equal 100 and the weights total one. (If the value weights are not adjusted to total one, the sum of the weighted indexes can be divided through by its base-year value to obtain the same values as in column 6 of the table.)

The composite index provides the executive with a summary measure of the growth in potential demand with which he can compare or predict his own sales.

A composite *price* index is constructed by this method in the same way as a quantity index. Table 18–4 illustrates the computation of a

Table 18–4

CONSTRUCTION OF COMPOSITE INDEX FOR THREE RETAIL MEAT PRICES
BY AVERAGE OF RELATIVES METHOD
(1967 = 100)

Period (1)	Simple Index (1967 = 100)			Weighted Index			Composite Index (Total, Columns 5–7) (8)
	Round Steak (2)	Smoked Ham (3)	Frying Chicken (4)	Steak (Col. 2 × .59) (5)	Ham (Col. 3 × .29) (6)	Chicken (Col. 4 × .12) (7)	
1967 Ave.	100	100	100	59	29	12	100
1970 Ave.	118	114	108	70	33	13	116
1971 Ave.	124	103	108	73	30	13	116
1972 Apr.	134	112	108	79	32	13	124

SOURCE OF PRICE DATA: U.S. Bureau of Labor Statistics, *Estimated Retail Food Prices by Cities.*

consumer price index for three types of meat in 1967 (the base period) and three later periods, using the price data in Table 18–5. Round steak is chosen as typical of all beef and veal prices in its price behavior, while smoked ham represents pork products and frying chicken represents poultry prices. The individual commodity price is then weighted in accordance with the importance of the whole commodity group it represents, rather than by its own individual importance. Of course, actual indexes involve hundreds of commodities and many dates. The steps are similar to those cited above:

1. Divide each price series by its price in the base period (1967 average) to express it as a simple index (Table 18–4, columns 2 to 4).
2. Measure the relative importance of each commodity group in dollars for some normal period. The relative weights in the heading of columns 5 to 7 are based on a hypothetical consumer survey which showed that for every dollar the typical family spent on meat, 59 cents went for beef and veal, 29 cents for pork products, and 12 cents for poultry. The weights preferably apply to the base period, but this is not always feasible. Thus, the U.S. Bureau of Labor Statistics formerly reported its Consumer Price Index with the base 1957–59 = 100, but after January 1964 it obtained its weights from a survey of consumer spending patterns made in 1960–61. (Note that *dollar values,* rather than prices or quantities, are used as weights in the weighted average of relatives method for computing either price or quantity indexes. Also, the weight must be held constant over a period of years; otherwise changes in the weight would affect the level of the index itself.)
3. Multiply the simple indexes (columns 2 to 4) by the weights to obtain the weighted indexes (columns 5 to 7).
4. Add the weighted indexes for each period to get the composite index (column 8). (If the weights are not adjusted to total 1, the last column must be divided by its base-period value to adjust this value to 100.)

Aggregative Method. The aggregative method is more direct than the average of relatives method in bypassing the calculation of simple indexes. Table 18–5 illustrates the construction of a *price* index by the aggregative method. The steps are:

Table 18–5

CONSTRUCTION OF COMPOSITE INDEX
FOR THREE RETAIL MEAT PRICES
BY AGGREGATIVE METHOD
(1967 = 100)

	Price per Pound, Dollars			Cost of Week's Supply, Dollars				(Composite Index (Col. 8 ÷ 9.40) (9)
Period (1)	Round Steak (2)	Smoked Ham (3)	Frying Chicken (4)	Steak (Col. 2 ×5 Lbs.) (5)	Ham (Col. 3 ×4 Lbs.) (6)	Chicken (Col. 4 ×3 Lbs.) (7)	Total (Cols. 5–7) (8)	
1967 Ave.	1.10	.69	.38	5.50	2.76	1.14	9.40	100
1970 Ave.	1.30	.79	.41	6.50	3.16	1.23	10.89	116
1971 Ave.	1.36	.71	.41	6.80	2.84	1.23	10.87	116
1972 Apr.	1.47	.77	.41	7.35	3.08	1.23	11.66	124

SOURCE OF PRICE DATA: U.S. Bureau of Labor Statistics, *Estimated Retail Food Prices by Cities.*

1. Choose as weights the physical *quantities* of each commodity produced or consumed in a typical period. In this case, it is the quantity of each of three food items consumed by an average family in a week: say, five pounds of beef and veal, four pounds of pork products, and three pounds of poultry.
2. Multiply each price (columns 2 to 4) by its weight to obtain the weighted prices (columns 5 to 7). The product of price times quantity gives the total cost of each commodity in the "market basket" as its price changes from time to time.
3. Total these products (column 8) to get the cost of the whole market basket.
4. Select a base period (1967 average) and divide the totals by the total in the base period ($9.40). The results (column 9) are aggregative index numbers. Here they indicate that in April 1972 the combined cost of the three commodity groups was about 124 percent of what it was in 1967.

As a more realistic sample of the aggregative method, Standard and Poor's constructs its price index of 500 stocks by multiplying the current market price of each stock by the number of shares outstanding in the base period (modified by later capitalization changes). This weighted price, or aggregate market value of the original shares, is then totaled for all 500 stocks and the grand total is divided by the aggregate market value in the base period to obtain the index.[5]

---

[5] The base is set at 1941–1943 = 10 in order to make the current index approximate the average price of all stocks listed on the New York Stock Exchange.

Quantity indexes are computed by the aggregative method in the same way as price indexes, except that quantity and price are interchanged. The varying quantities produced or consumed each month are multiplied by a fixed price in the base year or some other typical period. Hence, only changes in physical volume affect the movements of the index, and the fixed price serves to give each commodity its appropriate importance. Then the sum of the weighted quantities each month is divided by the sum in the average month of the base year to yield the weighted aggregative quantity index.

Dollar-value indexes (e.g., department store sales) reflect the movements of both price and quantity, so neither one need be held constant. Furthermore, the original data are already available in the form of dollar values. In the aggregative method, the estimated values for each component of the index are simply added each year. The totals themselves may then be reported, as in gross national product estimates, or they may be divided by a base-year value and reported as index numbers, as in the F. W. Dodge index of the value of construction contracts awarded.

The average of relatives method is used when the components are not comparable, as in bank debits and department store sales used in regional business indexes. Here the components are expressed as relatives and then multiplied by arbitrary weights to arrive at the final value indexes.

Formulas for Computing Composite Indexes

The two basic methods of computing weighted index numbers can be expressed in formulas using the following symbols:

For an individual commodity—

p_0 = Price in the base period (e.g., 1967).
p_n = Price in current year of the series (e.g., 1974, 1975, . . .).
q_0 = Quantity in the base period.
q_n = Quantity in current year of the series.
$\Sigma(p_n q_0)$ = Sum of (price of first commodity in current year times base-period quantity) plus (price of second commodity in current year times base-year quantity), etc.

The formulas are:[6]

---

[6] These formulas, which use base-year weights, are variants of Laspeyres' formula, as opposed to Paasche's formula, which uses current-year weights, or Irving Fisher's "ideal" index, which is the geometric mean of the two.

	Average of Relatives Method	Aggregative Method
Price index	$\dfrac{\Sigma(p_n/p_0)(p_0q_0)}{\Sigma(p_0q_0)}$	$\dfrac{\Sigma(p_nq_0)}{\Sigma(p_0q_0)}$
Quantity index	$\dfrac{\Sigma(q_n/q_0)(p_0q_0)}{\Sigma(p_0q_0)}$	$\dfrac{\Sigma(p_0q_n)}{\Sigma(p_0q_0)}$
Value index	$\dfrac{\Sigma(p_nq_n/p_0q_0)(p_0q_0)}{\Sigma(p_0q_0)}$	$\dfrac{\Sigma(p_nq_n)}{\Sigma(p_0q_0)}$

The two formulas in each row are identical when the base-period price, quantity, or value is used as weight. That is, multiplying prices by base-year quantities gives the same algebraic result as multiplying price relatives by the same year's value, etc. If some other period is used as weight, as is often the case, the results will differ somewhat. Thus, when the principal U.S. government indexes used the same 1957–59 base for comparability, the weights for the Consumer Price Index were determined from a survey of consumer expenditures in 1960–61; the weights for the Wholesale Price Index represented sales of commodities reported in the 1958 censuses; and the weights of the Federal Reserve Board Index of Industrial Production depended on the "value added" by the industry in 1957.

Formulas for quantity indexes are the same as for price indexes with p and q interchanged.

Comparison of Average of Relatives and Aggregative Methods

The average of relatives and aggregative methods often yield identical results, as described above. Then which is the better one to use?

The aggregative method is the simpler and the more easily understandable of the two, so it may be used whenever appropriate weights (i.e., quantities for a price index) are available and when only the composite index is needed.

The average of relatives method, on the other hand, must be used when:

1. It is desired to compare the individual components in the form of relatives, as in the Wholesale Price Index. The first step in this method produces these relatives directly.
2. The available weights are in value form, as in the Federal Reserve Board index, which applies the "value added by manufacture" for a group of related items as a weight for the production of a single representative item. It is usually easier to obtain dollar values as weights than it is to find quantities.

3. The component series are already in the form of relatives, as in combining several segments of the Federal Reserve Board Index of Industrial Production for comparison with a particular industry.

Since one or more of these conditions usually exist, the average of relatives method is more widely used than the aggregative method.

TESTS OF A GOOD INDEX NUMBER

A businessman must often refer to index numbers in gauging the state of the economy and in making necessary day-to-day decisions for the control and planning of his operations. Yet he cannot accept an index uncritically at its face value without inquiring into its characteristics and limitations. Appearances are deceiving, and the official names of indexes are often little more than general guides to their nature.

If one makes any regular use of an index, therefore, it is surely worthwhile to write the publisher for a description, or at least to check one of the publications at the end of this chapter that provide a critical analysis of the major indexes. One should also appraise the reliability and reputation of the compiler. For example, the leading federal statistical agencies have improved their indexes tremendously while, on the other hand, certain regional agencies publish rather crude indexes of business activity in their areas.

In studying the nature of an index it is particularly important to apply the following tests, which determine whether the index is suitable for your need: (1) the purpose of the index, (2) selection of the sample, (3) choice of the base period, (4) selection of weights, and (5) statistical adjustments.

Purpose of the Index

The exact purpose that an index number is intended to serve should be clearly understood by the reader. Thus, the Consumer Price Index is intended to measure the cost of a fixed bill of goods and services purchased by lower income urban workers; it does *not* claim to measure the cost of living of consumers generally, as is often misconstrued. Again, the Dow-Jones averages purport to measure the relative price changes of blue-chip market leaders, not the stock market generally. In similar fashion, the F. W. Dodge Corp. index of construction contracts awarded was developed to indicate relative changes in the value of contract building. It cannot be used to measure changes in the physical volume of construction nor changes in the value of construction put in place.

If a single index number proves inadequate, the use of several related indexes may fulfill a given need. For example, in analyzing monthly changes in regional business activity, it is useful to supplement a composite business index with indexes of employment, payrolls, construction contracts, retail sales, and the like that reflect changes in component elements of business.

Selection of the Sample

The second test of a good index number arises from the statistical requirement that the data must provide a representative sample, unless, of course, they cover the entire field. The principles for selecting a sample have been treated in Chapter 12. It is of the utmost importance that the data collected for constructing index numbers conform to these principles. Otherwise, no valid generalizations can be drawn from the results.

The following sampling plan is an effective one in selecting a sample of items to include in an index number.

First, divide the commodities into a large number of small groups or strata. Each group should comprise a closely related line of products that might be expected to move fairly uniformly in price, quantity, or value, as the case may be. Weights must be available for these groups. This stratification permits accurate weighting and flexible grouping into main categories as desired.

Then select from these groups a typical list of items to include not only all of the most important articles but also some that are typical of every category of goods in the group both in physical characteristics and price behavior, in the case of a price index. Of course, each item must be precisely identified. The prices are then weighted and the products totaled to form group indexes, and the latter are again combined to provide the overall index. The result may be called a highly *stratified judgment sample.*

In groups or parts of groups where there is little basis for selection, as when there are many items of minor or relatively equal importance, each 10th, 20th, or some other numbered item may be taken from the list.[7] This is a *systematic,* rather than a judgment, sample.

In any case, the proper selection of a typical cross section of items is the most crucial step in the entire process. Many regional "general business" indexes and others fail in this respect—they just do not measure what they purport to represent.

---

[7] Alternatively, the items may be selected with "probability proportional to size," size being defined as the relative weight of the item. See M. Wilkerson, *Sampling Aspects of the Revised CPI* (Washington, D.C.: U.S. Bureau of Labor Statistics, October 1, 1964), p. 12.

The number of items selected in each group may vary from 1 to 20 or more, depending on the group's importance and diversification. For all groups combined, several hundred items should be priced to constitute a sample of adequate size. The Bureau of Labor Statistics, for example, includes about 400 items in its Consumer Price Index,[8] while the Standard and Poor's index includes the prices of 500 common stocks. A smaller number might be used, however, for items that are fairly homogeneous as to type and price behavior.

Choice of a Base Period

The base of an index showing changes from time to time may be any period that provides the most suitable standard for comparison. There are a number of criteria for the selection of such a base. The most important of these are (1) normality of the period, (2) trustworthiness of the data in the period, (3) comparability with existing index numbers, and (4) inclusion of census years for bench-mark data.

Normality of Period. It is frequently held that the base period should be one that is "normal" or "average"; that is, a period when the level of the data is about midway between the peaks and troughs of business cycles in that era. A period of very high prices, for instance, should ordinarily be avoided as the base because the influence of the most inflated components would be disproportionately low in other periods. However, the peak year 1967 was chosen as base for government indexes because the industrial censuses that year provided base-year weights, as noted under "Inclusion of Census Years" below.

Trustworthiness of Data. Source materials have become generally more accurate and comprehensive in recent years, so that a recent period is more likely to provide a reliable base than an earlier period. For this reason, most government indexes have been revised in recent years to include new products and to embody new weights reflecting changed production and consumption patterns. At the same time the older base periods were replaced by a 1967 base, which more nearly encompasses both the recently developed products and the particular year for which the weights are available from census data.

Comparability with Other Index Numbers. The base for a new index number is often chosen to coincide with that of existing index numbers with which the new one is most likely to be compared. Index numbers are not directly comparable unless their base periods are identical. For this reason the Bureau of the Budget (now the Office of Man-

---

[8] On the other hand, some 2,500 items are included in the Bureau's Wholesale Price Index in order to insure the reliability of its many component indexes.

agement and Budget) has endeavored to standardize governmental indexes on a 1947–49, 1957–59, and 1967 base in these successive decades.

Inclusion of Census Years. Since it is preferable to use base-year weights as nearly as possible,[9] the base period should include census years for which bench-mark data are available as weights. Thus, the base year 1967 was selected for government indexes to coincide with the censuses of business, manufactures, mineral and construction industries, transportation, and other censuses taken that year.

Weights

Earlier in this chapter, weights were defined and used in calculating composite index numbers. Here the problems of selection of weights, type of weights, shifting weights, and weight bias are discussed.

Selection of Weights. Weights may be selected to represent either the importance of a specific commodity or the importance of the entire economic group of which it is typical. In the latter case, one might include in a production index of house furnishings the relative for a standard type of domestic wool rug weighted by the total value of all sorts of similar rugs rather than to include a large number of different rugs and weight each one according to its own specific importance. This group weighting system is used in the Federal Reserve Board Index of Industrial Production and the Bureau of Labor Statistics Consumer Price Index, as described later in this chapter.

Weights should also be appropriate to the purpose of an index. An average of relatives price index for a company's inventory, for example, should be weighted by inventory values; a price index of goods sold should be weighted by sales values; while a consumer price index should be weighted by consumer expenditures.[10]

Physical Quantities or Values as Weights. The factors used as weights for a given index number depend upon the method of construction and the kinds of data being employed. If it is an index number of prices and the aggregative method is used, that is, a method which adds the actual weighted prices, the weights must be *quantity* data of some kind, never value. Value includes the effect of price, since it equals price times quantity. Its use as a weight in an aggregative index would actually have the effect of squaring the prices, which would give undue

---

[9] U.S. Bureau of the Budget, Division of Statistical Standards, *Recommendations on Postwar Base Period for Index Numbers* (March 14, 1951), p. 2.

[10] Weights may be rounded off to two or three significant figures, or even one figure for minor items, since an appreciable difference in weights will affect an index but little.

importance to changes in the larger prices. Conversely, an aggregative quantity index would be weighted by prices. For an average of either price or quantity relatives, on the other hand, *value* weights should be used, as illustrated in Table 18–4.

Whether the weights used will be quantities or values may, however, depend upon the availability of data. For most kinds of commodities, exchange values in dollars are more likely to be available than quantities. Values must also be used for group weights where the items are in different units. In these cases, the weighted average of relatives method should be used.

Constant or Variable Weights.   Index numbers are designed to show changes only in the variable being measured—a price index, for instance, should isolate changes in price from changes which may be due to quality changes and other factors. None of the factors in the computation except prices should be allowed to fluctuate. The weights, therefore, should usually be kept constant for an extended period. If prices and weights were allowed to vary simultaneously, the resulting index numbers would reflect changes due to both factors, and no one could tell what part of the final result was due to variations in prices and what part was due to variations in the weights.

This raises the question: If the weights are to be held constant for extended periods, which specific period should they represent? In the examples used as illustrations of method, the weights were quantities or values in the period used as the base of the index numbers, but this is not necessarily the best procedure to follow in every case.

The importance of commodities may change during relatively short periods so that, if weights of an early period are used, there is a danger that the current index number will not accurately reflect the present relative importance of its several constituents. For instance, the cost of purchasing and maintaining a color television set is an important element in the present-day cost of living that did not exist a few years ago.

When it is definitely known that the constituents of the index are changing in importance, weights should be revised from time to time. Too frequent revisions, however, tend to impair the usefulness of an index number, so that ordinarily no change should be made as long as the weights are approximately correct. In long-established indexes the weights have been changed at intervals of about 10 years.

Bias Due to Weighting.   Bias due to methods of weighting is almost certain to occur in some degree. In this sense "bias" means that the index number tends to understate or overstate the degree of change because of the failure of the weights to represent accurately the relative importance of shifts in the items included. Price indexes are generally based on the cost of a fixed bill of goods, but people actually buy different quantities as prices change. The probable bias of any index due to shifts in consumption patterns and the like should be carefully considered before it is used in a major policy decision.

Statistical Adjustments

Most composite monthly indexes should be adjusted statistically to show the cycles and the long-term trend in the underlying data and to eliminate seasonal and irregular movements. (These adjustments will be discussed in Chapter 20.) That is, (1) the data should be adjusted for seasonal and calendar variations if necessary; (2) the resulting figures should be smoothed by moving averages (described in Chapter 20), so that the series will show more consistent trend-cycle changes from month to month than meaningless zigzag irregularities; and (3) a dollar value series should be deflated by a price index if it is desired to show physical volume changes (Chapter 19). It is also desirable to determine whether the index is typically a leading, coincident, or lagging indicator at business cycle turning points. (See U.S. Department of Commerce, *Business Conditions Digest,* monthly.)

Monthly business indexes should also be checked against more complete annual data or quinquennial censuses of manufactures and other censuses in order to adjust the general trend of the monthly series to these more accurate bench marks. Otherwise, a monthly index based on sample data will develop a cumulative upward or downward bias over the years which will destroy its validity for long-term comparisons.

REVISIONS OF INDEX NUMBERS

Substitution of Items

Changes in production, distribution, habits of consumption, and a variety of other economic factors sometimes necessitate substitutions in the items included in an index, in its list of respondents, or in the specifications of the items included. For example, in 1971 the Bureau of Labor Statistics added 86 items (such as a diesel farm tractor) and dropped 36 items (such as a wagon chassis and a drawn plow) in computing its Wholesale Price Index. The availability of new and better data may also make it desirable to revise established index numbers, as described above. When interpreting the movement of index numbers it is essential that these changes be kept in mind, for the particular method of revision may make a great deal of difference in the final result.

Changing the Base Period

The base period of an index number may need to be changed in either of the following situations: (1) When index numbers based on different periods are to be compared, it is necessary to shift one index to the same base period as the other, so that changes in the two will be measured from the same point in time. (2) It may be desired to shift the base of a series to some arbitrary reference date such as 1972 in order to compare subsequent changes with conditions at that time.

A series can be shifted to a new base by multiplying each of its index numbers by $100/X$, where X is the index number for the period selected as the new base. That is, $X \cdot 100/X = 100$. Since each of the indexes is multiplied by the same constant factor, the *relative* fluctuations of the series remain unchanged.

To illustrate, in Table 18–6 the base period for prices paid by farmers

Table 18–6

SHIFTING THE BASE PRICES PAID BY FARMERS
FROM 1910–14 TO 1967 FOR COMPARISON
WITH THE CONSUMER PRICE INDEX

| | Prices Paid by Farmers for Family Living Items | | Consumer Price Index |
	1910–14 = 100 (1)	1967 = 100* (2)	1967 = 100 (3)
1967	302	100	100
1968	310	103	104
1969	324	107	110
1970	336	111	116
1971	352	117	121

* Obtained by multiplying column 1 by 100/302 to shift the 302 value for the 1967 average to the 100 level.
Source: *Survey of Current Business.*

for family living items has been shifted from 1910–14 to 1967 for comparison with changes in the Consumer Price Index since that period. Since the original index of prices paid by farmers averaged 302 in 1967, the whole series has been multiplied by $100/302 = .3311$ to shift the 1967 average to 100 (column 2), the same as for the Consumer Price Index. The last two columns show that from 1967 to 1971, prices paid by farmers advanced only 17 percent as compared with 21 percent for consumer prices generally, even though the original farm price index increased by more points than the Consumer Price Index.

Splicing Two Series

It is often necessary to splice two series to form a continuous series, as when the specifications of a commodity in a price index are changed. Any two series may be spliced, provided they are both available for the same year. For example, the BLS Wholesale Price Index might be said to include everything but the kitchen sink. This is not true. It includes an enameled steel sink, but the price of a new reporting company was added to its sample in November 1958. As a result, the typical price had to be shifted from $13.39 (or an index of 100.8 on the 1957–59 base) to $13.13 in that month. Table 18–7 shows how to continue the

Table 18–7

SPLICING TWO PRICE SERIES
REPRESENTING AN ENAMELED STEEL SINK

(Prices in Dollars; Indexes on 1957–59 Base)

	Original Sample of Reporting Companies		Enlarged Sample of Reporting Companies		Spliced Series Index
	Price (1)	Index (2)	Price (3)	Index (4)	(5)
September 1958	$13.194	99.4			99.4
November 1958	$13.39	100.8	$13.13	100.8	100.8
June 1959			$12.71	97.6	97.6

SOURCE: U.S. Department of Labor, *Wholesale Prices and Price Indexes, 1958*, Bulletin No. 1257, (July 1959) pp. 225 and 230 (item #1053–11), shifted to 1957–59 base.

original price index (column 2) for the sink by splicing the new price (column 3) onto it. The new price of $13.13 in the overlapping month November 1958 must be shifted not to 100 but to 100.8, the index for that month. The new price series, therefore, is multiplied by 100.8/$13.13, as shown in column 4. The spliced series in column 5 (combining columns 2 and 4) now shows enameled steel sink prices continuously throughout the period, although the actual sample price shifts in November 1958.

As another example, the new car component of the Consumer Price Index (based on a standard-sized Chevrolet, Ford, and Plymouth) became outmoded in 1960 with the widespread introduction of compact cars, whose price behavior differed from that of standard-sized models. Hence, the Bureau of Labor Statistics introduced the prices of four small cars (Rambler, Falcon, Valiant, and Corvair), linking the new series onto the old in October 1960 so that the level of the index was not affected by the lower price of the compact cars.[11]

Strictly speaking, an index which is being shifted to a new base should be composed of the same items during the whole period of the index. Yet the most common use of base shifting is to link a current index containing one group of items to an earlier-period index containing a similar but not identical group of items. This procedure is legitimate if the old and new groups of items may be considered to be representative of the same population. This is true of the above example. In case the components of an index have changed more radically from time to time, however, as in the Cleveland Trust Company index of business activity from 1790 to date, the index loses its homogeneous character.

SOME IMPORTANT INDEXES

There are many more business indexes in common use than can be treated here. Hundreds of these are described in the readings at the end of the chapter. We will discuss only three major indexes—their construction, uses, and limitations—to illustrate the typical problems involved. These are the consumer and wholesale price indexes of the U.S. Bureau of Labor Statistics and the industrial production index of the Federal Reserve Board. The base period for all these indexes is 1967 = 100.

Consumer Price Index

"The Consumer Price Index (CPI) is a statistical measure of changes in prices of goods and services bought by urban wage earners and clerical workers, including families and single persons."[12]

The index is computed by the weighted average of relatives method[13] using constant weights. Prices are measured monthly or quarterly, and the aggregate cost of a fixed bill of goods and services is compared with that in the base period 1967. Since the quantities represent not only consumption of the 400 goods and services actually priced but also consumption of related items for which prices are not obtained, the total

---

[11] O. A. Larsgaard and L. J. Mack, "Compact Cars in the Consumer Price Index," *Monthly Labor Review* (May 1961).

[12] See U.S. Department of Labor, *The Consumer Price Index, A Short Description,* 1971, for further details.

[13] Three variants of this method are actually used: (1) the "average of price relatives for reporting outlets," (2) the "relative of average prices for identical outlets," and (3) the "relative of average prices for all reporting outlets." See M. Wilkerson, *Sampling Aspects of the Revised CPI* (U.S. Bureau of Labor Statistics, October 1, 1964).

cost of the "market basket" represents a broad sector of total consumer spending for goods and services.

The prices collected for this index are retail prices charged to consumers for "food, clothing, automobiles, homes, house furnishings, household supplies, fuel, drugs, and recreation goods; fees to doctors, lawyers, beauty shops; rent, repair costs, transportation fares, public utility rates, etc." These prices include sales and excise taxes as well as real property taxes, but not income or personal property taxes.

The 400 goods and services comprising the "market basket" of items sampled are representative of the typical goods and services purchased by urban wage and clerical worker families and single individuals living in urban areas with a 1960 population of 2,500 or more persons. These families and single workers comprised about 56 percent of the people living in urban places and about 40 percent of the total U.S. population in 1960. The index is designed to measure *only* changes in prices of the same "market basket" through time, *not* to measure changes in the composition of different market baskets or changes in consumers' standards of living.

Periodically, the bureau conducts Consumer Expenditure Surveys to determine the pattern of expenditures for goods and services by wage earners and clerical workers. The last survey was conducted in 72 urban areas which were chosen to represent all urban places in the 50 states. From the data collected, the bureau revised the quantity weights used to compute the index and objectively select the 400 items to be included.

All items purchased by wage earners and clerical workers were grouped or stratified into "expenditure classes." The items included in each of the 52 expenditure classes, which define the sampling strata, were primarily determined by grouping items which in a general way serve the same human needs. Items were selected with probability proportional to their relative importance as compared with total expenditures for all items. In relative importance, housing comprised 33 percent of the total index in December 1963, food 22 percent, health and recreation almost 20 percent, transportation 14 percent, and apparel 11 percent.

The urban places in which the bureau collects price data for the CPI also were selected by probability sampling. The primary sampling units were 50 standard metropolitan statistical areas. These units were stratified by broad region and by size of population into 12 strata.

The relative importance of each area in the CPI is determined by the proportion of total wage-earner and clerical-worker population it represents to the total for all areas represented in the CPI, based on 1960

Census data. Chart 18–1 shows the changes in the index and in three major components for 1966 to 1972. In addition to the overall index, a separate index is published for each of 23 Standard Metropolitan Statistical Areas.

Uses of the Consumer Price Index.   A major use of the index is to aid unions and management in adjusting wages to take account of changes in consumer prices. The most important impetus to the use of the index for this purpose was its designation as a basis of wage-rate

Chart 18–1
CONSUMER PRICES

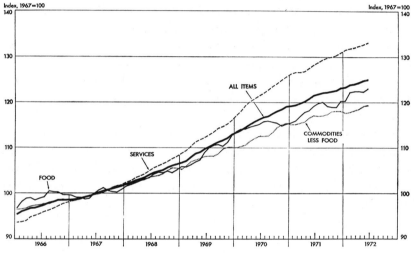

SOURCE: Department of Labor, Council of Economic Advisers.

escalation in the contract signed by the United Automobile Workers and the General Motors Corporation in May 1948. Since then the agreement has been extended several times.[14] After each major agreement, many other contracts have been signed on the same basis, frequently without any examination of the reasonableness of the relationship of wage-rate changes to index changes in each particular situation, or without full realization of the effects of arbitrarily accepting a ratio based on some other firm's or union's experience. Whatever the type of escalator employed, however, it is important to both sides in a bargaining group that the procedure be adjusted to each particular situation.

Escalator clauses based on the CPI are used not only to adjust wage

---

[14] See *Major Collective Bargaining Agreements: Deferred Wage Increase and Escalator Clauses,* U.S. Department of Labor Bulletin No. 1425–4 (January 1966).

payments but also to adjust rents, pensions, alimony, fiduciary payments, and many other types of contracts. Finally, the CPI is widely
cited as an indicator of inflation as it affects the consumer. It serves,
therefore, to measure the purchasing power of the consumer's dollar.

The Consumer Price Index also has limitations which should be
carefully considered: (1) It measures changes only in a fixed bill of
goods and services, but not changes in the standard or manner of living.
(2) It does not always reflect gains due to the improvement in the
quality of manufactured products. Hence, it is claimed to overstate the
true rate of inflation.[15] Conversely, in wartime conditions of material
shortages, it fails to reflect the full inflationary effect of black-market
prices, quality deterioration, and substitution of more expensive grades
for cheaper grades of products. (3) While it measures changes in
consumer prices from time to time, it cannot be used to compare prices
between different places at a single point in time. Geographic differences may be measured by comparing the individual prices compiled for
the Consumer Price Index, but not the index itself. (4) The index
measures changes in prices only for the worker group in urban areas. It
should not be used without modification for other income groups, for
families living in nonurban areas, nor for any individual family. (5)
The CPI is subject to sampling errors and faulty reports from respondents. However, it is believed to be accurate enough for most practical
uses.

Wholesale Price Index

The Wholesale Price Index of the Bureau of Labor Statistics measures the average rate and direction of movements in commodity prices
as primary-market levels—that is, at the point of the first commercial
transaction for each commodity—and specific price changes for individual commodities and groups of commodities.[16] The prices used in the
index are those representing all sales of goods by or to manufacturers
or producers, or those in effect on organized commodity exchanges.
Therefore, it represents producers' prices or primary-market prices
rather than those charged by wholesalers.

Prices for approximately 2,500 separate specifications of commodities
are included in the index. To obtain "real" or "pure" price changes not
influenced by changes in quality, identical lists of commodities defined

---

[15] See W. Allen Wallis, *Journal of the American Statistical Association* (March 1966),
pp. 1–10; also, *Monthly Labor Review* (September and November 1961), articles by
Milton Gilbert and Ethel Hoover, respectively.

[16] See U.S. Department of Labor, *Wholesale Prices and Price Indexes for January
1971* (July 1971), pp. 104–109.

by precise specifications are priced from month to month. Prices are adjusted for trade and quantity discounts, as well as cash and seasonal discounts when these are customary. Excise taxes are excluded. These prices are obtained from some 2,000 companies which are asked to quote the prices they actually charge for a specific commodity to a given type of buyer on a particular day, usually the Tuesday of the week including the 13th of the month. Some quotations from trade journals, organized exchanges, and other government agencies are also used.

Because the commodity population is so large, the index is based on a sample of commodities, a sample of specifications for the commodities, and a sample of reporting sources. The individual items are selected as the most important in each field and as those believed to represent the price movements of other closely related commodities. The sample is thus a highly stratified, selected group, rather than a random sample. The broad coverage of 2,500 items permits the development of reliable subindexes for many small subdivisions of the economy.

The index is calculated fundamentally as a weighted average of price relatives in which the weights are based on net sales values of commodities reported by the 1963 industrial censuses. Each item has a weight which includes its own weight based on its sales in 1963 and the weight of the other items it represents in the index.

The overall index is divided into the broad categories of industrial commodities, and farm and food products, as shown in Chart 18–2. Special wholesale price indexes are reported by stage of processing and by durability of product. In addition, separate indexes are published each month for many major groups and subgroups, hundreds of product classes, and for most of the individual series.

The Bureau of Labor Statistics also prepares a Weekly Wholesale Price Index based on actual weekly prices of a sample of several hundred of the commodities included in the monthly index and on estimates of the prices of the other commodities. This index may be used to give interim estimates of the monthly index.

Uses of the Wholesale Price Index. The Wholesale Price Index is one of the basic business barometers used to measure the economic health of the nation. It is also used as a price deflator or as a purchasing-power index, reflecting changes in the value of the dollar. The important application of price indexes in deflating value series is described in Chapter 19.

This index, or any of its component indexes, may be used for comparison with series of individual business data. For example, the General Electric Company provides its purchasing offices with a price index of commodities purchased by the company, weighted by their importance to the company, and compares this with the BLS wholesale price index for industrial commodities.

One of the most frequent uses of the Wholesale Price Index is as an escalator—that is, as the basis for adjusting contractual payments or values for changes in the value of the dollar. Long-term production contracts include escalator clauses as guarantees against losses due to

Chart 18–2

WHOLESALE PRICES

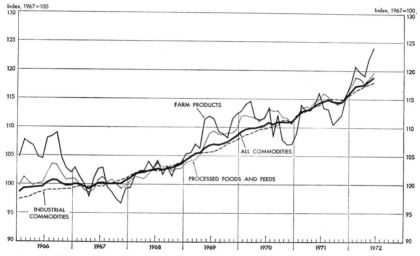

SOURCE: Department of Labor, Council of Economic Advisers.

increases in the prices of materials and other costs. Rentals on long-term leases are also often adjusted by this index.[17]

There are limitations to the wholesale price indexes which must be kept in mind when using them: (1) They measure primary-market prices, not wholesalers' prices as the name implies. (2) Most of the indexes relate to national coverage and hence should be used with caution in interpreting local or regional data. (3) Since they relate to changes of a given specification, they cannot be used with retail price indexes to calculate margins. (4) The indexes do not include any services, such as rent, transportation, or communications.

---

[17] See "The Use of Price Indexes in Escalator Clauses," *Monthly Labor Review* (August 1963).

Industrial Production Index

The monthly Federal Reserve Board Index of Industrial Production is one of the most widely used of the country's economic indicators. It measures changes in the physical volume of output of factories, mines, and gas and electric utilities from 1919 to date.[18]

The industrial production index includes 227 series expressed in physical terms—units, tons, yards, board feet, and the like—reflecting the production of American industries or data which represent such series. Where physical output data are lacking, other series which are believed to fluctuate in the same way as output data are substituted. Such series include volume of shipments, production-worker man-hours, materials consumed in production, etc. About one third of the monthly index is based on electric power consumption, and one fifth is based on man-hour data adjusted for estimated changes in output per man-hour. The balance represents actual production, shipments, and the like.

The component series of the index are combined with weights based on value added by manufacture (or gross value in some cases) in 1967. Monthly indexes are adjusted annually to the more detailed figures of the Census *Annual Survey of Manufactures*. The composite index is calculated as a weighted average of relatives. It is expressed in terms of the 1967 average as base, for comparability with other index numbers. The index is published for several broad "market groupings" having the following relative importance in 1967: final products, 48 percent (including consumer goods, 28 percent, and business and defense equipment, 20 percent); intermediate products (including construction), 13 percent; and materials, 39 percent. A separate classification is made into the major industry groupings of durable manufacturing, nondurable manufacturing, mining, and utilities. Indexes are also reported for hundreds of individual industrial groups, following the Standard Industrial Classification system. This great number of industry series permits flexible grouping for most desired comparisons.

The monthly production series are adjusted to levels shown by bench-mark production indexes based on the Censuses of Manufactures and Minerals and for interbench-mark years, mainly Census Annual Surveys. These adjustments are made periodically, and usually during a revision of the index. Between revisions, the levels of the monthly indexes are checked against independently compiled data, such as deflated manufacturers' shipments adjusted for inventory change and electric power used by the manufacturing and mining industries.

---

[18] See *Industrial Production, 1971 Edition,* Federal Reserve Board (November 1972).

Uses of the Industrial Production Index. The major use of the
Index of Industrial Production is as an indicator of the economy's output. It is the most sensitive and reliable indicator we have to answer
the questions "Is production increasing or decreasing?" and "In which
industries are major increases or decreases occurring?" Chart 18–3

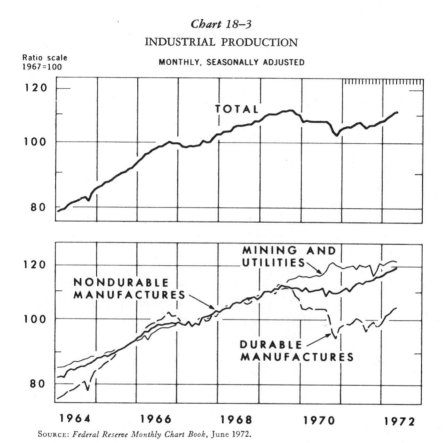

Chart 18–3

INDUSTRIAL PRODUCTION

MONTHLY, SEASONALLY ADJUSTED

Ratio scale
1967=100

Source: *Federal Reserve Monthly Chart Book*, June 1972.

shows the movements in total production and three major components
from 1964 to 1972. The index is widely used in conjunction with other
series for both forecasting and guidance in administrative decisions. For
example, it is compared with figures on unemployment to obtain estimates of the country's total number of unemployed workers that may
be associated with different levels of production. It is also compared
with data on inventories, new orders, manufacturers' shipments, and
retail sales.

The detailed industry indexes serve as very useful comparisons or bench marks in studying the production of individual companies. The individual indexes are also useful in comparing growth rates in different sectors of the economy.

One limitation of the industrial production index is its restriction to manufacturing, mining, and utilities, which keeps it from serving as a measure of total production. Agriculture, construction, transportation, communication, and other services are not included. Another limitation is that changes in electric power consumption, man-hours, and other indirect measures of industrial activity sometimes do not reflect accurately the changes in physical volume of production, particularly in times of war and postwar reconversion.

SUMMARY

Index numbers express the changes in a variable relative to some base taken as 100. They are particularly useful in comparing different series and in combining a group of series in a single summary figure. Most indexes are designed to show changes in price, quantity, or value (price times quantity), either from time to time or from place to place.

A simple index or relative is constructed by dividing a single series by its base figure and multiplying by 100.

Composite indexes should ordinarily be weighted arithmetic means of their components. A composite price or quantity index may be constructed by two methods: (1) In the weighted average of relatives method, the relatives are first computed for each series as described above and then multiplied by value weights expressed as decimal fractions of the total weight. The sum of the weighted relatives is the composite index. (2) In the aggregative method, the changing prices are multiplied by fixed quantity weights (or vice versa for a quantity index). The resulting products are then totaled, divided by the product in the base period or place, and multiplied by 100. The weights usually represent the importance of a component in the base years or some other normal period. In a value index the dollar values of each component are simply added in the aggregative method or else the components are expressed as relatives and multiplied by arbitrary weights before being totaled.

The aggregative method is the simpler of the two, but the average of relatives method is preferable when individual series are to be compared, when available weights are in value form, or when the component series are expressed as relatives.

The following tests of a good index should be applied in appraising the validity of an index for some specific use: (1) The purpose of the index should be clearly defined. (2) The items included must be specifically related to the purpose and must be a representative sample of the population being measured. (3) The base period should be a fairly normal one, adequate in length, easy to recall, and one used by comparable indexes. Trustworthy data and census bench marks should be available for this period. (4) Appropriate quantity weights should be used in an aggregative price index, and vice versa, or value weights in an average of relatives index. Weights must be held constant, but should be revised every decade or so as the importance of the components changes appreciably. The probable bias due to weighting should also be considered.

Items may be substituted for others in an index, as necessary, by proper "linking." An index number may be changed to a new base or spliced onto a similar series by multiplying or dividing by a constant factor without changing the relative movements of the index in any way.

The construction, uses, and limitations of three major indexes are discussed to illustrate typical examples. The consumer and wholesale price indexes of the Bureau of Labor Statistics represent broad samples of prices at the retail level and the primary market level, respectively. They are widely used as economic indicators, as deflators of value series, and as escalators in contracts. The proper use of the Consumer Price Index in wage contracts is particularly important.

The Federal Reserve Board Index of Industrial Production is an important and sensitive measure of general industrial activity. It represents the physical volume of production, shipments, or man-hours in the manufacturing, mining, and utility industries.

Many other indexes are described in the Selected Readings below.

PROBLEMS

1. *a*) Briefly describe three broad types of index numbers that are used to measure changes in business and economics.
   b) In your opinion, what is the one most important use of (1) simple index numbers and (2) composite indexes? Give reasons for your choice in each case.
   c) Cite the principal limitations of index numbers.

2. *a*) Compute a composite index of grain prices for the data below by the average of relatives method, with 1969 = 100, using base-year weights.
   b) Compute a composite price index by the aggregative method, using the same base.
   c) Compare the merits of the two methods in this case.

	Price per Bushel		Billions of Bushels	
	Wheat	Corn	Wheat	Corn
1969	$1.75	$1.19	1.46	4.58
1970	1.79	1.33	1.37	4.10
1971	1.72	1.36	1.64	5.54

Note: Price is wholesale, average, all grades; production is crop estimate.

SOURCE: *Survey of Current Business*, February and June 1972.

3. Using the data in Problem 2:
   a) Compute a composite index of grain *production* by the average of relatives method, with 1969 = 100, using base-year weights.
   b) Compute a composite production index by the aggregative method, on the same base.
   c) Compute an index of the *value* of grain production, on the same base.

4. As a purchasing agent for the Erie Steel Supply Company, you wish to compile a composite price index for iron and steel purchased, based on the following data (pig iron and steel scrap are in long tons, steel billets in short tons):

	Price per Ton			Thousands of Tons Purchased		
	Pig Iron	Steel Scrap	Steel Billets	Pig Iron	Steel Scrap	Steel Billets
1970..................	$61	$54	$81	10.0	3.0	5.0
1972..................	66	38	94	11.0	2.1	5.5
1974..................	66	34	95	10.7	3.6	2.7

   a) Compute a composite index for iron and steel prices each year by the average of relatives method with 1970 = 100, using value purchased in 1970 as weights.
   b) Compute a composite price index by the aggregative method, using the same year for the base and for weights as above.
   c) How do the indexes obtained in (a) and (b) differ? Explain. What is the chief advantage of each method in this case?

5. a) Compute a composite index of the quantity of iron and steel purchased each year, from the table above, using the average of relatives method. Take 1970 as base, and use 1970 values as weights.
   b) Compute a composite index of the dollar *value* of iron and steel purchased each year, with 1974 = 100.
   c) Explain the significance of the quantity and value indexes computed above, as opposed to the price index.

6. As a cost analyst with a petroleum company, you are asked to compile an annual index of oil well drilling costs beginning in 1966, with 1967 as a

base. You determine that the cost of drilling an oil well is made up of approximately 60 percent labor and 40 percent material, and you decide that the following data adequately represent these elements.

Year	Average Hourly Earnings, Petroleum and Coal Workers	Wholesale Price Index, Metals and Metal Products (1967 = 100)
1966	$3.41	98.8
1967	3.58	100.0
1968	3.75	102.6
1969	4.00	108.5
1970	4.28	116.7
1971	4.58	119.0

SOURCE: *Survey of Current Business,* June 1972, and supplement, *Business Statistics, 1971.*

a) List the indexes of drilling costs, along with any columns of computations needed.

b) What was the percentage increase in drilling costs from 1966 to 1971? If 1971 were the base of the drilling cost index, what would the 1966 index be? If labor and materials each made up half of drilling costs, would the index be higher or lower in 1971 than that shown? Why?

c) What more refined indexes might you be able to find, to replace those used here, to provide a better index of your company's drilling costs?

7. The Bureau of Business Research of the University of Texas published a monthly *Index of Texas Business Activity* in the 1950s with the following description: "1947–49 average = 100. Components: Retail sales, industrial electric power consumption, miscellaneous freight carloadings, building authorized, crude petroleum production, ordinary life insurance sales, crude oil runs to stills, total electric power consumption (weighted 46.8, 14.6, 10.0, 9.4, 8.1, 4.2, 3.9, and 3.0, respectively, and adjusted seasonally)." Each component was expressed as an index with 1947–49 = 100 before being weighted. Apply our tests of a good index number to give an appraisal of this index, listing its good and bad points.

8. Index numbers are ordinarily based on samples, so that care must be exercised to insure that the items included in the index are typical of the population.
    a) Describe the population represented by (i) an index of prices received by farmers, (ii) an index of industrial building costs, (iii) an index of manufacturing production, and (iv) an index of retail sales in urban areas, for the United States in each case.
    b) Samples used in index numbers are usually stratified. Why?
    c) Compare the advantages of random, systematic, and judgment sampling in selecting items for a price index representing a comprehensive list of women's apparel items.

9. If you were to choose a new base period to replace the 1967 base for federal government indexes, what year or years would you choose? Appraise the merits and drawbacks of this period according to the four criteria given in this chapter for choice of a base period.

10. *a*) Convert the American Appraisal Co. index of construction costs, below, to the 1967 average as base.
    b) Compare the changes in construction costs since 1967, as shown by the Engineering News-Record and American Appraisal Co. indexes.
    c) If in early 1973 the only available construction cost index for 1972 was the Department of Commerce figure of 140, compared with 131 for 1971, use these figures to estimate the American Appraisal Co. index (1967 = 100) for 1972.

	Engineering News-Record (1967 = 100)	American Appraisal Co. (1913 = 100)
1967	100.0	909
1968	107.8	970
1969	118.7	1,050
1970	128.9	1,132
1971	146.7	1,258

SOURCE: *Survey of Current Business*, June 1972, and supplement, *Business Statistics, 1971*.

11. Find an article in *Monthly Labor Review* or elsewhere reporting on the Bureau of Labor Statistics' latest program of revising the Consumer Price Index or the Wholesale Price Index. Describe the principal steps in this program and explain how the resulting improvements justify the considerable expense involved.

12. The Ford Motor Company's agreement of September 1958 with the UAW–CIO unions called for a quarterly cost of living allowance of approximately 1 cent per hour in straight-time hourly earnings for each 0.5 point change in the Bureau of Labor Statistics Consumer Price Index (1947–49 = 100) above, but not below, the base index level of 119.1, beginning with 1 cent for index 119.2 to 119.6. (The November 1958 index was 123.7.)

   In another case, the H Company reached an agreement with the Metal Workers' Union stating that if the Consumer Price Index increased or decreased by 5 percent or more in any semiannual period, wages would be adjusted upward or downward by the same percentage.

   Compare the merits of these two agreements as to:
   a) Adjusting wages at all levels by 1 cent per hour for each 0.5 point change in the Consumer Price Index versus adjusting wages by the same percentage as the change in the Consumer Price Index.
   b) Adjusting wages in little jumps (i.e., quarterly for each 0.5 point change in the Consumer Price Index) versus big jumps (i.e., semiannually by 5 percent or more provided the Consumer Price Index has changed that much).

c) Setting a minimum level of wages 4.6 cents an hour below the September 1958 rate, as indicated in the first paragraph, versus adjusting wages upward or downward without limit, in line with the Consumer Price Index.

13. Why is the Bureau of Labor Statistics Wholesale Price Index, excluding farm products and foods, frequently used in place of the All Commodities Index as a measure of general price changes?

14. If you were the economist of a national chain of drugstores and wished to compare the prices you pay with the Bureau of Labor Statistics Wholesale Price Index:
    a) Which subgroups of this index would you combine to meet your needs?
    b) What method, arithmetically, would you employ to combine them?

15. Is the following procedure appropriate? If not, suggest improvements. In order to allow for changes in the cost of living, a wage contract is set up by the Ajax Machine Tool Company of Houston, Texas, providing that machine tool workers' wages will be adjusted upward or downward each month by 1 cent per hour for each one-point change in the Wholesale Price Index.

16. What subindex or group of subindexes of the Federal Reserve Board Index of Industrial Production is appropriate for comparisons with the physical volume of production of:
    a) A large integrated oil company?
    b) A manufacturer of home laundry and kitchen appliances?
    c) A household furniture factory?

17. Present a critical analysis of a composite business or economic index of interest to you (other than the Bureau of Labor Statistics price indexes or the Federal Reserve Board Index of Industrial Production), describing its (a) purpose, (b) method of construction, and (c) limitations. (See Selected Readings, below, for sources.)

18. Considering the economic characteristics of your own state or metropolitan area:
    a) List four business indicators that are most significant for this state or area, giving exact sources.
    b) Describe and appraise a general business index published for this state or area.

19. What published indexes or indicators are most appropriate for use in the following situations?
    a) You wish to set a price at which to sell your frame house, which you bought new for $25,000 four years ago.
    b) The manager of a wool textile mill is anxious to learn if the expansion

in his volume of production over the past 18 months has kept pace with that of the industry.

c) The controller of a gas and electric company needs an adjustment factor with which to revise the basic level of pension payments, set up 10 years ago, for the company's retired workers.

d) An agricultural implement manufacturer needs information on recent trends in farmers' operating margins.

e) The president of a chain of department stores desires a monthly measure of changes in consumer purchasing power. He intends to compare this with the sales of his stores.

20. Justify or criticize the following actions. If an action is incorrect, indicate what should be done instead.

a) An oil company economist is asked to compare the growth of his industry since 1960 with that of industry in general. He prepares a chart showing total dollar sales of the oil industry each year, expressed as index numbers on a 1967 base, together with the Federal Reserve Board Index of Industrial Production.

b) An executive in Kansas City is offered a job in Cleveland and wishes to compare the cost of living in the two cities. The latest Consumer Price Index is 115.3 for Kansas City and 108.1 for Cleveland. Therefore, he concludes that living costs are somewhat lower in Cleveland.

c) The purchasing agent for a chain of automobile accessory stores who buys his major items direct from manufacturers needs a summary measure of general price changes each month with which to compare his costs. He chooses the Bureau of Labor Statistics Wholesale Price Index for this purpose.

d) A newspaper writer observes that gross national product has increased from $251 billion in 1948 to $1,047 billion in 1971. Therefore, he reports that the nation's output of goods and services has increased about fourfold over this period.

21. You wish to construct an index for 1964–71 representing the selling price of the merchandise of your retail clothing chain. You have constructed a price index since 1969. You also find in the company files an index computed by a former sales manager. This index was discontinued in 1966, but appears to have been correctly constructed for the years in which it was used. Since your index begins in 1969, you decide that the Bureau of Labor Statistics Consumer Price Index (CPI) for apparel would be satisfactory for the intervening years only. Construct the required index by splicing together the three series. Keep 1969 as the base year.

	Previous Price Index (1964 = 100)	CPI—Apparel Index (1967 = 100)	Your Price Index (1969 = 100)
1964	100.0	92.7	
1965	102.0	93.7	
1966	105.2	96.1	
1967		100.0	
1968		105.4	
1969		111.5	100.0
1970		116.1	105.8
1971		119.8	110.0

SOURCE: Consumer Price Index for apparel from *Survey of Current Business*, June 1972, and *Business Statistics, 1971*.

SELECTED READINGS

COMAN, EDWIN T. *Sources of Business Information.* Berkeley: University of California Press, 1964.

A comprehensive guide to publications providing a wide range of business indicators.

DOODY, FRANCIS S. *Introduction to the Use of Economic Indicators.* New York: Random House, 1965.

A guide to economic measurement and forecasting, with exercises in the use of major indicators.

MAUNDER, W. F. (ed.). *Bibliography of Index Numbers.* London: Athlone Press, 1970.

A computer listing of 2,600 indexes published in many countries through 1967.

MOORE, GEOFFREY H., and SHISHKIN, JULIUS. *Indicators of Business Expansions and Contractions.* New York: National Bureau of Economic Research, Columbia University Press, 1967.

Analyses and charts of the National Bureau of Economic Research indicators, with classification into leading, coincident, and lagging series.

U.S. BOARD OF GOVERNORS OF THE FEDERAL RESERVE SYSTEM. *Industrial Production, 1971 Edition.* Washington, D.C.: Board of Governors of the Federal Reserve System, 1972.

An authoritative treatment of principles and methods of constructing a quantity index.

U.S. BUREAU OF LABOR STATISTICS. *Major Programs* (annual). Washington, D.C.: United States Department of Labor, 1972 *et seq.*

Contains descriptions of data collection and methods of preparing 50 major Bureau of Labor Statistics series.

U.S. BUREAU OF THE BUDGET. *Statistical Services of the United States Government.* Rev. ed. Washington, D.C.: U.S. Government Printing Office, 1968.

Part III describes the principal statistical publications of federal agencies.

U.S. CONGRESS, JOINT ECONOMIC COMMITTEE. *1967 Supplement to Economic Indicators.* Washington, D.C.: U.S. Government Printing Office, 1967.

Contains brief descriptions of the series regularly included in *Economic Indicators* and describes uses and limitations of each.

U.S. DEPARTMENT OF COMMERCE. *Business Statistics,* biennial supplement to the *Survey of Current Business.* Washington, D.C.: U.S. Government Printing Office, 1971 *et seq.*

The "Explanatory Notes to the Statistical Series" referred to in the footnotes of the tables cover 2,500 monthly or quarterly series.

19. TIME SERIES ANALYSIS:
SECULAR TREND

MODERN BUSINESS and economic affairs are intensely dynamic in nature, and the analyst must be alert to interpret the significance of the passing scene. The changes are of many types. The long-term growth of industrial production, the residential building cycle, seasonal swings in department store sales, the daily movements of stock prices, and countless other elements in the dynamics of enterprise must be measured and appraised as an aid in understanding current developments and in formulating future policy. The importance of dynamic fluctuations, as opposed to static analysis, is reflected by the fact that the great bulk of data in business and economic publications (e.g., *Survey of Current Business, Economic Indicators*) is in the form of time series rather than being primarily classified by size, space, or other qualitative criteria at a given point of time.

TYPES OF BUSINESS FLUCTUATIONS

It is not sufficient for a businessman to observe merely the overall behavior of an economic indicator. There are various factors at work, the combined effect of which produced this result. Suppose a company's sales increased 6 percent over last month. Was this increase attributable to normal growth, a cyclical business boom, a pickup in seasonal demand, or an advertising campaign? What action should be taken as a result? Analysis of the data involves segregation of these factors so that their separate importance can be understood. The first necessity, then, is to know what factors are present in a time series. Next, how can the effect of each force be measured? Finally, how can it be predicted as an aid to forward planning?

The principal component fluctuations in a time series are as follows:

1. Secular trend.
2. Cyclical fluctuations.
3. Seasonal variation.
4. Irregular movements.

To illustrate, Chart 19–1 shows the monthly production of chemicals over a 15-year period, broken down into a rising trend, the wavelike cycles having a period of three to five years, the seasonal movement repeating its pattern each 12 months, and the small irregular residual. The trend value is measured in the original unit of the series (an index

Chart 19–1

THE ANATOMY OF A TIME SERIES
PRODUCTION OF CHEMICALS AND RELATED PRODUCTS

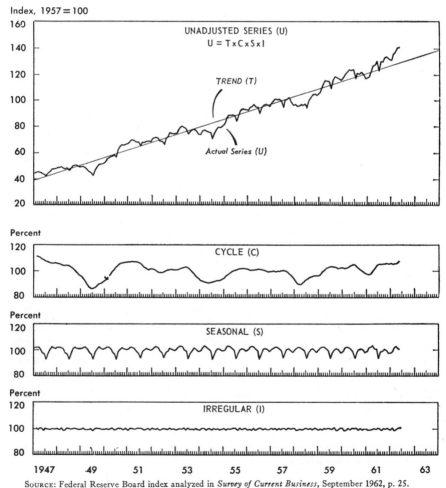

SOURCE: Federal Reserve Board index analyzed in *Survey of Current Business*, September 1962, p. 25.

number in this case), while the other three components are expressed in percentages. The product of the four components makes up the actual series.

Some time series contain all of the foregoing elements; others contain only some of them. Certain series are so largely controlled by one type of fluctuation that it is easily recognized from the original data. Thus, the production of synthetic fibers and frozen foods have a strong upward trend, durable goods suffer wide cyclical swings, department store sales are predominantly seasonal, and manufacturers' purchased material inventories move irregularly. Usually, however, the several components are not separately recognizable in the original data, but the businessman or economist needs to know the influence of each in order to understand the forces at work and the probable future behavior of the series. Therefore, the analyst's problem in dealing with time series is to identify the components and measure them separately.

The work of analysis can be divided into three parts: (1) fitting a secular trend curve, (2) measuring seasonal variation, and (3) analyzing cyclical-irregular residuals.

This chapter and the next one contain an explanation of the most useful methods for carrying out these three steps in the analysis of time series. In a particular application, only one or perhaps two of the steps may be needed, depending on the importance of the component and the purpose of the study.

SECULAR TREND

Secular trend is the gradual growth or decline of a series over a long period of time. The growth is ordinarily one of physical volume, like biological change; it does not strictly apply to long-term movements in prices, which do not grow in the biological sense. Hence, secular trend analysis usually applies to physical volume series and "deflated" dollar value series, expressed in constant dollars, rather than to dollar value or price series. However, trend curves are sometimes used to describe long-term movements in prices, even though the rational basis of growth is absent.

The tremendous expansion of population and technology in recent decades has stimulated widespread interest in the problem of measuring and predicting economic growth. Long-range planning has become a must for progressive companies, and trends must be projected as the first step in making a complete forecast and in setting a viable goal for future operations. It is particularly important to gauge the growth trends for individual industries and products, since they vary so widely, from the explosive growth of computers to the dismal decline of the railway passenger business. Most industries will also vary in their own rate of growth over a long period of years.

The variations in the nature of the secular trend component can be seen in the three curves of Chart 19–2. Gross national product in constant dollars represents the physical volume of total production, aluminum production typifies a young industry, and bituminous coal an older one. The data have been plotted on identical ratio scales, and smooth trend curves have been fitted by the National Industrial Conference Board to indicate average growth tendencies. The slopes of these curves show how the percentage rates of change differ in each case.

Gross national product has maintained nearly a straight line or uniform percentage rate of growth since 1890. Aluminum production, on the other hand, has shot up much more rapidly throughout its short life, although the trend curvature indicates that the rate of growth is slackening. The older bituminous coal industry developed at a more gradual rate from 1890 until World War I; since then it has matured and leveled off. Its course, however, has been steadier than that of aluminum. The three production series therefore exhibit marked differences in (1) shape of trend curve; (2) steepness of curve, or rate of growth; and (3) instability, measured in deviations from the curve. Trend analysis is most useful and reliable when growth is steady and steep and when the deviations about the trend curve are small. In this case the trend curve may even be projected into the future as a forecast if the factors affecting past growth are expected to continue.

The trend types in Chart 19–2 illustrate an application of a useful growth hypothesis popularly called the "law of growth." According to this principle, "If the population is expanding freely over unoccupied country, the percent rate of increase is constant. If it is growing in a limited area, the percentage rate of increase must tend to get less and less as population grows . . ."[1] until it finally levels off as an upper limit is approached. The constant rate of growth is characteristic not only of young industries (e.g., aluminum) but of total production (e.g., GNP), which is a cumulation of individual growth curves. The "law of growth" principle will be applied to the measurement of industrial trends later in the chapter.

These examples are sufficient evidence that the growth factor may be described by a simple curve, although it differs for each series. The

---

[1] P. F. Verhulst, "Recherches mathematiques sur la loi d'accroissement de la population," *Nouveaux memoires de l'Academie Royale de Sciences et Belles-Lettres de Bruxelles,* Tome XVIII (1845).

Chart 19–2

GROWTH PATTERNS IN AMERICAN INDUSTRY, 1890–71

GROSS NATIONAL PRODUCT
(Constant Dollars)

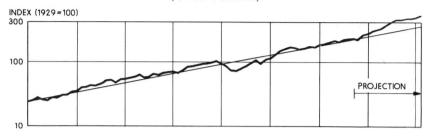

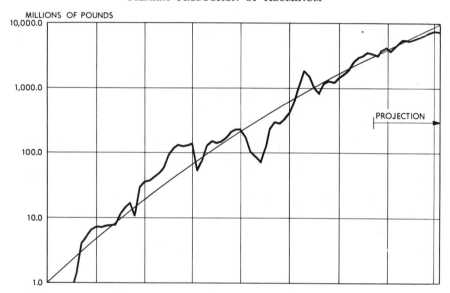

PRIMARY PRODUCTION OF ALUMINUM

BITUMINOUS COAL PRODUCTION

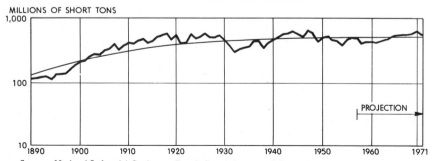

SOURCE: National Industrial Conference Board, *Growth Patterns: A Reexamination*, pp. 53, 40, 42.

problem of trend measurement, however, is not merely one of fitting a curve to the data; it also requires a knowledge of the industry under consideration. With this insight, one can apply methods of time series analysis that are not only mechanically correct but logical as well.

Purposes of Measuring Trend

There are three principal purposes of measuring secular trend:

1. The first purpose is to study the basic growth tendency of a series, ignoring short-term fluctuations due to business cycles, seasons, wars, or other causes. The trend curve answers such questions as: Has the company maintained its historic rate of expansion in recent years or is this rate tapering off? Has the company kept pace with its competitors or with the industry as a whole?

2. The second and most important purpose of measuring secular trend is to project the curve as a long-term forecast. If the past growth has been steady and if the conditions that determine this growth may reasonably be expected to persist, a trend curve may be projected over 5 to 10 years into the future as a preliminary forecast. Then regression analysis can be applied (Chapters 16–17), and a qualitative study of other factors, such as business cycles and specific demand and supply conditions, should be made to modify the trend forecast.

A long-term forecast is desirable in making a decision to take a job with a given company or to invest in its stock. It is even more essential in the management's decision to expand its plant, develop a new product, or enter a new regional market in order to justify the capital expansion. The projection of trend curves into the future is subject to considerable error and is deplored by many because of its inexactness and dependence on subjective judgment. Nevertheless it is a necessary expedient, since any major business decision affecting future operations involves a forecast, whether explicit or implicit; and an explicit projection is preferable, at least as a first step in planning.

3. The third purpose of measuring secular trend is to eliminate it, in order to clarify the cycles and other short-term movements in the data. Dividing the data by the trend values yields ratios which make the curve fluctuate around a horizontal line, thus bringing the cycles into clear relief. The Cleveland Trust Company Index of American Business Activity since 1790 is an example. However, these cyclical relatives may be affected arbitrarily by the type of trend curve used. Also, cycles can usually be discerned without trend adjustment, so trend is not often eliminated in current practice.

Period of Years Selected

The following rules should be observed in selecting the period of years to be used in fitting a trend curve:

1. The period should be as long as possible, preferably at least 15 years. In a long period the trend curve is but little affected by short-term episodes such as booms and depressions, whereas in a short period the trend curve may be distorted by these factors.
2. If the nature of a product or industry is abruptly changed by war, the introduction of a new product, or some other fundamental force, the series should be broken at this point and separate curves fitted to each segment. An examination of the graph of the data will be helpful in revealing such changes.
3. Each end of the series should represent the same phase of the business cycle. Thus, if recent years are prosperous, the series should begin with a prosperous period. If the series began in a depressed period, the trend line would be tilted upward by the recession at the beginning and prosperity at the end of the period so that it would exaggerate the true basic growth.

Serious errors have occurred through fitting trend curves to short periods of years dominated by cycles and other temporary disturbances.

Chart 19–3 shows trends fitted to various periods of years in output per man-hour, an important factor determining "productivity" or "improvement factor" increases in wage-rate contracts. Over short periods the average "trend" has varied from a growth of 4.1 percent per year to a decline of more than 3 percent. In particular, the United Auto Workers have cited the average annual growth of over 3 percent since 1947 to support their demands for future wage-rate increases. On the other hand, the long-term growth since 1909 has averaged only 2.2 percent per year, according to the Joint Economic Committee statisticians.

Price Deflation

Many series on the volume of sales, production, and other economic activities are available only in the form of dollar values. These values are affected not only by the physical quantity of goods involved but also by their prices, and prices have varied widely over the years. For many purposes it is necessary to know how much of the dollar value change represents a real change in physical quantity and how much is due to mere markups or markdowns in price tags. Physical quantities may be

Chart 19–3

ANNUAL RATES OF CHANGE IN OUTPUT PER MAN-HOUR
IN THE TOTAL PRIVATE ECONOMY
(1947 = 100)

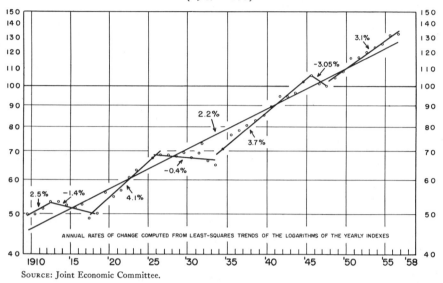

Source: Joint Economic Committee.

estimated by dividing the dollar values by the prices of the goods represented to eliminate the effect of price changes. (Price data are widely available.) That is, since value equals price times quantity, then value divided by price equals quantity. This adjustment is called price deflation or expressing a series in terms of constant dollars.

For example, suppose the sales in a shoe department increase from $20,000 in April to $20,900 in May. What was the change in physical volume? If we ascertain that the average price of shoes increased from $20 to $22 a pair in this period, we may divide the value by the price to learn that there has been an actual decline in shoes sold from 1,000 to 950 pairs, as shown below:

	April	May
1. Dollar sales	$20,000	$20,900
2. Average price per pair	$ 20	$ 22
3. Estimated number of pairs sold (1 ÷ 2)	1,000	950

Similarly, money wages may be deflated to find "real" wages, that is, wages in terms of the actual goods and services which can be purchased for a given amount of money.

The deflating process is a very simple one; the major problem is the selection of the proper price index. The rule to be followed is "Use an index number computed from the prices of the commodities whose values are to be deflated." For example, hardware store sales should be deflated by an index of hardware prices, not by a general price index.

In deflating dollar values that represent a variety of commodities, an appropriate price index may be pieced together from available sources to represent this particular "mix." For example, a mutual fund manager may desire to study the long-term growth of Sears, Roebuck & Company. The secular trend curve should be fitted to the physical volume of sales, since the price changes reflected in dollar sales follow no consistent pattern and merely obscure the real growth. The dollar sales therefore should be divided by a price index of goods sold by the company.

Such an index might be constructed by pricing a sample of important items sold by the store and weighting these prices by the sales volume of the departments represented. It is simpler, however, and adequate for

Table 19–1

SEARS, ROEBUCK ANNUAL NET SALES, 1953–71

Year*	Net Sales** (Billions of Dollars)	Price Index† (1967 = 100)	Deflated Net Sales‡ (Billions of 1967 Dollars)
1953.	2.982	88.6	3.366
1954.	2.965	88.3	3.358
1955.	3.307	87.6	3.775
1956.	3.556	88.3	4.027
1957.	3.601	90.1	3.997
1958.	3.721	90.4	4.116
1959.	4.036	91.1	4.430
1960.	4.134	92.1	4.489
1961.	4.268	92.4	4.619
1962.	4.578	92.6	4.944
1963.	5.093	93.5	5.447
1964.	5.716	94.1	6.074
1965.	6.357	94.7	6.713
1966.	6.769	96.6	7.007
1967.	7.296	100.0	7.296
1968.	8.178	104.8	7.803
1969.	8.844	110.0	8.040
1970.	9.251	114.5	8.079
1971. . . . . .10.006		118.8	8.423

* Fiscal years beginning February 1.
** From stockholders' reports.
† Constructed from U.S. Department of Commerce Consumer Price Index for apparel (weight 40%) and house furnishings (weight 60%).
‡ Net sales divided by price index times 100.

the purpose, to use existing retail price indexes. The Consumer Price Index itself is not suitable, since it contains elements such as foods, rents, and personal services not sold by the store; but the apparel and house furnishings components of this index may be appropriate. An analysis of Sears, Roebuck sales indicates that roughly 40 percent of the sales are in apparel and other soft goods, and 60 percent in household furnishings, appliances, and other durable goods. We can therefore multiply the CPI apparel component by .40, multiply the house furnishings component by .60, and add these two to get a combined price index appropriate for Sears, Roebuck sales. We can keep the 1967 base (expressing sales in dollars of 1967 purchasing power) for comparability with other indexes. Dividing reported net sales by this index gives deflated sales (actually "inflated" prior to 1967).

Table 19–1 compares the actual and deflated sales along with the price index from 1953 to 1971. The physical volume of business has increased more gradually than reported sales because of price inflation. In particular, the apparent gain in sales from 1956 to 1957 and from 1969 to 1970 was due to price markups; there was little change in "real" sales. On the other hand, nearly the entire rise in sales from 1957 to 1965 represented a real increase in physical volume, since prices were fairly stable during this period. Several types of secular trend curves will be fitted to the deflated sales in the next section.

METHODS OF MEASURING TREND

Trend analysis may be considered a special case of simple regression, in which time is the independent variable X. Thus, we can correlate Sears, Roebuck sales with U.S. personal income as in Chapter 16, or else fit a trend line to Sears' sales, in which the years (X) serve as a "proxy variable" representing the combined effect on sales of the growth in U.S. personal income, number of Sears stores, and other factors. Hence, we can use much the same graphic and least-squares methods of fitting a trend curve as in fitting a regression curve.

However, in trend analysis, the assumptions underlying the least-squares method may not be valid, as described in "Regression of Time Series," Chapter 16. The residuals from the trend line $(Y - Y_c)$ are not independent but are autocorrelated because of cycles. Also, the residuals may not have a uniform scatter; and they are not normally distributed, but often reflect marked irregularities such as booms, wars, and crises. These extreme values may have an exaggerated influence on regression measures, since the effect is proportional to the squares of the residuals.

Hence, while the trend line itself may be valid, the standard error of estimate and related measures are subject to the same limitations as described on pages 477–81. These standard errors serve some purposes, such as comparing the goodness of fit between two curves, but they are generally not valid for setting up confidence intervals or testing hypotheses in terms of probabilities. (A possible exception might be fitting trends to yearly percentage changes to reduce autocorrelation, as on page 482). Therefore, we will not repeat the discussion of these measures. Rather, attention will be concentrated on the rational types of trend curves and their validity in forecasting.

The graphic method may be preferable for preliminary analysis or for highly irregular series (since extreme values can be ignored), or for series whose growth does not follow a simple mathematical function.[2] On the other hand, the method of least squares has the advantage of being objective and precise, and it is easily carried out on the computer, using a regression program. In either case the statistical technique must be supplemented by a knowledge of the economic forces involved and the rational nature of the growth factor represented.

The series should first be plotted on a graph to determine the appropriate type of the trend curve. The trend can then be computed and plotted to check its fit. The arithmetic scale is appropriate for fitting trend equations to the natural values of the data by least squares.

For trend analysis in general, however, it is recommended that the data be plotted on a ratio scale, since this grid shows the two most important types of trend curves in their simplest form: (1) The exponential curve, with a constant percentage rate of growth, appears as a straight line. This logarithmic straight line characterizes many young industries and affords easy comparison of average rates of change in different series. (2) The "growth" curve, with a decreasing rate of gain, appears as a simple curve bending over to the right, rather than as an elongated S on an arithmetic scale (Chart 19–5).

Annual data are ordinarily used in secular trend analysis, rather than quarterly or monthly figures, because short-term movements are usually insignificant in measuring the broad sweep of an industry's growth or

---

[2] As Simon Kuznets puts it: "We must bear in mind the essential uncertainty of the whole process of separation or we shall be unduly influenced by mechanical methods of fitting. The method of least squares may save the investigator the trouble of decision in fitting to selected points and may seem more objective in the sense that identical results will be reached by different investigators. But mechanical arbitrariness is no whit better for being mechanical, and the method of least squares does not assure satisfaction of the two most obvious criteria of goodness of fit; namely, the balance and the minimizing of relative deviations from trend within each cycle." *Secular Movements in Production and Prices* (New York: Houghton Mifflin, 1930), p. 62.

decline and because the use of such detailed data involves much extra work. However, the methods applied in this chapter to annual data can be easily adapted to quarterly or monthly figures if desired.

Graphic Measurement

A simple method of fitting a trend curve is to draw it through the center of the plotted data by inspection.[3] If the general tendency of the data roughly follows a straight line, a transparent ruler or a piece of string may be used to locate the approximate central trend. If the trend is curved, a large-size transparent French curve or flexible spline rule may be used.

The trend line or curve should be drawn through the graph of the data in such a way that the vertical (not perpendicular) deviations above and below the trend are equal. They should be exactly equal for the series as a whole and approximately equal for the first half and last half of the series separately and for each major cycle. These deviations may be marked off cumulatively on the edge of a strip of paper, one above the other, for comparison.

Use of Group Averages. The average values of groups of data may be plotted as guide points in drawing a smooth trend curve. These averages may be computed for successive three- or five-year periods, or they may be computed for each cycle, marked off from trough to trough and plotted at the center year of the cycle. The trend is then drawn as a smooth curve between the plotted averages, but not necessarily through each one.

An Example: Fitting and Projecting Graphic Curves. Chart 19–4 shows two secular trend curves fitted by the graphic method to Sears, Roebuck deflated sales from 1926 to 1956. Sales for the next 15 years, through 1971, have then been plotted as a check on the validity of the trend projections that might have been made in 1957 as long-range forecasts. The ratio scale is chosen because the percentage rate of growth has been nearly constant during this period and so can be represented by a simple straight line, whereas the trend would curve up more and more steeply on an arithmetic chart.

The period of years is long enough so that the trend growth dominates the short-term cyclical-irregular movements. This period also balances the high-level prosperity levels of 1926–29 and 1952–56 at its

---

[3] For a more precise but detailed method of fitting a straight line, see S. I. Askovitz, "A Short-Cut Graphic Method for Fitting the Best Straight Line to a Series of Points According to the Criterion of Least Squares," *Journal of the American Statistical Association* (March 1957), pp. 13–17.

Chart 19–4

TREND CURVES FITTED BY GRAPHIC METHOD
SEARS, ROEBUCK DEFLATED SALES, 1926–56, PROJECTED TO 1971

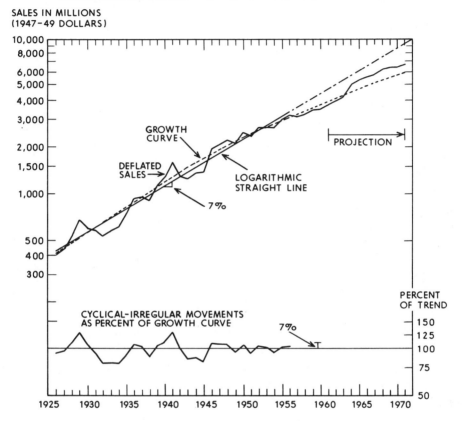

two extremes. Finally, it represents the entire era of the company's expansion in urban department stores, the first one of which was established in 1925.

Since the general growth tendency was nearly linear in 1926–56, a "logarithmic straight line" was drawn through the data with a transparent ruler so as to approximately bisect each of the major cycles. Then the vertical deviations above and below the line were cumulated and the line adjusted slightly to equalize the sum of these deviations for the two halves of the series.

The average annual rate of growth was then measured as follows: the vertical rise in the trend line in any year (see 1940–41 in Chart 19–4) was laid off on the right-hand percentage scale of the chart.

This distance extends from 100 percent upward to 107 percent, indicating an average growth of 7 percent per year in deflated sales over this period. This rate may be compared directly with that in deflated sales of other stores or real personal income, if desired.

The graphic measurement of average growth rate is subject to errors in drawing the slope of the trend line and in reading the result off the chart. The error in slope is small, however, if the trend is linear and the deviations from the trend line small. The error in reading values from a chart is also small if the scale is large.

The straight line indicates that Sears, Roebuck expanded at a fairly sustained rate over this 30-year period, although some flattening out is evident after 1947. A "growth" function was therefore drawn with a French curve to embody a decreasing rate of gain. In this case, the growth curve appears to describe the trend of sales somewhat better than the straight line. The growth curve may also be preferable for long-term projection into the future, since it follows the retardation-of-growth principle characteristic of many industries.

A logarithmic straight line may be projected for a limited period—say, 5 or 10 years—since the rate of expansion may be nearly constant for such a period, and the troublesome problem of curvature is avoided. In the very long run, however, the logarithmic straight line becomes too optimistic, since it increases indefinitely at a geometric rate.

The 1957–71 sales plotted on Chart 19–4 show how the trend projections would have worked out for these years. The extended growth curve predicted the average rate of increase in sales fairly well, while the straight line was consistently too high, as it had indicated it might be, by rising above the actual curve in 1954–56. On the other hand, a logarithmic straight line fitted only to the postwar years 1947–56 would have forecast 1957–71 sales reasonably well. This trend type is fitted by least squares to the post–Korean War years later in the chapter. Of course, trend projections do not forecast cyclical and irregular fluctuations, such as the company's expansion in new stores. These factors must be analyzed separately.

Eliminating Trend. The growth component of Sears, Roebuck sales may be eliminated graphically on the ratio chart for the purpose of isolating cyclical-irregular movements as follows: Draw a horizontal line at some convenient level away from the original curve—say opposite the lower printed number 2. Then mark a percent scale with 50, 100, and 150 percent opposite the printed scale numbers 1, 2, and 3, respectively. Caption this scale "Percent of Trend." Now take the *vertical* distances from each point to the original trend (the growth curve in Chart 19–4) with a paper strip, and lay these distances off in the same years above and below the horizontal 100 percent line. Connect these points with straight lines.

The resulting curve represents the cyclical-irregular movements in sales, since the trend is eliminated or flattened out. (There are no seasonal fluctuations in annual data.) The sales are now "adjusted for trend" or expressed as percentages of the trend values. This graphic adjustment is a short-cut method of dividing the sales data by the corresponding trend values and plotting the results.

The cyclical peak in 1929, the depression trough in 1932–34, the 1941 peak, the period of World War II shortages, and the mild post-war cycles are all clearly shown. The cyclical levels at the ends of the series, however, are somewhat uncertain, since the trend curve has a larger error where nearby past or future data are not known.

Growth Curves. Growth curves may be fitted either graphically (as described above) or mathematically to three selected points. The procedure will not be presented here (the equations of growth curves are too complex to be easily fitted by the least-squares method).[4] These curves are useful for representing both past trends and probable future tendencies, since they embody the rational "law of growth" principle described above. That is, an industry or population tends to grow at a nearly constant percentage rate during its youth; but as it matures, this rate tends to diminish.

There are several types of growth curves—the logistic (Pearl-Reed) and Gompertz being the most common—but all have the general characteristics illustrated in Chart 19–5. Here the same logistic curve is plotted on an arithmetic scale in panel A and a ratio scale in panel B. During the period shown, the curve rises from 1 to 99 and approaches an upper limit of 100.

The elongated S curve in panel A shows the growth of a typical industry or product in absolute units. The first stage is one of experimentation and slow initial growth. Second, there is a period of rapid exploitation of the product, and third, a leveling off of growth with maturity and saturation of demand. The relative age of different industries may be determined by locating them on this curve. Thus, the electronics and atomic energy industries would be located near its beginning, flour milling and railroads near the saturation level.

The same curve plotted on a ratio scale (panel B) is simpler in form,

---

[4] See F. E. Croxton, D. J. Cowden, and B. W. Bolch, *Practical Business Statistics* (4th ed.; Englewood Cliffs, N.J.: Prentice-Hall, 1969), pp. 327–38, for a description of mathematical methods of fitting logistic, Gompertz, and other growth curves. See also W. A. Spurr and D. R. Arnold, "A Short-Cut Method at Fitting a Logistic Curve," *Journal of the American Statistical Association*, March 1948, pp. 127–34. This article presents a graph on which a logistic curve can be drawn as a straight line.

being concave downward throughout its length. This is the grid that best illustrates the growth principle of a nearly constant percentage rate of change at first, followed by smaller and smaller percentage gains as the industry ages.

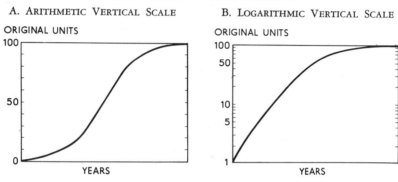

Chart 19–5

THE LOGISTIC GROWTH CURVE

A. ARITHMETIC VERTICAL SCALE   B. LOGARITHMIC VERTICAL SCALE

Before fitting a growth curve, two conditions should be satisfied: (1) The process represented should have the characteristics of biological growth to justify the use of this curve on logical grounds. Prices, ratios, business failures, or unemployment series would not qualify. (2) The data, when plotted on a ratio scale, must show a declining rate of growth or decline (i.e., must tend to flatten out) empirically, like this: growing series, $\int$; declining series $\searrow$. Otherwise, a growth function cannot be fitted.

Chart 19–2 shows Gompertz curves fitted mathematically by the National Industrial Conference Board to three series for more than a half century through 1958. We have plotted the actual data through 1971 and extended the trend curves to test their validity as projections. GNP exceeded its trend extrapolation substantially in the 1960s and early 1970s, but aluminum and coal continued surprisingly close to their projected growth curves.

The Method of Least Squares

In choosing the type of trend equation which best fits the data, one can judge goodness of fit in several ways. For example, one might like to have the average trend values equal the corresponding averages of the data not only for the series as a whole but also for selected parts (e.g., halves or thirds), or one might prefer to have the fitted curve pass through certain key points, such as cycle averages.

The most widely used criterion, however, is that of least squares. This criterion states that the best fitting curve of a given type is the one from which the sum of the *squared* deviations of the data is least. Also, the sum of the deviations of the data (Y) above and below the trend line (Y_c) must equal zero.

The method of least squares is applied here to the arithmetic straight line, the parabola, and the logarithmic straight line in turn. The sum of the squared deviations from the least-squares straight line is less than that from any other straight line. Similarly, the sum of the squared deviations from the least-squares parabola is less than that from any other curve described by a polynomial in X and X^2. Since the logarithmic straight line is fitted to the logarithms of the data, the sum of squares of logarithmic deviations is minimized. These usually correspond closely to percentage or relative deviations from trend rather than to the absolute deviations.

The method of least squares is most appropriate for data having a uniform variance of deviations along the trend line, few extreme deviations, and deviations that are independent of each other, especially in adjacent periods. As noted above, these conditions do not hold in time series. The deviations from trend are cyclical-irregular rather than random. Hence, one should attribute no special virtues to the method of least squares for fitting trends except simplicity from a practical point of view.

No matter what method is used to fit a trend, the equation type should be capable of describing the basic tendency of the series. Straight lines are often fitted to series having curved trends, with ridiculous results. Even if a straight line or parabola fits the past growth accurately, it is a purely empirical description and will not necessarily fit future growth. There should be some logical justification for curves used in forecasting, such as the tendency of many industries to grow at a nearly constant percentage rate in their youth and at a decreasing rate as they mature. These tendencies are described by logarithmic straight lines and growth curves, respectively.

Arithmetic Straight Line. The general equation of an arithmetic straight line trend is $Y_c = a + bX$, where Y_c is the computed or trend value of the time series Y in the year numbered X. The constant a is the value of Y_c when $X = 0$, and the constant b is the slope of the trend line—the change in Y_c per unit change in X. In the method of least squares, the trend line is fitted by finding the values of a and b that minimize the sum of the squared deviations from the trend line. To do this, we can use a computer program for regression analysis, placing the X origin at any point, such as the first year of the series. Or, we can solve the same normal equations as in simple regression. These equations are:

$$\Sigma Y = na + b\Sigma X$$
$$\Sigma XY = a\Sigma X + b\Sigma X^2$$

where n is the number of items in the series.

The short-cut method in Chapter 16 (using small x's and y's) can be simplified somewhat in trend analysis by choosing an odd number of years with the X origin at the midpoint in time. Then the negative values of X balance out the positive values, so that $\Sigma X = 0$. In other words, the time variable is measured as a deviation from its mean. Accordingly, X is changed to the small letter x, where $x = X - \overline{X}$. Since $\Sigma X = 0$, the terms containing ΣX drop out of the normal equations, which become:

$$a = \frac{\Sigma Y}{n}$$

$$b = \frac{\Sigma xY}{\Sigma x^2}$$

where x is measured from the middle year as origin. Here, the constant a is the arithmetic mean of the series and b is a simple ratio.

A straight-line trend can now be fitted by the method of least squares as follows:

1. Set up a table with columns for the year (x), the value of the time series (Y), the product xY, and x^2 for each year. (The column for x^2 may be omitted, if desired, by looking up Σx^2 in Appendix K.)
2. Add the columns and substitute the totals ΣY, ΣxY, and Σx^2 in the above formulas to find the constants a and b of the trend equation $Y_c = a + bx$.
3. Take any two widely separated values of x, find the value Y_c from the trend equation in each case, plot the corresponding points, and draw a straight line through them. This is the trend line.

If a trend curve must be fitted to an even number of years, the x origin must be placed midway between the two middle years in order to make $\Sigma X = 0$. From this origin it is $\frac{1}{2}$ year to the middle of the next year, $1\frac{1}{2}$ years to the middle of the following year, and so on. In order to avoid fractions, therefore, let the x unit equal six months. Then mark the x values of the years following the origin 1, 3, 5, . . . , and the x values going back from the origin -1, -3, -5. . . . The computation proceeds as above, and Σx^2 may be found in Appendix K. Then a is again the trend value at the origin, but b is the increase in the trend in six months rather than in a year.

The trend values (Y_c) can be listed for each year, if desired, by computing the value for the first year and adding the b value successively to get the other trend values. Note that $\Sigma Y_c = \Sigma Y$ as a check.

If it is desired to eliminate trend, in order to clarify cyclical-irregular movements, compute and plot Y/Y_c for each year. As in other statistical adjustments, dividing by a factor ($Y_c =$ trend) eliminates the influence of that factor.

As an example, an arithmetic straight line is fitted to Sears, Roebuck deflated sales in Table 19–2. In our graphic analysis of sales trends from 1926 to 1971 (Chart 19–4), we noted that the rate of growth in Sears, Roebuck sales had declined slightly since 1947. Thereafter,

Table 19–2

ARITHMETIC STRAIGHT LINE FITTED BY LEAST SQUARES
SEARS, ROEBUCK DEFLATED NET SALES, 1953–71

Year (1)	x (2)	Deflated Sales (Billions) — Y (3)	xY (4)	x^2 (5)
1953	-9	$ 3.366	-30.294	81
1954	-8	3.358	-26.864	64
1955	-7	3.775	-26.425	49
1956	-6	4.027	-24.162	36
1957	-5	3.997	-19.985	25
1958	-4	4.116	-16.464	16
1959	-3	4.430	-13.290	9
1960	-2	4.489	-8.978	4
1961	-1	4.619	-4.619	1
1962	0	4.944	0.000	0
1963	1	5.447	5.447	1
1964	2	6.074	12.148	4
1965	3	6.713	20.139	9
1966	4	7.007	28.028	16
1967	5	7.296	36.480	25
1968	6	7.803	46.818	36
1969	7	8.040	56.280	49
1970	8	8.079	64.632	64
1971	9	8.423	75.807	81
Total	0	106.003	174.698	570

SOURCE: Table 19–1.

the Korean war years were marked by erratic buying scares. Therefore, we now measure the post–Korean war trend from 1953 to 1971. This 19-year period is long enough for the growth factor to dominate cyclical-irregular influences. Also, cyclical forces are balanced by recessions in 1954 and 1970 next to the end years (see Table 20–5) and no marked irregularities are evident. Hence, the period of years chosen is a reasonable one.

To compute the trend equation, mark off the x values as integers from the middle year 1962 as origin, let $Y =$ sales, compute xY and x^2 (or look up Σx^2 in Appendix K), and total these columns. Then

$$a = \frac{\Sigma Y}{n} = \frac{106.003}{19} = 5.5791 \quad \text{(average sales in billions of dollars)}$$

$$b = \frac{\Sigma xY}{\Sigma x^2} = \frac{174.698}{570} = .30649 \quad \text{(average increase per year in billions of dollars)}$$

and the trend equation is $Y_c = 5.5791 + .30649x$. This equation is plotted in Chart 19–6. It is not a good fit; the line is too high throughout the middle years 1957–64 and too low in the end years. Its extension into the past goes below zero in 1943!

The indiscriminate use of the arithmetic straight line is a common error in trend analysis. For example, a large steel company featured this "standard" trend equation in a full-page magazine advertisement to emphasize the growth in per capita production of light steel products since 1901. The result was similar to that in Chart 19–6: The production data curved more and more steeply upward, while the straight trend line touched this curve at only two points and was far below it at the ends. An arithmetic straight line is a valid measure of trend for a series that tends to increase by constant absolute increments, but it cannot describe the long-term growth of an industry that expands by bigger increments as the industry itself increases in size. A type of trend curve must be chosen that will follow the tendency of a series throughout its course and will pass as nearly as possible through the center of individual cycles.

Parabola. The parabola is more flexible than the straight line as a measure of trend because of its curvature. To fit a parabola, its equation, $Y_c = a + bX + cX^2$ can be expressed as $Y_c = a + b_1X_1 + b_2X_2$ by transforming X^2 into X_2, a second independent variable. The values of a, b_1, and b_2 can then be determined by running a computer program for linear multiple regression.

Alternatively, the equation of the parabola can be written $Y_c = a + bx + cx^2$ when the x origin is placed at the middle year. The three

Chart 19–6

STRAIGHT LINE AND PARABOLA FITTED BY LEAST SQUARES

SEARS, ROEBUCK DEFLATED SALES, 1953–71, PROJECTED TO 1975

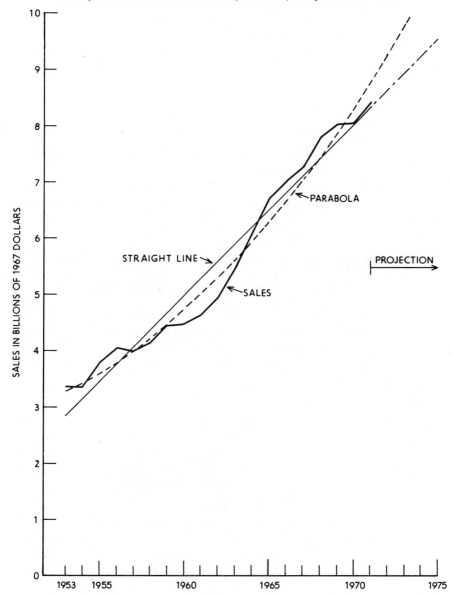

constants, a, b, and c, may then be found as follows: first, compute b by the same formula as in the straight line:

$$b = \frac{\Sigma xY}{\Sigma x^2} = \frac{174.698}{570} = .30649$$

Then find a and c by solving the following normal equations simultaneously:

$$\Sigma Y = na + c\Sigma x^2 \tag{1}$$
$$\Sigma x^2Y = a\Sigma x^2 + c\Sigma x^4 \tag{2}$$

In addition to the totals shown in Table 19–2, we need Σx^2Y (column 2 × column 4, not shown in detail) and Σx^4 (from Appendix K). Here, $\Sigma x^2Y = 3,304.59$ and $\Sigma x^4 = 30,666$.

Substitute these values in the above equations, multiply Equation 1 by 30 to equalize the coefficients of a, and subtract Equation 1 from Equation 2 to find $c = .009177$. Insert this value in Equation 1 to find $a = 5.3038$.

Hence, the equation of the parabola fitted to Sears, Roebuck sales is

$$Y_c = 5.3038 + .30649x + .009177x^2 \quad \text{(origin 1962)}$$

Finally, compute Y_c at three-year intervals and plot as on Chart 19–6. Here, a is the height of the curve at the origin (but not the arithmetic mean); b is the slope of the curve at this point only; and c determines the amount and direction of curvature. The numerical values are in billions of dollars at 1967 prices.

We can compare the goodness of fit between the parabola and the straight line (or between any two curves) by inspecting the chart, perhaps cumulating the vertical deviations from each curve on a paper strip to determine which sum is smaller. More precisely, we could compute the standard error of estimate (S_{YX}) or coefficient of determination (r^2) as described in Chapter 16, pages 462–67. The curve with the lower S_{YX} or higher r^2 is the better fit.

The parabola in Chart 19–6 is seen to be a better fit than the straight line. On the other hand, the parabola may be unduly influenced by extreme values, and it is not very logical for sales to increase with the square of time. Finally, the curve tends to become unreasonably steep (or to turn down, if c is negative) when it is projected far into the future.

Logarithmic Straight Line. A straight line drawn on a ratio chart (sometimes called an exponential or compound-interest curve) is often more useful for trend analysis than either the arithmetic straight line or parabola described above. Many younger industries tend to expand at a constant percentage rate of growth rather than at a constant amount of growth per year which appears as a straight line on an arithmetic chart. Furthermore, the arithmetic straight line is often illogical in that the constant amount of growth each year is independent of the size of the industry itself. Finally, the slopes of logarithmic straight lines show average percentage rates of growth, and so they are comparable for series of different units or widely different size, whereas the slopes of trend lines on arithmetic scales are not comparable in such cases.

Even if the rate of growth tends to diminish over a long period, the logarithmic straight line can be used to average the rate over some shorter interval, when the rate of change may be nearly constant.

A logarithmic straight line may be fitted either graphically or by the method of least squares. The graphic method was applied to Sears, Roebuck sales in Chart 19–4 for the first 30 years of its department-store expansion period, 1926–56. However, because of the retardation in the rate of growth after the Korean war, it appeared desirable to fit a separate straight line to the postwar period 1953–71. This line is fitted by least squares below.

First, a computer can be instructed to transform Y into $\log Y$ and solve the equation $\log Y_c = a + bX$ in linear fashion. Otherwise, look up the logarithms of the sales, then fit the equation $\log Y_c = a + bx$ in the same way as the arithmetic straight line, where x is measured from the middle year, and $\log Y$ replaces Y.

In Table 19–3, the years (x) are listed in column 2 with the origin centered in 1962, sales are shown in column 3 in billions, the logarithms of the sales $(\log Y)$ appear in column 4, and the product for each year $(x \log Y)$ appears in column 5. Columns 4 and 5 are then totaled, and Σx^2 is found from Appendix K. To determine a and b (which are both logarithms in this equation),

$$a = \frac{\Sigma \log Y}{n} = \frac{13.7941}{19} = .7260$$

$$b = \frac{\Sigma x \log Y}{\Sigma x^2} = \frac{13.7654}{570} = .02415$$

The trend equation is therefore

$$\log Y_c = .7260 + .02415x \quad \text{(origin 1962)}$$

To graph the trend on a ratio chart, plot any two widely separated points, using natural values of Y_c, and draw a straight line through them, as in Chart 19–7.

Table 19–3

LOGARITHMIC STRAIGHT LINE FITTED BY LEAST SQUARES

To Sears, Roebuck Deflated Net Sales, 1953–71

Year (1)	x (2)	Deflated Sales* (Billions) Y (3)	log Y (4)	x log Y (5)	log Y_c (6)	Trend Y_c (7)	Adjustment for Trend Y/Y_c (8)
1953	−9	3.366	0.5271	−4.7439	0.5087	3.226	1.043
1954	−8	3.358	0.5261	−4.2088	0.5328	3.410	0.985
1955	−7	3.775	0.5769	−4.0383	0.5570	3.605	1.047
1956	−6	4.027	0.6050	−3.6300	0.5811	3.812	1.057
1957	−5	3.997	0.6017	−3.0085	0.6053	4.030	0.992
1958	−4	4.116	0.6145	−2.4580	0.6294	4.260	0.966
1959	−3	4.430	0.6464	−1.9392	0.6536	4.504	0.984
1960	−2	4.489	0.6521	−1.3044	0.6777	4.761	0.943
1961	−1	4.619	0.6645	−0.6645	0.7019	5.033	0.918
1962	0	4.944	0.6941	0.0000	0.7260	5.321	0.929
1963	1	5.447	0.7362	0.7362	0.7501	5.625	0.968
1964	2	6.074	0.7835	1.5670	0.7743	5.947	1.021
1965	3	6.713	0.8269	2.4807	0.7984	6.287	1.068
1966	4	7.007	0.8455	3.3820	0.8226	6.646	1.054
1967	5	7.296	0.8631	4.3155	0.8467	7.026	1.038
1968	6	7.803	0.8923	5.3538	0.8709	7.428	1.050
1969	7	8.040	0.9053	6.3371	0.8950	7.853	1.024
1970	8	8.079	0.9074	7.2592	0.9192	8.302	0.973
1971	9	8.423	0.9255	8.3295	0.9433	8.777	0.960
Total	0		13.7941	13.7654			

* Sales in billions of 1967 dollars, years beginning February 1, from Table 19–1.

In 1953, $x = -9$,

$$\log Y_c = .7260 - .2173 = .5087 \qquad \text{so } Y_c = 3.226$$

In 1971, $x = +9$,

$$\log Y_c = .7260 + .2173 = .9433 \qquad \text{so } Y_c = 8.777$$

As a forecast for 1975, $x = 13$, $\log Y_c = 1.0400$, and the trend forecast Y_c is \$10.965 billion, in 1967 dollars. The slope of the least squares trend line is the logarithm b. This means that the *ratio* of each year's trend value to the preceding year's is antilog b, or 1.057. The average rate of growth for 1953–71 is then $1.057 - 1 = .057$, or 5.7 percent. This compares with the 7 percent growth rate determined graphically for the 1926–56 period.

Chart 19–7

LOGARITHMIC STRAIGHT LINE FITTED BY LEAST SQUARES

To Sears, Roebuck Deflated Sales, 1953–71, Projected to 1975

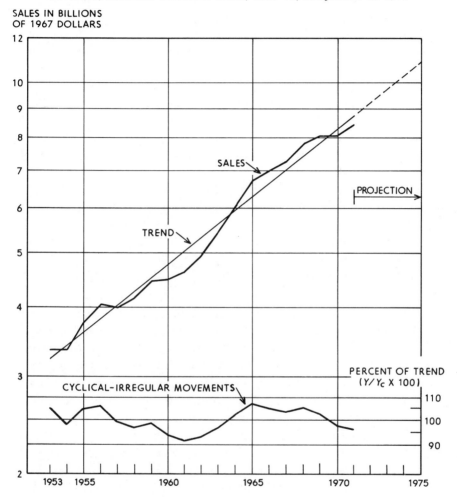

The trend may be eliminated, if desired, by computing and plotting Y/Y_c, or antilog (log Y — log Y_c), for each year. The computations are shown in Table 19–3, columns 6 to 8. The resulting curve resembles the graphically adjusted curve at the bottom of Chart 19–4, except that the trend base is the logarithmic straight line rather than the growth curve.

The parabola and logarithmic straight line appear to fit the trend of Sears, Roebuck sales about equally well over the period 1953–71. The latter, though, is generally preferable to the parabola because it is more rational in expressing growth as a constant percentage per year rather than as an arithmetic function of both time (x) and the square of time (x^2).

The graphic and least-squares methods of fitting a logarithmic straight line give nearly the same results. The graphic method is recommended for quick, approximate results and as a check on other methods, while the least-squares method is preferable for detailed, objective study, where computational assistance is available. The logarithmic least-squares method has the same merits and limitations as the arithmetic least-squares method, except that the logarithmic straight line is more likely to be distorted by extremely low values than by extremely high values.

In summary, the trend analysis in Chart 19–7 shows that (1) Sears, Roebuck real sales have increased at an average rate of 5.7 percent per year from 1953 to 1971; (2) there is no recent evidence that the rate of growth is slowing down (despite the 1970–71 recession); (3) there were two long cycles, with troughs in 1954, 1961, and 1971, and peaks in 1956 and 1965, though the amplitude was mild; (4) real sales can be projected over the next few years at an annual increase of 5.7 percent if the forces that made for past growth can be expected to persist.

The trend projection for 1975 is $10.965 billion at 1967 prices, as noted above. But this is only the first step in long-range forecasting. Assume that we forecast a *cyclical* recovery (as described later) from 96 percent of trend in 1971 (Table 19–3, column 8) to 105 percent of trend in 1975. The trend-cycle forecast is then $10.965 \times 1.05 = $11.513 billion at 1967 prices. Finally, if a forecast in current dollars is needed, prices must be projected. Thus, if we predicted an annual increase of 3 percent in Sears, Roebuck prices, based on a separate analysis, the price index in Table 19–1 would increase to 133.7 by 1975, and the forecast would be $11.513 \times 1.337 = $15,393 billion at current prices. (This last step is often omitted because of the difficulties of projecting price changes; forecasts are then expressed in terms of constant dollars.)

The actual forecast of the cyclical-irregular element requires the analysis of prospective changes in population and its age composition;[5] the regression of sales with disposable personal income and other economic factors (as described in Chapter 16), together with available

---

[5] See U.S. Bureau of the Census, *Current Population Reports, Population Estimates and Projections,* Series P-25, No. 470 (1971) *et seq.* for projections to 2020.

forecasts of the latter;[6] changes in consumer preferences; and the company's own expansion policy. The trend projection must therefore be modified by a thorough study of all pertinent economic factors.[7]

SUMMARY

An understanding of the nature and causes of business fluctuations is essential in a dynamic economy. These fluctuations may best be understood by analyzing economic time series into their principal components—secular trend, seasonal variations, cyclical fluctuations, and irregular movements.

The trend and seasonal components are measured directly, while cyclical-irregular movements are usually treated as a residual in combined form.

Secular trend is the gradual long-term increase or decrease in a series resulting from such basic factors as the growth of population, technology, and productivity. This development can be represented by a smooth trend curve fitted to the plotted data. Different series vary greatly in the shape and steepness of these trends, as well as in the variations of the data from the trend curve. Young industries and total production tend to grow at a constant percentage rate. The rate of growth is often retarded as an industry matures, following the "law of growth" principle, and eventually tends to level off or even turn down.

Secular trend may be measured for three purposes: (1) the appraisal of recent trends, (2) long-term forecasting, and (3) the elimination of trend to isolate cycles. The period of years selected for trend analysis should be as long as possible in order to minimize short-term disturbances; it should be broken at points of abrupt change; and it should begin and end at the same stage of the business cycle.

Price deflation is the process of dividing a dollar value series by a pertinent price index in order to reveal physical volume changes, expressed in "constant dollars." An appropriate price index may be compiled from segments of existing indexes, properly weighted, as in the Sears, Roebuck example. Price deflation is particularly necessary in

---

[6] See Predicasts, Inc., *Predicasts* (quarterly) for forecasts to 1985 or beyond for personal income, other GNP components, and many industry figures.

[7] See W. F. Butler and R. A. Karesh, *How Business Economists Forecast* (Englewood Cliffs, N.J.: Prentice-Hall, 1966); H. D. Wolfe, *Business Forecasting Methods* (New York: Holt, Rinehart & Winston, 1966); H. O. Stekler, *Economic Forecasting* (New York: Praeger, 1970); R. K. Chisholm and G. R. Whitaker, Jr., *Forecasting Methods* (Homewood, Ill.: Richard D. Irwin, 1971); or the sources listed in J. B. Woy, *Business Trends and Forecasting* (New York: Gale Research, 1965) for further study of forecasting methods.

times of wide price changes, since the "real" changes in output may differ drastically from the reported dollar figures.

Trend may be measured by either the graphic method or by least squares, as in regression analysis. In fact, trend may be considered a special case of regression in which the year X represents all the forces affecting Y. Also, the standard error of estimate and coefficient of determination may not be valid in probability terms (see "Regression of Time Series" in Chapter 16) but may serve to compare the goodness of fit between two curves.

In fitting trends, annual data are usually used, preferably plotted on a ratio chart.

1. To fit a trend curve by the graphic method, draw it with a transparent ruler or French curve so as to equalize the vertical deviations above and below each major segment of the curve. Averages of groups of years may be plotted as aids in locating the trend. The average growth rate of a logarithmic straight line can be read off the percentage scale on the chart. To eliminate trend, lay off the vertical deviations from the trend line about a horizontal line on the ratio chart and label the scale "Percent of Trend."

Graphic methods are quick, flexible, and afford a check on computations, while mathematical methods are more objective and often more accurate; the latter can be performed by clerical labor or by computers, and the results can be expressed in concise form. The two methods may be combined for optimum effectiveness.

Growth curves of the logistic or Gompertz type represent the rational tendency of many industries and populations to grow at a declining percentage rate as they mature. A curve of this type can be drawn graphically, using a transparent French curve or flexible rule, concave downward on a ratio chart. A growth curve can also be fitted mathematically to three selected points, but this procedure is not described here. Such curves provide a logical basis for forecasting.

2. The method of least squares fits a mathematical curve to the data such that the total of the squared deviations from the curve is less than that for any similar curve. The plus and minus deviations themselves total zero. This method is objective and reasonably accurate, provided the data follow the equation type chosen and are not too erratic. Unfortunately, however, the optimum conditions for the least-squares method do not occur in time series.

A trend equation can be determined by using a linear regression program on a computer, transforming X^2 into another variable X_2 in the case of a parabola, or transforming Y into log Y for a logarithmic straight line. Alternatively, to fit a straight line by least squares, center the X origin at the middle year; set up a table of x, Y, xY, and x^2; and substitute the column totals in the given equations to find a and b in the equation $Y_c = a + bx$. To eliminate trend and isolate cyclical-irregular movements, compute and plot Y/Y_c for each year. A straight line is simple, but may be illogical in that the constant growth increments are independent of the variable Y itself.

To fit a parabola, add columns for x^2Y and x^4 to the foregoing and substitute the totals in three equations to find a, b, and c in the equation $Y_c = a + bx + cx^2$. This is usually a better fit than a straight line, although it may be unduly affected by extreme values. It is also somewhat illogical as a function of the square of time, and its projection ahead may be too steep.

The logarithmic straight line is superior to the other two in describing a rational growth tendency of young industries and in comparing relative rates of change. It may be drawn graphically as a straight line on a ratio chart or computed by the method of least squares. The least-squares procedure is the same as in the arithmetic straight line, except that log Y is used in place of Y. The projection of this function is often a reasonable first step in making medium-range forecasts for a few years in the future. In the long run, however, a growth curve projection may be preferable, since the percentage rate of growth is likely to decline.

PROBLEMS

1. *a*) If you were an economist with the Eastman Kodak Company, manufacturers of camera and film (or other selected company), what would be the principal purpose of separating the company's monthly dollar sales into its component fluctuations? Give reasons to support your opinion.
   b) Briefly describe the causes of the four major components of this particular time series.
   c) Plot the company's annual sales for the past 15 or 20 years or make a photocopy of them from an available chart.
   d) Describe the trend characteristics of this series: Is the trend a straight line, concave upward, or concave downward? What does this mean in terms of growth? Is the growth steady or erratic?

2. Select in the *Survey of Current Business* a price index that might be appropriate for deflating the gross revenues of each of the following:
   a) A manufacturer of drugs and pharmaceuticals.
   b) A Cleveland building contractor.
   c) A clothing store.
   d) A grocery supermarket.

3. The U.S. Department of Commerce publication *Business Statistics, 1971* gives the following data on disposable personal income (in billions) and average hourly earnings of manufacturing production workers. These are shown with the Consumer Price Index for the same years:

	Disposable Income	Hourly Earnings	CPI (1967 = 100)
1950.............	$206.9	$1.44	72.1
1955.............	275.3	1.86	80.2
1960.............	350.0	2.26	88.7
1965.............	473.2	2.61	94.5
1970.............	687.8	3.36	116.3

a) Deflate disposable personal income by the consumer price index and list the results.

b) Plot actual and deflated income on a small chart.

c) Explain the significance of the deflated data and compare the trends of the two curves.

4. As a labor union economist, you wish to prepare a report summarizing the changes in real hourly earnings in manufacturing industries from 1950 to 1970, by five-year intervals. Besides eliminating changes in living costs, you feel that the results would be most meaningful if expressed in dollars of 1970 purchasing power, since recent price levels are most easily remembered. Using the figures in Problem 3:

a) Compute real hourly earnings in 1970 dollars.

b) Compare the 1950–70 percentage increase in average hourly earnings with that in the real purchasing power of these earnings.

c) To buy the same amount of goods and services that the 1970 worker could earn in one hour, how many hours would his father have had to work in 1950?

5. *a*) Under what conditions is it valid to forecast by extrapolating a trend curve fitted to past data? Discuss briefly.

b) Why may the particular purpose of measuring trend affect the choice of a trend curve?

c) What factors determine the period of years used in fitting a secular trend curve to an industry's sales?

d) Describe the use of group averages in trend fitting.

e) What is the one chief advantage of mathematical methods and of graphic methods, respectively, in trend analysis? Justify your selection.

6. *a*) Explain the "law of growth" principle implicit in the use of growth curves.

b) Describe briefly the graphic method of fitting a growth curve.

    c) What is the logical justification, if any, of fitting and projecting such a curve as a 10-year forecast of aluminum production (Chart 19–2)?

7. As part of a planning study for Kraftco Corporation, you are asked to analyze and project the growth trend in the output of manufactured food products, as measured by the Federal Reserve Index of Production in Food Manufactures shown in Chapter 16, Problem 18.

    a) Plot this series on an arithmetic chart. Since the growth is roughly linear, fit a straight-line trend by the method of least squares.

    b) State the average annual growth from 1957 to 1971 (give the unit). Compute Y/Y_c for 1971 to find the cyclical-irregular component, or the value "adjusted for trend," in that year (give unit).

    c) Plot the trend line on the chart and extend it beyond 1971 to the latest year for which the index is available. Multiply the projected trend value by the cyclical-irregular component for 1971 (assuming that this factor continues unchanged) to obtain a forecast. Find the actual food manufactures index for this year and give the percentage error of the forecast. Explain the probable causes of this error.

8. *a*) Perhaps a parabola would fit the data in Problem 7 better than a straight line. Fit a parabola to the food products indexes for 1957–71 by least squares, and plot on your arithmetic chart.

    b) Which is the better fit—the parabola or the straight line? To answer this, instead of comparing the standard errors of estimate $\sqrt{(Y - Y_c)^2/(n - k)}$ as in Chapter 16, simply cumulate the vertical (not perpendicular) deviations from each trend curve with a paper strip and find $\Sigma|Y - Y_c|/(n - k)$, where n is 15 years and k is the number of constants (2 for a straight line and 3 for a parabola). The curve with the smaller mean deviation is the better fit, by this criterion.

9 to 12. In studying trend forecasting, it is sometimes desirable to hold out the last few years and use them to test the projection of trends fitted to the earlier years. Suppose, therefore, that you were an economist in the chemical industry in 1966, and wished to make a six-year projection of chlorine production (shown here in millions of short tons) based on the postwar trends from 1947 through 1965. Then in 1972 you wish to check how your projection came out for the years 1966–71.

Year	Chlorine Production	Year	Chlorine Production	Year	Chlorine Production
1947	1.45	1957	3.95	1966	7.20
1948	1.64	1958	3.60	1967	7.68
1949	1.77	1959	4.35	1968	8.44
1950	2.08	1960	4.64	1969	9.41
1951	2.52	1961	4.60	1970	9.76
1952	2.61	1962	5.14	1971	9.35
1953	2.80	1963	5.46		
1954	2.90	1964	5.94		
1955	3.42	1965	6.44		
1956	3.80				

SOURCE: *Survey of Current Business*, June 1972, and *Business Statistics, 1971.*

9. *a*) Plot the 1947–65 figures only on a one-cycle ratio chart, with the time scale extended to 1971.

   b) Draw a smooth growth curve (slightly concave downward) through the data by inspection, and adjust it so that the vertical deviations above and below are about equal for each major segment (the deviations may be cumulated on a paper strip). Extend the curve to 1971 as a forecast on the assumption that some retardation in growth rate will occur after 1965.

   c) Draw a logarithmic straight line through the data from 1951 to 1965 by inspection, and extend it to 1971 on the more optimistic assumption that the average 1951–65 rate of growth will continue unchanged. Find the average annual rate of growth graphically and state it as a percentage.

   d) Forecast chlorine production to 1971, using (1) the trend in (*b*) or (*c*) that appears the more reasonable and (2) a cyclical-irregular adjustment (either as a percentage of trend or as a vertical distance laid off on the chart) based on 1965 production relative to trend, modified by your best judgment. Explain the reasons for your procedure.

   e) Plot actual chlorine production for 1966–71 on the chart, and note the percentage error in your forecast for 1971. What is the probable reason for this error?

10. *a*) Eliminate the trend through 1965 in Problem 9 graphically (using the trend curve you prefer), and plot the cyclical-irregular relatives in the lower part of the chart.

    b) Describe the cyclical timing and amplitude of chlorine production, and the principal irregular forces at work during this period.

11. *a*) Plot chlorine production for 1951–65 on an arithmetic chart.

    b) Fit either a straight line or parabola by least squares, depending on which appears to be a better fit.

    c) Using this trend, project chlorine production to 1971 and compare with actual results, as described in Problem 9 (*d*) and (*e*) above.

12. *a*) Fit a logarithmic straight line by least squares to chlorine production for 1951–65, and extend it to 1971. What is the percentage error in the trend forecast for 1971?

    b) Find the average annual rate of growth, using logarithms.

    c) Compare the goodness of fit of the logarithmic straight line fitted graphically with that fitted by least squares.

13 to 16. Problems 13 to 16 may be assigned either for full-length analysis, as given, or as short illustrative exercises covering only the seven years beginning in 1959.

   As noted above, we can test different types of trend forecasting by holding out the latest years to determine the future accuracy of a trend curve fitted to earlier years. Assume you are a utility analyst, therefore, and wish to test several types of trends fitted to U.S. electricity production (in billions of kilowatt-hours) fitted to the years 1947–65, against the actual results for 1966–71.

Year	Electricity Production	Year	Electricity Production	Year	Electricity Production
1947	256	1957	632	1966	1,144
1948	283	1958	645	1967	1,214
1949	291	1959	710	1968	1,329
1950	329	1960	753	1969	1,442
1951	371	1961	792	1970	1,530
1952	399	1962	852	1971	1,614
1953	443	1963	914		
1954	472	1964	984		
1955	547	1965	1,055		
1956	601				

SOURCE: *Survey of Current Business,* June 1972, and *Business Statistics, 1971.*

13. *a*) Plot the 1947–65 figures only on a one-cycle ratio chart, with the vertical scale beginning at 200 billion kilowatt-hours and the horizontal scale extended to 1971.
   b) Draw a smooth freehand trend line or curve through the data, plotting group averages as guides and equalizing the deviations above and below the trend as described in the text. Project the trend to 1971.
   c) Describe the nature of growth in this industry. What was the average annual percentage rate of growth from 1959 to 1965? (Show on the chart how this value was obtained.)

14. *a*) Plot electricity production on an arithmetic chart with the time scale extended to 1971, and compute an arithmetic straight line by the method of least squares. Fit it either to the seven years (1959–65) that electricity production has grown at a nearly constant rate or to the whole postwar period 1947–65, as assigned. Show computations and trend equation. Plot this curve on the arithmetic chart and project it to 1971.
   b) Then plot actual data for 1966–71 as a check. What is the percentage error in the trend projection for 1971? Explain this error.

15. *a*) Fit a logarithmic straight line to the electricity data by least squares, either for 1959–65 or 1947–65 as assigned; plot it with the data on a ratio chart; and extend the trend line to 1971.
   b) How does the least-squares criterion of goodness of fit differ in its application to the arithmetic straight line and the logarithmic straight line?
   c) Explain the meaning of the constants *a* and *b* in each of these equations.
   d) Plot the actual 1966–71 data on the chart to test your projection. Compute the trend forecast for 1971. What is its percentage error compared with the actual value of 1,614 billion kilowatt-hours? Explain the probable reason for this error.

16. *a*) Compare the goodness of fit of the freehand trend, the arithmetic straight line, and the logarithmic straight line in describing the growth of electricity production.
   b) Which of these three curves is the most logical for use in forecasting? Why?

17. As an economist with Pacific Gas & Electric Co. you are asked to project the company's future natural gas requirements, based on the following rec-

ord of natural gas purchased, in millions MCF, taken from the company's 1971 annual report. (The period of years is short, but appears adequate in this case because of the steadiness of the growth rate.)

Year	Gas Purchased	Year	Gas Purchased
1961	581	1967	802
1962	612	1968	888
1963	654	1969	878
1964	737	1970	951
1965	749	1971	1,005
1966	808		

a) Fit a logarithmic straight line by least squares to the data.

b) Find the average percentage rate of growth (from antilog *b*).

c) Project this trend to the latest year available, and compare with actual gas purchased (obtainable from the company's annual report).

18. You wish to forecast U.S. gasoline demand for 1971, using the following data, in billions of barrels. (Prior figures were not comparable; see *Business Statistics, 1971*, p. 167 notes.)

Year	Gasoline Demand	Increase Amount	Increase Percent
1964	1.658		
1965	1.720	62	3.7
1966	1.793	73	4.2
1967	1.843	50	2.8
1968	1.956	113	6.1
1969	2.042	96	4.4
1970	2.131	89	4.4

a) Since demand is growing at a fairly level percentage rate, as shown above, fit a logarithmic straight line by least squares to the gasoline demand series. (If the *amounts* of increase had been more level, rather than increasing, an arithmetic straight line would have fit better.)

b) Find the average annual percentage increase from log *b*. (You cannot average the above percentages, as the base of each is different.)

c) Forecast 1971 demand by applying this percentage increase to 1970 demand. Actual 1971 demand was 2,213 million barrels. What is the percentage error?

d) Forecast 1971 demand by extending the trend line. Why does this estimate differ from that in part (*c*)? Which is better logically? Why?

SELECTED READINGS

Readings for this chapter have been included in the list which appears on pages 655–56.

20. SEASONAL AND CYCLICAL VARIATIONS

OF THE PRINCIPAL types of fluctuations in economic activities, trend analysis was discussed in Chapter 19. In this chapter we consider seasonal and cyclical fluctuations.

In trend analysis, annual data are usually used. For the study of shorter term seasonal and cyclical movements, however, quarterly, monthly, or weekly data are needed. Monthly figures are most common.

NATURE OF SEASONALITY

Seasonal variations are of two kinds: (1) those resulting from natural forces and (2) those arising from man-made conventions. For example, in the northern United States and Canada, construction work is greatly curtailed during the winter season. Hence, data concerning road construction, building activity, and the like have seasonal variations that are directly related to the weather. On the other hand, department store sales expand before Easter and Christmas, a circumstance related to man-made festivals rather than to the weather.

Seasonal variations affect nearly all economic activities. The impact of seasonal influences is likely to be greatest at the point of origin and the point of consumption and less in the intervening manufacturing process. The cotton crop, for example, is seasonal, and so are retail sales of cotton goods (in a different pattern), but textile mills manage to operate at a more stable rate by manufacturing for stock in the slack seasons. In some industries, however, only the supply is markedly seasonal (e.g., wheat versus bread) or the demand (consumer durable goods) or the fabrication process itself (building construction). Inventories in general are more seasonal, and prices less seasonal, than production or sales. The typical seasonal pattern includes either one peak

611

and trough per year, as in building construction, or else peaks in both spring and fall and troughs in midwinter and midsummer, as in retail trade generally.

The latter pattern is illustrated by the monthly sales of Sears, Roebuck shown in Chart 20–2. The year starts with the midwinter slump, followed by a brisk spring trade, a June dip, a fall pickup, and a big Christmas rush. Accurate measures of seasonal behavior by products are invaluable to the management of such a firm in planning purchasing, inventory control, and selling programs.

Two important features of the seasonal rhythm should be noted: (1) it recurs year after year with a fixed period and (2) the increases and decreases of sales occur at about the same time and in about the same proportion each year.[1] The seasonal rhythm therefore has a fixed period and a fairly regular amplitude, whereas the cyclical rhythm is variable in both respects. Seasonal movements, consequently, may be measured and projected into the future much more accurately than cycles.

Calendar Variation

One cause of "seasonal" disturbances in monthly and weekly data is neither the weather nor customs but the eccentricity of the calendar itself. The months not only vary from 28 to 31 days in length, but some have four Saturdays and Sundays, others have five. Some also have one or several holidays, others have none. Further, certain series of data arise from activities which operate five days a week, others $5\frac{1}{2}$, 6 or even 7 days. All these factors cause spurious movements in monthly data which cannot be entirely eliminated by most seasonal adjustment methods.

It is usually desirable, therefore, to eliminate the effect of calendar variation as a preliminary step before measuring regular seasonal movements. The method of adjusting for calendar variation is to divide each monthly total by the number of operating days in that month to reduce it to a uniform average daily basis. The general rule is to count the number of days that the particular activity was carried on during the month. In some cases this will mean all of the days in the month; in others, Sundays or Saturdays, Sundays, and holidays will be excluded. If one day in the week is unusually light or heavy in volume, it may be weighted accordingly. Thus, the Federal Reserve Board formerly weighted Sunday as $1\frac{1}{2}$ days in adjusting monthly newspaper output—

---

[1] Two notable exceptions occur because (1) the date of Easter varies and (2) automobile production and sales are affected by the variable dates of offering new models. These irregularities require special corrections in seasonal measurement.

a component of the Industrial Production Index.[2] Different holidays are also observed in the several fields of business activity and in different areas.

Chart 20–1 shows the effect of calendar adjustment on a city's monthly bank clearings in a leap year when banks closed Sundays and 11 holidays. The monthly totals are divided by the number of operating days per month (bottom curve) to yield the daily averages

Chart 20–1

ADJUSTMENT FOR CALENDAR VARIATION

MONTHLY BANK CLEARINGS
(Millions of Dollars)

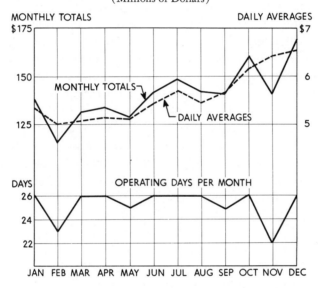

(dashed line, right scale). It is evident that most of the month-to-month fluctuations in total clearings—particularly the dips in February and November—were due merely to the erratic calendar and not to any significant change in banking activity.

The method of reducing to a daily average basis should be used only for quantities that cumulate during the month, such as bank clearings, production, or sales. These series all add up to larger amounts in long months than in short months. On the other hand, series such as bank

---

[2] The Census II method is currently used, as described later. See also A. Young, *Estimating Trading-Day Variation in Monthly Economic Time Series* (Technical Paper No. 12. Washington, D.C.: U.S. Bureau of the Census, 1964).

deposits, prices, employment, or other "point data" should not be reduced to an average daily basis, because they do not cumulate or build up to larger values in longer months. Yearly and quarterly data in general are not adjusted for the calendar either, since the irregularity is negligible in these longer periods.

In the case of weekly data the number of weekdays is constant, and only holidays cause irregularities. These may be corrected by (1) adjusting weeks containing holidays to a full-time basis (e.g., adding one fourth to the figure for a four-day week to make it comparable with data for five-day weeks) or (2) plotting curves for one year over the other on a tier chart so that weeks containing a given holiday are lined up vertically for direct comparability in different years, as in Chart 20–5.

When data are to be adjusted for seasonal variation, as described below, the calendar adjustment may sometimes be omitted, since the seasonal correction eliminates the difference between the *average* number of operating days in January and those in February. However, it does not smooth out the differences in operating days between one January and the next. Thus, if one January had 26 days and the next had 27 days, and we divided the two January totals by the same seasonal index, the adjusted data would still show a spurious difference due to the calendar. Calendar adjustment is incorporated in certain computer programs—such as Census II described later in the chapter—which automatically allow for this factor.

Other Rhythms

Many economic activities exhibit rhythmic movements having a shorter period than seasonal variations. Quarterly dividend and income tax payments and monthly payrolls cause regular fluctuations in the flow of funds through banks and in consumers' expenditures. Weekly rhythms may be illustrated by the sales in a department store. Monday is apt to be light, except after a long holiday weekend; then trade builds up gradually during the week to a peak on Saturday. The average sales on a number of Mondays may be compared with the averages for other weekdays (with separate norms for days before and after holidays) and a normal pattern of weekly variation worked out to aid in the timing of purchasing, advertising, and hiring of extra help.

Daily rhythms occur in such data as the hourly number of messages crossing a telephone switchboard, the hourly number of riders on buses, or the hourly use of electric power. These and many similar series have such regular fluctuations that engineers use them to determine the amount of equipment to be kept in service each hour of the day and night.

The rhythms having a shorter period than the seasonal, therefore, may be worth analyzing as an aid to short-term programming. Since they do not require the use of statistical techniques beyond averages, however, no further attention will be given to them here.

WHY SEASONAL ANALYSIS?

There are three principal purposes of measuring seasonal movements: (1) to analyze current seasonal behavior, (2) to predict seasonal movements as an aid in short-term planning, and (3) to eliminate seasonality in order to reveal cyclical movements.

1. Measures of typical seasonal behavior in production, sales, inventories, and prices are indispensable in controlling the operations of a business during the year through better understanding of current figures. Seasonal indexes serve to answer such questions as: Was the decline in sales last month more or less than the usual seasonal amount? How much does the price of a given product usually decline between July and August? What is the normal variation in inventories from month to month?

2. Seasonal measures are also useful in forecasting and planning operations over the next year or two. Every successful business concern operates on a budget, in which the coming year's income and expense items are estimated and later checked against actual results. By means of seasonal indexes, next year's budget items may be allocated by months. Seasonal indexes are also particularly useful in scheduling purchases, personnel requirements, seasonal financing, and advertising programs. Seasonal movements, like cycles, are wasteful because the men and equipment needed in the peak season are idle in the slack season. An accurate knowledge of seasonal behavior is an aid in mitigating and ironing out seasonal movements through business policy. This may be done by introducing diversified products having different seasonal peaks, accumulating stocks in slack seasons in order to manufacture at a more regular rate, cutting prices in slack seasons, and advertising off-seasonal uses for products.

3. Another purpose of measuring seasonal variations is to get rid of them. Business cycles are of critical importance, but these cycles are frequently obscured by large seasonal movements. The latter must ordinarily be measured and eliminated to reveal the former. Many monthly statistical series in economic publications are "adjusted for seasonal variation" for this purpose. The *Survey of Current Business,* for example, lists the following data and many others on a seasonally adjusted or simply "adjusted" basis: gross national product, industrial production, business sales and inventories, manufacturers' orders, new

construction, retail sales, and employment. A knowledge of seasonal adjustment therefore is essential for the economic analyst.

METHODS OF MEASURING SEASONAL VARIATION

Seasonal variation is a rhythmic movement which recurs each year with about the same relative intensity. This movement may be summarized by a seasonal pattern which is assumed to be typical of any year of a series or which changes gradually from year to year. The pattern consists of 12 monthly indexes (or four quarterly indexes) whose average is 100 percent. The problem of measuring seasonal variation is then one of determining these indexes for a given series.

A great many methods have been advanced for computing seasonal indexes. Essentially, however, most refined methods arrive at a seasonal index for a given month by averaging its ratios to a trend-cycle base in several years (or fitting a trend curve to these ratios) to cancel out the nonseasonal factors.

In any case, the series is first plotted on a chart to show the general nature of the seasonal pattern and to aid in further analysis. Unless a fairly pronounced and regular rhythm is apparent, seasonal measurement may not be worthwhile. A ratio scale must be used in the graphic method described below and is usually desirable in other methods as well, since seasonal movements in most economic data are more stable as percentages than in absolute amounts. Hence, seasonal indexes themselves are expressed as percentages.

The period of time covered should be at least six or seven years for series having a regular seasonal pattern, and longer for irregular data, in order to average out the peculiarities in individual years. The normal seasonal rhythm may be disrupted by wars, strikes, government edicts, depressions, and abrupt changes in business policy. Such erratic periods should be excluded, as far as possible. Sometimes the seasonal nature of a series will change gradually over the years. In this case a relatively long period of years should be used, as in trend analysis, and "changing" indexes of seasonal variation should be computed as described later in the chapter.

Graphic Method

In the graphic shortcut method, most of the steps are performed directly on the chart. This technique will be applied to monthly sales of Sears, Roebuck from 1966 to 1971.[3] The steps are:

---

[3] Sears, Roebuck and Co. sales have not been adjusted for calendar variation because the seasonal indexes themselves will reflect the difference in average length of months and

1. Plot the data on a ratio chart, preferably with a one-cycle scale. The large scale makes measurements more accurate than on two-cycle paper, and the ratio scale permits measuring and averaging percentages on the graph. As shown in Chart 20–2, Sears, Roebuck sales have a

Chart 20–2

GRAPHIC SEASONAL METHOD

SEARS, ROEBUCK SALES, 1966–71 WITH FORECAST FOR 1972
(Ratio Chart)

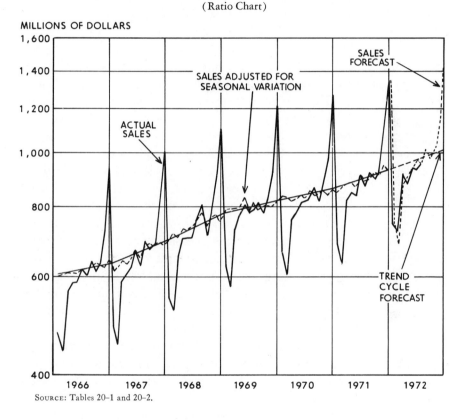

SOURCE: Tables 20–1 and 20–2.

pronounced seasonal rhythm, so that seasonal analysis is worthwhile.

2. Plot the annual average of monthly sales at the middle of each year (between June and July) and draw a freehand trend-cycle curve

correct for this in the adjusted data. Slight variations due to the varying number of weekdays between one January and the next, etc., remain, and should be corrected by a separate calendar adjustment in a more refined study, unless a computer program is used that makes this adjustment.

It is not necessary to deflate sales for price changes in seasonal analysis, since they have little effect on the seasonal rhythm and tend to cancel out in the averaging process.

through these points by inspection. The curve should follow not only the trend but also cyclical and extended irregular movements such as those caused by war. A knowledge of economic conditions in this period will also help in locating the peaks and troughs of cycles. Thus, the period 1966–71 was marked by continuous expansion except for a general business recession from a peak in November 1969 to a trough in November 1970.[4]

The trend-cycle curve in Chart 20–2 is drawn with a French curve through the annual averages on a rising trend, with only a slight flattening out during the recession of 1970. The fitting of this curve involves a subjective error, but part of the error is canceled in subsequent operations,[5] and the curve can be altered later to improve the fit, if necessary, as described under "Revision for Greater Accuracy" below.

3. Take another sheet of one-cycle ratio graph paper, and lay off a percentage scale on its right margin, as illustrated, marking 100 percent with a red arrow opposite the number "5" printed on the graph paper, 120 percent opposite "6," 80 percent opposite "4," and the other numbers in the same proportion. Cut out a vertical measuring strip containing these values. Find the percentage of sales to the trend-cycle base for each month by placing the 100 percent red arrow of the measuring strip on the trend-cycle curve of the sales chart and reading off the value on the measuring strip opposite the plotted sales. Tabulate the percentages, as in Table 20–1. Dividing sales by the trend-cycle base eliminates most of the influence of trend and cycle, so that the percentages reflect primarily the effect of seasonal and irregular movements. When these percentages are averaged for a given month (step 4), the irregular factors tend to cancel out and the average itself reflects the seasonal influence alone.

MEASURING STRIP (PERCENT)

160—
150—
140—
130—
120—
110—
100 ▶
90—
80—
70—
60—

4. Compute a "modified" mean of the percentages for each month in the different years, omitting the highest and lowest values as being unduly influenced by irregular factors (such as a strike or a stock market break).

In Table 20–1 the highest figure and the lowest figure in each column are crossed out and the remaining four items are totaled and di-

---

[4] According to the National Bureau of Economic Research reference dates of business cycle turning points shown in Table 20–5.

[5] The error cancels out either if the average level of the freehand curve is too high or too low (since the seasonal indexes are adjusted to average 100 percent) or if its positive and negative errors are equal (since the ratios for each month are averaged).

Table 20–1

PERCENTAGES OF GRAPHIC TREND-CYCLE CURVE AND COMPUTATION OF SEASONAL INDEXES

SEARS, ROEBUCK SALES, 1966–71

	Jan.	Feb.	Mar.	Apr.	May	June	July	Aug.	Sept.	Oct.	Nov.	Dec.	Total
1966	79	72	92	96	96	101	97	102	98	102	114	149	
1967	77	71	92	95	96	101	94	104	100	101	116	146	
1968	80	74	92	98	97	96	100	107	94	102	117	144	
1969	81	73	93	97	101	98	99	100	96	100	112	147	
1970	81	73	91	95	97	98	98	102	97	102	113	147	
1971	78	72	93	96	95	101	96	101	99	100	118	146	
Total, middle four	318	290	369	384	386	398	390	409	390	405	460	586	
Mean, middle four	79.5	72.5	92.2	96.0	96.5	99.5	97.5	102.2	97.5	101.2	115.0	146.5	1,196.1
Seasonal index	79.8	72.7	92.6	96.3	96.8	99.8	97.8	102.5	97.8	101.5	115.4	147.0	1,200.0

vided by four to give the modified means shown in the next to the bottom row. These means are preliminary seasonal indexes. They should average 100 percent, or total 1,200 for 12 months, by definition. The total in Table 20–1, however, is 1,196.1 because extreme values have been dropped before averaging the rest.

5. Therefore, multiply each of the 12 modified means by the quotient of 1,200 over their total to yield the final *seasonal indexes*. Here, each mean is multiplied by 1,200/1,196.1 and the resulting indexes are listed in the last row. They total 1,200 and hence average 100 percent.

The individual percentages and seasonal indexes in Table 20–1 are plotted in Chart 20–3, the seasonal indexes being connected by straight lines.

These indexes of seasonal variation provide a quantitative measure of typical seasonal behavior and a basis for future planning. The slump in January and February, the autumn rise, and the December peak are clearly evident. The volume ranges from a low of 73 percent of the average month, in February, to double that volume, 147 percent, in December. The normal seasonal rise from November to December is 28 percent, that is, (147–115)/115; the decline from December to January is 46 percent; and so on. (The seasonal indexes are rounded here, since they are only accurate to the nearest percent.)

The irregularities in seasonal behavior are shown by the scatter of the percentages of trend-cycle for a given month in Chart 20–3. If the percentages are closely bunched, it means that the seasonal standing of the month is regular from year to year and the seasonal index is reliable for

Chart 20–3

SEASONAL INDEXES AND PERCENTAGES OF TREND-CYCLE—
GRAPHIC METHOD

SEARS, ROEBUCK SALES, 1966–71

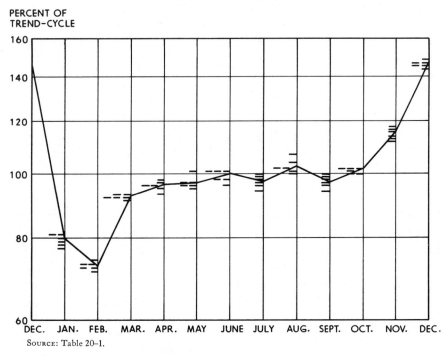

Source: Table 20–1.

use in forecasting. If all the scatters were centered about the 100 percent line, as in October, there would be no significant seasonality. In
this case, however, the average seasonal movement shown by the dis-
placement of the clusters away from the base line is unmistakable.

6. If it is desired to adjust the data to eliminate seasonal variation,
mark the January seasonal index on the measuring strip, place this mark
on each January sales point of Chart 20–2, and plot the adjusted value
on the chart opposite the 100 percent arrow of the measuring strip. This
has the effect of dividing actual sales by the seasonal index (e.g., for
January 1971, 681 ÷ 79.8 percent = 853). Do this for all months,
raising the values for months with seasonal indexes below 100 and
lowering those with indexes above 100. (The span between the seasonal index and 100 can be laid off on a blank strip for convenience in
adjusting different months.)

The adjusted sales for all months, drawn as a dashed line in Chart

20–2, reflect the trend, cycle, and irregular movements of the data, eliminating only the typical seasonal rhythm. This curve shows a steady rise in Sears, Roebuck sales, with a mild cycle marked by only a slight leveling off during the general business decline from November 1969 to November 1970. The month-to-month irregularities are due to calendar variation, the changing date of Easter, unusual weather conditions, special sales, and numerous unidentifiable causes. These irregularities can be smoothed out graphically or by a short-term moving average, as described later in this chapter, to clarify the trend-cycle pattern of sales.

Revision for Greater Accuracy. The graphic method can be refined for more accurate results as follows: Draw a revised trend-cycle curve on the ratio chart so as to bisect the *seasonally adjusted* data, following the cyclical drift and ignoring only the month-to-month zigzag movements. The revised trend-cycle curve is shown in Chart 20–6. Then repeat steps 3 to 5 (and step 6 if the data are to be adjusted for seasonality), using the new curve. The revised trend-cycle curve is more sensitive to the cyclical positions of individual months than the original curve. Hence, the seasonal indexes are better. The correction in this case, however, does not seem to justify a revision. The same procedure can be used to improve the results of the 12-month moving-average method described below.

Moving-Average Method

The moving-average method of measuring seasonal variation involves the same basic steps as the graphic method except that the steps are performed arithmetically. This method will be illustrated by the same Sears, Roebuck sales data as before. The steps are as follows:

1. Plot the series either on an arithmetic scale, for easier plotting, or on a ratio scale, to show seasonal swings of more uniform amplitude.

2. Compute a 12-month moving average to represent the trend-cycle base. This is simply a yearly average moved up a month at a time. A 12-month average includes both the high and low seasonal months during the year, and so the seasonal influences cancel out and the trend and cycle remain. The 12-month moving average is more objective than the freehand trend-cycle curve, although it tends to cut corners at cyclical turning points.[6]

---

[6] The 12-month moving average does not show the true trend-cycle position of its middle month but rather the average level of 12 adjoining months. Hence, it cannot reach the peaks, valleys, and extremities of a series; it errs in the direction of curvature in either trend or cycle, and distorts the 12 months centered on a point of abrupt change.

To compute a 12-month moving average, first find the moving *total* as follows: add the first 12 figures on an adding machine, list the total with the "subtotal" key on the tape, then add the next month and subtract the first month, list the subtotal again, and so on throughout the series. Check the last subtotal against an independent total of the last 12 months to verify all totals.

List each total in a table opposite the *seventh* of its 12 months.[7] Then divide the totals by 12 to get the moving averages. This may be done most easily by entering the reciprocal of 12—.083333—in a calculating machine and multiplying it successively by each of the totals without clearing the machine.[8]

In Table 20–2, Sears, Roebuck sales are listed from July 1965 to May 1972 to determine the moving averages for the six-year period January 1966 to December 1971, since they cannot be computed for the end months. The total for the first 12 months, July 1965–June 1966, is listed in column 3 opposite the seventh month, January 1966. Moving up a month, the next 12-month total for August 1965–July 1966 is computed as $7,222 + 601 - 563 = 7,260$ and listed opposite the seventh month, February 1966, and so on. These totals are then multiplied by $\frac{1}{12} = .083333$ with a calculating machine. The resulting moving averages are listed in Table 20–2, column 4.

3. Divide each monthly item of original data by the corresponding 12-month moving average, and list the quotient as "Percent of Moving Average." In Table 20–2, column 2 divided by column 4 equals column 5. Division is preferable to subtraction here because seasonal variation tends to repeat itself from year to year with the same *relative* intensity. That is, a normal seasonal rise in a given month tends to remain at the

---

[7] A 12-month total or average can be centered on either the sixth or seventh month, but the latter is a month more up to date. The exact center is midway between the two, so that sometimes two adjoining 12-month moving totals are themselves averaged in order to center exactly on a given month. Thus, a total of July 1971–June 1972 and August 1971–July 1972 would center precisely on January 1972. The steps are as follows: (1) Compute a 12-month moving total, listing the first item opposite the sixth month. (2) Compute a two-item moving total of these totals, entering the first item opposite the seventh month of the original data. (3) Divide by 24. This is the centered moving average. However, since the moving average is only a rough approximation of trend-cycle at best, this very minor refinement in timing does not appear to justify the considerable extra labor.

[8] Twelve-month moving averages are used here to clarify the method, but the moving totals themselves can more easily be used in subsequent steps to save the labor of multiplying through by $\frac{1}{12}$, as follows: (1) Divide each month's sales by the moving total. The results will be just $\frac{1}{12}$ the percentages of moving averages. (2) Compute the modified mean of these ratios for each month and total the 12 means. (3) Multiply each mean by 1,200 over this total to arrive at seasonal indexes identical with those in the text, the final multiplication factors being just 12 times those in the text method.

Table 20–2

COMPUTATION OF 12-MONTHS MOVING AVERAGES
SEARS, ROEBUCK SALES, 1966–71

Month	Sales (Millions)	12-Month Moving Total	12-Month Moving Average	Percent of Moving Average (Col. 2 ÷ Col. 4)	Month	Sales (Millions)	12-Month Moving Total	12-Month Moving Average	Percent of Moving Average (Col. 2 ÷ Col. 4)
(1)	(2)	(3)	(4)	(5)	(1)	(2)	(3)	(4)	(5)
1965:					**1969:**				
July	563	....	...	...	Jan.	628	9,318	776.5	80.9
Aug.	590	...	...	...	Feb.	575	9,372	781.0	73.6
Sept.	595	...	...	...	Mar.	731	9,386	782.2	93.5
Oct.	611	...	...	...	Apr.	769	9,458	788.2	97.6
Nov.	682	...	...	...	May	804	9,509	792.4	101.5
Dec.	908	...	...	...	June	784	9,542	795.2	98.6
1966:					July	797	9,651	804.2	99.1
Jan.	478	7,222	601.8	79.4	Aug.	817	9,693	807.7	101.1
Feb.	439	7,260	605.0	72.6	Sept.	781	9,725	810.4	96.4
Mar.	563	7,304	608.7	92.5	Oct.	823	9,754	812.8	101.3
Apr.	586	7,320	610.0	96.1	Nov.	926	9,779	814.9	113.6
May	588	7,343	611.9	96.1	Dec.	1,216	9,787	815.6	149.1
June	619	7,378	614.8	100.7	**1970:**				
July	601	7,416	618.0	97.2	Jan.	670	9,828	819.0	81.8
Aug.	634	7,427	618.9	102.4	Feb.	607	9,863	821.9	73.9
Sept.	611	7,443	620.2	98.5	Mar.	760	9,907	825.6	92.1
Oct.	634	7,471	622.6	101.8	Apr.	794	9,950	829.2	95.8
Nov.	717	7,492	624.3	114.8	May	812	10,001	833.4	97.4
Dec.	946	7,532	627.7	150.7	June	825	10,053	837.7	98.5
1967:					July	832	10,111	842.6	98.7
Jan.	489	7,581	631.7	77.4	Aug.	861	10,122	843.5	102.1
Feb.	455	7,604	633.7	71.8	Sept.	824	10,146	845.5	97.5
Mar.	591	7,668	639.0	92.5	Oct.	874	10,205	850.4	102.8
Apr.	607	7,730	644.2	94.2	Nov.	978	10,268	855.7	114.3
May	628	7,785	648.7	96.8	Dec.	1,274	10,304	858.7	148.4
June	668	7,868	655.7	101.9	**1971:**				
July	624	7,926	660.5	94.5	Jan.	681	10,397	866.4	78.6
Aug.	698	7,991	665.9	104.8	Feb.	631	10,442	870.2	72.5
Sept.	673	8,058	671.5	100.2	Mar.	819	10,504	875.3	93.6
Oct.	689	8,121	676.7	101.8	Apr.	857	10,588	882.3	97.1
Nov.	800	8,220	685.0	116.8	May	848	10,642	886.8	95.6
Dec.	1,004	8,300	691.7	145.2	June	918	10,765	897.1	102.3
1968:					July	877	10,858	904.8	96.9
Jan.	554	8,342	695.2	79.7	Aug.	923	10,925	910.4	101.4
Feb.	522	8,461	705.1	74.0	Sept.	908	11,012	917.7	98.9
Mar.	654	8,566	713.8	91.6	Oct.	928	11,119	926.6	100.2
Apr.	706	8,602	716.8	98.5	Nov.	1,101	11,143	928.6	118.6
May	708	8,685	723.7	97.8	Dec.	1,367	11,242	936.8	145.9
June	710	8,778	731.5	97.1	**1972:**				
July	743	8,881	740.1	100.4	Jan.	748	...	...	...
Aug.	803	8,955	746.2	107.6	Feb.	718	...	...	...
Sept.	709	9,008	750.7	94.4	Mar.	926	...	...	...
Oct.	772	9,085	757.1	102.0	Apr.	881	...	...	...
Nov.	893	9,148	762.3	117.1	May	947	...	...	...
Dec.	1,107	9,244	770.3	143.7					

same percentage as the enterprise grows, even though the dollar value rise in this month increases with the size of the business. Since the 12-month moving average roughly describes the path of the trend and cyclical fluctuations combined, the percentages of the original data divided by this average represent primarily the seasonal-irregular components, as in the graphic method. That is, actual sales = trend $(T) \times$ cycle $(C) \times$ seasonal $(S) \times$ irregular (I) components in our time series model. (Trend is expressed in the original unit, such as dollars, while the other components are stated as percentages.) Then, in step 3, $TCSI/TC = SI$, and averaging the SI ratios in the same month for different years (step 4) cancels out most of the I factor.

4. Compute the modified mean of the percentages of moving averages for a given month in the various years, omitting the highest and lowest values as being dominated by irregular factors, exactly as in the graphic method.

The percentages in Table 20–2, column 5, are grouped in Table 20–3. The highest and lowest figures in each column are then crossed out as before, and the remaining four values are totaled and divided by four to give the modified means or preliminary seasonal indexes.

5. Since the 12 modified means total 1202.0 rather than 1,200 (last column), each one is multiplied by 1,200/1202.0 to yield the final seasonal indexes shown in the row below. These indexes total 1,200 and therefore average 100 percent.

Since steps 4 and 5 are both the same as in the graphic method, Table

Table 20–3

PERCENTAGES OF 12-MONTH MOVING AVERAGES
AND COMPUTATION OF SEASONAL INDEXES
SEARS, ROEBUCK SALES, 1966–71

	Jan.	Feb.	Mar.	Apr.	May	June	July	Aug.	Sept.	Oct.	Nov.	Dec.	Total
1966	79.4	72.6	92.5	96.1	96.1	100.7	97.2	102.4	98.5	101.8	114.8	~~150.7~~	
1967	~~77.1~~	~~71.8~~	92.5	~~94.2~~	96.8	101.9	~~94.5~~	104.8	~~100.2~~	101.8	116.8	145.2	
1968	79.7	~~74.0~~	~~91.6~~	~~98.5~~	97.8	~~97.1~~	~~100.4~~	~~107.6~~	~~94.4~~	102.0	117.1	~~143.7~~	
1969	80.9	73.6	93.5	97.6	~~101.5~~	98.6	99.1	~~101.1~~	96.4	101.3	~~113.6~~	149.1	
1970	~~81.8~~	73.9	92.1	95.8	97.4	98.5	98.7	102.1	97.5	~~102.8~~	114.3	148.4	
1971	78.6	72.5	~~93.6~~	97.1	~~95.6~~	~~102.3~~	96.9	101.4	98.9	~~100.2~~	~~118.6~~	145.9	
Total, middle four	318.6	292.6	370.6	386.6	388.1	399.7	391.9	410.7	391.3	406.9	463.0	588.6	
Mean, middle four	79.6	73.1	92.6	96.6	97.0	99.9	98.0	102.7	97.8	101.7	115.8	147.2	1,202.0
Seasonal index	79.5	73.0	92.5	96.5	96.9	99.7	97.8	102.5	97.6	101.5	115.6	146.9	1,200.0
Seasonal index (graphic)*	79.8	72.7	92.6	96.3	96.8	99.8	97.8	102.5	97.8	101.5	115.4	147.0	1,200.0
Difference	−0.3	0.3	−0.1	0.2	0.1	−0.1	0.0	0.0	−0.2	0.0	0.2	−0.1	

* From Table 20–1.

20–3 is quite similar to Table 20–1, and a graph of the figures in Table 20–3 (not shown here) would show nearly the same pattern of seasonal indexes and seasonal irregularities as in Chart 20–3. The seasonal indexes obtained by the two methods are compared at the bottom of Table 20–3. The average absolute difference between the two is only 0.1 point for the 12 months, which is trivial, since seasonal indexes are only accurate to within about one point, unless more refined methods are used.

6. In order to adjust the data for seasonal variation (to eliminate its effects), divide the actual sales by the seasonal indexes. Thus, in December 1971, actual sales of $1,367 million (Table 20–2) divided by 147 percent (Table 20–3) give $930 million as the sales *adjusted for seasonal variation.* That is, $TCSI/S = TCI$. These figures are not listed here, since their graph would be almost identical with the dashed line in Chart 20–2 showing sales adjusted by the graphic method.

Changing Seasonality

Seasonal rhythm may change gradually over a period of years. The changes may be due to business policy or consumer habits. Thus, new customs, such as increasing vacation travel in summer, stimulate many activities in this season. This gradual change in seasonal behavior is called changing (moving or progressive) seasonality, as opposed to the "constant" seasonality discussed above.

Changing seasonality may be measured as follows in either the graphic or moving-average method: (1) Set up 12 small charts with the vertical scale marked "Percent of Trend-Cycle" or "Percent of 12-Month Moving Average," and mark the years on the horizontal scale. Either arithmetic or ratio charts may be used. Plot the January percentages from Table 20–1 or Table 20–3 in the first chart as a time series, the February percentages in the second chart, and so on. Then, if the January points show a sustained upward or downward drift over the years, draw a smooth, freehand trend curve through the plotted points. Now, read off the preliminary seasonal indexes from the trend curve, a different index for January in each year. Correct the 12 indexes in each calendar year to average 100 percent if necessary, as in step 5 above.

It is necessary to use a considerable number of years to determine reliable trends in the seasonal indexes for a given month. Chart 20–4 therefore contrasts the trends in October and November, relative to the 12-month moving averages, throughout the 12-year period 1960–71. October appears to drift downward, while November follows a rising trend. Therefore, we have drawn sloping freehand curves through these

Chart 20–4

CHANGING SEASONALITY

SEARS, ROEBUCK SALES, 1960–71

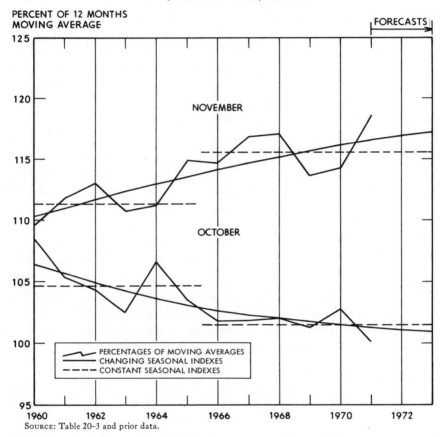

SOURCE: Table 20–3 and prior data.

panels to smooth out the irregularities and thus determine the pre-liminary changing seasonal indexes over the years. The index is read from this curve each year, rather than using the constant seasonal indexes for the two six-year periods, which are drawn as horizontal lines. The curves have been projected ahead to 1973 for use in forward planning.

This trend fit is justifiable provided there is some known explanation for the shift and a long enough period of years is included to be sure that our slope does not represent merely a random run. In this case, customers may be putting off beginning their Christmas shopping from October to November; but a special study would be required to verify the reason for this shift.

To check the changes in seasonality over a longer period, Table 20–4 presents constant seasonal indexes for four six-year periods since World War II, all computed by the moving-average method. September and October have consistently declined in importance, while gains have been experienced in July and November (since the 1950s). This confirms the shorter term trends in Chart 20–4. Other months do not

Table 20–4

CHANGES IN SEASONAL PATTERN OF SEARS, ROEBUCK SALES
CONSTANT SEASONAL INDEXES IN FOUR PERIODS, 1946–71

Period	Jan.	Feb.	Mar.	Apr.	May	June	July	Aug.	Sept.	Oct.	Nov.	Dec.
1946–51	81.8	71.9	93.5	98.7	98.7	98.9	87.1	97.5	105.7	109.9	114.9	141.4
1953–58	77.0	70.2	86.4	96.8	104.8	105.8	94.4	102.3	101.1	107.0	109.6	144.8
1960–65	77.1	70.1	88.6	96.7	100.6	102.5	96.3	103.5	99.5	104.7	111.4	149.0
1966–71	79.5	73.0	92.5	96.5	96.9	99.7	97.8	102.5	97.6	101.5	115.6	146.9

show persistent trends. For further analysis, we should extend Chart 20–4 to cover all 12 months plotted yearly for the whole period since 1946.

Changing seasonal measurement is recommended for refined analysis, since it takes into account gradual changes in seasonal behavior. However, it still does not allow fully for cyclical changes in seasonality, such as the pickup in the slack season during cyclical booms, or abrupt changes, such as those caused by war. Disruptions can best be avoided by simply omitting the abnormal periods in computing the seasonal indexes. Furthermore, changing seasonal indexes are cumbersome because they differ for each month of each year. For ordinary purposes, therefore, the use of constant seasonal indexes for homogeneous periods of years should be adequate.

Use of Computers

Electronic computer programs for measuring seasonal variation have been developed in recent years to speed the computations and permit various refinements of technique. The principal methods include the Census II Seasonal Adjustment Program,[9] the BLS Seasonal Factor Method,[10] and the use of multiple regression with a "dummy" variable

[9] See U.S. Bureau of the Census, "The X-11 Variant of the Census II Seasonal Adjustment Program," Technical Paper No. 15 (Rev., February 1967). Includes a sample printout and bibliography. See also Lawrence Salzman, *Computerized Economic Analysis* (New York: McGraw-Hill, 1968).

[10] U.S. Bureau of Labor Statistics, May 1966. Like Census II, the BLS method is based on ratios-to-12-month moving averages, and provides changing seasonal indexes, with numerous refinements. This program is adapted to many computer systems.

for each month.[11] The first of these methods is summarized below.

The Census II program is based on the ratio-to-12-month-moving-average method, using changing seasonal indexes, but the program offers a variety of optional refinements, summary measures, and tests of significance. This program is available in FORTRAN IV language, which can be used on many medium- and large-scale computers. The typical run will require less than five minutes of computer time.

The Census II method (Variant X-11) has these important features: (1) The preliminary calendar correction can be performed by correlating the original series with the number of times each day of the week occurs in each month, rather than by having to introduce explicitly the number of working days in the month. (2) The series is then adjusted for seasonal variation by the ratio-to-centered-12-month-average method. (3) The adjusted series (TCI) is then smoothed by a weighted moving average of 9, 13, or 23 terms (depending on how irregular the series is), in order to smooth out irregularities and provide a revised trend-cycle curve. This type of trend-cycle curve is much more sensitive to cyclical movements than the original 12-month moving average. (4) The original daily averages are then divided by this new trend-cycle base and the seasonal measurement process is repeated as before. (5) The seasonal-irregular ratios for a given month in different years are smoothed by a weighted moving average (obtained by taking a three-term average of a five-term moving average) to estimate the changing seasonal indexes. (6) Extreme values are given reduced weight or no weight, depending on how many standard deviations they depart from the norm. (7) A set of summary measures is prepared, such as the percentage contributions of the trend-cycle, calendar, seasonal, and irregular factors in a time series, and the ratio of the average irregular component in month-to-month changes to the average trend-cycle component. Various tests of significance are also provided. (8) The results are printed out graphically on a chart.

Therefore, the computer carries the ratio-to-moving-average method through more refinements than would otherwise be feasible. Furthermore, seasonality can be analyzed in far more economic time series than was formerly possible.

Electronic computers cannot handle certain problems such as abrupt changes in plantwide vacation schedules or the shifting dates for offer-

---

[11] See Michael C. Lovell, "Seasonal Adjustment of Economic Time Series and Multiple Regression Analysis," *Journal of the American Statistical Association*, Vol. 58 (1963), pp. 993–1010. This method is appropriate in some circumstances, but the ratio-to-moving-average computer methods are more generally preferred.

ing new automobile models. These situations should be adjusted by hand before the data are put into the computer, or else the series should be broken at the point of discontinuity and the two segments analyzed separately. Computers provide speed and precision of results in the hands of the skilled analyst, but they still do not take the place of human judgment.

Which Method to Use?

The following suggestions may be helpful in selecting an appropriate method for measuring seasonal variation:

1. The graphic method is recommended as a shortcut, since it substitutes graphic measurements for the three laborious steps (2, 3, and 6) of the moving-average method. The freehand trend-cycle curve can follow cyclical movements more closely than the 12-month moving average, if drawn with skill and judgment, particularly when revised to follow the seasonally adjusted data. The graph also affords a visual check on each step, revealing irregularities in the data and allowing necessary variations in technique.

2. The moving-average method has the advantage of being a standard, objective procedure than can be performed by clerical labor with a hand calculator and adding machine. It is the most commonly used of the many simple arithmetic methods proposed for analyzing seasonality. Like the graphic method, its results are usually accurate enough for most purposes.

3. Electronic computer methods provide both the greatest time saving and the most accurate seasonal measurement, when many series are to be analyzed, and the program and computer are available. Such programs, however, are complex and require a sophisticated analyst to select the appropriate options and to interpret results.

Other Methods of Taking Seasonality into Account

There are several commonly used methods of allowing for seasonality without actually measuring it:

1. Seasonal movements are sometimes referred to merely in directional terms. For example, "Retail sales made a seasonal gain in September over the August level." This statement, however, does not say whether the gain was more or less than the normal seasonal amount and how much it differed. It would be more meaningful to say: "Retail sales gained 8 percent in September over the August level, after allowance for the usual seasonal increase."

2. The common practice of comparing a month with the same

month a year ago serves to eliminate the seasonal factor common to both months. This usage, however, may still distort the cyclical picture for either of two reasons: (*a*) The current month is judged in comparison with a single historic month that might have been erratic itself. Thus, the statement "Production in March was 3 percent above a year ago" appears favorable, but it might represent an unfavorable situation if March last year was unduly depressed. (*b*) The comparison with a year ago ignores the trends of the past 11 months. For example, Sears, Roebuck sales in November 1970 were 6 percent above those of November 1969. This report appears favorable, but it would have been more significant to note that seasonally adjusted sales had failed to gain since July 1970, as shown in Chart 20–2.

3. Plotting weekly or monthly data for several years above each other on a tier chart with the horizontal scale extending from January to December enables one to compare current tendencies with those in the same seasons of other years without any calculations. But the comparison with several such years is apt to be confusing and offers no precise adjustment for the seasonal factor. In Chart 20–5, for example,

Chart 20–5

ELECTRIC POWER PRODUCTION

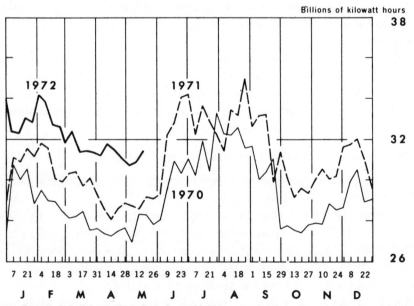

SOURCE: *Federal Reserve Chart Book*, June 1972. This source also charts the seasonally adjusted data, which clarify nonseasonal movements.

the general level of 1972 electric power production is obviously above that of the two previous years, but the weekly nonseasonal comparisons are not clear. In particular, was the decline in production from February to May 1972 more or less than the usual seasonal amount?

These methods are sometimes useful for simple presentation. For careful analysis, however, seasonal indexes should be computed as described earlier in the chapter.

SEASONAL FORECASTING

Seasonal indexes play an important part in short-term business planning. Chart 20–2 shows how Sears, Roebuck sales can be forecast (at the end of 1971) for each month of 1972 by projecting the trend-cycle curve and multiplying these values by the seasonal indexes.

The trend-cycle curve has been projected from the regression between Sears annual net sales and disposable income (Table 16–5), as well as the trend projection in deflated sales (Chart 19–4), combined with a prediction of Sears' price index (Table 19–1) and the cyclical outlook based on statistical indicators (Chart 20–8) and other methods described later in the chapter.

We can then forecast monthly sales by multiplying the values on the extended trend-cycle curve by the seasonal indexes (i.e., $TC \times S = TCS$). Or, using the graphic method, we can place the 100 percent mark of the measuring strip on the trend-cycle curve, and lay off the seasonal indexes from Table 20–1 above and below it to predict the combined trend-cycle-seasonal effects. (The irregular element cannot be estimated.)

In Chart 20–2 this projection is plotted as a dotted line through 1972 and compared with actual sales through July 1972 as a check on its accuracy. Thus the January forecast is the trend-cycle value of $946 million times 79.8 percent (the seasonal index) or $755 millions, compared with actual sales of $748 millions, an error of 1 percent. (The February forecast could be improved by allowing for the extra trading day in Leap Year, and the March and April figures could be adjusted for the changing date of Easter, in further analysis.)

The error of the forecast includes that of the trend-cycle projection (which increases with time) and that of the irregularity in the seasonal element itself, which can be estimated from the scatter of the arrays in Chart 20–3. When seasonal fluctuations are large and regular, and short-term cyclical movements are mild, as in retail trade generally, short-term forecasting is relatively accurate.

CYCLICAL VARIATIONS

Cyclical swings, or alternations between expansion and contraction, are of prime importance in short-term business analysis and planning.

Business cycles are a type of fluctuation found in the aggregate economic activity of nations that organize their work mainly in business enterprises: a cycle consists of expansions occurring at about the same time in many economic activities, followed by similarly general recessions, contractions, and revivals which merge into the expansion phase of the next cycle; this sequence of changes is recurrent but not periodic; in duration business cycles vary from more than one year to ten or twelve years.[12]

Business cycles have developed in modern industrialized countries having closely integrated business structures. The cycles are affected by factors outside business, such as wars, acts of government, and the size of crops, but it is the conditions within the business system itself that cause a protracted expansion to give way to contraction and vice versa, in a roughly rhythmic fashion. Nearly all economic activities are affected by cyclical forces, but heavy industrial production and finance are most susceptible, and retail trade, personal service, and agricultural production are least affected.

The average length of peacetime business cycles since 1945 has been about $3\frac{1}{2}$ years, of which the expansion phase has averaged over three times as long as the contraction phase. Table 20–5 lists the turning points of the general business cycle, averaged from thousands of individual series by the National Bureau of Economic Research. Despite the Korean and Vietnam wars, the amplitude of cycles in this period has been much milder than in earlier times.

Cycles in individual series also differ markedly in these respects from the general business cycle. Consider the cyclical swings of gross national product, aluminum, and coal production in Chart 19–2 as the major deviations from the trend lines. Gross national product is relatively insensitive to the cycle, since it contains many stable types of expenditures, such as interest payments, while aluminum production is extremely volatile, and coal is both moderate in amplitude and more sensitive to general business conditions than is aluminum. All three series, however, reflect the booms of the two world wars and the depressions of 1921 and 1932. The study of cycles is more crucial in cyclical or sensitive industries than in stable activities.

---

[12] This definition of Wesley C. Mitchell is used as the point of departure in the National Bureau of Economic Research studies in business cycles. See Arthur F. Burns and Wesley C. Mitchell, *Measuring Business Cycles* (New York: National Bureau of Economic Research, 1946), p. 3. See also Wesley C. Mitchell, *What Happens during Business Cycles: A Progress Report* (New York: National Bureau of Economic Research, 1951).

Table 20–5

TURNING POINTS OF BUSINESS CYCLES
IN THE UNITED STATES, 1945–70

		Number of Months		
Trough	Peak	Contraction (Trough from Previous Peak)	Expansion (Trough to Peak)	Total Cycle (Peak from Previous Peak)
October 1945. . . . . . .November 1948		*8*	37	45
October 1949. . . . . . .July 1953		11	*45*	*56*
August 1954. . . . . . .July 1957		*13*	35	48
April 1958. . . . . .May 1960		9	25	34
February 1961. . . . . .November 1969*		9	*105*	*114*
November 1970*. 12				
Average, 5 cycles, 1945–70. 11			49	59†
Average, 3 peacetime cycles, 1945–61. 10			32	42‡

Note: Italics indicate wartime expansions, postwar contractions, and full cycles that include wartime expansions.
   * Tentative.
   † 5 cycles, 1945–69.
   ‡ 3 cycles, 1945–60.
   Source: National Bureau of Economic Research, reported in *Business Conditions Digest*, Appendix E, February 1973. This source also gives earlier turning points, beginning in 1854.

Irregular fluctuations in economic time series are caused by such forces as government expenditures, taxes, unusual weather, labor strife, war, and all forms of unpredictable events. These forces are of two types. The first group serves as "originating forces" in inducing or altering business cycle movements. War and its aftermath, for example, tend to produce the familiar boom and bust phases of a major peacetime cycle. A government public-works program may stimulate a similar cycle on a smaller scale. A protracted steel strike, on the other hand, creates a condition similar to cyclical depression in that industry. These forces are generally unpredictable, although many Washington "services" advise business on what the government is likely to do and whether there will be war, strikes, large or small crops, etc., with partial success.

The second group of irregular factors comprises the host of miscellaneous forces that act in a more or less random fashion to give a plotted curve its familiar zigzag contour. These factors are usually numerous, unidentifiable, and unpredictable. The random element varies widely in different series, from nothing in the Federal Reserve rediscount rate to a major influence in the value of building permits issued.

The irregular component in a time series represents the residue of fluctuations after secular trend, cyclical, and seasonal movements have been accounted for. In practice, however, the cycle itself is so erratic and is so interwoven with irregular movements that it is impossible to separate them, except in smoothing out some of the random factors of the second type.

Reasons for Measuring Cycles

Three important purposes are served by isolating the cyclical, or cyclical-irregular, component in a time series.

1. Measures of typical cyclical behavior are valuable aids in controlling the operations of a business. These measures will answer such questions as: How sensitive is this business to general cyclical influences? What is the typical timing, amplitude, and general cyclical pattern of the company's production, sales, inventories, or raw material prices? How do these factors compare with those of other companies or with the industry as a whole? Are there leads or lags compared with other series that would aid in forecasting?

The study of business cycles is also one of the major branches of economics. Today economists generally recognize the need not only of theory but also of accurate statistical measures in order to gain a clear understanding of this phenomenon. Hence, the National Bureau of Economic Research and other agencies have devoted decades of study to this measurement.

2. Successful businessmen plan ahead; planning requires forecasting; and forecasting involves a knowledge of both typical and recent cyclical behavior. Measures of *typical* cycles are used in the "economic rhythm" school of forecasting, which projects past cycles ahead in periodic fashion. Such measures also appear in the "specific historical analogy" method of relating present conditions to those in a comparable period of the past and anticipating similar developments. Measures of *recent* cyclical behavior are necessary as a starting point in any kind of forecast. Articles abound in business journals, particularly around the first of the year, containing forecasts based on cyclical indicators.

3. Cyclical measures are useful tools in formulating policy aimed at stabilizing the level of business activity. Major efforts are being made by the federal government and by business to iron out the business cycle, since depressions are disastrous for the economy. The President's Council of Economic Advisers and the congressional Joint Economic Committee are important agencies that evaluate cyclical indicators as aids in devising safeguards against depression. Accurate cyclical measures are as

necessary in planning preventive action as in anticipating what will happen without such action.

Despite the importance of business cycles, they are the most difficult type of economic fluctuation to predict. This is because successive cycles vary so widely in timing, amplitude (percentage rise and fall), and pattern, and because the cyclical rhythm is inextricably mixed with irregular factors.

HOW TO MEASURE CYCLES

The standard method of isolating cycles in economic data is to eliminate seasonal, secular, and irregular movements as far as possible and to plot the residuals to show the cyclical fluctuations.[13] Not all of these movements, however, need to be eliminated in practice. The more pronounced a noncyclical factor, the more it tends to obliterate the cyclical pattern and the greater the need for its elimination. Thus, a wide seasonal swing, a steep trend, or a violently zigzag irregular contour requires adjustment more than if each of these factors were neutral. Ordinarily, the seasonal adjustment is the most important of the three. Frequently, *only* this adjustment is made in the data, together with some smoothing of random-type irregularities. This is because the secular trend does not ordinarily obscure short-term cycles, and the adjustment for trend introduces an error arising from the fitting of the trend curve itself. Furthermore, cycles cannot be separated successfully from the sustained irregular movements caused by originating forces.

Annual data need be adjusted only for secular trend, since seasonal and short-term irregular fluctuations tend to cancel out in the yearly totals. Charts 19–4 and 19–7 show the yearly deflated sales of Sears, Roebuck, adjusted for trend. The cycles in the annual data were described on pages 591 and 601–2. However, since cycles are of short-term duration, monthly data are usually needed to give a more detailed picture.

Graphic Adjustment

Cycles may be isolated graphically as follows:

1. Adjust the data for seasonal variation as described above. To illustrate, Chart 20–6 is reproduced from Chart 20–2 to show Sears, Roebuck sales adjusted for seasonality by the graphic method (dashed line).

---

[13] A method of *averaging* the cycles in seasonally adjusted data is described in Burns and Mitchell, *op. cit.,* chap. 2; see also Mitchell, *op. cit.*

Chart 20–6

TREND-CYCLE MOVEMENTS IN SEARS, ROEBUCK SALES, 1966–72
GRAPHIC METHOD
Ratio Chart

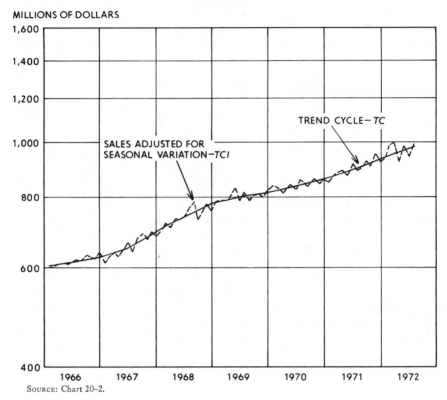

SOURCE: Chart 20-2.

2. Draw a freehand curve through the adjusted data, if necessary, to smooth out the zigzag irregularities and bring out the trend-cycle component in clear relief. The deviations above the curve should equal those below. This trend-cycle curve itself usually suffices for cycle analysis. Thus, the trend-cycle curve of Sears, Roebuck sales (Chart 20–6) gave warning of a possible recession by leveling off in the latter part of both 1969 and 1970, whereas the unadjusted sales in Chart 20–2 might have misled management, since they rose sharply from September to December each year because of seasonal influences. No cyclical downturn occurred in dollar sales, however, despite the general business recession of 1970. (This curve can also be used in place of the preliminary freehand trend-cycle curve or 12-month moving average in

recomputing the seasonal indexes, as described on page 621 under "Revision for Greater Accuracy.")

3. The trend-cycle curve in Chart 20–6 can be adjusted further for trend by fitting a smooth trend curve (e.g., a logarithmic straight line) and laying off the vertical deviations of the trend-cycle curve from the trend around a horizontal line. The result is the cyclical component expressed as a percentage of trend. This procedure is not shown here, since it was illustrated for Sears, Roebuck annual sales in Chart 19–7, and the trend adjustment is not usually necessary for short-term analysis.

Arithmetic Adjustment

Cycles can also be isolated arithmetically in three steps:

1. Adjust the data for calendar and seasonal variation as described in the ratio-to-12-month-moving-average method.

2. Compute a three-month moving average, if necessary, to smooth out short-term irregular movements. That is, the January–March average is plotted in the middle month, February; the February–April average is used for March; and so on. If the data are extremely erratic, a five-month moving average may be preferable. This results in a smoother curve but one which is less sensitive to month-to-month movements than the three-month moving average. Of course, irregular movements do not exactly offset each other every three or five months, so some of the irregularities remain in the smoothed curve. Ordinarily, the resulting trend-cycle values can be used for cycle analysis without further adjustment.

3. If it is desired to adjust for trend, fit an appropriate trend curve to the monthly data by least squares and divide the seasonally adjusted data by the trend values before computing the three- or five-month moving averages. (However, the order of operations makes little or no difference.) That is, assuming that sales represent the product of $T \times C \times S \times I$,[14] the seasonal adjustment is $TCSI/S = TCI$; dividing by the trend value gives $TCI/T = CI$; and a three- or five-month moving average cancels out part of the irregular movements to leave C as a residual. All steps can be performed by hand calculators.

We will not illustrate the arithmetic method of isolating cycles in Sears, Roebuck sales here, as we have already described step 1; step 2 is laborious; step 3 is usually unnecessary; and the TCI and TC curves

---

[14] This is TCSI not $T + C + S + I$ because C, S, and even I tend to be more constant as percentages than as absolute amounts. However, these factors can be added (or subtracted) on a ratio chart, since this operation is equivalent to adding the logarithms or multiplying the natural values.

resulting from steps 1 and 2, respectively, would be quite similar to those shown in Chart 20–6. The chief difference is that the short-term moving average would be somewhat more irregular, though more objective, than the freehand TC curve.

Computer Methods

The electronic computer programs described above not only adjust monthly or quarterly data for seasonality but also smooth out irregularities by means of a short-term moving average. An average of from one to six months is used in the Census II method, depending on the relative amplitude of the month-to-month irregular changes as compared with the cyclical changes in a series. That is, the number of "months for cyclical dominance" is computed as $MCD = \bar{I}/\bar{C}$, where $\bar{I}$ is the average absolute irregular movement per month and $\bar{C}$ is the average absolute cyclical change.[15] This is the span of months in which the cumulative cyclical element in the series typically exceeds the irregular element for one month. In a very irregular series such as liabilities of business failures, a six-month moving average is required for the cyclical element to dominate over the irregular movements. On the other hand, a single month's change in the Federal Reserve Board Index of Industrial Production typically contains a larger cyclical than irregular element, so the actual monthly figures are used without averaging several months.

Chart 20–7 illustrates the elimination of seasonality and the smoothing of irregularities in the number of unemployed men from 1948 to 1965, using the Bureau of Labor Statistics computer method. The top panel shows the actual data and the final trend-cycle component, after eliminating the changing seasonal pattern and the irregularities depicted separately in the lower panels. Note how clearly the cycles of unemployment emerge in the trend-cycle curve as compared with the actual data, which are dominated by strong seasonal-irregular influences. In particular, the peaks and troughs of the unemployment cycle occur at quite different times from those which appear in the actual data.

CYCLICAL FORECASTING

We can forecast monthly changes in a series for the next year by combining their trend, seasonal, and cyclical components. Projecting the trend and seasonal elements is a straightforward statistical process, but

---

[15]$\bar{C}$ includes the trend component, but this is negligible in one month. See *Business Conditions Digest* for a more detailed explanation.

Chart 20-7

TREND-CYCLE, SEASONAL AND IRREGULAR COMPONENTS
UNEMPLOYED MEN* IN THE UNITED STATES, APRIL 1948–JUNE 1965

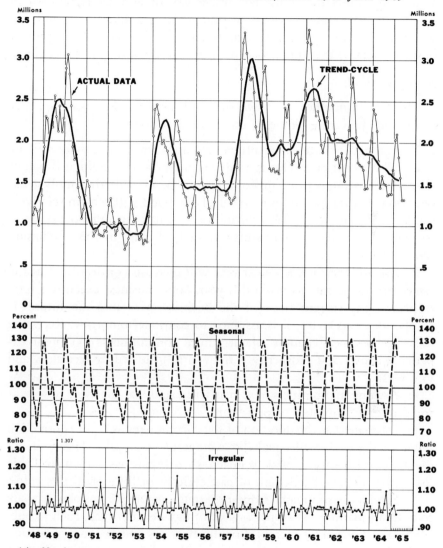

* Age 20 and over.
SOURCE: U.S. Bureau of Labor Statistics, *The BLS Seasonal Factor Method (1966)*, p. 2.

foretelling cyclical changes is much more difficult. Cycles are recurrent but not periodic; their expansion or contraction periods may be reversed at turning points that must be anticipated, or at least recognized in passing, for successful business planning. Also, unlike trends and seasonal movements, cycles in specific series are influenced by the general business cycle, so their prediction requires a study of the entire economy.

Naïve Methods

There are a number of "naïve" methods used explicitly or implicitly to foretell the future. Some of these are as follows:

1. Assume that the most probable future level of activity will be that of the recent past. This is a fallacy; the normal condition is one of change. For example, an investor buys a bond with the implicit expectation that the purchasing power of the bond will remain relatively stable during its life. If the probabilities point to an inflationary trend in prices, however, he may suffer a costly decline in the purchasing power of his bond.

2. Assume that business next year will increase (or decrease) at the same percentage rate as it did this year. Some business executives tend to project the current stage of the business cycle into the future. If prosperity exists today, it is assumed to continue tomorrow. Recession today makes men cautious about future commitments. Yet past experience shows that prosperity is frequently followed by recession, and vice versa.

3. Assume that business next year will expand at the average secular trend rate of a number of past years.

4. Estimate that the duration of the current expansion or contraction phase of the cycle will equal the average of past cycles. However, individual cycles vary so widely in length of phase, as shown in Table 20–5, that the average length of past cycles is of little predictive value.

5. Send a questionnaire requesting opinions on the business outlook to a broad mailing list of persons who may be interested, such as subscribers to *Fortune* or members of the Business and Economics Section of the American Statistical Association. Thus, from a quantity of casual replies one hopes to distill a precise forecast. The use of surveys to elicit a consensus of opinions and guesses is a widespread pastime in economic, political, and social affairs.

Some of these methods, particularly 1 and 3, prove more often right than wrong, since the usual estimate of continued rise reflects the long-term growth of the economy and the fact that cyclical expansions last longer than contractions. But such success is illusory. More sophisticated statistical analysis is needed to provide an adequate basis for planning future operations.

Exponentially Weighted Moving Averages

A simple computer program can be used to forecast sales of a large number of products a few months ahead, for short-term planning and inventory control. The estimate is a moving average of past months, with weights declining exponentially. That is, the latest month is given the heaviest weight, and the weight for each preceding month is reduced by a constant percentage. (The weights must total one.) Such a procedure seems cumbersome, but it is actually simple for the computer, since all prior data can be summarized in a single number and only the latest month added to bring the moving average up to date. The result is often a reasonable estimate for the coming month, since the moving average gives greatest weight to the latest month but still smooths out most irregularities by averaging a number of prior values. Trend and seasonal adjustments can also be incorporated in the program.[16]

The foregoing methods have the limitation of being based essentially on *past* trends rather than on future prospects. The most important function of business cycle forecasting, however, is not to predict a continuance of the current phase, but rather to recognize the turning points. The following methods may be useful for this purpose.

Lead and Lag Indicators

Most business processes move up and down roughly concurrently in the business cycle, but some are more sensitive than others, or represent earlier stages in production, and so reach their peak and troughs before the aggregate indicators. Thus, the average work week of production workers in manufacturing responds more promptly to economic stimuli than does total nonagricultural employment. New orders for durable goods and construction contracts precede actual business expenditures for new plant and equipment. Common stock prices anticipate future changes in profits. Finally, sensitive commodity prices such as steel scrap move more promptly than composite nonfarm wholesale prices.

The Natural Bureau of Economic Research has selected a number of monthly and quarterly series that tend to lead the general business cycle

---

[16] See Peter R. Winters, "Forecasting Sales by Exponentially Weighted Moving Averages," in F. M. Bass *et al., Mathematical Models and Methods of Marketing* (Homewood, Ill.: Richard D. Irwin, 1961), pp. 482–514. See also Robert G. Brown, *Smoothing, Forecasting, and Prediction of Discrete Time Series* (Englewood Cliffs, N.J.: Prentice-Hall, 1963), chaps. 7 and 12.

at its turning points, another group that are roughly coincident in timing with general business, and some indicators that tend to lag.[17] These are adjusted for seasonal variation and irregularities by computer methods, and are reported monthly in *Business Conditions Digest*. Thus, during a cyclical expansion, a marked down-turn by a majority of the leading indexes gives a warning of a possible impending downturn in general business. If most of the coincident indexes then also turn down, this confirms the movements of the leaders, and if the lagging indicators follow suit, a general business contraction is almost certainly in progress.

Unfortunately, none of these indicators is consistent in timing, and while most of them in fact have reversed direction at actual business peaks and troughs, they often give false signals because of minor intermediate movements. They must be used with caution.

Diffusion Indexes

A diffusion index is also based on the principle that different processes in business reach their peaks and troughs at different times, but

Chart 20–8

COMPOSITE INDEXES

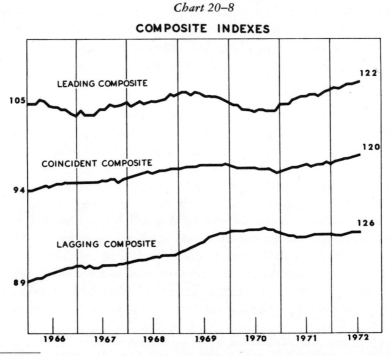

---

[17] For a further description, see G. H. Moore and J. Shishkin, *Indicators of Business Expansions and Contractions* (New York: Columbia University Press, 1967).

Chart 20–8—Continued
PERCENTAGE EXPANDING

LEADING INDICATORS

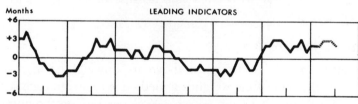

COINCIDENT INDICATORS

LAGGING INDICATORS

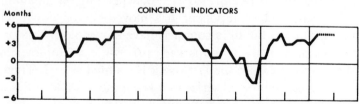

AVERAGE MONTHLY DURATION

LEADING INDICATORS

COINCIDENT INDICATORS

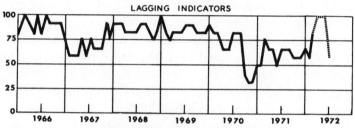

LAGGING INDICATORS

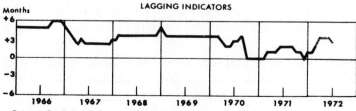

SOURCE: Statistical Indicator Associates, North Egremont, Mass.

this device does not require identifying *which* particular series lead and which lag. A diffusion index is simply the percentage of all seasonally adjusted series that are rising in a given month. (Sometimes a six- or nine-month span is used.) Thus, if 60 out of 100 series increased in October over September, and 40 were stationary or declining, the diffusion index would be 60.

During the midexpansion period, perhaps 80 percent or more of all series are rising. But at the peak of aggregate activity, about half of the indicators of business volume will have turned down, while the other half are still rising, so that the diffusion index will cross the 50 percent line on the way down. Similarly, in midcontraction the diffusion index may drop as low as 20 percent. But at the trough of general business, about half the series of business volume will have turned up while the other half are still declining, and the diffusion index will have risen to about 50 percent. Hence, a diffusion index signals a peak or trough in general business activity by crossing the 50 percent line on the way down or up. Theoretically, therefore, a diffusion index can lead the aggregates on which it is based by perhaps a quarter-cycle. Diffusion indexes are shown monthly for many industries (e.g., new orders for durable goods in 36 industries) in *Business Conditions Digest*. Like the lead and lag indicators themselves, diffusion indexes usually mark actual business cycle turning points very well, but often give false signals in crossing the 50 percent line because of short-term irregular movements.

Average Duration of Run

The diffusion indexes described above are unweighted in that each series counts the same. One weighting method is to assign each series in a given month a number from +6 to —6, depending on the number of months its trend-cycle component has moved up or down without interruption. Thus, if building contracts have moved up for six or more months through January it is marked +6, while if employment has declined two months since the last rise it is counted as —2. Then these numbers are averaged for all series in a given month, and the resulting "average duration of run" series is plotted. It signalizes a peak or trough in business when it crosses the zero line, going downward or upward, respectively, just as the diffusion index does when it crosses the 50 percent line.

Chart 20–8 shows a composite group of leading, coincident, and lagging indicators, diffusion indexes ("Percentage Expanding") and average monthly duration, as compiled by Statistical Indicator Associates. In 1966 the leading indicators gave a preliminary warning of a business downturn, but this was not confirmed by the coincident indicators, and no recession occurred. In 1969, however, the drop in the three leading indicators was confirmed by most coincident indicators to signal the business contraction of November 1969–November 1970 (Table 20–5). The indicators also gave an early signal of the following recovery.

Surveys of Anticipations Data

This method is based on the premise that businessmen, and to a lesser extent consumers, make forward plans for the expenditure of capital goods, and that a survey of these intentions will have forecasting significance. The surveys of businessmen's plans for new plant and equipment expenditures conducted by the U.S. Department of Commerce–Securities and Exchange Commission and by McGraw-Hill are widely followed. The National Industrial Conference Board surveys capital appropriations of large firms. The University of Michigan Survey Research Center and the U.S. Bureau of the Census canvass consumers' plans to purchase houses, cars, and durable equipment.[18]

Surveys of professional forecasters' opinions of course are valuable, as opposed to the surveys of general mailing lists which were classified under naïve methods above. Thus, the National Industrial Conference Board publishes the conclusions of an annual conference of leading forecasters. United Business Service summarizes the views of eight other financial services each month. The Federal Reserve Banks of Philadelphia and Richmond select and compile hundreds of forecasts early in the year. If you are confused by the multiplicity of expert opinions, just follow the consensus.

A RÉSUMÉ OF STATISTICAL METHODS IN FORECASTING

At this point we may summarize the statistical methods useful in business forecasting. Sample survey methods (Chapter 12) are needed to poll the expectations of businessmen and consumers for the near-term future. The regression analysis of time series (Chapters 16–17) will enable us to correlate our own process (e.g., a company or industry sales) to some aggregate series (e.g., personal income) for which projections are available. Thus, *Predicasts* compiles forecasts for many economic aggregates and industry totals for 20 or more years in the future from many sources. Index numbers (Chapter 18) serve to sum-

---

[18] See National Bureau of Economic Research, *The Quality and Economic Significance of Anticipations Data* (Princeton, N.J.: Princeton University Press, 1960), for appraisal of these methods. *Business Conditions Digest* reports current data.

marize economic aggregates and their characteristics (e.g., diffusion indexes) as well as to make disparate series comparable. Finally, time series analysis (Chapters 19–20) provide a means of projecting the secular trends, seasonal movements, and cycles of a business series to achieve a composite forecast.

Not all the statistical methods used in short-term forecasting are needed in long-term forecasting. A long-term forecast, extending perhaps 5 or 10 years into the future, typically involves a secular trend projection and regression analysis, to compare the series with basic economic aggregates. The long-term forecast is not concerned, however, with seasonal variation, nor is it possible to forecast the phase of the business cycle more than a year or two ahead. Surveys of anticipations or expectations, also, are generally not valid in the long run.

In short-term forecasting, which usually involves monthly estimates for the coming year, all the above statistical methods are applicable. In particular, it is useful to extrapolate the trend and seasonal movements of a monthly series by the methods described above, and then estimate by statistical and economic analysis whether the current phase of the business cycle is likely to continue or whether a turning point is in prospect. Finally, the cyclical components of an individual series (e.g., industry sales) can be correlated with the cyclical elements in some basic series (such as personal income) for which estimates are available. All the above methods can be carried out efficiently and comprehensively by electronic computers in large-scale analysis.

While statistical methods are necessary tools in business forecasting, they are not sufficient in themselves to complete the job. It is necessary to supplement the statistical results with a thorough economic analysis of cyclical and growth factors at the national, industry, and company levels. Accordingly, the corporate staff specialist responsible for forecasting is more often called a business economist than a statistician. The economics of forecasting, of course, lies beyond the scope of this book.[19]

SUMMARY

Seasonal variations are regular rhythmic movements within a period of one year resulting from the weather and from man-made conventions such as holidays. They affect nearly all economic processes in varying

[19] See W. F. Butler and R. A. Kavesh, *How Business Economists Forecast* (Englewood Cliffs, N.J.: Prentice-Hall, 1966); H. D. Wolfe, *Business Forecasting Methods* (New York: Holt, Rinehart & Winston, 1966); H. O. Stekler, *Economic Forecasting* (New York: Praeger, 1970); R. K. Chisholm and G. R. Whitaker, Jr., *Forecasting Methods* (Homewood, Ill.: Richard D. Irwin, 1971); or the sources listed in J. B. Woy, *Business Trends and Forecasting* (New York: Gale Research, 1965) for further study.

degrees, particularly at the point of origin and the point of consumption. Seasonal variations may change in character over the years. However, seasonal fluctuations are much more regular than cycles, and so they can be measured and projected more accurately. Regular rhythms also occur within a quarterly, monthly, weekly, or daily period. Finally, the calendar itself causes quasi-seasonal variations in monthly and weekly data, since the number of operating days varies from one month or week to the next.

Adjustment for calendar variation is made as a preliminary step in seasonal measurement in order to eliminate fluctuations in the data caused by the varying length of the working month. The data are divided by the number of operating days in each month to place the series on a uniform daily average basis.

Seasonal variation is measured for the purpose of controlling current operations, forecasting, or adjusting data in order to reveal cycles. The seasonal pattern is best described by seasonal indexes that represent the average value for each month related to the average of all 12 months as 100 percent. The period analyzed should be long enough to average out peculiarities in individual years, but abnormal periods should be omitted.

Several methods of computing seasonal indexes are described. The graphic and moving-average methods are summarized in the table, with symbols to indicate how the trend (T), cycle (C), and irregular (I) factors are eliminated to isolate the seasonal index (S).

Results can be improved by redrawing the trend-cycle curve through

Step	Graphic Method	Moving-Average Method	Shows
1.	Plot on ratio chart.	Plot on ratio chart.	TCSI
2.	Draw freehand TC curve.	Compute 12-month moving average.	TC
3.	Read ratios of data to TC from measuring strip.	Divide data by moving average.	SI
4.	Average ratios for each month.	Average ratios for each month.	S (preliminary)
5.	Multiply indexes by 1,200 over their sum.	Multiply indexes by 1,200 over their sum.	S
6.	To adjust for seasonality, shift plotted data the distance from seasonal index to base line of measuring strip.	To adjust for seasonality, divide data by seasonal indexes.	TCI

the seasonally adjusted data and repeating steps 3 to 5 (and 6 if desired).

If the seasonal pattern changes over the years, changing or moving seasonal indexes can be computed in either method by plotting the ratios for each month in step 3 chronologically and reading the preliminary indexes from freehand trend curves drawn through these plots.

Computer programs such as Census II greatly speed the necessary calculations and permit several refinements in technique, such as calendar adjustment from internal evidence, improved trend-cycle estimates using weighted moving averages, reduced weights for extreme items, computation of changing seasonal indexes, and various summary measures and tests of significance.

The methods compare as follows: the graphic method is quick, flexible, and affords a continuous check on operations, while the moving-average method is objective and can be performed by clerical labor on hand calculators. Electronic computer programs are recommended where many series are to be treated, since they give fast and accurate results in the hands of a skilled analyst.

Seasonality is sometimes taken into account without actual measurement by means of (1) qualitative description, (2) comparing a month with the same month a year ago, or (3) plotting several years on a tier chart with the same monthly time scale. These devices are useful for simple presentation, but seasonal indexes are needed for refined analysis.

To make a short-term forecast, project the trend-cycle curve (see cyclical forecasting) and multiply these values by the seasonal indexes each month (i.e., $TC \times S = TCS$) or lay off these indexes from the TC curve with the measuring strip graphically.

Cyclical variations are the rhythmic movements of alternating expansion and contraction that have developed in industrialized economies. Cycles vary widely in timing, pattern, and amplitude, both from one cycle to the next and from industry to industry. Major expansions and contractions, however, affect nearly all economic activities.

Irregular fluctuations are the residual component in a time series after secular trend, cyclical, and seasonal movements have been accounted for. It is usually impossible, however, to separate cyclical and irregular fluctuations satisfactorily. The irregular factors may be major "exogenous forces" (such as wars and acts of government) that influence business cycles, or they may be miscellaneous unknown and unpredictable factors of a random nature.

Measures of business cycles are important in the study of current cyclical behavior, in forecasting business activity, and in planning stabilization policy. Cycles can be isolated (1) by eliminating seasonality (and perhaps trend) by division or graphic adjustment, and (2) by smoothing irregularities by a short-term moving average or freehand curve. The cyclical component remains as a residual. Sometimes only the seasonal adjustment is necessary. Computer programs such as Census II eliminate the calendar and seasonal components in successive steps and then smooth the residuals with a moving average of from one to six months, depending on the irregularity of the data, to arrive at the trend-cycle component. Trend is left in, since it does not obscure the short-term cyclical pattern.

It is important to forecast the cyclical swings of business, particularly at turning points. A number of statistical forecasting methods are discussed: (1) various naïve methods in common use, (2) exponentially weighted moving averages, (3) lead and lag indicators, (4) diffusion indexes, (5) average duration of run, and (6) surveys of anticipations data. Statistical methods, however, must be supplemented by careful economic analysis to achieve an adequate forecast.

The statistical forecaster should be familiar with the materials in Chapters 12 and 16 to 20 of this book, as well as appropriate economics texts, as a basis for becoming adept in the strategic art of business forecasting.

PROBLEMS

1. *a*) Make a photocopy of a published graph of monthly data dominated by seasonal movements. Do not use textbook examples.

    b) Describe the seasonal characteristics: Is the seasonal amplitude wide or narrow? Is the seasonal pattern regular or irregular? What are the high and low months and the seasonal tendency of other months? Give reasons for these movements.

2. *a*) Make a photocopy of a published graph of monthly data dominated by cyclical-irregular fluctuations rather than by secular or seasonal movements. Do not use textbook examples.

    b) Describe its cyclical characteristics: Is the amplitude wide or narrow? How does the timing of peaks and troughs compare with the timing of turning points in general business (Table 20–5)? What is the current phase of the cycle—expansion or contraction?

    c) Describe the irregular movements: What was the behavior of this series during recent wars? What other major nonbusiness influences appear to have caused extended irregular movements? Are the month-to-month zigzag random forces marked or mild?

3. Which of the following should be changed to an average daily basis, and which should not? Explain in each case.

a) Monthly data on average sales per salesperson in a chain of women's apparel stores.

b) A monthly record of the stocks of a department store.

c) The total loans of a commercial bank on the last day of each month.

4. *a*) List, from Moody's or Standard and Poor's reports, Sears, Roebuck sales for the first five periods of four or five weeks each for this year or last year.

b) Adjust these sales to a daily average basis, counting Saturday as 1½ days and omitting Sundays, January 1, and Memorial Day. (See calendar.)

c) Plot the actual sales and daily average sales on a small chart, using two scales.

d) How does the calendar adjustment affect month-to-month movements?

5. *a*) Define "seasonal index." Distinguish between constant and changing seasonal indexes.

b) Having computed seasonal indexes, describe briefly how to make a seasonal forecast.

c) A chart is captioned "Adjusted for Seasonal Variation." Explain.

d) Why is it sometimes necessary to adjust monthly data for calendar variation before measuring seasonality?

6. Seasonal indexes of sales for the Holloway Company are January, 97; February, 89; March, 101; April, 104; May, 120; etc.

a) Company sales increased from $2,910,000 in January 1973 to $2,964,000 in April of the same year. What was the percentage change in the seasonally adjusted sales between January and April?

b) The company treasurer has forecast sales of $36 million for the next calendar year. He believes that by May the trend-cycle component should be about 5 percent above the average monthly level. Based upon his assumptions, what is the treasurer's sales forecast for the month of May?

7. The following table shows production of Portland cement (in thousands of barrels) by the Quik-Dry Concrete Company for 1968–72:

Year	First	Second	Third	Fourth	Annual Average
1968	100.3	148.5	147.6	128.7	131.3
1969	111.5	162.9	164.6	147.2	146.6
1970	142.5	171.2	170.8	162.5	161.8
1971	151.0	174.8	167.6	155.1	162.1
1972	147.3	168.8	167.7	153.6	159.4
Total	652.6	826.2	818.3	747.1	761.2
Quarterly average	130.5	165.2	163.7	149.4	152.2

    a) Compute indexes of seasonal variation for the cement production data by the graphic method.

    b) Adjust this series graphically for seasonal variation.

    c) Forecast cement production graphically for the four quarters of 1973, extending your trend-cycle curve freehand.

8. *a*) Compute indexes of seasonal variation for the cement production data in Problem 7 by the moving-average method, centering the moving average on the third quarter. Use these additional production figures: 1967, third quarter, 156.0 thousand barrels; fourth quarter, 132.2; and 1973, first quarter, 137.3 thousand barrels.

    b) How much do these indexes differ from those of the graphic method? Give reasons for the differences.

    c) Adjust this series arithmetically for seasonal variation and plot the results. What is the purpose of this adjustment?

    d) Forecast cement production in the second quarter of 1973, assuming a trend-cycle decline of 2 percent from the first quarter.

9. Using the data in Problems 7 and 8:

    a) What factors determine whether constant or changing seasonal indexes should be computed?

    b) How does the computation of a changing seasonal index differ from that of a constant seasonal index?

    c) Is there evidence of changing seasonality in cement production? Present small charts of each of the four quarters to support your answer.

10. As an analyst with the Extron Petroleum Company, you wish to measure the seasonal variation in the company's gasoline sales by the graphic method, using the following data:

GASOLINE SALES, DAILY AVERAGES IN HUNDREDS OF BARRELS

	1967	1968	1969	1970	1971	1972	1973
January	252	264	269	274	330	327	361
February	271	263	278	295	330	355	398
March	264	283	298	318	336	348	382
April	287	300	320	334	357	397	407
May	287	307	321	359	374	398	406
June	317	340	351	368	406	410	452
July	298	328	342	377	399	429	438
August	320	335	355	376	408	428	
September	304	342	344	367	380	416	
October	298	298	319	348	401	411	
November	275	311	320	332	349	376	
December	296	292	308	324	344	387	
Average	289	305	319	339	368	390	

    a) Plot the data on a one-cycle ratio chart; draw a trend-cycle curve through the 1967–72 annual averages (extended through 1973), and determine the 12 seasonal indexes by means of a measuring strip.

b) Describe briefly the typical seasonal behavior in the company's sales. Is the seasonality regular or irregular?

c) Forecast gasoline demand for the next four months (August–November 1973) by laying off the seasonal indexes from your measuring strip above or below the extended trend-cycle curve on the chart. Plot your forecast as a dashed line, and the actual figures below (determined later) as a solid line to compare the results. Actual sales were: August, 433; September, 438; October, 411; November, 392.

d) Adjust the data for seasonal variation graphically and plot the results in red. Describe the principal nonseasonal movements in gasoline demand over this period. Which of these movements dominates the adjusted series—trend, cycles, or irregular fluctuations?

11. In order to analyze the factors affecting gasoline sales of the Extron Petroleum Company, you decide to compute indexes of seasonal variation for the data in Problem 10 by the moving-average method. You first compute a 12-month moving average for each month, and then divide the original sales by these averages, obtaining the following percentages:

MONTHLY GASOLINE SALES AS PERCENTAGES
OF 12-MONTH MOVING AVERAGES

	1967	1968	1969	1970	1971	1972	1973
January	91.5	89.0	86.1	83.1	92.9	86.7	89.4
February	97.8	88.0	88.7	88.9	92.2	93.5	98.1
March	94.5	94.1	94.7	95.4	93.4	91.0	91.5
April	101.8	99.2	101.2	99.3	98.4	103.4	101.6
May	101.0	101.0	101.3	106.4	102.4	103.1	98.5
June	110.3	111.3	110.2	108.7	110.7	105.5	108.1
July	102.8	107.5	107.4	110.2	108.4	109.6	108.1
August	110.6	109.4	111.1	108.9	110.8	108.5	
September	104.8	111.1	107.1	105.5	102.6	104.6	
October	102.1	96.3	98.9	99.6	107.6	103.0	
November	93.8	100.2	98.4	94.6	93.1	93.9	
December	100.4	93.6	94.1	91.7	91.5	96.2	

a) If the original data represent $T \times C \times S \times I$ (trend $\times$ cycle $\times$ seasonal $\times$ irregular forces), what types of fluctuations do the data in the above table primarily represent? How were these elements derived from the original figures?

b) Compute a modified mean of these percentages for each of the 12 months (first omitting the highest and lowest percentage in each case as being the most erratic) to average out the irregular elements. Then multiply these means by (1,200/their total), if necessary, so that they will average 100. List the resulting seasonal indexes rounded to the nearest whole number.

c) In July 1973 the company economist predicts that a cyclical recession during the balance of the year will offset the usual secular trend growth. On this assumption, forecast daily average gasoline sales for November 1973 based on the normal seasonal change from July (the latest month available). Give the percentage error of forecast, compared with the actual figure of 392 thousand barrels daily average in November.

   d) You wish to analyze the change in gasoline sales between February and July 1973. Actual sales increased from 398 to 438, or 10 percent, in this period. Adjust the data in these two months for seasonal variation and compute the percentage change in the adjusted figures.

   e) Show how the adjusted February and July figures were derived, in terms of the *TCSI* concept, and explain the significance of the change in the adjusted demand.

12. Gasoline demand is said to be less seasonal than formerly, since people in colder areas who once stored their cars during the winter now drive the year round, and vacation trips that were formerly confined to the summer months are now made throughout the year. Do the figures in Problems 10 and 11 confirm this claim? That is, does gasoline demand in a winter month, expressed as a ratio to the average month, tend to rise, and does the ratio for a summer month fall correspondingly over the years? Test this hypothesis of changing seasonality for the two months February and June as follows:

   a) Plot the February and June percentages of moving averages from Problem 11 on two panels of an arithmetic chart.

   b) Draw a freehand trend line through each of these diagrams, ignoring erratic points.

   c) Do these charts support the claim that gasoline demand is becoming less seasonal? Explain.

   d) Read off from your trend lines and list the changing seasonal indexes for February and June 1973.

13. *a*) Cite the one chief advantage of graphic methods and of arithmetic methods, respectively, in seasonal analysis, and explain your choice.

   b) In what type of study may the electronic calculator method be preferable?

   c) How could you measure the *irregularity* of seasonal fluctuations in your business?

14. *a*) Find a series of recent monthly data that is published both with and without seasonal adjustment in *Survey of Current Business* or some other source. Discuss the latest monthly figure in terms of (1) the percentage change in the unadjusted value over a year ago and (2) the relation of the seasonally adjusted value to those of recent months. Compare these two methods of taking seasonality into account.

   b) Find a weekly business indicator presented as a tier chart for the past several years and describe its recent behavior, indicating what component types of fluctuations can be distinguished. (One source is the *Federal Reserve Chart Book*.)

15. *a*) List any peaks and troughs in general business that have occurred since November 1970 (from *Business Conditions Digest,* Appendix E) to update Table 20–5.

b) How did the National Bureau of Economic Research arrive at these "reference dates"?

16. Which of the three purposes of measuring cycles is most important, in your opinion, for (*a*) the business executive and (*b*) the President's Council of Economic Advisers? Explain your choices.

17. *a*) Outline both the graphic and the alternative arithmetic steps necessary to isolate the trend-cycle component of a time series.
    b) Just how do these procedures eliminate seasonal and irregular influences? What traces of these elements are likely to remain in the trend-cycle residuals?

18. Cycles in monthly series are usually studied by examining data that are adjusted only for seasonal variation, since secular trend rarely obscures short-term cycles, and cyclical-irregular movements cannot be completely separated from each other. In your analysis of gasoline sales (Problems 10 and 11), however, the cycles in the seasonally adjusted data, Problem 10 (*e*), were obscured by secular trend and irregular elements. You therefore decide to eliminate these factors, as far as possible, in order to determine the nature of the cycle, if any, that may exist in this industry.
    a) Trace the seasonally adjusted gasoline demand curve from Problem 10 (*e*), onto another ratio chart and fit a straight trend line (since the trend is practically linear) by inspection, using the annual averages as guides.
    b) Adjust the series for secular trend by laying off the vertical (not perpendicular) deviations above or below the trend line, with a paper strip, around the horizontal line printed with "2" on the chart. Mark a "Percent of Trend" vertical scale with 50, 100, and 150 opposite the lines printed 1, 2, and 3, respectively. The curve is now adjusted for both seasonality and trend, so that it represents the estimated cyclical-irregular fluctuations in gasoline demand.
    c) Draw a flexible freehand curve through the adjusted series to smooth out the month-to-month zigzags, but make it follow closely the short-term cyclical swings. This curve approximates the cycle itself (including extended irregular influences).
    d) Describe the cyclical fluctuations, if any, in gasoline demand. In what months did cyclical peaks or troughs occur?

19. If a computer program is available (e.g., Census II, Variant X–11), analyze Sears, Roebuck sales in Table 20–2 to:
    a) Adjust for calendar and seasonal variation.
    b) Smooth out irregularities with a short-term moving average, to isolate the trend-cycle component.
    c) Interpret all results on your printout sheet.

20. Analyze the gasoline sales in Problem 10, using the computer method as outlined in Problem 19*a*, *b*, and *c*.

21. Estimate the percentage change in gross national product this year compared with last year, using any three of the five "naïve" methods of cyclical forecasting described in the text. Comment briefly on the validity of the results.

22. Find an article on the use of exponentially weighted moving averages in short-term forecasting and prepare a short report explaining this method (going beyond the textbook outline), together with its pros and cons.

23. What is the present stage of the general business cycle—expansion or contraction? Is a turning point in prospect? Cite evidence supporting or modifying your view from:
    a) Lead and lag indicators.
    b) Diffusion indexes.
    c) A survey of anticipations data (e.g., businessmen's plans for new plant and equipment expenditures).

24. Select a leading indicator from *Business Conditions Digest* (as assigned) and:
    a) Explain on logical grounds why this indicator should lead general business at cyclical turning points.
    b) Describe its performance and reliability in recent years as a barometer of business.

25. Prepare a critical review on the use of diffusion indexes (including average duration of run) as cyclical forecasting devices. Explanation should go beyond that in this text. See National Bureau of Economic Research publications, *Statistical Indicator Reports*, or *Business Conditions Digest*.

26. Select a survey of anticipations data, as assigned, and report on its validity as a forecasting tool. Cite not only the original source but an outside critical study of its efficacy.

SELECTED READINGS

ANDERSON, T. W. *The Statistical Analysis of Time Series.* New York: John Wiley, 1971.
    A mathematical text on regression, trends, and cycles.
BOX, G. E. P., and JENKINS, G. M. *Time Series Analysis, Forecasting and Control.* San Francisco: Holden-Day, 1970.
    An advanced treatment of time series models and their use in forecasting.
BRY, G., and BOSCHAN, C. *Cyclical Analysis of Time Series: Selected Procedures and Computer Programs.* New York: Columbia University Press, 1971.
    Summarizes the National Bureau of Economic Research programmed analysis of business cycles.
CROXTON, F. E.; COWDEN, D. J.; and BOLCH, B. W. *Practical Business Statistics.* 4th ed. Englewood Cliffs, N.J.: Prentice-Hall, 1969, chaps. 19–21.
    Explores several methods of isolating seasonal and cyclical fluctuations and trends, including the use of polynomials and growth curves.

HICKMAN, B. G. (ed.). *Econometric Models of Cyclical Behavior,* 2 vols. New York: National Bureau of Economic Research and Social Science Research Council, Columbia University Press, 1972.

Various papers on econometric models of business cycles and the evaluation of forecasts.

NETER, J.; WASSERMAN, W.; and WHITMORE, G. A. *Fundamental Statistics for Business and Economics.* 4th ed. Boston: Allyn & Bacon, 1973.

Covers time series analysis for forecasting, planning, and control.

SHISKIN, J. *Signals of Recession and Recovery: An Experiment with Monthly Reporting.* New York: National Bureau of Economic Research, 1961.

Introduces the monthly indicators reported currently in *Business Conditions Digest.*

——— *et al. The X-11 Variant of the Census Method II Seasonal Adjustment Program.* U.S. Bureau of the Census, Technical Paper No. 15, rev. 1967.

The latest census method, summarized in *Business Cycle Developments,* October, 1965.

ZARNOWITZ, V. (ed.). *The Business Cycle Today.* New York: National Bureau of Economic Research, Columbia University Press, 1972.

A colloquium on measures of recent business cycles, forecasting, and the use of econometric models.

APPENDIXES

APPENDIXES

A. GLOSSARY OF SYMBOLS

$\backsim$	Not.
$\approx$	Approximately equals.
$\mid$	Given.
$\mid\ \mid$	Absolute value of enclosed symbol, ignoring sign.
$!$	Factorial: $n! = 1 \times 2 \times 3 \times \cdots \times n$.
$>$	Greater than.
$\geqq$	Greater than or equal to.
$<$	Less than.
$\leqq$	Less than or equal to.
a, A	Value of Y_c in trend or regression equation when all X's $= 0$; a is for sample, A is for population.
A'	Transpose of matrix A.
A^{-1}	Inverse of square matrix A.
a	Level of significance (alpha).
b, B	Slope of trend line; slope of higher degree curve at Y axis; simple regression coefficient, where b is for sample, B is for population.
$b_1, B_1, \ldots$	Net regression coefficients in multiple regression; where $b_1, b_2, \ldots$ are for sample, $B_1, B_2 \ldots$ are for population.
$\beta_1, \beta_2 \ldots$	Standardized values (betas) of $b_1, b_2 \ldots$.
C	Number of combinations; cyclical component in time series, expressed as percentage of trend (T); column effect in analysis of variance.
$C(n)$	Cost of sample of size n.
c	Cost per unit; constant determining curvature in second-degree equation.
D	Factor used in determining unit normal loss function (Appendix E).
d	Deviation of class midpoint from assumed mean of frequency distribution in class interval units; expected value of χ^2 distribution; Durban-Watson statistic.
d, df	Degrees of freedom.

659

GLOSSARY OF SYMBOLS (*Continued*)

E	Expected value; expected frequency in a category of χ^2 distribution.
EMV	Expected monetary value.
ENGS	Expected net gain from sampling.
EOL	Expected opportunity loss.
EVPI	Expected value of perfect information.
EVSI	Expected value of sample information.
e	The constant 2.718 . . . , the base of natural logarithms.
ϵ	Residual or unexplained variation (epsilon). $\epsilon = Y - Y_c$.
F	Cumulative frequency for all classes below median or quartile interval; ratio of two (χ^2/df) variables in F distribution.
f	Frequency or number of items in any class $(\Sigma f = n)$.
fpc	Finite population correction.
H	Statistic used in rank-sum test for several independent samples.
I	Irregular component in time series expressed as percentage of $T \times C \times S$; identity matrix $(I = A \times A^{-1})$; interaction effect in analysis of variance.
i	Class interval in a frequency distribution; subscript denoting the ith item in a set of items.
K	Break-even value.
k	Number of constants in an equation; sampling interval in systematic selection; number of replicated samples drawn from population.
L	Lower limit of class containing median or quartile in a frequency distribution.
LCL	Lower control limit in quality control chart.
$L_N(D)$	Unit normal loss function (find D in Appendix E).
$L(X)$	Opportunity loss.
λ	Average number of arrivals per unit of time in exponential distribution (lambda).
log	Logarithm.
M	Number of secondary units in population—cluster sampling.
M_i	Number of items in ith stratum of stratified sample; number of secondary units in ith primary unit in cluster sampling.
MCD	Months for cyclical dominance: MCD $= \bar{I}/\bar{C}$.
MD	Mean (average) deviation.
Md	Median.
M_0, M_1	Mean of decision maker's prior (M_0) or posterior (M_1) betting distribution about unknown mean μ.
m	Mean (or variance) of Poisson distribution; number of independent variables in matrix solution of multiple regression.
m_i	Sample size of ith stratum in stratified sample; number of secondary units sampled in ith primary unit in cluster sample.
μ	Arithmetic mean of a population (mu).
μ_h	Hypothetical population mean.
N	Number of items in a population; number of primary units in population—cluster sampling.

GLOSSARY OF SYMBOLS (*Continued*)

n	Number of items in a sample; number of primary units in cluster sample.
O	Observed frequency in a category of χ^2 distribution.
OC	Operating characteristic.
P	Probability.
$P(A)$	Probability that event A will occur.
$P(A|B)$	Conditional probability of A given B.
$P(A, B)$	Joint probability of A and B.
$P(p)$	Prior probability that fraction defective is p.
p	Population proportion; probability of a success; price.
p_h, q_h	Hypothetical population proportions.
p_s	Sample proportion.
π	The constant 3.14159 . . . ; profit (pi).
Q	Quartile deviation.
Q_1, Q_3	First and third quartiles (Q_2 = median).
q	Population proportion, probability of a failure, where $q = 1 - p$; quantity (e.g., units stocked).
q_s	Sample proportion, where $q_s = 1 - p_s$.
R	Ratio; coefficient of multiple correlation;† row effect in analysis of variance.
R^2	Coefficient of multiple determination.†
r	Coefficient of simple correlation;† number of successes (e.g., defectives) in sample.
r^2	Coefficient of simple determination.†
r_s	Coefficient of simple correlation for a sample.
S	Seasonal index.
S_f	Standard error of an individual forecast.
Sk	Coefficient of skewness.
S_0, S_1	Standard deviation of decision maker's prior (S_0) or posterior (S_1) betting distribution about unknown mean μ.
S_{YX}	Standard error of estimate in simple regression.
S^2_{YX}	Unexplained variance in simple regression.
$S_{Y\cdot 12}$ . . .	Standard error of estimate in multiple regression.
$S^2_{Y\cdot 12}$ . . .	Unexplained variance in multiple regression.
S^2_*	Reduction of prior variance as a result of taking sample.
SSD	Sum of squared deviations.
Σ	Sum, total (capital sigma).
s	Standard deviation.†
s^2	Variance.†
s_{po}	Pooled or common estimate of standard deviation from several samples.
$s_{\bar{x}}$ etc.	Standard error of the mean;† s is used with other subscripts for standard errors of other measures.
$s^2_{Yc-\bar{Y}}$	Explained variance.

---

† Population value as estimated from a sample (except computer printout of R in Chapter 17, which is not adjusted for sample bias).

GLOSSARY OF SYMBOLS (*Continued*)

σ	Standard deviation (small sigma) of a population; or, in quality control, its estimated value.
$\sigma_{\bar{x}}$, etc.	Standard error of the mean; σ is used with other subscripts for standard errors of other measures.
T	Total (population); trend ordinate in time series ($T = Y_c$); sum of ranks in rank-sum test.
T_i	Total estimated for ith cluster, in cluster sampling.
t	Deviation of sample mean, etc., from population mean, expressed in standard error units; the t distribution applies to small samples. Slope of opportunity loss function. Time between successive arrivals in exponential distribution.
UCL	Upper control limit in quality control chart.
u	Utility value.
w_i	Weight of ith stratum in stratified sampling.
X, Y	Independent and dependent variables measured from zero.
$\bar{X}, \bar{Y}$	Arithmetic means of X and Y in a sample. (Subscripts 1, 2, etc., refer to different samples.)
$\bar{\bar{X}}$	Arithmetic mean of several sample means.
$\bar{X}_a$	Assumed mean of X.
$X_1, X_2 \cdots$	Independent variables in multiple regression.
x, y	Variables measured from means; e.g., $x = X - \bar{X}$, $y = Y - \bar{Y}$.
χ^2	A variable representing the sums of squared normal random variables in χ^2 (chi-square) distribution.
Y	Dependent variable in regression or trend analysis.
Y'	Value of Y adjusted for X_1 in graphic method of multiple regression.
Y_c	Value of Y computed from regression or trend equation.
$\bar{Y}_c$	Population mean estimated from cluster sample.
$\bar{Y}_i$	Sample mean of ith stratum in stratified sample; sample mean of secondary units in ith primary unit of cluster sample.
$\bar{Y}_j$	Mean of replicated sample.
$\bar{Y}_R$	Ratio estimate of true mean μ_Y.
$\bar{Y}_s$	Estimate of overall mean from stratified sample.
z	Standard normal deviate: $z = (X - \mu)/\sigma = x/\sigma$ in normal distribution.

B. LOGARITHMS

HOW TO USE THE TABLE OF LOGARITHMS

LOGARITHMS are used to simplify the operations of multiplication, division, raising numbers to powers, and extracting roots. They are especially valuable in constructing ratio charts, and in fitting certain types of regression or secular trend curves.

The common logarithm of a number is the power of 10 which is equal to that number. For example, the third power of 10 is 1,000, so

$$\log 1,000 = 3$$

That is, the logarithm of 1,000 is 3 because $10^3 = 1,000$. Similarly, log $100 = 2$, log $10 = 1$, log $1 = 0$, log $.1 = -1$, log $.01 = -2$, etc. For intermediate numbers the logarithm is a whole number, as above, followed by a decimal fraction.

The whole number part of a logarithm (to the left of the decimal point) is called the *characteristic,* and the fractional part (to the right of the decimal point) is called the *mantissa.* To find the logarithm of any number, determine the characteristic from the following rules and look up the mantissa in the accompanying table.

Rules for Determining the Characteristic

1. The characteristics of the logarithms of all numbers greater than one are positive, and their numerical values are one unit less than the number of digits to the left of the decimal point in the numbers themselves.

Examples:

Number	Characteristic of Logarithm
286.	2
12,769.	4
1,008.73	3
1.827	0

2. The characteristics of the logarithms of all numbers between zero and one are negative, and their numerical values are one unit greater than the number of zeros between the decimal point and the first significant digit of the numbers themselves. A negative value is indicated either by a minus sign written above the characteristic or a positive number followed by −10, as shown below.

Examples:

Number	Characteristic of Logarithm
.764	$\bar{1}$, or 9. ...−10
.031	$\bar{2}$, or 8. ...−10
.02793	$\bar{2}$, or 8. ...−10
.00004	$\bar{5}$, or 5. ...−10

3. The number zero and negative numbers have no logarithms.

How to Find the Mantissa

The table at the end of this appendix shows four-place mantissas of logarithms for three-digit numbers. This table is accurate enough for most business and economic data. For convenience in printing, decimal points are omitted, but each entry in the table must be interpreted as a four-place decimal. Mantissas are always positive. The mantissa of any number of three digits or less can be read directly from the table. The first two digits are found in the column labeled "N" at the left of the page, and the third digit is found at the top of the page. Thus, to find the logarithm of 316, write down the characteristic 2 from Rule 1 above, followed by the mantissa .4997 from the table. This is found by moving down the column on the left to 31 and going to the right under the column headed 6. The log of 316 therefore is 2.4997.

Examples:

$$\log 3.160 \ = 0.4997$$
$$\log 0.316 \ = \bar{1}.4997, \ \text{or} \ 9.4997 - 10$$
$$\log 180,000 = 5.2553$$
$$\log 0.031 \ = \bar{2}.4914, \ \text{or} \ 8.4914 - 10$$

The logarithms of four-place numbers may be determined by interpolation. Thus, to find the log of 3.162, go two tenths of the way from log 3.160 (i.e., 0.4997) to log 3.170 (i.e., 0.5011). This is 0.4997 + 0.2 × 0.0014 = 0.5000.

Antilogarithms

To find the antilogarithm or natural number corresponding to a logarithm, find the nearest logarithm in the table and read the first two digits of the corresponding natural number from the left-hand column and the third digit from the top row. Thus, to get the antilog of 3.3101, find 3096, the nearest mantissa in the table, and read across and up to the number 204. Then from the rules on characteristics the answer is 2,040. This value may also be interpolated if four-place accuracy is desired.

Rules for Using Logarithms

1. To multiply numbers add their logarithms. Then look up the antilogarithm of their sum. The fact that numbers may be multiplied by adding their logarithms is the most basic property of logarithms.

Example: Multiply 19 by 28:

$$\begin{array}{lll}\log 19 & = & 1.2788 \\ \log 28 & = & 1.4472 \\ \log \text{product} = & & \overline{2.7260} \\ \text{product} & = \text{antilog } 2.7260 = & 532 \end{array}$$

2. To divide one number by another, subtract the logarithm of the latter from that of the former. Then look up the antilogarithm of the difference.

Example: Divide 532 by 28.

$$\begin{array}{lll}\log 532 & = & 2.7259 \\ \log 28 & = & 1.4472 \\ \log \text{difference} = & & \overline{1.2787} \\ \text{quotient} & = \text{antilog } 1.2787 = & 19.0 \end{array}$$

3. To raise a number to a given power, multiply the logarithm of the number by the exponent of the power and look up the antilogarithm of the product.

4. To extract any root of a number, divide its logarithm by the index of the root and look up the antilogarithm of the quotient.

FOUR-PLACE LOGARITHMS

N	0	1	2	3	4	5	6	7	8	9
10	0000	0043	0086	0128	0170	0212	0253	0294	0334	0374
11	0414	0453	0492	0531	0569	0607	0645	0682	0719	0755
12	0792	0828	0864	0899	0934	0969	1004	1038	1072	1106
13	1139	1173	1206	1239	1271	1303	1335	1367	1399	1430
14	1461	1492	1523	1553	1584	1614	1644	1673	1703	1732
15	1761	1790	1818	1847	1875	1903	1931	1959	1987	2014
16	2041	2068	2095	2122	2148	2175	2201	2227	2253	2279
17	2304	2330	2355	2380	2405	2430	2455	2480	2504	2529
18	2553	2577	2601	2625	2648	2672	2695	2718	2742	2765
19	2788	2810	2833	2856	2878	2900	2923	2945	2967	2989
20	3010	3032	3054	3075	3096	3118	3139	3160	3181	3201
21	3222	3243	3263	3284	3304	3324	3345	3365	3385	3404
22	3424	3444	3464	3483	3502	3522	3541	3560	3579	3598
23	3617	3636	3655	3674	3692	3711	3729	3747	3766	3784
24	3802	3820	3838	3856	3874	3892	3909	3927	3945	3962
25	3979	3997	4014	4031	4048	4065	4082	4099	4116	4133
26	4150	4166	4183	4200	4216	4232	4249	4265	4281	4298
27	4314	4330	4346	4362	4378	4393	4409	4425	4440	4456
28	4472	4487	4502	4518	4533	4548	4564	4579	4594	4609
29	4624	4639	4654	4669	4683	4698	4713	4728	4742	4757
30	4771	4786	4800	4814	4829	4843	4857	4871	4886	4900
31	4914	4928	4942	4955	4969	4983	4997	5011	5024	5038
32	5051	5065	5079	5092	5105	5119	5132	5145	5159	5172
33	5185	5198	5211	5224	5237	5250	5263	5276	5289	5302
34	5315	5328	5340	5353	5366	5378	5391	5403	5416	5428
35	5441	5453	5465	5478	5490	5502	5514	5527	5539	5551
36	5563	5575	5587	5599	5611	5623	5635	5647	5658	5670
37	5682	5694	5705	5717	5729	5740	5752	5763	5775	5786
38	5798	5809	5821	5832	5843	5855	5866	5877	5888	5899
39	5911	5922	5933	5944	5955	5966	5977	5988	5999	6010
40	6021	6031	6042	6053	6064	6075	6085	6096	6107	6117
41	6128	6138	6149	6160	6170	6180	6191	6201	6212	6222
42	6232	6243	6253	6263	6274	6284	6294	6304	6314	6325
43	6336	6345	6355	6365	6375	6385	6395	6405	6415	6425
44	6435	6444	6454	6464	6474	6484	6493	6503	6513	6522
45	6532	6542	6551	6561	6571	6580	6590	6599	6609	6618
46	6628	6637	6646	6656	6665	6675	6684	6693	6702	6712
47	6721	6730	6739	6749	6758	6767	6776	6785	6794	6803
48	6812	6821	6830	6839	6848	6857	6866	6875	6884	6893
49	6902	6911	6920	6928	6937	6946	6955	6964	6972	6981
50	6990	6998	7007	7016	7024	7033	7042	7050	7059	7067
51	7076	7084	7093	7101	7110	7118	7126	7135	7143	7152
52	7160	7168	7177	7185	7193	7202	7210	7218	7226	7235
53	7243	7251	7259	7267	7275	7284	7292	7300	7308	7316
54	7324	7332	7340	7348	7356	7364	7372	7380	7388	7396

FOUR-PLACE LOGARITHMS (*Continued*)

N	0	1	2	3	4	5	6	7	8	9
55	7404	7412	7419	7427	7435	7443	7451	7459	7466	7474
56	7482	7490	7497	7505	7513	7520	7528	7536	7543	7551
57	7559	7566	7574	7582	7589	7597	7604	7612	7619	7627
58	7634	7642	7649	7657	7664	7672	7679	7686	7694	7701
59	7709	7716	7723	7731	7738	7745	7752	7760	7767	7774
60	7782	7789	7796	7803	7810	7818	7825	7832	7839	7846
61	7853	7860	7868	7875	7882	7889	7896	7903	7910	7917
62	7924	7931	7938	7945	7952	7959	7966	7973	7980	7987
63	7993	8000	8007	8014	8021	8028	8035	8041	8048	8055
64	8062	8069	8075	8082	8089	8096	8102	8109	8116	8122
65	8129	8136	8142	8149	8156	8162	8169	8176	8182	8189
66	8195	8202	8209	8215	8222	8228	8235	8241	8248	8254
67	8261	8267	8274	8280	8287	8293	8299	8306	8312	8319
68	8325	8331	8338	8344	8351	8357	8363	8370	8376	8382
69	8388	8395	8401	8407	8414	8420	8426	8432	8439	8445
70	8451	8457	8463	8470	8476	8482	8488	8494	8500	8506
71	8513	8519	8525	8531	8537	8543	8549	8555	8561	8567
72	8573	8579	8585	8591	8597	8603	8609	8615	8621	8627
73	8633	8639	8645	8651	8657	8663	8669	8675	8681	8686
74	8692	8698	8704	8710	8716	8722	8727	8733	8739	8745
75	8751	8756	8762	8768	8774	8779	8785	8791	8797	8802
76	8808	8814	8820	8825	8831	8837	8842	8848	8854	8859
77	8865	8871	8876	8882	8887	8893	8899	8904	8910	8915
78	8921	8927	8932	8938	8943	8949	8954	8960	8965	8971
79	8976	8982	8987	8993	8998	9004	9009	9015	9020	9025
80	9031	9036	9042	9047	9053	9058	9063	9069	9074	9079
81	9085	9090	9096	9101	9106	9112	9117	9122	9128	9133
82	9138	9143	9149	9154	9159	9165	9170	9175	9180	9186
83	9191	9196	9201	9206	9212	9217	9222	9227	9232	9238
84	9243	9248	9253	9258	9263	9269	9274	9279	9284	9289
85	9294	9299	9304	9309	9315	9320	9325	9330	9335	9340
86	9345	9350	9355	9360	9365	9370	9375	9380	9385	9390
87	9395	9400	9405	9410	9415	9420	9425	9430	9435	9440
88	9445	9450	9455	9460	9465	9469	9474	9479	9484	9489
89	9494	9499	9504	9509	9513	9518	9523	9528	9533	9538
90	9542	9547	9552	9557	9562	9566	9571	9576	9581	9586
91	9590	9595	9600	9605	9609	9614	9619	9624	9628	9633
92	9638	9643	9647	9652	9657	9661	9666	9671	9675	9680
93	9685	9689	9694	9699	9703	9708	9713	9717	9722	9727
94	9731	9736	9741	9745	9750	9754	9759	9763	9768	9773
95	9777	9782	9786	9791	9795	9800	9805	9809	9814	9818
96	9823	9827	9832	9836	9841	9845	9850	9854	9859	9863
97	9868	9872	9877	9881	9886	9890	9894	9899	9903	9908
98	9912	9917	9921	9926	9930	9934	9939	9943	9948	9952
99	9956	9961	9965	9969	9974	9978	9983	9987	9991	9996

C. SQUARES, SQUARE ROOTS, AND RECIPROCALS 1-1,000

HOW TO FIND A SQUARE ROOT

SQUARE ROOTS can be read from the following table by any of three methods:

1. For any whole number from 1 to 1,000 listed in the N column, find the square root in the same row of the $\sqrt{N}$ column. Thus, the square root of 458 (in the N column) is 21.4+ (in the $\sqrt{N}$ column).

2. For any multiple of 10 from 10 to 10,000, move the decimal point one place to the left, look up this number in the N column, and find the square root in the $\sqrt{10N}$ column. For example, to get the square root of 8,670, look up 867 in the N column and find 93.1+ in the $\sqrt{10N}$ column.

3. When a problem calls for the square root of a number not given in the N column, it may be possible to find that number in the N^2 column. If the number is located in the N^2 column, its square root is given in the N column. Thus, to obtain the square root of 1,225, find this number under N^2 and read the square root, 35, to the left in the N column.

The square root of other numbers may also be read from the table in any of these methods by observing the rule that moving the decimal point two places to the left or right in the number moves it one place in the square root. As an example, the number 123,201 is given in the N^2 column of the table. Then,

$$\text{The square root of } 123,201. = 351.$$
$$\text{The square root of } 1,232.01 = 35.1$$
$$\text{The square root of } 12.3201 = 3.51$$

The square root of any number not shown in the table may be estimated by interpolating between values which are included. For example, the square root of 65.12 must be between the square root of 65 and the square root of 66. Since 65.12 stands at a point .12 of the way from 65 to 66, its square root should be approximately .12 of the way from the square root of 65 to the square root of 66. The following procedure is used:

Number	*Root*
66.............................	.8.124
65.12...........................	?
65.............................	.8.062
Difference......................	.062

$$\sqrt{65.12} = 8.062 + .12(.062) = 8.069+$$

More detailed values of square roots may be obtained without interpolation by the use of *Barlow's Tables,* published by the Chemical Publishing Co., Inc., 234 King Street, Brooklyn, New York, which gives the squares, cubes, square roots, cube roots, and reciprocals of all integer numbers up to 12,500.

THE USE OF RECIPROCALS

Multiplication and division can often be facilitated by the use of reciprocals. Instead of multiplying, one can divide one number by the reciprocal of the second, if the reciprocal is a simple number. For example:

$$1,582 \times 25 \quad = \frac{158,200}{4} = 39,550$$

$$220 \times 50 \quad = \frac{22,000}{2} = 11,000$$

$$17,228 \times 125 = \frac{17,228,000}{8} = 2,153,500$$

Similarly, instead of dividing, it may be easier to multiply the numerator by the reciprocal of the denominator. Thus,

$$5,725 \div 25 \ = 5,725 \times .04 = 229$$
$$280,400 \div 50 \ = 2,804 \times 2 = 5,608$$
$$245,925 \div 125 = 245.925 \times 8 = 1,967.4$$

This shortcut is particularly useful in computing a series of percentages on a common base, such as percentages of various asset accounts to total assets in a balance sheet. Simply place the reciprocal of the base (e.g., total assets) in a calculating machine, and *multiply* by each of the other items in turn, without clearing the machine. Reciprocals may be found in the last column of the following table.

SQUARES, SQUARE ROOTS, AND RECIPROCALS 1–1,000*

N	N^2	$\sqrt{N}$	$\sqrt{10N}$	$1/N$	N	N^2	$\sqrt{N}$	$\sqrt{10N}$	$1/N$ .0
					50	2 500	7.071 068	22.36068	2000000
1	1	1.000 000	3.162 278	1.0000000	51	2 601	7.141 428	22.58318	1960784
2	4	1.414 214	4.472 136	.5000000	52	2 704	7.211 103	22.80351	1923077
3	9	1.732 051	5.477 226	.3333333	53	2 809	7.280 110	23.02173	1886792
4	16	2.000 000	6.324 555	.2500000	54	2 916	7.348 469	23.23790	1851852
5	25	2.236 068	7.071 068	.2000000	55	3 025	7.416 198	23.45208	1818182
6	36	2.449 490	7.745 967	.1666667	56	3 136	7.483 315	23.66432	1785714
7	49	2.645 751	8.366 600	.1428571	57	3 249	7.549 834	23.87467	1754386
8	64	2.828 427	8.944 272	.1250000	58	3 364	7.615 773	24.08319	1724138
9	81	3.000 000	9.486 833	.1111111	59	3 481	7.681 146	24.28992	1694915
10	100	3.162 278	10.00000	.1000000	60	3 600	7.745 967	24.49490	1666667
11	121	3.316 625	10.48809	.09090909	61	3 721	7.810 250	24.69818	1639344
12	144	3.464 102	10.95445	.08333333	62	3 844	7.874 008	24.89980	1612903
13	169	3.605 551	11.40175	.07692308	63	3 969	7.937 254	25.09980	1587302
14	196	3.741 657	11.83216	.07142857	64	4 096	8.000 000	25.29822	1562500
15	225	3.872 983	12.24745	.06666667	65	4 225	8.062 258	25.49510	1538462
16	256	4.000 000	12.64911	.06250000	66	4 356	8.124 038	25.69047	1515152
17	289	4.123 106	13.03840	.05882353	67	4 489	8.185 353	25.88436	1492537
18	324	4.242 641	13.41641	.05555556	68	4 624	8.246 211	26.07681	1470588
19	361	4.358 899	13.78405	.05263158	69	4 761	8.306 624	26.26785	1449275
20	400	4.472 136	14.14214	.05000000	70	4 900	8.366 600	26.45751	1428571
21	441	4.582 576	14.49138	.04761905	71	5 041	8.426 150	26.64583	1408451
22	484	4.690 416	14.83240	.04545455	72	5 184	8.485 281	26.83282	1388889
23	529	4.795 832	15.16575	.04347826	73	5 329	8.544 004	27.01851	1369863
24	576	4.898 979	15.49193	.04166667	74	5 476	8.602 325	27.20294	1351351
25	625	5.000 000	15.81139	.04000000	75	5 625	8.660 254	27.38613	1333333
26	676	5.099 020	16.12452	.03846154	76	5 776	8.717 798	27.56810	1315789
27	729	5.196 152	16.43168	.03703704	77	5 929	8.774 964	27.74887	1298701
28	784	5.291 503	16.73320	.03571429	78	6 084	8.831 761	27.92848	1282051
29	841	5.385 165	17.02939	.03448276	79	6 241	8.888 194	28.10694	1265823
30	900	5.477 226	17.32051	.03333333	80	6 400	8.944 272	28.28427	1250000
31	961	5.567 764	17.60682	.03225806	81	6 561	9.000 000	28.46050	1234568
32	1 024	5.656 854	17.88854	.03125000	82	6 724	9.055 385	28.63564	1219512
33	1 089	5.744 563	18.16590	.03030303	83	6 889	9.110 434	28.80972	1204819
34	1 156	5.830 952	18.43909	.02941176	84	7 056	9.165 151	28.98275	1190476
35	1 225	5.916 080	18.70829	.02857143	85	7 225	9.219 544	29.15476	1176471
36	1 296	6.000 000	18.97367	.02777778	86	7 396	9.273 618	29.32576	1162791
37	1 369	6.082 763	19.23538	.02702703	87	7 569	9.327 379	29.49576	1149425
38	1 444	3.164 414	19.49359	.02631579	88	7 744	9.380 832	29.66479	1136364
39	1 521	6.244 998	19.74842	.02564103	89	7 921	9.433 981	29.83287	1123596
40	1 600	6.324 555	20.00000	.02500000	90	8 100	9.486 833	30.00000	1111111
41	1 681	6.403 124	20.24846	.02439024	91	8 281	9.539 392	30.16621	1098901
42	1 764	6.480 741	20.49390	.02380952	92	8 464	9.591 663	30.33150	1086957
43	1 849	6.557 439	20.73644	.02325581	93	8 649	9.643 651	30.49590	1075269
44	1 936	6.633 250	20.97618	.02272727	94	8 836	9.695 360	30.65942	1063830
45	2 025	6.708 204	21.21320	.02222222	95	9 025	9.746 794	30.82207	1052632
46	2 116	6.782 330	21.44761	.02173913	96	9 216	9.797 959	30.98387	1041667
47	2 209	6.855 655	21.67948	.02127660	97	9 409	9.848 858	31.14482	1030928
48	2 304	6.928 203	21.90890	.02083333	98	9 604	9.899 495	31.30495	1020408
49	2 401	7.000 000	22.13594	.02040816	99	9 801	9.949 874	31.46427	1010101
50	2 500	7.071 068	22.36068	.02000000	100	10 000	10.00000	31.62278	1000000

* From Frederick E. Croxton and Dudley J. Cowden, *Practical Business Statistics*, © 1948, pp. 524–33. Reprinted by permission of Prentice-Hall, Inc., Englewood Cliffs, N.J.

SQUARES, SQUARE ROOTS, AND RECIPROCALS 1–1,000 (*Continued*)

N	N²	√N	√10N	1/N .0	N	N²	√N	√10N	1/N .00
100	10 000	10.00000	31.62278	10000000	150	22 500	12.24745	38.72983	6666667
101	10 201	10.04988	31.78050	09900990	151	22 801	12.28821	38.85872	6622517
102	10 404	10.09950	31.93744	09803922	152	23 104	12.32883	38.98718	6578947
103	10 609	10.14889	32.09361	09708738	153	23 409	12.36932	39.11521	6535948
104	10 816	10.19804	32.24903	09615385	154	23 716	12.40967	39.24283	6493506
105	11 025	10.24695	32.40370	09523810	155	24 025	12.44990	39.37004	6451613
106	11 236	10.29563	32.55764	09433962	156	24 336	12.49000	39.49684	6410256
107	11 449	10.34408	32.71085	09345794	157	24 649	12.52996	39.62323	6369427
108	11 664	10.39230	32.86335	09259259	158	24 964	12.56981	39.74921	6329114
109	11 881	10.44031	33.01515	09174312	159	25 281	12.60952	39.87480	6289308
110	12 100	10.48809	33.16625	09090909	160	25 600	12.64911	40.00000	6250000
111	12 321	10.53565	33.31666	09009009	161	25 921	12.68858	40.12481	6211180
112	12 544	10.58301	33.46640	08928571	162	26 244	12.72792	40.24922	6172840
113	12 769	10.63015	33.61547	08849558	163	26 569	12.76715	40.37326	6134969
114	12 996	10.67708	33.76389	08771930	164	26 896	12.80625	40.49691	6097561
115	13 225	10.72381	33.91165	08695652	165	27 225	12.84523	40.62019	6060606
116	13 456	10.77033	34.05877	08620690	166	27 556	12.88410	40.74310	6024096
117	13 689	10.81665	34.20526	08547009	167	27 889	12.92285	40.86563	5988024
118	13 924	10.86278	34.35113	08474576	168	28 224	12.96148	40.98780	5952381
119	14 161	10.90871	34.49638	08403361	169	28 561	13.00000	41.10961	5917160
120	14 400	10.95445	34.64102	08333333	170	28 900	13.03840	41.23106	5882353
121	14 641	11.00000	34.78505	08264463	171	29 241	13.07670	41.35215	5847953
122	14 884	11.04536	34.92850	08196721	172	29 584	13.11488	41.47288	5813953
123	15 129	11.09054	35.07136	08130081	173	29 929	13.15295	41.59327	5780347
124	15 376	11.13553	35.21363	08064516	174	30 276	13.19091	41.71331	5747126
125	15 625	11.18034	35.35534	08000000	175	30 625	13.22876	41.83300	5714286
126	15 876	11.22497	35.49648	07936508	176	30 976	13.26650	41.95235	5681818
127	16 129	11.26943	35.63706	07874016	177	31 329	13.30413	42.07137	5649718
128	16 384	11.31371	35.77709	07812500	178	31 684	13.34166	42.19005	5617978
129	16 641	11.35782	35.91657	07751938	179	32 041	13.37909	42.30839	5586592
130	16 900	11.40175	36.05551	07692308	180	32 400	13.41641	42.42641	5555556
131	17 161	11.44552	36.19392	07633588	181	32 761	13.45362	42.54409	5524862
132	17 424	11.48913	36.33180	07575758	182	33 124	13.49074	42.66146	5494505
133	17 689	11.53256	36.46917	07518797	183	33 489	13.52775	42.77850	5464481
134	17 956	11.57584	36.60601	07462687	184	33 856	13.56466	42.89522	5434783
135	18 225	11.61895	36.74235	07407407	185	34 225	13.60147	43.01163	5405405
136	18 496	11.66190	36.87818	07352941	186	34 596	13.63818	43.12772	5376344
137	18 769	11.70470	37.01351	07299270	187	34 969	13.67479	43.24350	5347594
138	19 044	11.74734	37.14835	07246377	188	35 344	13.71131	43.35897	5319149
139	19 321	11.78983	37.28270	07194245	189	35 721	13.74773	43.47413	5291005
140	19 600	11.83216	37.41657	07142857	190	36 100	13.78405	43.58899	5263158
141	19 881	11.87434	37.54997	07092199	191	36 481	13.82027	43.70355	5235602
142	20 164	11.91638	37.68289	07042254	192	36 864	13.85641	43.81780	5208333
143	20 449	11.95826	37.81534	06993007	193	37 249	13.89244	43.93177	5181347
144	20 736	12.00000	37.94733	06944444	194	37 636	13.92839	44.04543	5154639
145	21 025	12.04159	38.07887	06896552	195	38 025	13.96424	44.15880	5128205
146	21 316	12.08305	38.20995	06849315	196	38 416	14.00000	44.27189	5102041
147	21 609	12.12436	38.34058	06802721	197	38 809	14.03567	44.38468	5076142
148	21 904	12.16553	38.47077	06756757	198	39 204	14.07125	44.49719	5050505
149	22 201	12.20656	38.60052	06711409	199	39 601	14.10674	44.60942	5025126
150	22 500	12.24745	38.72983	06666667	200	40 000	14.14214	44.72136	5000000

SQUARES, SQUARE ROOTS, AND RECIPROCALS 1–1,000 (*Continued*)

N	N^2	$\sqrt{N}$	$\sqrt{10N}$	$1/N$ .00	N	N^2	$\sqrt{N}$	$\sqrt{10N}$	$1/N$ .00
200	40 000	14.14214	44.72136	5000000	250	62 500	15.81139	50.00000	4000000
201	40 401	14.17745	44.83302	4975124	251	63 001	15.84298	50.09990	3984064
202	40 804	14.21267	44.94441	4950495	252	63 504	15.87451	50.19960	3968254
203	41 209	14.24781	45.05552	4926108	253	64 009	15.90597	50.29911	3952569
204	41 616	14.28286	45.16636	4901961	254	64 516	15.93738	50.39841	3937008
205	42 025	14.31782	45.27693	4878049	255	65 025	15.96872	50.49752	3921569
206	42 436	14.35270	45.38722	4854369	256	65 536	16.00000	50.59644	3906250
207	42 849	14.38749	45.49725	4830918	257	66 049	16.03122	50.69517	3891051
208	43 264	14.42221	45.60702	4807692	258	66 564	16.06238	50.79370	3875969
209	43 681	14.45683	45.71652	4784689	259	67 081	16.09348	50.89204	3861004
210	44 100	14.49138	45.82576	4761905	260	67 600	16.12452	50.99020	3846154
211	44 521	14.52584	45.93474	4739336	261	68 121	16.15549	51.08816	3831418
212	44 944	14.56022	46.04346	4716981	262	68 644	16.18641	51.18594	3816794
213	45 369	14.59452	46.15192	4694836	263	69 169	16.21727	51.28353	3802281
214	45 796	14.62874	46.26013	4672897	264	69 696	16.24808	51.38093	3787879
215	46 225	14.66288	46.36809	4651163	265	70 225	16.27882	51.47815	3773585
216	46 656	14.69694	46.47580	4629630	266	70 756	16.30951	51.57519	3759398
217	47 089	14.73092	46.58326	4608295	267	71 289	16.34013	51.67204	3745318
218	47 524	14.76482	46.69047	4587156	268	71 824	16.37071	51.76872	3731343
219	47 961	14.79865	46.79744	4566210	269	72 361	16.40122	51.86521	3717472
220	48 400	14.83240	46.90416	4545455	270	72 900	16.43168	51.96152	3703704
221	48 841	14.86607	47.01064	4524887	271	73 441	16.46208	52.05766	3690037
222	49 284	14.89966	47.11688	4504505	272	73 984	16.49242	52.15362	3676471
223	49 729	14.93318	47.22288	4484305	273	74 529	16.52271	52.24940	3663004
224	50 176	14.96663	47.32864	4464286	274	75 076	16.55295	52.34501	3649635
225	50 625	15.00000	47.43416	4444444	275	75 625	16.58312	52.44044	3636364
226	51 076	15.03330	47.53946	4424779	276	76 176	16.61325	52.53570	3623188
227	51 529	15.06652	47.64452	4405286	277	76 729	16.64332	52.63079	3610108
228	51 984	15.09967	47.74935	4385965	278	77 284	16.67333	52.72571	3597122
229	52 441	15.13275	47.85394	4366812	279	77 841	16.70329	52.82045	3584229
230	52 900	15.16575	47.95832	4347826	280	78 400	16.73320	52.91503	3571429
231	53 361	15.19868	48.06246	4329004	281	78 961	16.76305	53.00943	3558719
232	53 824	15.23155	48.16638	4310345	282	79 524	16.79286	53.10367	3546099
233	54 289	15.26434	48.27007	4291845	283	80 089	16.82260	53.19774	3533569
234	54 756	15.29706	48.37355	4273504	284	80 656	16.85230	53.29165	3521127
235	55 225	15.32971	48.47680	4255319	285	81 225	16.88194	53.38539	3508772
236	55 696	15.36229	48.57983	4237288	286	81 796	16.91153	53.47897	3496503
237	56 169	15.39480	48.68265	4219409	287	82 369	16.94107	53.57238	3484321
238	56 644	15.42725	48.78524	4201681	288	82 944	16.97056	53.66563	3472222
239	57 121	15.45962	48.88763	4184100	289	83 521	17.00000	53.75872	3460208
240	57 600	15.49193	48.98979	4166667	290	84 100	17.02939	53.85165	3448276
241	58 081	15.52417	49.09175	4149378	291	84 681	17.05872	53.94442	3436426
242	58 564	15.55635	49.19350	4132231	292	85 264	17.08801	54.03702	3424658
243	59 049	15.58846	49.29503	4115226	293	85 849	17.11724	54.12947	3412969
244	59 536	15.62050	49.39636	4098361	294	86 436	17.14643	54.22177	3401361
245	60 025	15.65248	49.49747	4081633	295	87 025	17.17556	54.31390	3389831
246	60 516	15.68439	49.59839	4065041	296	87 616	17.20465	54.40588	3378378
247	61 009	15.71623	49.69909	4048583	297	88 209	17.23369	54.49771	3367003
248	61 504	15.74802	49.79960	4032258	298	88 804	17.26268	54.58938	3355705
249	62 001	15.77973	49.89990	4016064	299	89 401	17.29162	54.68089	3344482
250	62 500	15.81139	50.00000	4000000	300	90 000	17.32051	54.77226	3333333

SQUARES, SQUARE ROOTS, AND RECIPROCALS 1–1,000 (*Continued*)

N	N²	√N	√10N	1/N .00	N	N²	√N	√10N	1/N .00
300	90 000	17.32051	54.77226	3333333	350	122 500	18.70829	59.16080	2857143
301	90 601	17.34935	54.86347	3322259	351	123 201	18.73499	59.24525	2849003
302	91 204	17.37815	54.95453	3311258	352	123 904	18.76166	59.32959	2840909
303	91 809	17.40690	55.04544	3300330	353	124 609	18.78829	59.41380	2832861
304	92 416	17.43560	55.13620	3289474	354	125 316	18.81489	59.49790	2824859
305	93 025	17.46425	55.22681	3278689	355	126 025	18.84144	59.58188	2816901
306	93 636	17.49286	55.31727	3267974	356	126 736	18.86796	59.66574	2808989
307	94 249	17.52142	55.40758	3257329	357	127 449	18.89444	59.74948	2801120
308	94 864	17.54993	55.49775	3246753	358	128 164	18.92089	59.83310	2793296
309	95 481	17:57840	55.58777	3236246	359	128 881	18.94730	59.91661	2785515
310	96 100	17.60682	55.67764	3225806	360	129 600	18.97367	60.00000	2777778
311	96 721	17.63519	55.76737	3215434	361	130 321	19.00000	60.08328	2770083
312	97 344	17.66352	55.85696	3205128	362	131 044	19.02630	60.16644	2762431
313	97 969	17.69181	55.94640	3194888	363	131 769	19.05256	60.24948	2754821
314	98 596	17.72005	56.03570	3184713	364	132 496	19.07878	60.33241	2747253
315	99 225	17.74824	56.12486	3174603	365	133 225	19.10497	60.41523	2739726
316	99 856	17.77639	56.21388	3164557	366	133 956	19.13113	60.49793	2732240
317	100 489	17.80449	56.30275	3154574	367	134 689	19.15724	60.58052	2724796
318	101 124	17.83255	56.39149	3144654	368	135 424	19.18333	60.66300	2717391
319	101 761	17.86057	56.48008	3134796	369	136 161	19.20937	60.74537	2710027
320	102 400	17.88854	56.56854	3125000	370	136 900	19.23538	60.82763	2702703
321	103 041	17.91647	56.65686	3115265	371	137 641	19.26136	60.90977	2695418
322	103 684	17.94436	56.74504	3105590	372	138 384	19.28730	60.99180	2688172
323	104 329	17.97220	56.83309	3095975	373	139 129	19.31321	61.07373	2680965
324	104 976	18.00000	56.92100	3086420	374	139 876	19.33908	61.15554	2673797
325	105 625	18.02776	57.00877	3076923	375	140 625	19.36492	61.23724	2666667
326	106 276	18.05547	57.09641	3067485	376	141 376	19.39072	61.31884	2659574
327	106 929	18.08314	57.18391	3058104	377	142 129	19.41649	61.40033	2652520
328	107 584	18.11077	57.27128	3048780	378	142 884	19.44222	61.48170	2645503
329	108 241	18.13836	57.35852	3039514	379	143 641	19.46792	61.56298	2638522
330	108 900	18.16590	57.44563	3030303	380	144 400	19.49359	61.64414	2631579
331	109 561	18.19341	57.53260	3021148	381	145 161	19.51922	61.72520	2624672
332	110 224	18.22087	57.61944	3012048	382	145 924	19.54483	61.80615	2617801
333	110 889	18.24829	57.70615	3003003	383	146 689	19.57039	61.88699	2610966
334	111 556	18.27567	57.79273	2994012	384	147 456	19.59592	61.96773	2604167
335	112 225	18.30301	57.87918	2985075	385	148 225	19.62142	62.04837	2597403
336	112 896	18.33030	57.96551	2976190	386	148 996	19.64688	62.12890	2590674
337	113 569	18.35756	58.05170	2967359	387	149 769	19.67232	62.20932	2583979
338	114 244	18.38478	58.13777	2958580	388	150 544	19.69772	62.28965	2577320
339	114 921	18.41195	58.22371	2949853	389	151 321	19.72308	62.36986	2570694
340	115 600	18.43909	58.30952	2941176	390	152 100	19.74842	62.44998	2564103
341	116 281	18.46619	58.39521	2932551	391	152 881	19.77372	62.52999	2557545
342	116 964	18.49324	58.48077	2923977	392	153 664	19.79899	62.60990	2551020
343	117 649	18.52026	58.56620	2915452	393	154 449	19.82423	62.68971	2544529
344	118 336	18.54724	58.65151	2906977	394	155 236	19.84943	62.76942	2538071
345	119 025	18.57418	58.73670	2898551	395	156 025	19.87461	62.84903	2531646
346	119 716	18.60108	58.82176	2890173	396	156 816	19.89975	62.92853	2525253
347	120 409	18.62794	58.90671	2881844	397	157 609	19.92486	63.00794	2518892
348	121 104	18.65476	58.99152	2873563	398	158 404	19.94994	63.08724	2512563
349	121 801	18.68154	59.07622	2865330	399	159 201	19.97498	63.16645	2506266
350	122 500	18.70829	59.16080	2857143	400	160 000	20.00000	63.24555	2500000

SQUARES, SQUARE ROOTS, AND RECIPROCALS 1–1,000 (*Continued*)

N	N²	$\sqrt{N}$	$\sqrt{10N}$	1/N .00	N	N²	$\sqrt{N}$	$\sqrt{10N}$	1/N .00
400	160 000	20.00000	63.24555	2500000	450	202 500	21.21320	67.08204	2222222
401	160 801	20.02498	63.32456	2493766	451	203 401	21.23676	67.15653	2217295
402	161 604	20.04994	63.40347	2487562	452	204 304	21.26029	67.23095	2212389
403	162 409	20.07486	63.48228	2481390	453	205 209	21.28380	67.30527	2207506
404	163 216	20.09975	63.56099	2475248	454	206 116	21.30728	67.37952	2202643
405	164 025	20.12461	63.63961	2469136	455	207 025	21.33073	67.45369	2197802
406	164 836	20.14944	63.71813	2463054	456	207 936	21.35416	67.52777	2192982
407	165 649	20.17424	63.79655	2457002	457	208 849	21.37756	67.60178	2188184
408	166 464	20.19901	63.87488	2450980	458	209 764	21.40093	67.67570	2183406
409	167 281	20.22375	63.95311	2444988	459	210 681	21.42429	67.74954	2178649
410	168 100	20.24846	64.03124	2439024	460	211 600	21.44761	67.82330	2173913
411	168 921	20.27313	64.10928	2433090	461	212 521	21.47091	67.89698	2169197
412	169 744	20.29778	64.18723	2427184	462	213 444	21.49419	67.97058	2164502
413	170 569	20.32240	64.26508	2421308	463	214 369	21.51743	68.04410	2159827
414	171 396	20.34699	64.34283	2415459	464	215 296	21.54066	68.11755	2155172
415	172 225	20.37155	64.42049	2409639	465	216 225	21.56386	68.19091	2150538
416	173 056	20.39608	64.49806	2403846	466	217 156	21.58703	68.26419	2145923
417	173 889	20.42058	64.57554	2398082	467	218 089	21.61018	68.33740	2141328
418	174 724	20.44505	64.65292	2392344	468	219 024	21.63331	68.41053	2136752
419	175 561	20.46949	64.73021	2386635	469	219 961	21.65641	68.48357	2132196
420	176 400	20.49390	64.80741	2380952	470	220 900	21.67948	68.55655	2127660
421	177 241	20.51828	64.88451	2375297	471	221 841	21.70253	68.62944	2123142
422	178 084	20.54264	64.96153	2369668	472	222 784	21.72556	68.70226	2118644
423	178 929	20.56696	65.03845	2364066	473	223 729	21.74856	68.77500	2114165
424	179 776	20.59126	65.11528	2358491	474	224 676	21.77154	68.84766	2109705
425	180 625	20.61553	65.19202	2352941	475	225 625	21.79449	68.92024	2105263
426	181 476	20.63977	65.26868	2347418	476	226 576	21.81742	68.99275	2100840
427	182 329	20.66398	65.34524	2341920	477	227 529	21.84033	69.06519	2096436
428	183 184	20.68816	65.42171	2336449	478	228 484	21.86321	69.13754	2092050
429	184 041	20.71232	65.49809	2331002	479	229 441	21.88607	69.20983	2087683
430	184 900	20.73644	65.57439	2325581	480	230 400	21.90890	69.28203	2083333
431	185 761	20.76054	65.65059	2320186	481	231 361	21.93171	69.35416	2079002
432	186 624	20.78461	65.72671	2314815	482	232 324	21.95450	69.42622	2074689
433	187 489	20.80865	65.80274	2309469	483	233 289	21.97726	69.49820	2070393
434	188 356	20.83267	65.87868	2304147	484	234 256	22.00000	69.57011	2066116
435	189 225	20.85665	65.95453	2298851	485	235 225	22.02272	69.64194	2061856
436	190 096	20.88061	66.03030	2293578	486	236 196	22.04541	69.71370	2057613
437	190 969	20.90454	66.10598	2288330	487	237 169	22.06808	69.78539	2053388
438	191 844	20.92845	66.18157	2283105	488	238 144	22.09072	69.85700	2049180
439	192 721	20.95233	66.25708	2277904	489	239 121	22.11334	69.92853	2044990
440	193 600	20.97618	66.33250	2272727	490	240 100	22.13594	70.00000	2040816
441	194 481	21.00000	66.40783	2267574	491	241 081	22.15852	70.07139	2036660
442	195 364	21.02380	66.48308	2262443	492	242 064	22.18107	70.14271	2032520
443	196 249	21.04757	66.55825	2257336	493	243 049	22.20360	70.21396	2028398
444	197 136	21.07131	66.63332	2252252	494	244 036	22.22611	70.28513	2024291
445	198 025	21.09502	66.70832	2247191	495	245 025	22.24860	70.35624	2020202
446	198 916	21.11871	66.78323	2242152	496	246 016	22.27106	70.42727	2016129
447	199 809	21.14237	66.85806	2237136	497	247 009	22.29350	70.49823	2012072
448	200 704	21.16601	66.93280	2232143	498	248 004	22.31591	70.56912	2008032
449	201 601	21.18962	67.00746	2227171	499	249 001	22.33831	70.63993	2004008
450	202 500	21.21320	67.08204	2222222	500	250 000	22.36068	70.71068	2000000

SQUARES, SQUARE ROOTS, AND RECIPROCALS 1–1,000 (*Continued*)

N	N²	√N	√10N	1/N .00	N	N²	√N	√10N	1/N .00
500	250 000	22.36068	70.71068	2000000	550	302 500	23.45208	74.16198	1818182
501	251 001	22.38303	70.78135	1996008	551	303 601	23.47339	74.22937	1814882
502	252 004	22.40536	70.85196	1992032	552	304 704	23.49468	74.29670	1811594
503	253 009	22.42766	70.92249	1988072	553	305 809	23.51595	74.36397	1808318
504	254 016	22.44994	70.99296	1984127	554	306 916	23.53720	74.43118	1805054
505	255 025	22.47221	71.06335	1980198	555	308 025	23.55844	74.49832	1801802
506	256 036	22.49444	71.13368	1976285	556	309 136	23.57965	74.56541	1798561
507	257 049	22.51666	71.20393	1972387	557	310 249	23.60085	74.63243	1795332
508	258 064	22.53886	71.27412	1968504	558	311 364	23.62202	74.69940	1792115
509	259 081	22.56103	71.34424	1964637	559	312 481	23.64318	74.76630	1788909
510	260 100	22.58318	71.41428	1960784	560	313 600	23.66432	74.83315	1785714
511	261 121	22.60531	71.48426	1956947	561	314 721	23.68544	74.89993	1782531
512	262 144	22.62742	71.55418	1953125	562	315 844	23.70654	74.96666	1779359
513	263 169	22.64950	71.62402	1949318	563	316 969	23.72762	75.03333	1776199
514	264 196	22.67157	71.69379	1945525	564	318 096	23.74868	75.09993	1773050
515	265 225	22.69361	71.76350	1941748	565	319 225	23.76973	75.16648	1769912
516	266 256	22.71563	71.83314	1937984	566	320 356	23.79075	75.23297	1766784
517	267 289	22.73763	71.90271	1934236	567	321 489	23.81176	75.29940	1763668
518	268 324	22.75961	71.97222	1930502	568	322 624	23.83275	75.36577	1760563
519	269 361	22.78157	72.04165	1926782	569	323 761	23.85372	75.43209	1757469
520	270 400	22.80351	72.11103	1923077	570	324 900	23.87467	75.49834	1754386
521	271 441	22.82542	72.18033	1919386	571	326 041	23.89561	75.56454	1751313
522	272 484	22.84732	72.24957	1915709	572	327 184	23.91652	75.63068	1748252
523	273 529	22.86919	72.31874	1912046	573	328 329	23.93742	75.69676	1745201
524	274 576	22.89105	72.38784	1908397	574	329 476	23.95830	75.76279	1742160
525	275 625	22.91288	72.45688	1904762	575	330 625	23.97916	75.82875	1739130
526	276 676	22.93469	72.52586	1901141	576	331 776	24.00000	75.89466	1736111
527	277 729	22.95648	72.59477	1897533	577	332 929	24.02082	75.96052	1733102
528	278 784	22.97825	72.66361	1893939	578	334 084	24.04163	76.02631	1730104
529	279 841	23.00000	72.73239	1890359	579	335 241	24.06242	76.09205	1727116
530	280 900	23.02173	72.80110	1886792	580	336 400	24.08319	76.15773	1724138
531	281 961	23.04344	72.86975	1883239	581	337 561	24.10394	76.22336	1721170
532	283 024	23.06513	72.93833	1879699	582	338 724	24.12468	76.28892	1718213
533	284 089	23.08679	73.00685	1876173	583	339 889	24.14539	76.35444	1715266
534	285 156	23.10844	73.07530	1872659	584	341 056	24.16609	76.41989	1712329
535	286 225	23.13007	73.14369	1869159	585	342 225	24.18677	76.48529	1709402
536	287 296	23.15167	73.21202	1865672	586	343 396	24.20744	76.55064	1706485
537	288 369	23.17326	73.28028	1862197	587	344 569	24.22808	76.61593	1703578
538	289 444	23.19483	73.34848	1858736	588	345 744	24.24871	76.68116	1700680
539	290 521	23.21637	73.41662	1855288	589	346 921	24.26932	76.74634	1697793
540	291 600	23.23790	73.48469	1851852	590	348 100	24.28992	76.81146	1694915
541	292 681	23.25941	73.55270	1848429	591	349 281	24.31049	76.87652	1692047
542	293 764	23.28089	73.62065	1845018	592	350 464	24.33105	76.94154	1689189
543	294 849	23.30236	73.68853	1841621	593	351 649	24.35159	77.00649	1686341
544	295 936	23.32381	73.75636	1838235	594	352 836	24.37212	77.07140	1683502
545	297 025	23.34524	73.82412	1834862	595	354 025	24.39262	77.13624	1680672
546	298 116	23.36664	73.89181	1831502	596	355 216	24.41311	77.20104	1677852
547	299 209	23.38803	73.95945	1828154	597	356 409	24.43358	77.26578	1675042
548	300 304	23.40940	74.02702	1824818	598	357 604	24.45404	77.33046	1672241
549	301 401	23.43075	74.09453	1821494	599	358 801	24.47448	77.39509	1669449
550	302 500	23.45208	74.16198	1818182	600	360 000	24.49490	77.45967	1666667

SQUARES, SQUARE ROOTS, AND RECIPROCALS 1–1,000 (*Continued*)

N	N^2	$\sqrt{N}$	$\sqrt{10N}$	$1/N$ .00	N	N^2	$\sqrt{N}$	$\sqrt{10N}$	$1/N$ .00
600	360 000	24.49490	77.45967	1666667	650	422 500	25.49510	80.62258	1538462
601	361 201	24.51530	77.52419	1663894	651	423 801	25.51470	80.68457	1536098
602	362 404	24.53569	77.58866	1661130	652	425 104	25.53429	80.74652	1533742
603	363 609	24.55606	77.65307	1658375	653	426 409	25.55386	80.80842	1531394
604	364 816	24.57641	77.71744	1655629	654	427 716	25.57342	80.87027	1529052
605	366 025	24.59675	77.78175	1652893	655	429 025	25.59297	80.93207	1526718
606	367 236	24.61707	77.84600	1650165	656	430 336	25.61250	80.99383	1524390
607	368 449	24.63737	77.91020	1647446	657	431 649	25.63201	81.05554	1522070
608	369 664	24.65766	77.97435	1644737	658	432 964	25.65151	81.11720	1519757
609	370 881	24.67793	78.03845	1642036	659	434 281	25.67100	81.17881	1517451
610	372 100	24.69818	78.10250	1639344	660	435 600	25.69047	81.24038	1515152
611	373 321	24.71841	78.16649	1636661	661	436 921	25.70992	81.30191	1512859
612	374 544	24.73863	78.23043	1633987	662	438 244	25.72936	81.36338	1510574
613	375 769	24.75884	78.29432	1634321	663	439 569	25.74879	81.42481	1508296
614	376 996	24.77902	78.35815	1628664	664	440 896	25.76820	81.48620	1506024
615	378 225	24.79919	78.42194	1626016	665	442 225	25.78759	81.54753	1503759
616	379 456	24.81935	78.48567	1623377	666	443 556	25.80698	81.60882	1501502
617	380 689	24.83948	78.54935	1620746	667	444 889	25.82634	81.67007	1499250
618	381 924	24.85961	78.61298	1618123	668	446 224	25.84570	81.73127	1497006
619	383 161	24.87971	78.67655	1615509	669	447 561	25.86503	81.79242	1494768
620	384 400	24.89980	78.74008	1612903	670	448 900	25.88436	81.85353	1492537
621	385 641	24.91987	78.80355	1610306	671	450 241	25.90367	81.91459	1490313
622	386 884	24.93993	78.86698	1607717	672	451 584	25.92296	81.97561	1488095
623	388 129	24.95997	78.93035	1605136	673	452 929	25.94224	82.03658	1485884
624	389 376	24.97999	78.99367	1602564	674	454 276	25.96151	82.09750	1483680
625	390 625	25.00000	79.05694	1600000	675	455 625	25.98076	82.15838	1481481
626	391 876	25.01999	79.12016	1597444	676	456 976	26.00000	82.21922	1479290
627	393 129	25.03997	79.18333	1594896	677	458 329	26.01922	82.28001	1477105
628	394 384	25.05993	79.24645	1592357	678	459 684	26.03843	82.34076	1474926
629	395 641	25.07987	79.30952	1589825	679	461 041	26.05763	82.40146	1472754
630	396 900	25.09980	79.37254	1587302	680	462 400	26.07681	82.46211	1470588
631	398 161	25.11971	79.43551	1584786	681	463 761	26.09598	82.52272	1468429
632	399 424	25.13961	79.49843	1582278	682	465 124	26.11513	82.58329	1466276
633	400 689	25.15949	79.56130	1579779	683	466 489	26.13427	82.64381	1464129
634	401 956	25.17936	79.62412	1577287	684	467 856	26.15339	82.70429	1461988
635	403 225	25.19921	79.68689	1574803	685	469 225	26.17250	82.76473	1459854
636	404 496	25.21904	79.74961	1572327	686	470 596	26.19160	82.82512	1457726
637	405 769	25.23886	79.81228	1569859	687	471 969	26.21068	82.88546	1455604
638	407 044	25.25866	79.87490	1567398	688	473 344	26.22975	82.94577	1453488
639	408 321	25.27845	79.93748	1564945	689	474 721	26.24881	83.00602	1451379
640	409 600	25.29822	80.00000	1562500	690	476 100	26.26785	83.06624	1449275
641	410 881	25.31798	80.06248	1560062	691	477 481	26.28688	83.12641	1447178
642	412 164	25.33772	80.12490	1557632	692	478 864	26.30589	83.18654	1445087
643	413 449	25.35744	80.18728	1555210	693	480 249	26.32489	83.24662	1443001
644	414 736	25.37716	80.24961	1552795	694	481 636	26.34388	83.30666	1440922
645	416 025	25.39685	80.31189	1550388	695	483 025	26.36285	83.36666	1438849
646	417 316	25.41653	80.37413	1547988	696	484 416	26.38181	83.42661	1436782
647	418 609	25.43619	80.43631	1545595	697	485 809	26.40076	83.48653	1434720
648	419 904	25.45584	80.49845	1543210	698	487 204	26.41969	83.54639	1432665
649	421 201	25.47548	80.56054	1540832	699	488 601	26.43861	83.60622	1430615
650	422 500	25.49510	80.62258	1538462	700	490 000	26.45751	83.66600	1428571

SQUARES, SQUARE ROOTS, AND RECIPROCALS 1–1,000 (*Continued*)

N	N^2	$\sqrt{N}$	$\sqrt{10N}$	$1/N$ .00	N	N^2	$\sqrt{N}$	$\sqrt{10N}$	$1/N$ .00
700	490 000	26.45751	83.66600	1428571	750	562 500	27.38613	86.60254	1333333
701	491 401	26.47640	83.72574	1426534	751	564 001	27.40438	86.66026	1331558
702	492 804	26.49528	83.78544	1424501	752	565 504	27.42262	86.71793	1329787
703	494 209	26.51415	83.84510	1422475	753	567 009	27.44085	86.77557	1328021
704	495 616	26.53300	83.90471	1420455	754	568 516	27.45906	86.83317	1326260
705	497 025	26.55184	83.96428	1418440	755	570 025	27.47726	86.89074	1324503
706	498 436	26.57066	84.02381	1416431	756	571 536	27.49545	86.94826	1322751
707	499 849	26.58947	84.08329	1414427	757	573 049	27.51363	87.00575	1321004
708	501 264	26.60827	84.14274	1412429	758	574 564	27.53180	87.06320	1319261
709	502 681	26.62705	84.20214	1410437	759	576 081	27.54995	87.12061	1317523
710	504 100	26.64583	84.26150	1408451	760	577 600	27.56810	87.17798	1315789
711	505 521	26.66458	84.32082	1406470	761	579 121	27.58623	87.23531	1314060
712	506 944	26.68333	84.38009	1404494	762	580 644	27.60435	87.29261	1312336
713	508 369	26.70206	84.43933	1402525	763	582 169	27.62245	87.34987	1310616
714	509 796	26.72078	84.49852	1400560	764	583 696	27.64055	87.40709	1308901
715	511 225	26.73948	84.55767	1398601	765	585 225	27.65863	87.46428	1307190
716	512 656	26.75818	84.61678	1396648	766	586 756	27.67671	87.52143	1305483
717	514 089	26.77686	84.67585	1394700	767	588 289	27.69476	87.57854	1303781
718	515 524	26.79552	84.73488	1392758	768	589 824	27.71281	87.63561	1302083
719	516 961	26.81418	84.79387	1390821	769	591 361	27.73085	87.69265	1300390
720	518 400	26.83282	84.85281	1388889	770	592 900	27.74887	87.74964	1298701
721	519 841	26.85144	84.91172	1386963	771	594 441	27.76689	87.80661	1297017
722	521 284	26.87006	84.97058	1385042	772	595 984	27.78489	87.86353	1295337
723	522 729	26.88866	85.02941	1383126	773	597 529	27.80288	87.92042	1293661
724	524 176	26.90725	85.08819	1381215	774	599 076	27.82086	87.97727	1291990
725	525 625	26.92582	85.14693	1379310	775	600 625	27.83882	88.03408	1290323
726	527 076	26.94439	85.20563	1377410	776	602 176	27.85678	88.09086	1288660
727	528 529	26.96294	85.26429	1375516	777	603 729	27.87472	88.14760	1287001
728	529 984	26.98148	85.32292	1373626	778	605 284	27.89265	88.20431	1285347
729	531 441	27.00000	85.38150	1371742	779	606 841	27.91057	88.26098	1283697
730	532 900	27.01851	85.44004	1369863	780	608 400	27.92848	88.31761	1282051
731	534 361	27.03701	85.49854	1367989	781	609 961	27.94638	88.37420	1280410
732	535 824	27.05550	85.55700	1366120	782	611 524	27.96426	88.43076	1278772
733	537 289	27.07397	85.61542	1364256	783	613 089	27.98214	88.48729	1277139
734	538 756	27.09243	85.67380	1362398	734	614 656	28.00000	88.54377	1275510
735	540 225	27.11088	85.73214	1360544	785	616 225	28.01785	88.60023	1273885
736	541 696	27.12932	85.79044	1358696	786	617 796	28.03569	88.65664	1272265
737	543 169	27.14774	85.84870	1356852	787	619 369	28.05352	88.71302	1270648
738	544 644	27.16616	85.90693	1355014	788	620 944	28.07134	88.76936	1269036
739	546 121	27.18455	85.96511	1353180	789	622 521	28.08914	88.82567	1267427
740	547 600	27.20294	86.02325	1351351	790	624 100	28.10694	88.88194	1265823
741	549 081	27.22132	86.08136	1349528	791	625 681	28.12472	88.93818	1264223
742	550 564	27.23968	86.13942	1347709	792	627 264	28.14249	88.99438	1262626
743	552 049	27.25803	86.19745	1345895	793	628 849	28.16026	89.05055	1261034
744	553 536	27.27636	86.25543	1344086	794	630 436	28.17801	89.10668	1259446
745	555 025	27.29469	86.31338	1342282	795	632 025	28.19574	89.16277	1257862
746	556 516	27.31300	86.37129	1340483	796	633 616	28.21347	89.21883	1256281
747	558 009	27.33130	86.42916	1338688	797	635 209	28.23119	89.27486	1254705
748	559 504	27.34959	86.48699	1336898	798	636 804	28.24889	89.33085	1253133
749	561 001	27.36786	86.54479	1335113	799	638 401	28.26659	89.38680	1251564
750	562 500	27.38613	86.60254	1333333	800	640 000	28.28427	89.44272	1250000

SQUARES, SQUARE ROOTS, AND RECIPROCALS 1–1,000 (*Continued*)

N	N^2	$\sqrt{N}$	$\sqrt{10N}$	$1/N$ .00	N	N^2	$\sqrt{N}$	$\sqrt{10N}$	$1/N$ .00
800	640 000	28.28427	89.44272	1250000	850	722 500	29.15476	92.19544	1176471
801	641 601	28.30194	89.49860	1248439	851	724 201	29.17190	92.24966	1175088
802	643 204	28.31960	89.55445	1246883	852	725 904	29.18904	92.30385	1173709
803	644 809	28.33725	89.61027	1245330	853	727 609	29.20616	92.35800	1172333
804	646 416	28.35489	89.66605	1243781	854	729 316	29.22328	92.41212	1170960
805	648 025	28.37252	89.72179	1242236	855	731 025	29.24038	92.46621	1169591
806	649 636	28.39014	89.77750	1240695	856	732 736	29.25748	92.52027	1168224
807	651 249	28.40775	89.83318	1239157	857	734 449	29.27456	92.57429	1166861
808	652 864	28.42534	89.88882	1237624	858	736 164	29.29164	92.62829	1165501
809	654 481	28.44293	89.94443	1236094	859	737 881	29.30870	92.68225	1164144
810	656 100	28.46050	90.00000	1234568	860	739 600	29.32576	92.73618	1162791
811	657 721	28.47806	90.05554	1233046	861	741 321	29.34280	92.79009	1161440
812	659 344	28.49561	90.11104	1231527	862	743 044	29.35984	92.84396	1160093
813	660 969	28.51315	90.16651	1230012	863	744 769	29.37686	92.89779	1158749
814	662 596	28.53069	90.22195	1228501	864	746 496	29.39388	92.95160	1157407
815	664 225	28.54820	90.27735	1226994	865	748 225	29.41088	93.00538	1156069
816	665 856	28.56571	90.33272	1225490	866	749 956	29.42788	93.05912	1154734
817	667 489	28.58321	90.38805	1223990	867	751 689	29.44486	93.11283	1153403
818	669 124	28.60070	90.44335	1222494	868	753 424	29.46184	93.16652	1152074
819	670 761	28.61818	90.49862	1221001	869	755 161	29.47881	93.22017	1150748
820	672 400	28.63564	90.55385	1219512	870	756 900	29.49576	93.27379	1149425
821	674 041	28.65310	90.60905	1218027	871	758 641	29.51271	93.32738	1148106
822	675 684	28.67054	90.66422	1216545	872	760 384	29.52965	93.38094	1146789
823	677 329	28.68798	90.71935	1215067	873	762 129	29.54657	93.43447	1145475
824	678 976	28.70540	90.77445	1213592	874	763 876	29.56349	93.48797	1144165
825	680 625	28.72281	90.82951	1212121	875	765 625	29.58040	93.54143	1142857
826	682 276	28.74022	90.88454	1210654	876	767 376	29.59730	93.59487	1141553
827	683 929	28.75761	90.93954	1209190	877	769 129	29.61419	93.64828	1140251
828	685 584	28.77499	90.99451	1207729	878	770 884	29.63106	93.70165	1138952
829	687 241	28.79236	91.04944	1206273	879	772 641	29.64793	93.75500	1137656
830	688 900	28.80972	91.10434	1204819	880	774 400	29.66479	93.80832	1136364
831	690 561	28.82707	91.15920	1203369	881	776 161	29.68164	93.86160	1135074
832	692 224	28.84441	91.21403	1201923	882	777 924	29.69848	93.91486	1133787
833	693 889	28.86174	91.26883	1200480	883	779 689	29.71532	93.96808	1132503
834	695 556	28.87906	91.32360	1199041	884	781 456	29.73214	94.02127	1131222
835	697 225	28.89637	91.37833	1197605	885	783 225	29.74895	94.07444	1129944
836	698 896	28.91366	91.43304	1196172	886	784 996	29.76575	94.12757	1128668
837	700 569	28.93095	91.48770	1194743	887	786 769	29.78255	94.18068	1127396
838	702 244	28.94823	91.54234	1193317	888	788 544	29.79933	94.23375	1126126
839	703 921	28.96550	91.59694	1191895	889	790 321	29.81610	94.28680	1124859
840	705 600	28.98275	91.65151	1190476	890	792 100	29.83287	94.33981	1123596
841	707 281	29.00000	91.70605	1189061	891	793 881	29.84962	94.39280	1122334
842	708 964	29.01724	91.76056	1187648	892	795 664	29.86637	94.44575	1121076
843	710 649	29.03446	91.81503	1186240	893	797 449	29.88311	94.49868	1119821
844	712 336	29.05168	91.86947	1184834	894	799 236	29.89983	94.55157	1118568
845	714 025	29.06888	91.92388	1183432	895	801 025	29.91655	94.60444	1117318
846	715 716	29.08608	91.97826	1182033	896	802 816	29.93326	94.65728	1116071
847	717 409	29.10326	92.03260	1180638	897	804 609	29.94996	94.71008	1114827
848	719 104	29.12044	92.08692	1179245	898	806 404	29.96665	94.76286	1113586
849	720 801	29.13760	92.14120	1177856	899	808 201	29.98333	94.81561	1112347
850	722 500	29.15476	92.19544	1176471	900	810 000	30.00000	94.86833	1111111

SQUARES, SQUARE ROOTS, AND RECIPROCALS 1–1,000 (*Concluded*)

N	N^2	$\sqrt{N}$	$\sqrt{10N}$	$1/N$ .00	N	N^2	$\sqrt{N}$	$\sqrt{10N}$	$1/N$ .00
900	810 000	30.00000	94.86833	1111111	950	902 500	30.82207	97.46794	1052632
901	811 801	30.01666	94.92102	1109878	951	904 401	30.83829	97.51923	1051525
902	813 604	30.03331	94.97368	1108647	952	906 304	30.85450	97.57049	1050420
903	815 409	30.04996	95.02631	1107420	953	908 209	30.87070	97.62172	1049318
904	817 216	30.06659	95.07891	1106195	954	910.116	30.88689	97.67292	1048218
905	819 025	30.08322	95.13149	1104972	955	912 025	30.90307	97.72410	1047120
906	820 836	30.09983	95.18403	1103753	956	913 936	30.91925	97.77525	1046025
907	822 649	30.11644	95.23655	1102536	957	915 849	30.93542	97.82638	1044932
908	824 464	30.13304	95.28903	1101322	958	917 764	30.95158	97.87747	1043841
909	826 281	30.14963	95.34149	1100110	959	919 681	30.96773	97.92855	1042753
910	828 100	30.16621	95.39392	1098901	960	921 600	30.98387	97.97959	1041667
911	829 921	30.18278	95.44632	1097695	961	923 521	31.00000	98.03061	1040583
912	831 744	30.19934	95.49869	1096491	962	925 444	31.01612	98.08160	1039501
913	833 569	30.21589	95.55103	1095290	963	927 369	31.03224	98.13256	1038422
914	835 396	30.23243	95.60335	1094092	964	929 296	31.04835	98.18350	1037344
915	837 225	30.24897	95.65563	1092896	965	931 225	31.06445	98.23441	1036269
916	839 056	30.26549	95.70789	1091703	966	933 156	31.08054	98.28530	1035197
917	840 889	30.28201	95.76012	1090513	967	935 089	31.09662	98.33616	1034126
918	842 724	30.29851	95.81232	1089325	968	937 024	31.11270	98.38699	1033058
919	844 561	30.31501	95.86449	1088139	969	938 961	31.12876	98.43780	1031992
920	846 400	30.33150	95.91663	1086957	970	940 900	31.14482	98.48858	1030928
921	848 241	30.34798	95.96874	1085776	971	942 841	31.16087	98.53933	1029866
922	850 084	30.36445	96.02083	1084599	972	944 784	31.17691	98.59006	1028807
923	851 929	30.38092	96.07289	1083424	973	946 729	31.19295	98.64076	1027749
924	853 776	30.39737	96.12492	1082251	974	948 676	31.20897	98.69144	1026694
925	855 625	30.41381	96.17692	1081081	975	950 625	31.22499	98.74209	1025641
926	857 476	30.43025	96.22889	1079914	976	952 576	31.24100	98.79271	1024590
927	859 329	30.44667	96.28084	1078749	977	954 529	31.25700	98.84331	1023541
928	861 184	30.46309	96.33276	1077586	978	956 484	31.27299	98.89388	1022495
929	863 041	30.47950	96.38465	1076426	979	958 441	31.28898	98.94443	1021450
930	864 900	30.49590	96.43651	1075269	980	960 400	31.30495	98.99495	1020408
931	866 761	30.51229	96.48834	1074114	981	962 361	31.32092	99.04544	1019368
932	868 624	30.52868	96.54015	1072961	982	964 324	31.33688	99.09591	1018330
933	870 489	30.54505	96.59193	1071811	983	966 289	31.35283	99.14636	1017294
934	872 356	30.56141	96.64368	1070664	984	968 256	31.36877	99.19677	1016260
935	874 225	30.57777	96.69540	1069519	985	970 225	31.38471	99.24717	1015228
936	876 096	30.59412	96.74709	1068376	986	972 196	31.40064	99.29753	1014199
937	877 969	30.61046	96.79876	1067236	987	974 169	31.41656	99.34787	1013171
938	879 844	30.62679	96.85040	1066098	988	976 144	31.43247	99.39819	1012146
939	881 721	30.64311	96.90201	1064963	989	978 121	31.44837	99.44848	1011122
940	883 600	30.65942	96.95360	1063830	990	980 100	31.46427	99.49874	1010101
941	885 481	30.67572	97.00515	1062699	991	982 081	31.48015	99.54898	1009082
942	887 364	30.69202	97.05668	1061571	992	984 064	31.49603	99.59920	1008065
943	889 249	30.70831	97.10819	1060445	993	986 049	31.51190	99.64939	1007049
944	891 136	30.72458	97.15966	1059322	994	988 036	31.52777	99.69955	1006036
945	893 025	30.74085	97.21111	1058201	995	990 025	31.54362	99.74969	1005025
946	894 916	30.75711	97.26253	1057082	996	992 016	31.55947	99.79980	1004016
947	896 809	30.77337	97.31393	1055966	997	994 009	31.57531	99.84989	1003009
948	898 704	30.78961	97.36529	1054852	998	996 004	31.59114	99.89995	1002004
949	900 601	30.80584	97.41663	1053741	999	998 001	31.60696	99.94999	1001001
950	902 500	30.82207	97.46794	1052632	1000	1 000 000	31.62278	100.00000	1000000

D. AREAS UNDER THE NORMAL CURVE

EACH ENTRY in this table is the proportion of the total area under a normal curve which lies under the segment between the mean and x/σ or z standard deviations from the mean. Example: $x = X - \mu = 31$ and $\sigma = 20$, so $z = x/\sigma = 1.55$. Then the required area is .4394. The area in the tail beyond the point $x = 31$ is then $.5000 - .4394 = .0606$.

x/σ	.00	.01	.02	.03	.04	.05	.06	.07	.08	.09
0.0	.0000	.0040	.0080	.0120	.0160	.0199	.0239	.0279	.0319	.0359
0.1	.0398	.0438	.0478	.0517	.0557	.0596	.0636	.0675	.0714	.0753
0.2	.0793	.0832	.0871	.0910	.0948	.0987	.1026	.1064	.1103	.1141
0.3	.1179	.1217	.1255	.1293	.1331	.1368	.1406	.1443	.1480	.1517
0.4	.1554	.1591	.1628	.1664	.1700	.1736	.1772	.1808	.1844	.1879
0.5	.1915	.1950	.1985	.2019	.2054	.2088	.2123	.2157	.2190	.2224
0.6	.2257	.2291	.2324	.2357	.2389	.2422	.2454	.2486	.2518	.2549
0.7	.2580	.2612	.2642	.2673	.2704	.2734	.2764	.2794	.2823	.2852
0.8	.2881	.2910	.2939	.2967	.2995	.3023	.3051	.3078	.3106	.3133
0.9	.3159	.3186	.3212	.3238	.3264	.3289	.3315	.3340	.3365	.3389
1.0	.3413	.3438	.3461	.3485	.3508	.3531	.3554	.3577	.3599	.3621
1.1	.3643	.3665	.3686	.3708	.3729	.3749	.3770	.3790	.3810	.3830
1.2	.3849	.3869	.3888	.3907	.3925	.3944	.3962	.3980	.3997	.4015
1.3	.4032	.4049	.4066	.4082	.4099	.4115	.4131	.4147	.4162	.4177
1.4	.4192	.4207	.4222	.4236	.4251	.4265	.4279	.4292	.4306	.4319
1.5	.4332	.4345	.4357	.4370	.4382	.4394	.4406	.4418	.4429	.4441
1.6	.4452	.4463	.4474	.4484	.4495	.4505	.4515	.4525	.4535	.4545
1.7	.4554	.4564	.4573	.4582	.4591	.4599	.4608	.4616	.4625	.4633
1.8	.4641	.4649	.4656	.4664	.4671	.4678	.4686	.4693	.4699	.4706
1.9	.4713	.4719	.4726	.4732	.4738	.4744	.4750	.4756	.4761	.4767
2.0	.4772	.4778	.4783	.4788	.4793	.4798	.4803	.4808	.4812	.4817
2.1	.4821	.4826	.4830	.4834	.4838	.4842	.4846	.4850	.4854	.4857
2.2	.4861	.4864	.4868	.4871	.4875	.4878	.4881	.4884	.4887	.4890
2.3	.4893	.4896	.4898	.4901	.4904	.4906	.4909	.4911	.4913	.4916
2.4	.4918	.4920	.4922	.4925	.4927	.4929	.4931	.4932	.4934	.4936
2.5	.4938	.4940	.4941	.4943	.4945	.4946	.4948	.4949	.4951	.4952
2.6	.4953	.4955	.4956	.4957	.4959	.4960	.4961	.4962	.4963	.4964
2.7	.4965	.4966	.4967	.4968	.4969	.4970	.4971	.4972	.4973	.4974
2.8	.4974	.4975	.4976	.4977	.4977	.4978	.4979	.4979	.4980	.4981
2.9	.4981	.4982	.4982	.4983	.4984	.4984	.4985	.4985	.4986	.4986
3.0	.49865	.4987	.4987	.4988	.4988	.4989	.4989	.4989	.4990	.4990
3.1	.49903	.4991	.4991	.4991	.4992	.4992	.4992	.4992	.4993	.4993
3.2	.4993129	.4993	.4994	.4994	.4994	.4994	.4994	.4995	.4995	.4995
3.3	.4995166	.4995	.4995	.4996	.4996	.4996	.4996	.4996	.4996	.4997
3.4	.4996631	.4997	.4997	.4997	.4997	.4997	.4997	.4997	.4998	.4998
3.5	.4997674	.4998	.4998	.4998	.4998	.4998	.4998	.4998	.4998	.4998
3.6	.4998409	.4998	.4999	.4999	.4999	.4999	.4999	.4999	.4999	.4999
3.7	.4998922	.4999	.4999	.4999	.4999	.4999	.4999	.4999	.4999	.4999
3.8	.4999277	.4999	.4999	.4999	.4999	.4999	.4999	.4999	.5000	.5000
3.9	.4999519	.5000	.5000	.5000	.5000	.5000	.5000	.5000	.5000	.5000
4.0	.4999683									
4.5	.4999966									
5.0	.4999997133									

SOURCE: Frederick E. Croxton and Dudley J. Cowden, *Practical Business Statistics* (2d ed.; New York: Prentice-Hall, Inc., 1948), p. 511. Reprinted by permission of the publisher.
Through $x/\sigma = 2.99$, from Rugg's *Statistical Methods Applied to Education*, by arrangement with the publishers, Houghton Mifflin Company. A much more detailed table of normal curve areas is given in Federal Works Agency, Work Projects Administration for the City of New York, *Tables of Probability Functions* (New York: National Bureau of Standards, 1942), Vol. II, pp. 2–238. In this appendix values for $x/\sigma = 3.00$ through 5.00 were computed from the latter source.

E. UNIT NORMAL LOSS FUNCTION

THE VALUE $L_N(D)$ is the expected opportunity loss (or EVPI) for a linear loss function with slope one and a unit normal distribution. The value D represents the relative position of the break-even point.

When using $L_N(D)$ for a general normal distribution, the value D represents the absolute deviation of the break-even point K from the mean M_0, expressed in standard deviation, S_0, units. That is

$$D = \left| \frac{K - M_0}{S_0} \right|.$$

D	.00	.01	.02	.03	.04	.05	.06	.07	.08	.09
.0	.3989	.3940	.3890	.3841	.3793	.3744	.3697	.3649	.3602	.3556
.1	.3509	.3464	.3418	.3373	.3328	.3284	.3240	.3197	.3154	.3111
.2	.3069	.3027	.2986	.2944	.2904	.2863	.2824	.2784	.2745	.2706
.3	.2668	.2630	.2592	.2555	.2518	.2481	.2445	.2409	.2374	.2339
.4	.2304	.2270	.2236	.2203	.2169	.2137	.2104	.2072	.2040	.2009
.5	.1978	.1947	.1917	.1887	.1857	.1828	.1799	.1771	.1742	.1714
.6	.1687	.1659	.1633	.1606	.1580	.1554	.1528	.1503	.1478	.1453
.7	.1429	.1405	.1381	.1358	.1334	.1312	.1289	.1267	.1245	.1223
.8	.1202	.1181	.1160	.1140	.1120	.1100	.1080	.1061	.1042	.1023
.9	.1004	.09860	.09680	.09503	.09328	.09156	.08986	.08819	.08654	.08491
1.0	.08332	.08174	.08019	.07866	.07716	.07568	.07422	.07279	.07138	.06999
1.1	.06862	.06727	.06595	.06465	.06336	.06210	.06086	.05964	.05844	.05726
1.2	.05610	.05496	.05384	.05274	.05165	.05059	.04954	.04851	.04750	.04650
1.3	.04553	.04457	.04363	.04270	.04179	.04090	.04002	.03916	.03831	.03748
1.4	.03667	.03587	.03508	.03431	.03356	.03281	.03208	.03137	.03067	.02998
1.5	.02931	.02865	.02800	.02736	.02674	.02612	.02552	.02494	.02436	.02380
1.6	.02324	.02270	.02217	.02165	.02114	.02064	.02015	.01967	.01920	.01874
1.7	.01829	.01785	.01742	.01699	.01658	.01617	.01578	.01539	.01501	.01464
1.8	.01428	.01392	.01357	.01323	.01290	.01257	.01226	.01195	.01164	.01134
1.9	.01105	.01077	.01049	.01022	$.0^2 9957$	$.0^2 9698$	$.0^2 9445$	$.0^2 9198$	$.0^2 8957$	$.0^2 8721$
2.0	$.0^2 8491$	$.0^2 8266$	$.0^2 8046$	$.0^2 7832$	$.0^2 7623$	$.0^2 7418$	$.0^2 7219$	$.0^2 7024$	$.0^2 6835$	$.0^2 6649$
2.1	$.0^2 6468$	$.0^2 6292$	$.0^2 6120$	$.0^2 5952$	$.0^2 5788$	$.0^2 5628$	$.0^2 5472$	$.0^2 5320$	$.0^2 5172$	$.0^2 5028$
2.2	$.0^2 4887$	$.0^2 4750$	$.0^2 4616$	$.0^2 4486$	$.0^2 4358$	$.0^2 4235$	$.0^2 4114$	$.0^2 3996$	$.0^2 3882$	$.0^2 3770$
2.3	$.0^2 3662$	$.0^2 3556$	$.0^2 3453$	$.0^2 3352$	$.0^2 3255$	$.0^2 3159$	$.0^2 3067$	$.0^2 2977$	$.0^2 2889$	$.0^2 2804$
2.4	$.0^2 2720$	$.0^2 2640$	$.0^2 2561$	$.0^2 2484$	$.0^2 2410$	$.0^2 2337$	$.0^2 2267$	$.0^2 2199$	$.0^2 2132$	$.0^2 2067$
2.5	$.0^2 2005$	$.0^2 1943$	$.0^2 1883$	$.0^2 1826$	$.0^2 1769$	$.0^2 1715$	$.0^2 1662$	$.0^2 1610$	$.0^2 1560$	$.0^2 1511$
3.0	$.0^3 3822$	$.0^3 3689$	$.0^3 3560$	$.0^3 3436$	$.0^3 3316$	$.0^3 3199$	$.0^3 3087$	$.0^3 2978$	$.0^3 2873$	$.0^3 2771$
3.5	$.0^4 5848$	$.0^4 5620$	$.0^4 5400$	$.0^4 5188$	$.0^4 4984$	$.0^4 4788$	$.0^4 4599$	$.0^4 4417$	$.0^4 4242$	$.0^4 4073$
4.0	$.0^5 7145$	$.0^5 6835$	$.0^5 6538$	$.0^5 6253$	$.0^5 5980$	$.0^5 5718$	$.0^5 5468$	$.0^5 5227$	$.0^5 4997$	$.0^5 4777$

Reproduced with permission from Robert Schlaifer, *Introduction to Statistics for Business Decisions* (New York: McGraw-Hill 1961) pp. 370–71.

F. BINOMIAL DISTRIBUTION— INDIVIDUAL TERMS

THE TABLE presents individual binomial probabilities for the number of successes, r, in n trials, for selected values of p, the probability of a success on any one trial.

Examples and details in the use of this table for p greater than .50 are given on pages 133–34.

The symbol 0+ indicates a value that is positive but less than .0005.

BINOMIAL DISTRIBUTION—INDIVIDUAL TERMS

$$P(r) = {}_nC_r\, p^r q^{n-r}$$

The column headings .01 – .50 give values of p.

n	r	.01	.02	.04	.05	.06	.08	.10	.12	.14	.15	.16	.18	.20	.22	.24	.25	.30	.35	.40	.45	.50	r
2	0	980	960	922	902	884	846	810	774	740	722	706	672	640	608	578	562	490	422	360	302	250	0
	1	020	039	077	095	113	147	180	211	241	255	269	295	320	343	365	375	420	455	480	495	500	1
	2	0+	0+	002	002	004	006	010	014	020	022	026	032	040	048	058	062	090	122	160	202	250	2
3	0	970	941	885	857	831	779	729	681	636	614	593	551	512	475	439	422	343	275	216	166	125	0
	1	029	058	111	135	159	203	243	279	311	325	339	363	384	402	416	422	441	444	432	408	375	1
	2	0+	001	005	007	010	018	027	038	051	057	065	080	096	113	131	141	189	239	288	334	375	2
	3	0+	0+	0+	0+	0+	001	001	002	003	003	004	006	008	011	014	016	027	043	064	091	125	3
4	0	961	922	849	815	781	716	656	600	547	522	498	452	410	370	334	316	240	179	130	092	063	0
	1	039	075	142	171	199	249	292	327	356	368	379	397	410	418	421	422	412	384	346	299	250	1
	2	001	002	009	014	019	033	049	067	087	098	108	131	154	177	200	211	265	311	346	368	375	2
	3	0+	0+	0+	0+	001	002	004	006	009	011	014	019	026	033	042	047	076	111	154	200	250	3
	4	0+	0+	0+	0+	0+	0+	0+	0+	0+	001	001	001	002	002	003	004	008	015	026	041	062	4
5	0	951	904	815	774	734	659	590	528	470	444	418	371	328	289	254	237	168	116	078	050	031	0
	1	048	092	170	204	234	287	328	360	383	392	398	410	410	407	400	396	360	312	259	206	156	1
	2	001	004	014	021	030	050	073	098	125	138	152	179	205	230	253	264	309	336	346	337	312	2
	3	0+	0+	001	001	002	004	008	013	020	024	029	039	051	065	080	088	132	181	230	276	312	3
	4	0+	0+	0+	0+	0+	0+	0+	001	002	002	003	004	006	009	013	015	028	049	077	113	156	4
	5	0+	0+	0+	0+	0+	0+	0+	0+	0+	0+	0+	0+	0+	001	001	001	002	005	010	018	031	5
6	0	941	886	783	735	690	606	531	464	405	377	351	304	262	225	193	178	118	075	047	028	016	0
	1	057	108	196	232	264	316	354	380	395	399	401	400	393	381	365	356	303	244	187	136	094	1
	2	001	006	020	031	042	069	098	130	161	176	191	220	246	269	288	297	324	328	311	278	234	2
	3	0+	0+	001	002	004	008	015	024	035	041	049	064	082	101	121	132	185	235	276	303	312	3
	4	0+	0+	0+	0+	0+	001	001	002	004	005	007	011	015	021	029	033	060	095	138	186	234	4
	5	0+	0+	0+	0+	0+	0+	0+	0+	0+	0+	001	001	002	002	004	004	010	020	037	061	094	5
	6	0+	0+	0+	0+	0+	0+	0+	0+	0+	0+	0+	0+	0+	0+	0+	001	002	004	008	016		6
7	0	932	868	751	698	648	558	478	409	348	321	295	249	210	176	146	133	082	049	028	015	008	0
	1	066	124	219	257	290	340	372	390	396	396	393	383	367	347	324	311	247	185	131	087	055	1
	2	002	008	027	041	055	089	124	160	194	210	225	252	275	293	307	311	318	298	261	214	164	2
	3	0+	0+	002	004	006	013	023	036	053	062	071	092	115	138	161	173	227	268	290	292	273	3
	4	0+	0+	0+	0+	0+	001	003	005	009	011	014	020	029	039	051	058	097	144	194	239	273	4
	5	0+	0+	0+	0+	0+	0+	0+	0+	001	001	002	003	004	007	010	012	025	047	077	117	164	5
	6	0+	0+	0+	0+	0+	0+	0+	0+	0+	0+	0+	0+	0+	001	001	001	004	008	017	032	055	6
	7	0+	0+	0+	0+	0+	0+	0+	0+	0+	0+	0+	0+	0+	0+	0+	0+	001	002	004	008		7
8	0	923	851	721	663	610	513	430	360	299	272	248	204	168	137	111	100	058	032	017	008	004	0
	1	075	139	240	279	311	357	383	392	390	385	378	359	336	309	281	267	198	137	090	055	031	1
	2	003	010	035	051	070	109	149	187	222	238	252	276	294	305	311	311	296	259	209	157	109	2
	3	0+	0+	003	005	009	019	033	051	072	084	096	121	147	172	196	208	254	279	279	257	219	3
	4	0+	0+	0+	0+	001	002	005	009	015	018	023	033	046	066	077	087	136	188	232	263	273	4
	5	0+	0+	0+	0+	0+	0+	0+	001	002	003	003	006	009	014	020	023	047	081	124	172	219	5
	6	0+	0+	0+	0+	0+	0+	0+	0+	0+	0+	0+	001	001	002	003	004	010	022	041	070	109	6
	7	0+	0+	0+	0+	0+	0+	0+	0+	0+	0+	0+	0+	0+	0+	0+	0+	001	003	008	016	031	7
	8	0+	0+	0+	0+	0+	0+	0+	0+	0+	0+	0+	0+	0+	0+	0+	0+	0+	0+	001	002	004	8
9	0	914	834	693	630	573	472	387	316	257	232	208	168	134	107	085	075	040	021	010	005	002	0
	1	083	153	260	299	329	370	387	388	377	368	357	331	302	271	240	225	156	100	060	034	018	1
	2	003	013	043	063	084	129	172	212	245	260	272	291	302	306	304	300	267	216	161	111	070	2
	3	0+	001	004	008	013	026	045	067	093	107	121	149	176	201	224	234	267	272	251	212	164	3
	4	0+	0+	0+	001	001	003	007	014	023	028	035	049	066	085	106	117	172	219	251	260	246	4
	5	0+	0+	0+	0+	0+	001	001	002	004	005	007	011	017	024	033	039	074	118	167	213	246	5
	6	0+	0+	0+	0+	0+	0+	0+	0+	001	001	001	002	003	005	007	009	021	042	074	116	164	6
	7	0+	0+	0+	0+	0+	0+	0+	0+	0+	0+	0+	0+	0+	001	001	001	004	010	021	041	070	7
	8	0+	0+	0+	0+	0+	0+	0+	0+	0+	0+	0+	0+	0+	0+	0+	0+	0+	001	004	009	018	8
	9	0+	0+	0+	0+	0+	0+	0+	0+	0+	0+	0+	0+	0+	0+	0+	0+	0+	0+	0+	001	002	9
10	0	904	817	665	599	539	434	349	279	221	197	175	137	107	083	064	056	028	013	006	003	001	0
	1	091	167	277	315	344	378	387	380	360	347	333	302	268	235	203	188	121	072	040	021	010	1
	2	004	015	052	075	099	148	194	233	264	276	286	298	302	298	288	282	233	176	121	076	044	2
	3	0+	001	006	010	017	034	057	085	115	130	145	174	201	224	243	250	267	252	215	166	117	3
	4	0+	0+	001	001	002	005	011	020	033	040	048	067	088	111	134	146	200	238	251	238	205	4
	5	0+	0+	0+	0+	0+	001	001	003	006	008	011	018	026	037	051	058	103	154	201	234	246	5
	6	0+	0+	0+	0+	0+	0+	0+	0+	001	001	002	003	006	009	013	016	037	069	111	160	205	6
	7	0+	0+	0+	0+	0+	0+	0+	0+	0+	0+	0+	0+	001	001	002	003	009	021	042	075	117	7
	8	0+	0+	0+	0+	0+	0+	0+	0+	0+	0+	0+	0+	0+	0+	0+	0+	001	004	011	023	044	8
	9	0+	0+	0+	0+	0+	0+	0+	0+	0+	0+	0+	0+	0+	0+	0+	0+	0+	001	002	004	010	9

BINOMIAL DISTRIBUTION—INDIVIDUAL TERMS (*Continued*)

$$P(r) = {}_nC_r\, p^r q^{n-r}$$

Probability columns are headed by **P**.

n	r	.01	.02	.04	.05	.06	.08	.10	.12	.14	.15	.16	.18	.20	.22	.24	.25	.30	.35	.40	.45	.50	r
10	10	0+	0+	0+	0+	0+	0+	0+	0+	0+	0+	0+	0+	0+	0+	0+	0+	0+	0+	0+	0+	001	10
11	0	895	801	638	569	506	400	314	245	190	167	147	113	086	065	049	042	020	009	004	001	0+	0
	1	099	180	293	329	355	382	384	368	341	325	308	272	236	202	170	155	093	052	027	013	005	1
	2	005	018	061	087	113	166	213	251	277	287	293	299	295	284	268	258	200	140	089	051	027	2
	3	0+	001	008	014	022	043	071	103	135	152	168	197	221	241	254	258	257	225	177	126	081	3
	4	0+	0+	001	001	003	008	016	028	044	054	064	086	111	136	160	172	220	243	236	206	161	4
	5	0+	0+	0+	0+	0+	001	002	005	010	013	017	027	039	054	071	080	132	183	221	236	226	5
	6	0+	0+	0+	0+	0+	0+	0+	001	002	002	003	006	010	015	022	027	057	099	147	193	226	6
	7	0+	0+	0+	0+	0+	0+	0+	0+	0+	0+	0+	001	002	003	005	006	017	038	070	113	161	7
	8	0+	0+	0+	0+	0+	0+	0+	0+	0+	0+	0+	0+	0+	0+	001	001	004	010	023	046	081	8
	9	0+	0+	0+	0+	0+	0+	0+	0+	0+	0+	0+	0+	0+	0+	0+	0+	001	002	005	013	027	9
	10	0+	0+	0+	0+	0+	0+	0+	0+	0+	0+	0+	0+	0+	0+	0+	0+	0+	0+	001	002	005	10
	11	0+	0+	0+	0+	0+	0+	0+	0+	0+	0+	0+	0+	0+	0+	0+	0+	0+	0+	0+	0+	0+	11
12	0	886	785	613	540	476	368	282	216	164	142	123	092	069	051	037	032	014	006	002	001	0+	0
	1	107	192	306	341	365	384	377	353	320	301	282	243	206	172	141	127	071	037	017	008	003	1
	2	006	022	070	099	128	183	230	265	286	292	296	294	283	266	244	232	168	109	064	034	016	2
	3	0+	001	010	017	027	053	085	120	155	172	188	215	236	250	257	258	240	195	142	092	054	3
	4	0+	0+	001	002	004	010	021	037	057	068	080	106	133	159	183	194	231	237	213	170	121	4
	5	0+	0+	0+	0+	0+	001	004	008	015	019	025	037	053	072	092	103	158	204	227	222	193	5
	6	0+	0+	0+	0+	0+	0+	0+	001	003	004	005	010	016	024	034	040	079	128	177	212	226	6
	7	0+	0+	0+	0+	0+	0+	0+	0+	0+	001	001	002	003	006	009	011	029	059	101	149	193	7
	8	0+	0+	0+	0+	0+	0+	0+	0+	0+	0+	0+	0+	001	001	002	002	008	020	042	076	121	8
	9	0+	0+	0+	0+	0+	0+	0+	0+	0+	0+	0+	0+	0+	0+	0+	0+	001	005	012	028	054	9
	10	0+	0+	0+	0+	0+	0+	0+	0+	0+	0+	0+	0+	0+	0+	0+	0+	0+	001	002	007	016	10
	11	0+	0+	0+	0+	0+	0+	0+	0+	0+	0+	0+	0+	0+	0+	0+	0+	0+	0+	0+	001	003	11
	12	0+	0+	0+	0+	0+	0+	0+	0+	0+	0+	0+	0+	0+	0+	0+	0+	0+	0+	0+	0+	0+	12
13	0	878	769	588	513	447	338	254	190	141	121	104	076	055	040	028	024	010	004	001	0+	0+	0
	1	115	204	319	351	371	382	367	336	298	277	257	216	179	145	116	103	054	026	011	004	002	1
	2	007	025	080	111	142	199	245	275	291	294	293	285	268	245	220	206	139	084	045	022	010	2
	3	0+	002	012	021	033	064	100	138	174	190	205	229	246	254	254	252	218	165	111	066	035	3
	4	0+	0+	001	003	005	014	028	047	071	084	098	126	154	179	201	210	234	222	184	135	087	4
	5	0+	0+	0+	0+	001	003	006	012	021	027	033	050	069	091	114	126	180	215	221	199	157	5
	6	0+	0+	0+	0+	0+	0+	001	002	004	006	008	015	023	034	048	056	103	155	197	217	209	6
	7	0+	0+	0+	0+	0+	0+	0+	0+	001	001	002	003	006	010	015	019	044	083	131	177	209	7
	8	0+	0+	0+	0+	0+	0+	0+	0+	0+	0+	0+	001	001	002	004	005	014	034	066	109	157	8
	9	0+	0+	0+	0+	0+	0+	0+	0+	0+	0+	0+	0+	0+	0+	001	001	003	010	024	050	087	9
	10	0+	0+	0+	0+	0+	0+	0+	0+	0+	0+	0+	0+	0+	0+	0+	0+	001	002	006	016	035	10
	11	0+	0+	0+	0+	0+	0+	0+	0+	0+	0+	0+	0+	0+	0+	0+	0+	0+	0+	001	003	010	11
	12	0+	0+	0+	0+	0+	0+	0+	0+	0+	0+	0+	0+	0+	0+	0+	0+	0+	0+	0+	0+	002	12
	13	0+	0+	0+	0+	0+	0+	0+	0+	0+	0+	0+	0+	0+	0+	0+	0+	0+	0+	0+	0+	0+	13
14	0	869	754	565	488	421	311	229	167	121	103	087	062	044	031	021	018	007	002	001	0+	0+	0
	1	123	215	329	359	376	379	356	319	276	254	232	191	154	122	095	083	041	018	007	003	001	1
	2	008	029	089	123	156	214	257	283	292	291	287	272	250	223	195	180	113	063	032	014	006	2
	3	0+	002	015	026	040	074	114	154	190	206	219	239	250	252	246	240	194	137	085	046	022	3
	4	0+	0+	002	004	007	018	035	058	085	100	115	144	172	195	214	220	229	202	155	104	061	4
	5	0+	0+	0+	0+	001	003	008	016	028	035	044	063	086	110	135	147	196	218	207	170	122	5
	6	0+	0+	0+	0+	0+	0+	001	003	007	009	012	021	032	047	064	073	126	176	207	209	183	6
	7	0+	0+	0+	0+	0+	0+	0+	001	001	002	003	005	009	015	023	028	062	108	157	195	209	7
	8	0+	0+	0+	0+	0+	0+	0+	0+	0+	0+	0+	001	002	004	006	008	023	051	092	140	183	8
	9	0+	0+	0+	0+	0+	0+	0+	0+	0+	0+	0+	0+	0+	001	001	002	007	018	041	076	122	9
	10	0+	0+	0+	0+	0+	0+	0+	0+	0+	0+	0+	0+	0+	0+	0+	0+	001	005	014	031	061	10
	11	0+	0+	0+	0+	0+	0+	0+	0+	0+	0+	0+	0+	0+	0+	0+	0+	0+	001	003	009	022	11
	12	0+	0+	0+	0+	0+	0+	0+	0+	0+	0+	0+	0+	0+	0+	0+	0+	0+	0+	001	002	006	12
	13	0+	0+	0+	0+	0+	0+	0+	0+	0+	0+	0+	0+	0+	0+	0+	0+	0+	0+	0+	0+	001	13
	14	0+	0+	0+	0+	0+	0+	0+	0+	0+	0+	0+	0+	0+	0+	0+	0+	0+	0+	0+	0+	0+	14
15	0	860	739	542	463	395	286	206	147	104	087	073	051	035	024	016	013	005	002	0+	0+	0+	0
	1	130	226	339	366	378	373	343	301	254	231	209	168	132	102	077	067	031	013	005	002	0+	1
	2	009	032	099	135	169	227	267	287	290	286	279	258	231	201	171	156	092	048	022	009	003	2
	3	0+	003	018	031	047	086	129	170	204	218	230	245	250	246	234	225	170	111	063	032	014	3
	4	0+	0+	002	005	009	022	043	069	100	116	131	162	188	208	221	225	219	179	127	078	042	4

BINOMIAL DISTRIBUTION—INDIVIDUAL TERMS (*Continued*)

$$P(r) = {}_nC_r\, p^r q^{n-r}$$

The column headings .01–.50 are values of p.

n	r	.01	.02	.04	.05	.06	.08	.10	.12	.14	.15	.16	.18	.20	.22	.24	.25	.30	.35	.40	.45	.50	r
15	5	0+	0+	0+	001	001	004	010	021	036	045	055	078	103	129	154	165	206	212	186	140	092	5
	6	0+	0+	0+	0+	0+	001	002	005	010	013	017	029	043	061	081	092	147	191	207	191	153	6
	7	0+	0+	0+	0+	0+	0+	0+	001	002	003	004	008	014	022	033	039	081	132	177	201	196	7
	8	0+	0+	0+	0+	0+	0+	0+	0+	0+	001	001	002	003	006	010	013	035	071	118	165	196	8
	9	0+	0+	0+	0+	0+	0+	0+	0+	0+	0+	0+	0+	001	001	003	003	012	030	061	105	153	9
	10	0+	0+	0+	0+	0+	0+	0+	0+	0+	0+	0+	0+	0+	0+	0+	001	003	010	024	051	092	10
	11	0+	0+	0+	0+	0+	0+	0+	0+	0+	0+	0+	0+	0+	0+	0+	0+	001	002	007	019	042	11
	12	0+	0+	0+	0+	0+	0+	0+	0+	0+	0+	0+	0+	0+	0+	0+	0+	0+	0+	002	005	014	12
	13	0+	0+	0+	0+	0+	0+	0+	0+	0+	0+	0+	0+	0+	0+	0+	0+	0+	0+	0+	001	003	13
	14	0+	0+	0+	0+	0+	0+	0+	0+	0+	0+	0+	0+	0+	0+	0+	0+	0+	0+	0+	0+	0+	14
	15	0+	0+	0+	0+	0+	0+	0+	0+	0+	0+	0+	0+	0+	0+	0+	0+	0+	0+	0+	0+	0+	15
16	0	851	724	520	440	372	263	185	129	090	074	061	042	028	019	012	010	003	001	0+	0+	0+	0
	1	138	236	347	371	379	366	329	282	233	210	187	147	113	085	063	053	023	009	003	001	0+	1
	2	010	036	108	146	182	239	275	289	285	277	268	242	211	179	148	134	073	035	015	006	002	2
	3	0+	003	021	036	054	097	142	184	216	229	238	246	246	236	218	208	146	089	047	022	009	3
	4	0+	0+	003	006	011	027	051	081	114	131	147	177	200	216	224	225	204	155	101	057	028	4
	5	0+	0+	0+	001	002	006	014	027	045	056	067	093	120	146	170	180	210	201	162	112	067	5
	6	0+	0+	0+	0+	0+	001	003	007	013	018	023	037	055	076	098	110	165	198	198	168	122	6
	7	0+	0+	0+	0+	0+	0+	0+	001	003	005	006	012	020	030	044	052	101	152	189	197	175	7
	8	0+	0+	0+	0+	0+	0+	0+	0+	001	001	001	003	006	010	016	020	049	092	142	181	196	8
	9	0+	0+	0+	0+	0+	0+	0+	0+	0+	0+	0+	001	001	002	004	006	019	044	084	132	175	9
	10	0+	0+	0+	0+	0+	0+	0+	0+	0+	0+	0+	0+	0+	0+	001	001	006	017	039	075	122	10
	11	0+	0+	0+	0+	0+	0+	0+	0+	0+	0+	0+	0+	0+	0+	0+	0+	001	005	014	034	067	11
	12	0+	0+	0+	0+	0+	0+	0+	0+	0+	0+	0+	0+	0+	0+	0+	0+	0+	001	004	011	028	12
	13	0+	0+	0+	0+	0+	0+	0+	0+	0+	0+	0+	0+	0+	0+	0+	0+	0+	0+	001	003	009	13
	14	0+	0+	0+	0+	0+	0+	0+	0+	0+	0+	0+	0+	0+	0+	0+	0+	0+	0+	0+	001	002	14
	15	0+	0+	0+	0+	0+	0+	0+	0+	0+	0+	0+	0+	0+	0+	0+	0+	0+	0+	0+	0+	0+	15
	16	0+	0+	0+	0+	0+	0+	0+	0+	0+	0+	0+	0+	0+	0+	0+	0+	0+	0+	0+	0+	0+	16
17	0	843	709	500	418	349	242	167	114	077	063	052	034	023	015	009	008	002	001	0+	0+	0+	0
	1	145	246	354	374	379	358	315	264	213	189	167	128	096	070	051	043	017	006	002	001	0+	1
	2	012	040	118	158	194	249	280	288	278	267	255	225	191	158	128	114	058	026	010	004	001	2
	3	001	004	025	041	062	108	156	196	226	236	243	246	239	223	202	189	125	070	034	014	005	3
	4	0+	0+	004	008	014	033	060	094	129	146	162	189	209	221	223	221	187	132	080	041	018	4
	5	0+	0+	0+	001	002	007	017	033	054	067	080	108	136	162	183	191	208	185	138	087	047	5
	6	0+	0+	0+	0+	0+	001	004	009	018	024	031	047	068	091	116	128	178	199	184	143	094	6
	7	0+	0+	0+	0+	0+	0+	001	002	005	007	009	016	027	040	057	067	120	168	193	184	148	7
	8	0+	0+	0+	0+	0+	0+	0+	0+	001	001	002	004	008	014	023	028	064	113	161	188	185	8
	9	0+	0+	0+	0+	0+	0+	0+	0+	0+	0+	0+	001	002	004	007	009	028	061	107	154	185	9
	10	0+	0+	0+	0+	0+	0+	0+	0+	0+	0+	0+	0+	0+	001	002	002	009	026	057	101	148	10
	11	0+	0+	0+	0+	0+	0+	0+	0+	0+	0+	0+	0+	0+	0+	0+	001	003	009	024	052	094	11
	12	0+	0+	0+	0+	0+	0+	0+	0+	0+	0+	0+	0+	0+	0+	0+	0+	001	002	008	021	047	12
	13	0+	0+	0+	0+	0+	0+	0+	0+	0+	0+	0+	0+	0+	0+	0+	0+	0+	001	002	007	018	13
	14	0+	0+	0+	0+	0+	0+	0+	0+	0+	0+	0+	0+	0+	0+	0+	0+	0+	0+	0+	002	005	14
	15	0+	0+	0+	0+	0+	0+	0+	0+	0+	0+	0+	0+	0+	0+	0+	0+	0+	0+	0+	0+	001	15
	16	0+	0+	0+	0+	0+	0+	0+	0+	0+	0+	0+	0+	0+	0+	0+	0+	0+	0+	0+	0+	0+	16
	17	0+	0+	0+	0+	0+	0+	0+	0+	0+	0+	0+	0+	0+	0+	0+	0+	0+	0+	0+	0+	0+	17
18	0	835	695	480	397	328	223	150	100	066	054	043	028	018	011	007	006	002	0+	0+	0+	0+	0
	1	152	255	360	376	377	349	300	246	194	170	149	111	081	058	041	034	013	004	001	0+	0+	1
	2	013	044	127	168	205	258	284	285	268	256	241	207	172	139	109	096	046	019	007	002	001	2
	3	001	005	028	047	070	120	168	207	233	241	244	243	230	209	184	170	105	055	025	009	003	3
	4	0+	0+	004	009	017	039	070	106	142	159	175	200	215	221	218	213	168	110	061	029	012	4
	5	0+	0+	001	001	003	009	022	040	065	079	093	123	151	175	193	199	202	166	115	067	033	5
	6	0+	0+	0+	0+	0+	002	005	012	023	030	038	058	082	107	132	144	187	194	166	118	071	6
	7	0+	0+	0+	0+	0+	0+	001	003	006	009	013	022	035	052	071	082	138	179	189	166	121	7
	8	0+	0+	0+	0+	0+	0+	0+	001	001	002	003	007	012	020	031	038	081	133	173	186	167	8
	9	0+	0+	0+	0+	0+	0+	0+	0+	0+	0+	001	002	003	006	011	014	039	079	128	169	185	9
	10	0+	0+	0+	0+	0+	0+	0+	0+	0+	0+	0+	0+	001	002	003	004	015	038	077	125	167	10
	11	0+	0+	0+	0+	0+	0+	0+	0+	0+	0+	0+	0+	0+	0+	001	001	005	015	037	074	121	11
	12	0+	0+	0+	0+	0+	0+	0+	0+	0+	0+	0+	0+	0+	0+	0+	0+	001	005	015	035	071	12
	13	0+	0+	0+	0+	0+	0+	0+	0+	0+	0+	0+	0+	0+	0+	0+	0+	0+	001	004	013	033	13
	14	0+	0+	0+	0+	0+	0+	0+	0+	0+	0+	0+	0+	0+	0+	0+	0+	0+	0+	001	004	012	14
	15	0+	0+	0+	0+	0+	0+	0+	0+	0+	0+	0+	0+	0+	0+	0+	0+	0+	0+	0+	001	003	15
	16	0+	0+	0+	0+	0+	0+	0+	0+	0+	0+	0+	0+	0+	0+	0+	0+	0+	0+	0+	0+	001	16
	17	0+	0+	0+	0+	0+	0+	0+	0+	0+	0+	0+	0+	0+	0+	0+	0+	0+	0+	0+	0+	0+	17
	18	0+	0+	0+	0+	0+	0+	0+	0+	0+	0+	0+	0+	0+	0+	0+	0+	0+	0+	0+	0+	0+	18

BINOMIAL DISTRIBUTION—INDIVIDUAL TERMS (*Continued*)

$$P(r) = {}_nC_r\, p^r q^{n-r}$$

p heads the block of columns beginning at .15/.16.

n	r	.01	.02	.04	.05	.06	.08	.10	.12	.14	.15	.16	.18	.20	.22	.24	.25	.30	.35	.40	.45	.50	r
19	0	826	681	460	377	309	205	135	088	057	046	036	023	014	009	005	004	001	0+	0+	0+	0+	0
	1	159	264	364	377	374	339	285	228	176	153	132	096	068	048	033	027	009	003	001	0+	0+	1
	2	014	049	137	179	215	265	285	280	258	243	226	190	154	121	093	080	036	014	005	001	0+	2
	3	001	006	032	053	078	131	180	217	238	244	244	236	218	194	166	152	087	042	017	006	002	3
	4	0+	0+	005	011	020	045	080	118	155	171	186	207	218	219	210	202	149	091	047	020	007	4
	5	0+	0+	001	002	004	012	027	048	076	091	106	137	164	185	199	202	192	147	093	050	022	5
	6	0+	0+	0+	0+	001	002	007	015	029	037	047	070	095	122	146	157	192	184	145	095	052	6
	7	0+	0+	0+	0+	0+	0+	001	004	009	012	017	029	044	064	086	097	153	184	180	144	096	7
	8	0+	0+	0+	0+	0+	0+	0+	001	002	003	005	009	017	027	041	049	098	149	180	177	144	8
	9	0+	0+	0+	0+	0+	0+	0+	0+	0+	001	001	003	005	009	016	020	051	098	146	177	176	9
	10	0+	0+	0+	0+	0+	0+	0+	0+	0+	0+	0+	001	001	003	005	007	022	053	098	145	176	10
	11	0+	0+	0+	0+	0+	0+	0+	0+	0+	0+	0+	0+	0+	001	001	002	008	023	053	097	144	11
	12	0+	0+	0+	0+	0+	0+	0+	0+	0+	0+	0+	0+	0+	0+	0+	001	002	008	024	053	096	12
	13	0+	0+	0+	0+	0+	0+	0+	0+	0+	0+	0+	0+	0+	0+	0+	0+	001	002	008	023	052	13
	14	0+	0+	0+	0+	0+	0+	0+	0+	0+	0+	0+	0+	0+	0+	0+	0+	0+	001	002	008	022	14
	15	0+	0+	0+	0+	0+	0+	0+	0+	0+	0+	0+	0+	0+	0+	0+	0+	0+	0+	001	002	007	15
	16	0+	0+	0+	0+	0+	0+	0+	0+	0+	0+	0+	0+	0+	0+	0+	0+	0+	0+	0+	0+	002	16
	17	0+	0+	0+	0+	0+	0+	0+	0+	0+	0+	0+	0+	0+	0+	0+	0+	0+	0+	0+	0+	0+	17
	18	0+	0+	0+	0+	0+	0+	0+	0+	0+	0+	0+	0+	0+	0+	0+	0+	0+	0+	0+	0+	0+	18
	19	0+	0+	0+	0+	0+	0+	0+	0+	0+	0+	0+	0+	0+	0+	0+	0+	0+	0+	0+	0+	0+	19
20	0	818	668	442	358	290	189	122	078	049	039	031	019	012	007	004	003	001	0+	0+	0+	0+	0
	1	165	272	368	377	370	328	270	212	159	137	117	083	058	039	026	021	007	002	0+	0+	0+	1
	2	016	053	146	189	225	271	285	274	247	229	211	173	137	105	078	067	028	010	003	001	0+	2
	3	001	006	036	060	086	141	190	224	241	243	241	228	205	178	148	134	072	032	012	004	001	3
	4	0+	001	006	013	023	052	090	130	167	182	195	213	218	213	199	190	130	074	035	014	005	4
	5	0+	0+	001	002	005	015	032	057	087	103	119	149	175	192	201	202	179	127	075	036	015	5
	6	0+	0+	0+	0+	001	003	009	019	035	045	057	082	109	136	159	169	192	171	124	075	037	6
	7	0+	0+	0+	0+	0+	001	002	005	012	016	022	036	055	076	100	112	164	184	166	122	074	7
	8	0+	0+	0+	0+	0+	0+	0+	001	003	005	007	013	022	035	051	061	114	161	180	162	120	8
	9	0+	0+	0+	0+	0+	0+	0+	0+	001	001	002	004	007	013	022	027	065	116	160	177	160	9
	10	0+	0+	0+	0+	0+	0+	0+	0+	0+	0+	0+	001	002	004	008	010	031	069	117	159	176	10
	11	0+	0+	0+	0+	0+	0+	0+	0+	0+	0+	0+	0+	0+	001	002	003	012	034	071	119	160	11
	12	0+	0+	0+	0+	0+	0+	0+	0+	0+	0+	0+	0+	0+	0+	001	001	004	014	035	073	120	12
	13	0+	0+	0+	0+	0+	0+	0+	0+	0+	0+	0+	0+	0+	0+	0+	0+	001	004	015	037	074	13
	14	0+	0+	0+	0+	0+	0+	0+	0+	0+	0+	0+	0+	0+	0+	0+	0+	0+	001	005	015	037	14
	15	0+	0+	0+	0+	0+	0+	0+	0+	0+	0+	0+	0+	0+	0+	0+	0+	0+	0+	001	005	015	15
	16	0+	0+	0+	0+	0+	0+	0+	0+	0+	0+	0+	0+	0+	0+	0+	0+	0+	0+	0+	001	005	16
	17	0+	0+	0+	0+	0+	0+	0+	0+	0+	0+	0+	0+	0+	0+	0+	0+	0+	0+	0+	0+	001	17
	18	0+	0+	0+	0+	0+	0+	0+	0+	0+	0+	0+	0+	0+	0+	0+	0+	0+	0+	0+	0+	0+	18
	19	0+	0+	0+	0+	0+	0+	0+	0+	0+	0+	0+	0+	0+	0+	0+	0+	0+	0+	0+	0+	0+	19
	20	0+	0+	0+	0+	0+	0+	0+	0+	0+	0+	0+	0+	0+	0+	0+	0+	0+	0+	0+	0+	0+	20
21	0	810	654	424	341	273	174	109	068	042	033	026	015	010	005	003	002	001	0+	0+	0+	0+	0
	1	172	280	371	376	366	317	255	195	144	122	103	071	048	032	021	017	005	001	0+	0+	0+	1
	2	017	057	155	198	233	276	284	267	234	215	196	157	121	091	066	055	022	007	002	0+	0+	2
	3	001	007	041	066	094	152	200	230	242	241	236	218	192	162	132	117	058	024	009	003	001	3
	4	0+	001	008	016	027	059	100	141	177	191	202	215	216	205	187	176	113	059	026	009	003	4
	5	0+	0+	001	003	006	018	038	065	098	115	131	161	183	197	201	199	164	109	059	026	010	5
	6	0+	0+	0+	0+	001	004	011	024	043	054	067	094	122	148	169	177	188	156	105	057	026	6
	7	0+	0+	0+	0+	0+	001	003	007	015	020	027	044	065	089	114	126	172	180	149	101	055	7
	8	0+	0+	0+	0+	0+	0+	001	002	004	006	009	017	029	044	063	074	129	169	174	144	097	8
	9	0+	0+	0+	0+	0+	0+	0+	0+	001	002	002	005	010	018	029	036	080	132	168	170	140	9
	10	0+	0+	0+	0+	0+	0+	0+	0+	0+	0+	001	001	003	006	011	014	041	085	134	167	168	10
	11	0+	0+	0+	0+	0+	0+	0+	0+	0+	0+	0+	0+	001	002	003	005	018	046	089	137	168	11
	12	0+	0+	0+	0+	0+	0+	0+	0+	0+	0+	0+	0+	0+	0+	001	001	006	021	050	093	140	12
	13	0+	0+	0+	0+	0+	0+	0+	0+	0+	0+	0+	0+	0+	0+	0+	0+	002	008	023	053	097	13
	14	0+	0+	0+	0+	0+	0+	0+	0+	0+	0+	0+	0+	0+	0+	0+	0+	0+	002	009	025	055	14
	15	0+	0+	0+	0+	0+	0+	0+	0+	0+	0+	0+	0+	0+	0+	0+	0+	0+	001	003	009	026	15
	16	0+	0+	0+	0+	0+	0+	0+	0+	0+	0+	0+	0+	0+	0+	0+	0+	0+	0+	001	003	010	16
	17	0+	0+	0+	0+	0+	0+	0+	0+	0+	0+	0+	0+	0+	0+	0+	0+	0+	0+	0+	001	003	17
	18	0+	0+	0+	0+	0+	0+	0+	0+	0+	0+	0+	0+	0+	0+	0+	0+	0+	0+	0+	0+	001	18
	19	0+	0+	0+	0+	0+	0+	0+	0+	0+	0+	0+	0+	0+	0+	0+	0+	0+	0+	0+	0+	0+	19
	20	0+	0+	0+	0+	0+	0+	0+	0+	0+	0+	0+	0+	0+	0+	0+	0+	0+	0+	0+	0+	0+	20
	21	0+	0+	0+	0+	0+	0+	0+	0+	0+	0+	0+	0+	0+	0+	0+	0+	0+	0+	0+	0+	0+	21

BINOMIAL DISTRIBUTION—INDIVIDUAL TERMS (Continued)

$$P(r) = {}_nC_r\, p^r q^{n-r}$$

Column headings below (.01 … .50) are values of p.

n	r	.01	.02	.04	.05	.06	.08	.10	.12	.14	.15	.16	.18	.20	.22	.24	.25	.30	.35	.40	.45	.50	r
22	0	802	641	407	324	256	160	098	060	036	028	022	013	007	004	002	002	0+	0+	0+	0+	0+	0
	1	178	288	373	375	360	306	241	180	130	109	090	061	041	026	017	013	004	001	0+	0+	0+	1
	2	019	062	163	207	241	279	281	258	222	201	181	141	107	078	055	046	017	005	001	0+	0+	2
	3	001	008	045	073	103	162	208	235	241	237	230	207	178	146	116	102	047	018	006	002	0+	3
	4	0+	001	009	018	031	067	110	152	186	199	208	216	211	196	174	161	096	047	019	006	002	4
	5	0+	0+	001	003	007	021	044	075	109	126	143	170	190	199	197	193	149	091	046	019	006	5
	6	0+	0+	0+	001	001	005	014	029	050	063	077	106	134	159	177	183	139	086	043	018	006	6
	7	0+	0+	0+	0+	0+	001	004	009	019	025	033	053	077	102	128	139	177	171	131	081	041	7
	8	0+	0+	0+	0+	0+	0+	001	002	006	008	012	022	036	054	075	087	142	173	164	125	076	8
	9	0+	0+	0+	0+	0+	0+	0+	0+	001	002	004	007	014	024	037	045	095	145	170	159	119	9
	10	0+	0+	0+	0+	0+	0+	0+	0+	0+	001	001	002	005	009	015	020	053	101	148	169	154	10
	11	0+	0+	0+	0+	0+	0+	0+	0+	0+	0+	0+	001	001	003	005	007	025	060	107	151	168	11
	12	0+	0+	0+	0+	0+	0+	0+	0+	0+	0+	0+	0+	0+	001	002	002	010	029	066	113	154	12
	13	0+	0+	0+	0+	0+	0+	0+	0+	0+	0+	0+	0+	0+	0+	0+	001	003	012	034	071	119	13
	14	0+	0+	0+	0+	0+	0+	0+	0+	0+	0+	0+	0+	0+	0+	0+	0+	001	004	014	037	076	14
	15	0+	0+	0+	0+	0+	0+	0+	0+	0+	0+	0+	0+	0+	0+	0+	0+	001	005	016	041		15
	16	0+	0+	0+	0+	0+	0+	0+	0+	0+	0+	0+	0+	0+	0+	0+	0+	0+	001	006	018		16
	17	0+	0+	0+	0+	0+	0+	0+	0+	0+	0+	0+	0+	0+	0+	0+	0+	0+	0+	002	007		17
	18	0+	0+	0+	0+	0+	0+	0+	0+	0+	0+	0+	0+	0+	0+	0+	0+	0+	0+	0+	002		18
	19	0+	0+	0+	0+	0+	0+	0+	0+	0+	0+	0+	0+	0+	0+	0+	0+	0+	0+	0+	0+	0+	19
	20	0+	0+	0+	0+	0+	0+	0+	0+	0+	0+	0+	0+	0+	0+	0+	0+	0+	0+	0+	0+	0+	20
	21	0+	0+	0+	0+	0+	0+	0+	0+	0+	0+	0+	0+	0+	0+	0+	0+	0+	0+	0+	0+	0+	21
	22	0+	0+	0+	0+	0+	0+	0+	0+	0+	0+	0+	0+	0+	0+	0+	0+	0+	0+	0+	0+	0+	22
23	0	794	628	391	307	241	147	089	053	031	024	018	010	006	003	002	001	0+	0+	0+	0+	0+	0
	1	184	295	375	372	354	294	226	166	117	097	079	053	034	021	013	010	003	001	0+	0+	0+	1
	2	020	066	172	215	248	281	277	249	209	188	166	127	093	066	046	038	013	004	001	0+	0+	2
	3	001	009	050	079	111	171	215	237	238	232	222	195	163	131	101	088	038	014	004	001	0+	3
	4	0+	001	010	021	035	074	120	162	194	204	211	214	204	185	160	146	082	037	014	004	001	4
	5	0+	0+	002	004	009	025	051	084	120	137	153	179	194	198	192	185	133	076	035	013	004	5
	6	0+	0+	0+	001	002	006	017	034	059	073	087	118	145	168	182	185	171	122	070	032	012	6
	7	0+	0+	0+	0+	0+	001	005	011	023	031	040	063	088	115	139	150	178	160	113	064	029	7
	8	0+	0+	0+	0+	0+	0+	001	003	008	011	015	028	044	065	088	100	153	172	151	105	058	8
	9	0+	0+	0+	0+	0+	0+	0+	001	002	003	005	010	018	030	046	056	109	155	168	143	097	9
	10	0+	0+	0+	0+	0+	0+	0+	0+	0+	001	001	003	006	012	020	026	065	117	157	164	136	10
	11	0+	0+	0+	0+	0+	0+	0+	0+	0+	0+	0+	001	002	004	008	010	033	074	123	159	161	11
	12	0+	0+	0+	0+	0+	0+	0+	0+	0+	0+	0+	0+	0+	001	002	003	014	040	082	130	161	12
	13	0+	0+	0+	0+	0+	0+	0+	0+	0+	0+	0+	0+	0+	0+	0+	001	005	018	046	090	136	13
	14	0+	0+	0+	0+	0+	0+	0+	0+	0+	0+	0+	0+	0+	0+	0+	0+	002	007	022	053	097	14
	15	0+	0+	0+	0+	0+	0+	0+	0+	0+	0+	0+	0+	0+	0+	0+	0+	002	009	026	058		15
	16	0+	0+	0+	0+	0+	0+	0+	0+	0+	0+	0+	0+	0+	0+	0+	0+	001	003	011	029		16
	17	0+	0+	0+	0+	0+	0+	0+	0+	0+	0+	0+	0+	0+	0+	0+	0+	0+	001	004	012		17
	18	0+	0+	0+	0+	0+	0+	0+	0+	0+	0+	0+	0+	0+	0+	0+	0+	0+	0+	001	004		18
	19	0+	0+	0+	0+	0+	0+	0+	0+	0+	0+	0+	0+	0+	0+	0+	0+	0+	0+	0+	001		19
	20	0+	0+	0+	0+	0+	0+	0+	0+	0+	0+	0+	0+	0+	0+	0+	0+	0+	0+	0+	0+	0+	20
	21	0+	0+	0+	0+	0+	0+	0+	0+	0+	0+	0+	0+	0+	0+	0+	0+	0+	0+	0+	0+	0+	21
	22	0+	0+	0+	0+	0+	0+	0+	0+	0+	0+	0+	0+	0+	0+	0+	0+	0+	0+	0+	0+	0+	22
	23	0+	0+	0+	0+	0+	0+	0+	0+	0+	0+	0+	0+	0+	0+	0+	0+	0+	0+	0+	0+	0+	23
24	0	786	616	375	292	227	135	080	047	027	020	015	009	005	003	001	001	0+	0+	0+	0+	0+	0
	1	190	302	375	369	347	282	213	152	105	086	070	045	028	017	010	008	002	0+	0+	0+	0+	1
	2	022	071	180	223	255	282	272	239	196	174	153	114	081	056	038	031	010	003	001	0+	0+	2
	3	002	011	055	086	119	180	221	239	234	225	213	183	149	117	088	075	031	010	003	001	0+	3
	4	0+	001	012	024	040	082	129	171	200	209	213	211	196	173	146	132	069	029	010	003	001	4
	5	0+	0+	002	005	010	029	057	093	130	147	162	185	196	195	184	176	118	062	027	009	003	5
	6	0+	0+	0+	001	002	008	020	040	067	082	098	129	155	174	184	185	160	106	056	024	008	6
	7	0+	0+	0+	0+	0+	002	006	014	028	037	048	073	100	126	149	159	176	147	096	050	021	7
	8	0+	0+	0+	0+	0+	0+	001	004	010	014	019	034	053	076	100	112	160	168	136	087	044	8
	9	0+	0+	0+	0+	0+	0+	0+	001	003	004	007	013	024	038	056	067	122	161	161	126	078	9
	10	0+	0+	0+	0+	0+	0+	0+	0+	001	001	002	004	009	016	027	033	079	130	161	155	117	10
	11	0+	0+	0+	0+	0+	0+	0+	0+	0+	0+	0+	001	003	006	011	014	043	089	137	161	149	11
	12	0+	0+	0+	0+	0+	0+	0+	0+	0+	0+	0+	0+	001	002	004	005	020	052	099	143	161	12
	13	0+	0+	0+	0+	0+	0+	0+	0+	0+	0+	0+	0+	0+	0+	001	002	008	026	061	108	149	13
	14	0+	0+	0+	0+	0+	0+	0+	0+	0+	0+	0+	0+	0+	0+	0+	0+	003	011	032	069	117	14

BINOMIAL DISTRIBUTION—INDIVIDUAL TERMS (*Concluded*)

$$P(r) = {}_nC_r\, p^r q^{n-r}$$

n	r	.01	.02	.04	.05	.06	.08	.10	.12	.14	.15	.16	.18	.20	.22	.24	.25	.30	.35	.40	.45	.50	r
24	15	0+	0+	0+	0+	0+	0+	0+	0+	0+	0+	0+	0+	0+	0+	0+	0+	001	004	014	038	078	15
	16	0+	0+	0+	0+	0+	0+	0+	0+	0+	0+	0+	0+	0+	0+	0+	0+	0+	001	005	017	044	16
	17	0+	0+	0+	0+	0+	0+	0+	0+	0+	0+	0+	0+	0+	0+	0+	0+	0+	0+	002	007	021	17
	18	0+	0+	0+	0+	0+	0+	0+	0+	0+	0+	0+	0+	0+	0+	0+	0+	0+	0+	0+	002	008	18
	19	0+	0+	0+	0+	0+	0+	0+	0+	0+	0+	0+	0+	0+	0+	0+	0+	0+	0+	0+	001	003	19
	20	0+	0+	0+	0+	0+	0+	0+	0+	0+	0+	0+	0+	0+	0+	0+	0+	0+	0+	0+	0+	001	20
	21	0+	0+	0+	0+	0+	0+	0+	0+	0+	0+	0+	0+	0+	0+	0+	0+	0+	0+	0+	0+	0+	21
	22	0+	0+	0+	0+	0+	0+	0+	0+	0+	0+	0+	0+	0+	0+	0+	0+	0+	0+	0+	0+	0+	22
	23	0+	0+	0+	0+	0+	0+	0+	0+	0+	0+	0+	0+	0+	0+	0+	0+	0+	0+	0+	0+	0+	23
	24	0+	0+	0+	0+	0+	0+	0+	0+	0+	0+	0+	0+	0+	0+	0+	0+	0+	0+	0+	0+	0+	24
25	0	778	603	360	277	213	124	072	041	023	017	013	007	004	002	001	001	0+	0+	0+	0+	0+	0
	1	196	308	375	365	340	270	199	140	094	076	061	038	024	014	008	006	001	0+	0+	0+	0+	1
	2	024	075	188	231	260	282	266	228	183	161	139	101	071	048	031	025	007	002	0+	0+	0+	2
	3	002	012	060	093	127	188	226	239	229	217	203	170	136	104	076	064	024	008	002	0+	0+	3
	4	0+	001	014	027	045	090	138	179	205	211	213	206	187	161	132	118	057	022	007	002	0+	4
	5	0+	0+	002	006	012	033	065	103	140	156	170	190	196	190	175	165	103	051	020	006	002	5
	6	0+	0+	0+	001	003	010	024	047	076	092	108	139	163	179	184	183	147	091	044	017	005	6
	7	0+	0+	0+	0+	0+	002	007	017	034	044	056	083	111	137	158	165	171	133	080	038	014	7
	8	0+	0+	0+	0+	0+	0+	002	005	012	017	024	041	062	087	112	124	165	161	120	070	032	8
	9	0+	0+	0+	0+	0+	0+	0+	001	004	006	009	017	029	046	067	078	134	163	151	108	061	9
	10	0+	0+	0+	0+	0+	0+	0+	0+	001	002	003	006	012	021	034	042	092	141	161	142	097	10
	11	0+	0+	0+	0+	0+	0+	0+	0+	0+	0+	001	002	004	008	015	019	054	103	147	158	133	11
	12	0+	0+	0+	0+	0+	0+	0+	0+	0+	0+	0+	0+	001	003	005	007	027	065	114	151	155	12
	13	0+	0+	0+	0+	0+	0+	0+	0+	0+	0+	0+	0+	0+	001	002	002	011	035	076	124	155	13
	14	0+	0+	0+	0+	0+	0+	0+	0+	0+	0+	0+	0+	0+	0+	001	001	004	016	043	087	133	14
	15	0+	0+	0+	0+	0+	0+	0+	0+	0+	0+	0+	0+	0+	0+	0+	0+	001	006	021	052	097	15
	16	0+	0+	0+	0+	0+	0+	0+	0+	0+	0+	0+	0+	0+	0+	0+	0+	0+	002	009	027	061	16
	17	0+	0+	0+	0+	0+	0+	0+	0+	0+	0+	0+	0+	0+	0+	0+	0+	0+	001	003	012	032	17
	18	0+	0+	0+	0+	0+	0+	0+	0+	0+	0+	0+	0+	0+	0+	0+	0+	0+	0+	001	004	014	18
	19	0+	0+	0+	0+	0+	0+	0+	0+	0+	0+	0+	0+	0+	0+	0+	0+	0+	0+	0+	001	005	19
	20	0+	0+	0+	0+	0+	0+	0+	0+	0+	0+	0+	0+	0+	0+	0+	0+	0+	0+	0+	0+	002	20
	21	0+	0+	0+	0+	0+	0+	0+	0+	0+	0+	0+	0+	0+	0+	0+	0+	0+	0+	0+	0+	0+	21
	22	0+	0+	0+	0+	0+	0+	0+	0+	0+	0+	0+	0+	0+	0+	0+	0+	0+	0+	0+	0+	0+	22
	23	0+	0+	0+	0+	0+	0+	0+	0+	0+	0+	0+	0+	0+	0+	0+	0+	0+	0+	0+	0+	0+	23
	24	0+	0+	0+	0+	0+	0+	0+	0+	0+	0+	0+	0+	0+	0+	0+	0+	0+	0+	0+	0+	0+	24
	25	0+	0+	0+	0+	0+	0+	0+	0+	0+	0+	0+	0+	0+	0+	0+	0+	0+	0+	0+	0+	0+	25

G. BINOMIAL DISTRIBUTION— CUMULATIVE TERMS

THE TABLE presents the binomial probability for *r or more* successes in n trials for selected values of p, the probability of a success on any one trial.

Examples and details in the use of this table for p greater than .50 are given on pages 133–34.

The symbol $0+$ indicates a value that is positive but less than .0005.

The symbol $1-$ indicates a value that is less than 1 but greater than .9995.

BINOMIAL DISTRIBUTION—CUMULATIVE TERMS

Probability of r or more successes in n trials $= \sum\limits_{r}^{n} {}_nC_r p^r q^{n-r}$

n	r	.01	.02	.04	.05	.06	.08	.10	.12	.14	.15	.16	.18	.20	.22	.24	.25	.30	.35	.40	.45	.50	r
2	0	1	1	1	1	1	1	1	1	1	1	1	1	1	1	1	1	1	1	1	1	1	0
	1	020	040	078	098	116	154	190	226	260	278	294	328	360	392	422	438	510	578	640	698	750	1
	2	0+	0+	002	002	004	006	010	014	020	022	026	032	040	048	058	062	090	122	160	202	250	2
3	0	1	1	1	1	1	1	1	1	1	1	1	1	1	1	1	1	1	1	1	1	1	0
	1	030	059	115	143	169	221	271	319	364	386	407	449	488	525	561	578	657	725	784	834	875	1
	2	0+	001	005	007	010	018	028	040	053	061	069	086	104	124	145	156	216	282	352	425	500	2
	3	0+	0+	0+	0+	0+	001	001	002	003	003	004	006	008	011	014	016	027	043	064	091	125	3
4	0	1	1	1	1	1	1	1	1	1	1	1	1	1	1	1	1	1	1	1	1	1	0
	1	039	078	151	185	219	284	344	400	453	478	502	548	590	630	666	684	760	821	870	908	938	1
	2	001	002	009	014	020	034	052	073	097	110	123	151	181	212	245	262	348	437	525	609	688	2
	3	0+	0+	0+	0+	001	002	004	006	010	012	014	020	027	036	045	051	084	126	179	241	312	3
	4	0+	0+	0+	0+	0+	0+	0+	0+	0+	001	001	001	002	002	003	004	008	015	026	041	062	4
5	0	1	1	1	1	1	1	1	1	1	1	1	1	1	1	1	1	1	1	1	1	1	0
	1	049	096	185	226	266	341	410	472	530	556	582	629	672	711	746	763	832	884	922	950	969	1
	2	001	004	015	023	032	054	081	112	147	165	183	222	263	304	346	367	472	572	663	744	812	2
	3	0+	0+	001	001	002	005	009	014	022	027	032	044	058	074	093	104	163	235	317	407	500	3
	4	0+	0+	0+	0+	0+	0+	0+	001	002	002	002	003	004	007	010	013	016	031	054	087	131	188	4
	5	0+	0+	0+	0+	0+	0+	0+	0+	0+	0+	0+	0+	0+	001	001	001	002	005	010	018	031	5
6	0	1	1	1	1	1	1	1	1	1	1	1	1	1	1	1	1	1	1	1	1	1	0
	1	059	114	217	265	310	394	469	536	595	623	649	696	738	775	807	822	882	925	953	972	984	1
	2	001	006	022	033	046	077	114	156	200	224	247	296	345	394	442	466	580	681	767	836	891	2
	3	0+	0+	001	002	004	009	016	026	039	047	056	076	099	125	154	169	256	353	456	558	656	3
	4	0+	0+	0+	0+	0+	001	001	003	005	006	007	012	017	024	033	038	070	117	179	255	344	4
	5	0+	0+	0+	0+	0+	0+	0+	0+	0+	0+	001	001	002	003	004	005	011	022	041	069	109	5
	6	0+	0+	0+	0+	0+	0+	0+	0+	0+	0+	0+	0+	0+	0+	0+	0+	001	002	004	008	016	6
7	0	1	1	1	1	1	1	1	1	1	1	1	1	1	1	1	1	1	1	1	1	1	0
	1	068	132	249	302	352	442	522	591	652	679	705	751	790	824	854	867	918	951	972	985	992	1
	2	002	008	029	044	062	103	150	201	256	283	311	368	423	478	530	555	671	766	841	898	938	2
	3	0+	0+	002	004	006	014	026	042	062	074	087	115	148	184	223	244	353	468	580	684	773	3
	4	0+	0+	0+	0+	0+	001	003	005	009	012	015	023	033	046	062	071	126	200	290	392	500	4
	5	0+	0+	0+	0+	0+	0+	0+	0+	001	001	002	003	005	007	011	013	029	056	096	153	227	5
	6	0+	0+	0+	0+	0+	0+	0+	0+	0+	0+	0+	0+	0+	001	001	001	004	009	019	036	062	6
	7	0+	0+	0+	0+	0+	0+	0+	0+	0+	0+	0+	0+	0+	0+	0+	0+	001	002	004	008	7
8	0	1	1	1	1	1	1	1	1	1	1	1	1	1	1	1	1	1	1	1	1	1	0
	1	077	149	279	337	390	487	570	640	701	728	752	796	832	863	889	900	942	968	983	992	996	1
	2	003	010	038	057	079	130	187	248	311	343	374	437	497	554	608	633	745	831	894	937	965	2
	3	0+	0+	003	006	010	021	038	061	089	105	123	161	203	249	297	321	448	572	685	780	855	3
	4	0+	0+	0+	0+	001	002	005	010	017	021	027	040	056	076	100	114	194	294	406	523	637	4
	5	0+	0+	0+	0+	0+	0+	0+	001	002	003	004	007	010	016	023	027	058	106	174	260	363	5
	6	0+	0+	0+	0+	0+	0+	0+	0+	0+	0+	001	001	002	003	004	011	025	050	088	145	6
	7	0+	0+	0+	0+	0+	0+	0+	0+	0+	0+	0+	0+	0+	0+	001	001	004	009	018	035	7
	8	0+	0+	0+	0+	0+	0+	0+	0+	0+	0+	0+	0+	0+	0+	0+	0+	001	002	004	8
9	0	1	1	1	1	1	1	1	1	1	1	1	1	1	1	1	1	1	1	1	1	1	0
	1	086	166	307	370	427	528	613	684	743	768	792	832	866	893	915	925	960	979	990	995	998	1
	2	003	013	048	071	098	158	225	295	366	401	435	501	564	622	675	700	804	879	929	961	980	2
	3	0+	001	004	008	014	030	053	083	120	141	163	210	262	316	371	399	537	663	768	850	910	3
	4	0+	0+	0+	001	001	004	008	016	027	034	042	062	086	114	148	166	270	391	517	639	746	4
	5	0+	0+	0+	0+	0+	0+	001	002	004	006	007	012	020	029	042	049	099	172	267	379	500	5
	6	0+	0+	0+	0+	0+	0+	0+	0+	001	001	001	002	003	005	008	010	025	054	099	166	254	6
	7	0+	0+	0+	0+	0+	0+	0+	0+	0+	0+	0+	001	001	001	004	011	025	050	090	7
	8	0+	0+	0+	0+	0+	0+	0+	0+	0+	0+	0+	0+	0+	0+	0+	001	004	009	020	8
	9	0+	0+	0+	0+	0+	0+	0+	0+	0+	0+	0+	0+	0+	0+	0+	0+	0+	001	002	9
10	0	1	1	1	1	1	1	1	1	1	1	1	1	1	1	1	1	1	1	1	1	1	0
	1	096	183	335	401	461	566	651	721	779	803	825	863	893	917	936	944	972	987	994	997	999	1
	2	004	016	058	086	118	188	264	342	418	456	492	561	624	682	733	756	851	914	954	977	989	2
	3	0+	001	006	012	019	040	070	109	155	180	206	263	322	383	444	474	617	738	833	900	945	3
	4	0+	0+	0+	001	002	006	013	024	040	050	061	088	121	159	201	224	350	486	618	734	828	4
	5	0+	0+	0+	0+	0+	001	002	004	007	010	013	021	033	048	067	078	150	249	367	496	623	5
	6	0+	0+	0+	0+	0+	0+	0+	001	001	002	004	006	010	016	020	047	095	166	262	377	6
	7	0+	0+	0+	0+	0+	0+	0+	0+	0+	0+	001	002	003	004	011	026	055	102	172	7
	8	0+	0+	0+	0+	0+	0+	0+	0+	0+	0+	0+	0+	0+	002	005	012	027	055	8
	9	0+	0+	0+	0+	0+	0+	0+	0+	0+	0+	0+	0+	0+	0+	001	002	005	011	9

BINOMIAL DISTRIBUTION—CUMULATIVE TERMS (*Continued*)

Probability of *r* or more successes in *n* trials $= \sum\limits_{r}^{n} {}_{n}C_{r}\,p^{r}q^{n-r}$

Column headings .01 … .50 are values of *p* (the heading at .16 carries the label *p*).

n	r	.01	.02	.04	.05	.06	.08	.10	.12	.14	.15	.16	.18	.20	.22	.24	.25	.30	.35	.40	.45	.50	r
10	10	0+	0+	0+	0+	0+	0+	0+	0+	0+	0+	0+	0+	0+	0+	0+	0+	0+	0+	0+	0+	001	10
11	0	1	1	1	1	1	1	1	1	1	1	1	1	1	1	1	1	1	1	1	1	1	0
	1	105	199	362	431	494	600	686	755	810	833	853	887	914	935	951	958	980	991	996	999	1-	1
	2	005	020	069	102	138	218	303	387	469	508	545	615	678	733	781	803	887	939	970	986	994	2
	3	0+	001	008	015	025	052	090	137	191	221	252	316	383	449	513	545	687	800	881	935	967	3
	4	0+	0+	001	002	003	009	019	034	056	069	085	120	161	208	260	287	430	574	704	809	887	4
	5	0+	0+	0+	0+	0+	001	003	006	012	016	021	033	050	072	099	115	210	332	467	603	726	5
	6	0+	0+	0+	0+	0+	0+	0+	001	002	003	004	007	012	019	028	034	078	149	247	367	500	6
	7	0+	0+	0+	0+	0+	0+	0+	0+	0+	0+	0+	001	002	004	006	008	022	050	099	174	274	7
	8	0+	0+	0+	0+	0+	0+	0+	0+	0+	0+	0+	0+	0+	0+	001	002	004	012	029	061	113	8
	9	0+	0+	0+	0+	0+	0+	0+	0+	0+	0+	0+	0+	0+	0+	0+	0+	001	002	006	015	033	9
	10	0+	0+	0+	0+	0+	0+	0+	0+	0+	0+	0+	0+	0+	0+	0+	0+	0+	0+	001	002	006	10
	11	0+	0+	0+	0+	0+	0+	0+	0+	0+	0+	0+	0+	0+	0+	0+	0+	0+	0+	0+	0+	0+	11
12	0	1	1	1	1	1	1	1	1	1	1	1	1	1	1	1	1	1	1	1	1	1	0
	1	114	215	387	460	524	632	718	784	836	858	877	908	931	949	963	968	986	994	998	999	1-	1
	2	006	023	081	118	160	249	341	431	517	557	595	664	725	778	822	842	915	958	980	992	997	2
	3	0+	002	011	020	032	065	111	167	230	264	299	370	442	511	578	609	747	849	917	958	981	3
	4	0+	0+	001	002	004	012	026	046	075	092	111	155	205	261	320	351	507	653	775	866	927	4
	5	0+	0+	0+	0+	0+	002	004	009	018	024	031	049	073	102	138	158	276	417	562	696	806	5
	6	0+	0+	0+	0+	0+	0+	001	001	003	005	006	012	019	030	045	054	118	213	335	473	613	6
	7	0+	0+	0+	0+	0+	0+	0+	0+	0+	001	001	002	004	007	011	014	039	085	158	261	387	7
	8	0+	0+	0+	0+	0+	0+	0+	0+	0+	0+	0+	0+	001	001	002	003	009	026	057	112	194	8
	9	0+	0+	0+	0+	0+	0+	0+	0+	0+	0+	0+	0+	0+	0+	0+	0+	002	006	015	036	073	9
	10	0+	0+	0+	0+	0+	0+	0+	0+	0+	0+	0+	0+	0+	0+	0+	0+	0+	001	003	008	019	10
	11	0+	0+	0+	0+	0+	0+	0+	0+	0+	0+	0+	0+	0+	0+	0+	0+	0+	0+	0+	001	003	11
	12	0+	0+	0+	0+	0+	0+	0+	0+	0+	0+	0+	0+	0+	0+	0+	0+	0+	0+	0+	0+	0+	12
13	0	1	1	1	1	1	1	1	1	1	1	1	1	1	1	1	1	1	1	1	1	1	0
	1	122	231	412	487	553	662	746	810	859	879	896	924	945	960	972	976	990	996	999	1-	1-	1
	2	007	027	093	135	181	279	379	474	561	602	640	708	766	815	856	873	936	970	987	995	998	2
	3	0+	002	014	025	039	080	134	198	270	308	346	423	498	570	636	667	798	887	942	973	989	3
	4	0+	0+	001	003	006	016	034	061	097	118	141	194	253	316	382	416	579	722	831	907	954	4
	5	0+	0+	0+	0+	001	002	006	014	026	034	044	068	099	137	182	206	346	499	647	772	867	5
	6	0+	0+	0+	0+	0+	0+	001	002	005	008	010	018	030	046	068	080	165	284	426	573	709	6
	7	0+	0+	0+	0+	0+	0+	0+	0+	001	001	002	004	007	012	019	024	062	129	229	356	500	7
	8	0+	0+	0+	0+	0+	0+	0+	0+	0+	0+	0+	001	001	002	004	006	018	046	098	179	291	8
	9	0+	0+	0+	0+	0+	0+	0+	0+	0+	0+	0+	0+	0+	0+	001	001	004	013	032	070	133	9
	10	0+	0+	0+	0+	0+	0+	0+	0+	0+	0+	0+	0+	0+	0+	0+	0+	001	003	008	020	046	10
	11	0+	0+	0+	0+	0+	0+	0+	0+	0+	0+	0+	0+	0+	0+	0+	0+	0+	0+	001	004	011	11
	12	0+	0+	0+	0+	0+	0+	0+	0+	0+	0+	0+	0+	0+	0+	0+	0+	0+	0+	0+	001	002	12
	13	0+	0+	0+	0+	0+	0+	0+	0+	0+	0+	0+	0+	0+	0+	0+	0+	0+	0+	0+	0+	0+	13
14	0	1	1	1	1	1	1	1	1	1	1	1	1	1	1	1	1	1	1	1	1	1	0
	1	131	246	435	512	579	689	771	833	879	897	913	938	956	969	979	982	993	998	999	1-	1-	1
	2	008	031	106	153	204	310	415	514	603	643	681	747	802	847	884	899	953	979	992	997	999	2
	3	0+	002	017	030	048	096	158	232	311	352	393	474	552	624	689	719	839	916	960	983	994	3
	4	0+	0+	002	004	008	021	044	077	121	147	174	235	302	372	443	479	645	779	876	937	971	4
	5	0+	0+	0+	0+	001	004	009	020	036	047	059	091	130	176	230	258	416	577	721	833	910	5
	6	0+	0+	0+	0+	0+	001	001	004	008	012	016	027	044	066	095	112	219	359	514	663	788	6
	7	0+	0+	0+	0+	0+	0+	0+	001	001	002	003	006	012	020	031	038	093	184	308	454	605	7
	8	0+	0+	0+	0+	0+	0+	0+	0+	0+	0+	0+	001	002	005	008	010	031	075	150	259	395	8
	9	0+	0+	0+	0+	0+	0+	0+	0+	0+	0+	0+	0+	0+	001	002	002	008	024	058	119	212	9
	10	0+	0+	0+	0+	0+	0+	0+	0+	0+	0+	0+	0+	0+	0+	0+	0+	002	006	018	043	090	10
	11	0+	0+	0+	0+	0+	0+	0+	0+	0+	0+	0+	0+	0+	0+	0+	0+	0+	001	004	011	029	11
	12	0+	0+	0+	0+	0+	0+	0+	0+	0+	0+	0+	0+	0+	0+	0+	0+	0+	0+	001	002	006	12
	13	0+	0+	0+	0+	0+	0+	0+	0+	0+	0+	0+	0+	0+	0+	0+	0+	0+	0+	0+	0+	001	13
	14	0+	0+	0+	0+	0+	0+	0+	0+	0+	0+	0+	0+	0+	0+	0+	0+	0+	0+	0+	0+	0+	14
15	0	1	1	1	1	1	1	1	1	1	1	1	1	1	1	1	1	1	1	1	1	1	0
	1	140	261	458	537	605	714	794	853	896	913	927	949	965	976	984	987	995	998	1-	1-	1-	1
	2	010	035	119	171	226	340	451	552	642	681	718	781	833	874	906	920	965	986	995	998	1-	2
	3	0+	003	020	036	057	113	184	265	352	396	439	523	602	673	736	764	873	938	973	989	996	3
	4	0+	0+	002	005	010	027	056	096	148	177	209	278	352	427	502	539	703	827	909	958	982	4

BINOMIAL DISTRIBUTION—CUMULATIVE TERMS (*Continued*)

$$\text{Probability of } r \text{ or more successes in } n \text{ trials} = \sum_{r}^{n} {}_nC_r\, p^r q^{n-r}$$

Column headings are values of p.

n	r	.01	.02	.04	.05	.06	.08	.10	.12	.14	.15	.16	.18	.20	.22	.24	.25	.30	.35	.40	.45	.50	r
15	5	0+	0+	0+	001	001	005	013	026	048	062	078	117	164	219	281	314	485	648	783	880	941	5
	6	0+	0+	0+	0+	0+	001	002	006	012	017	023	039	061	090	127	148	278	436	597	739	849	6
	7	0+	0+	0+	0+	0+	0+	0+	001	002	004	005	010	018	030	046	057	131	245	390	548	696	7
	8	0+	0+	0+	0+	0+	0+	0+	0+	0+	001	001	002	004	008	013	017	050	113	213	346	500	8
	9	0+	0+	0+	0+	0+	0+	0+	0+	0+	0+	0+	0+	001	002	003	004	015	042	095	182	304	9
	10	0+	0+	0+	0+	0+	0+	0+	0+	0+	0+	0+	0+	0+	0+	001	001	004	012	034	077	151	10
	11	0+	0+	0+	0+	0+	0+	0+	0+	0+	0+	0+	0+	0+	0+	0+	0+	001	003	009	025	059	11
	12	0+	0+	0+	0+	0+	0+	0+	0+	0+	0+	0+	0+	0+	0+	0+	0+	0+	001	002	006	018	12
	13	0+	0+	0+	0+	0+	0+	0+	0+	0+	0+	0+	0+	0+	0+	0+	0+	0+	0+	0+	001	004	13
	14	0+	0+	0+	0+	0+	0+	0+	0+	0+	0+	0+	0+	0+	0+	0+	0+	0+	0+	0+	0+	0+	14
	15	0+	0+	0+	0+	0+	0+	0+	0+	0+	0+	0+	0+	0+	0+	0+	0+	0+	0+	0+	0+	0+	15
16	0	1	1	1	1	1	1	1	1	1	1	1	1	1	1	1	1	1	1	1	1	1	0
	1	149	276	480	560	628	737	815	871	910	926	939	958	972	981	988	990	997	999	1-	1-	1-	1
	2	011	040	133	189	249	370	485	588	677	716	751	811	859	897	925	937	974	990	997	999	1-	2
	3	001	004	024	043	067	131	211	300	393	439	484	570	648	717	777	803	901	955	982	993	998	3
	4	0+	0+	003	007	013	034	068	116	176	210	246	322	402	481	558	595	754	866	935	972	989	4
	5	0+	0+	0+	001	002	007	017	035	062	079	099	146	202	265	334	370	550	711	833	915	962	5
	6	0+	0+	0+	0+	0+	001	003	008	017	024	032	053	082	119	164	190	340	510	671	802	895	6
	7	0+	0+	0+	0+	0+	0+	001	002	004	006	008	015	027	043	066	080	175	312	473	634	773	7
	8	0+	0+	0+	0+	0+	0+	0+	0+	001	001	002	004	007	013	021	027	074	159	284	437	598	8
	9	0+	0+	0+	0+	0+	0+	0+	0+	0+	0+	0+	001	001	003	006	007	026	067	142	256	402	9
	10	0+	0+	0+	0+	0+	0+	0+	0+	0+	0+	0+	0+	0+	001	001	002	007	023	058	124	227	10
	11	0+	0+	0+	0+	0+	0+	0+	0+	0+	0+	0+	0+	0+	0+	0+	0+	002	006	019	049	105	11
	12	0+	0+	0+	0+	0+	0+	0+	0+	0+	0+	0+	0+	0+	0+	0+	0+	0+	001	005	015	038	12
	13	0+	0+	0+	0+	0+	0+	0+	0+	0+	0+	0+	0+	0+	0+	0+	0+	0+	0+	001	003	011	13
	14	0+	0+	0+	0+	0+	0+	0+	0+	0+	0+	0+	0+	0+	0+	0+	0+	0+	0+	0+	001	002	14
	15	0+	0+	0+	0+	0+	0+	0+	0+	0+	0+	0+	0+	0+	0+	0+	0+	0+	0+	0+	0+	0+	15
	16	0+	0+	0+	0+	0+	0+	0+	0+	0+	0+	0+	0+	0+	0+	0+	0+	0+	0+	0+	0+	0+	16
17	0	1	1	1	1	1	1	1	1	1	1	1	1	1	1	1	1	1	1	1	1	1	0
	1	157	291	500	582	651	758	833	886	923	937	948	966	977	985	991	992	998	999	1-	1-	1-	1
	2	012	045	147	208	272	399	518	622	710	748	781	838	882	915	940	950	981	993	998	999	1-	2
	3	001	004	029	050	078	150	238	335	432	480	527	613	690	758	812	836	923	967	988	996	999	3
	4	0+	0+	004	009	016	042	083	138	207	244	284	367	451	533	611	647	798	897	954	982	994	4
	5	0+	0+	0+	001	003	009	022	045	078	099	122	178	242	313	388	426	611	765	874	940	975	5
	6	0+	0+	0+	0+	0+	001	005	011	023	032	042	069	106	151	205	235	403	580	736	853	928	6
	7	0+	0+	0+	0+	0+	0+	001	002	006	008	012	022	038	060	089	107	225	381	552	710	834	7
	8	0+	0+	0+	0+	0+	0+	0+	0+	001	002	003	006	011	019	032	040	105	213	359	526	685	8
	9	0+	0+	0+	0+	0+	0+	0+	0+	0+	0+	0+	001	003	005	009	012	040	099	199	337	500	9
	10	0+	0+	0+	0+	0+	0+	0+	0+	0+	0+	0+	0+	0+	001	002	003	013	038	092	183	315	10
	11	0+	0+	0+	0+	0+	0+	0+	0+	0+	0+	0+	0+	0+	0+	0+	0+	003	012	035	083	166	11
	12	0+	0+	0+	0+	0+	0+	0+	0+	0+	0+	0+	0+	0+	0+	0+	0+	001	003	011	030	072	12
	13	0+	0+	0+	0+	0+	0+	0+	0+	0+	0+	0+	0+	0+	0+	0+	0+	0+	001	003	009	025	13
	14	0+	0+	0+	0+	0+	0+	0+	0+	0+	0+	0+	0+	0+	0+	0+	0+	0+	0+	0+	002	006	14
	15	0+	0+	0+	0+	0+	0+	0+	0+	0+	0+	0+	0+	0+	0+	0+	0+	0+	0+	0+	0+	001	15
	16	0+	0+	0+	0+	0+	0+	0+	0+	0+	0+	0+	0+	0+	0+	0+	0+	0+	0+	0+	0+	0+	16
	17	0+	0+	0+	0+	0+	0+	0+	0+	0+	0+	0+	0+	0+	0+	0+	0+	0+	0+	0+	0+	0+	17
18	0	1	1	1	1	1	1	1	1	1	1	1	1	1	1	1	1	1	1	1	1	1	0
	1	165	305	520	603	672	777	850	900	934	946	957	972	982	989	993	994	998	1-	1-	1-	1-	1
	2	014	050	161	226	294	428	550	654	740	776	808	861	901	931	952	961	986	995	999	1-	1-	2
	3	001	005	033	058	090	170	266	369	471	520	567	654	729	792	843	865	940	976	992	997	999	3
	4	0+	0+	005	011	020	051	098	162	238	280	323	411	499	582	659	694	835	922	967	988	996	4
	5	0+	0+	001	002	003	012	028	056	096	121	148	212	284	361	441	481	667	811	906	959	985	5
	6	0+	0+	0+	0+	0+	002	006	015	031	042	055	089	133	187	249	283	466	645	791	892	952	6
	7	0+	0+	0+	0+	0+	0+	001	003	008	012	017	031	051	080	117	139	278	451	626	774	881	7
	8	0+	0+	0+	0+	0+	0+	0+	001	002	003	004	009	016	028	046	057	141	272	437	609	760	8
	9	0+	0+	0+	0+	0+	0+	0+	0+	0+	001	001	002	004	008	015	019	060	139	263	422	593	9
	10	0+	0+	0+	0+	0+	0+	0+	0+	0+	0+	0+	0+	001	002	004	005	021	060	135	253	407	10
	11	0+	0+	0+	0+	0+	0+	0+	0+	0+	0+	0+	0+	0+	0+	001	001	006	021	058	128	240	11
	12	0+	0+	0+	0+	0+	0+	0+	0+	0+	0+	0+	0+	0+	0+	0+	0+	001	006	020	054	119	12
	13	0+	0+	0+	0+	0+	0+	0+	0+	0+	0+	0+	0+	0+	0+	0+	0+	0+	001	006	018	048	13
	14	0+	0+	0+	0+	0+	0+	0+	0+	0+	0+	0+	0+	0+	0+	0+	0+	0+	0+	001	005	015	14
	15	0+	0+	0+	0+	0+	0+	0+	0+	0+	0+	0+	0+	0+	0+	0+	0+	0+	0+	0+	001	004	15
	16	0+	0+	0+	0+	0+	0+	0+	0+	0+	0+	0+	0+	0+	0+	0+	0+	0+	0+	0+	0+	001	16
	17	0+	0+	0+	0+	0+	0+	0+	0+	0+	0+	0+	0+	0+	0+	0+	0+	0+	0+	0+	0+	0+	17
	18	0+	0+	0+	0+	0+	0+	0+	0+	0+	0+	0+	0+	0+	0+	0+	0+	0+	0+	0+	0+	0+	18

BINOMIAL DISTRIBUTION—CUMULATIVE TERMS (*Continued*)

Probability of r or more successes in n trials $= \sum_{r}^{n} {}_nC_r p^r q^{n-r}$

n	r	.01	.02	.04	.05	.06	.08	.10	.12	.14	.15	p .16	.18	.20	.22	.24	.25	.30	.35	.40	.45	.50	r
19	0	1	1	1	1	1	1	1	1	1	1	1	1	1	1	1	1	1	1	1	1	1	0
	1	174	319	540	623	691	795	865	912	943	954	964	977	986	991	995	996	999	1-	1-	1-	1-	1
	2	015	055	175	245	317	456	580	683	767	802	832	881	917	943	962	969	990	997	999	1-	1-	2
	3	001	006	038	067	102	191	295	403	509	559	606	691	763	822	869	889	954	983	995	998	1-	3
	4	0+	0+	006	013	024	060	115	187	271	316	362	455	545	628	703	737	867	941	977	992	998	4
	5	0+	0+	001	002	004	015	035	069	116	144	176	248	327	410	494	535	718	850	930	972	990	5
	6	0+	0+	0+	0+	001	003	009	020	040	054	070	111	163	225	295	332	526	703	837	922	968	6
	7	0+	0+	0+	0+	0+	0+	002	005	011	016	023	041	068	103	149	175	334	519	692	827	916	7
	8	0+	0+	0+	0+	0+	0+	0+	001	003	004	006	013	023	040	063	077	182	334	512	683	820	8
	9	0+	0+	0+	0+	0+	0+	0+	0+	001	001	001	003	007	013	022	029	084	185	333	506	676	9
	10	0+	0+	0+	0+	0+	0+	0+	0+	0+	0+	0+	001	002	003	007	009	033	087	186	329	500	10
	11	0+	0+	0+	0+	0+	0+	0+	0+	0+	0+	0+	0+	0+	001	002	002	011	035	088	184	324	11
	12	0+	0+	0+	0+	0+	0+	0+	0+	0+	0+	0+	0+	0+	0+	0+	0+	003	011	035	087	180	12
	13	0+	0+	0+	0+	0+	0+	0+	0+	0+	0+	0+	0+	0+	0+	0+	0+	001	003	012	034	084	13
	14	0+	0+	0+	0+	0+	0+	0+	0+	0+	0+	0+	0+	0+	0+	0+	0+	0+	001	003	011	032	14
	15	0+	0+	0+	0+	0+	0+	0+	0+	0+	0+	0+	0+	0+	0+	0+	0+	0+	0+	001	003	010	15
	16	0+	0+	0+	0+	0+	0+	0+	0+	0+	0+	0+	0+	0+	0+	0+	0+	0+	0+	0+	001	002	16
	17	0+	0+	0+	0+	0+	0+	0+	0+	0+	0+	0+	0+	0+	0+	0+	0+	0+	0+	0+	0+	0+	17
	18	0+	0+	0+	0+	0+	0+	0+	0+	0+	0+	0+	0+	0+	0+	0+	0+	0+	0+	0+	0+	0+	18
	19	0+	0+	0+	0+	0+	0+	0+	0+	0+	0+	0+	0+	0+	0+	0+	0+	0+	0+	0+	0+	0+	19
20	0	1	1	1	1	1	1	1	1	1	1	1	1	1	1	1	1	1	1	1	1	1	0
	1	182	332	558	642	710	811	878	922	951	961	969	981	988	993	996	997	999	1-	1-	1-	1-	1
	2	017	060	190	264	340	483	608	711	792	824	853	898	931	954	970	976	992	998	999	1-	1-	2
	3	001	007	044	075	115	212	323	437	545	595	642	725	794	849	891	909	965	988	996	999	1-	3
	4	0+	001	007	016	029	071	133	213	304	352	401	497	589	671	743	775	893	956	984	995	999	4
	5	0+	0+	001	003	006	018	043	083	137	170	206	285	370	458	544	585	762	882	949	981	994	5
	6	0+	0+	0+	0+	001	004	011	026	051	067	087	136	196	266	343	383	584	755	874	945	979	6
	7	0+	0+	0+	0+	0+	001	002	007	015	022	030	054	087	130	184	214	392	583	750	870	942	7
	8	0+	0+	0+	0+	0+	0+	0+	001	004	006	009	018	032	054	083	102	228	399	584	748	868	8
	9	0+	0+	0+	0+	0+	0+	0+	0+	001	001	002	005	010	019	032	041	113	238	404	586	748	9
	10	0+	0+	0+	0+	0+	0+	0+	0+	0+	0+	0+	001	003	005	010	014	048	122	245	409	588	10
	11	0+	0+	0+	0+	0+	0+	0+	0+	0+	0+	0+	0+	001	001	003	004	017	053	128	249	412	11
	12	0+	0+	0+	0+	0+	0+	0+	0+	0+	0+	0+	0+	0+	0+	001	001	005	020	057	131	252	12
	13	0+	0+	0+	0+	0+	0+	0+	0+	0+	0+	0+	0+	0+	0+	0+	0+	001	006	021	058	132	13
	14	0+	0+	0+	0+	0+	0+	0+	0+	0+	0+	0+	0+	0+	0+	0+	0+	0+	002	006	021	058	14
	15	0+	0+	0+	0+	0+	0+	0+	0+	0+	0+	0+	0+	0+	0+	0+	0+	0+	0+	002	006	021	15
	16	0+	0+	0+	0+	0+	0+	0+	0+	0+	0+	0+	0+	0+	0+	0+	0+	0+	0+	0+	002	006	16
	17	0+	0+	0+	0+	0+	0+	0+	0+	0+	0+	0+	0+	0+	0+	0+	0+	0+	0+	0+	0+	001	17
	18	0+	0+	0+	0+	0+	0+	0+	0+	0+	0+	0+	0+	0+	0+	0+	0+	0+	0+	0+	0+	0+	18
	19	0+	0+	0+	0+	0+	0+	0+	0+	0+	0+	0+	0+	0+	0+	0+	0+	0+	0+	0+	0+	0+	19
	20	0+	0+	0+	0+	0+	0+	0+	0+	0+	0+	0+	0+	0+	0+	0+	0+	0+	0+	0+	0+	0+	20
21	0	1	1	1	1	1	1	1	1	1	1	1	1	1	1	1	1	1	1	1	1	1	0
	1	190	346	576	659	727	826	891	932	958	967	974	985	991	995	997	998	999	1-	1-	1-	1-	1
	2	019	065	204	283	362	509	635	736	814	845	872	913	943	962	976	981	994	999	1-	1-	1-	2
	3	001	008	050	085	128	234	352	470	580	630	676	756	821	872	910	925	973	991	998	999	1-	3
	4	0+	001	009	019	034	082	152	240	338	389	440	538	630	710	779	808	914	967	989	997	999	4
	5	0+	0+	001	003	007	023	052	098	161	197	237	323	414	505	592	633	802	908	963	987	996	5
	6	0+	0+	0+	0+	001	005	014	033	063	083	106	162	231	308	391	433	637	799	904	961	987	6
	7	0+	0+	0+	0+	0+	001	003	009	020	029	039	068	109	160	222	256	449	643	800	904	961	7
	8	0+	0+	0+	0+	0+	0+	001	002	005	008	012	024	043	070	108	130	277	464	650	803	905	8
	9	0+	0+	0+	0+	0+	0+	0+	0+	001	002	003	007	014	026	044	056	148	294	476	659	808	9
	10	0+	0+	0+	0+	0+	0+	0+	0+	0+	0+	001	002	004	008	016	021	068	162	309	488	669	10
	11	0+	0+	0+	0+	0+	0+	0+	0+	0+	0+	0+	0+	001	002	005	006	026	077	174	321	500	11
	12	0+	0+	0+	0+	0+	0+	0+	0+	0+	0+	0+	0+	0+	001	001	002	009	031	085	184	332	12
	13	0+	0+	0+	0+	0+	0+	0+	0+	0+	0+	0+	0+	0+	0+	0+	0+	002	011	035	091	192	13
	14	0+	0+	0+	0+	0+	0+	0+	0+	0+	0+	0+	0+	0+	0+	0+	0+	001	003	012	038	095	14
	15	0+	0+	0+	0+	0+	0+	0+	0+	0+	0+	0+	0+	0+	0+	0+	0+	0+	001	004	013	039	15
	16	0+	0+	0+	0+	0+	0+	0+	0+	0+	0+	0+	0+	0+	0+	0+	0+	0+	0+	001	004	013	16
	17	0+	0+	0+	0+	0+	0+	0+	0+	0+	0+	0+	0+	0+	0+	0+	0+	0+	0+	0+	001	004	17
	18	0+	0+	0+	0+	0+	0+	0+	0+	0+	0+	0+	0+	0+	0+	0+	0+	0+	0+	0+	0+	001	18
	19	0+	0+	0+	0+	0+	0+	0+	0+	0+	0+	0+	0+	0+	0+	0+	0+	0+	0+	0+	0+	0+	19
	20	0+	0+	0+	0+	0+	0+	0+	0+	0+	0+	0+	0+	0+	0+	0+	0+	0+	0+	0+	0+	0+	20
	21	0+	0+	0+	0+	0+	0+	0+	0+	0+	0+	0+	0+	0+	0+	0+	0+	0+	0+	0+	0+	0+	21

BINOMIAL DISTRIBUTION—CUMULATIVE TERMS (Continued)

Probability of r or more successes in n trials $= \sum_r^n {}_nC_r\, p^r q^{n-r}$

Column heading row labelled p; the P marker appears above the .16 column.

n	r	.01	.02	.04	.05	.06	.08	.10	.12	.14	.15	.16	.18	.20	.22	.24	.25	.30	.35	.40	.45	.50	r
22	0	1	1	1	1	1	1	1	1	1	1	1	1	1	1	1	1	1-	1-	1-	1-	1-	0
	1	198	359	593	676	744	840	902	940	964	972	978	987	993	996	998	998	1-	1-	1-	1-	1-	1
	2	020	071	219	302	384	535	760	834	863	888	926	952	970	981	985	996	999	1-	1-	1-	1-	2
	3	001	009	056	095	142	256	380	502	612	662	707	785	846	892	926	939	979	994	998	1-	1-	3
	4	0+	001	011	022	040	094	172	267	372	425	477	578	668	746	810	838	932	975	992	998	1-	4
	5	0+	0+	002	004	009	027	062	115	186	226	270	362	457	550	637	677	835	928	973	992	998	5
	6	0+	0+	0+	001	002	006	018	041	077	100	127	191	267	351	439	483	687	837	928	973	992	6
	7	0+	0+	0+	0+	0+	001	004	012	026	037	050	085	133	193	263	301	506	698	842	929	974	7
	8	0+	0+	0+	0+	0+	0+	001	003	008	011	017	032	056	090	135	162	329	526	710	848	933	8
	9	0+	0+	0+	0+	0+	0+	0+	001	002	003	005	010	020	036	060	075	186	353	546	724	857	9
	10	0+	0+	0+	0+	0+	0+	0+	0+	0+	001	001	003	006	012	022	030	092	208	376	565	738	10
	11	0+	0+	0+	0+	0+	0+	0+	0+	0+	0+	0+	001	002	004	007	010	039	107	228	396	584	11
	12	0+	0+	0+	0+	0+	0+	0+	0+	0+	0+	0+	0+	0+	001	002	003	014	047	121	246	416	12
	13	0+	0+	0+	0+	0+	0+	0+	0+	0+	0+	0+	0+	0+	0+	0+	001	004	018	055	133	262	13
	14	0+	0+	0+	0+	0+	0+	0+	0+	0+	0+	0+	0+	0+	0+	0+	0+	001	006	021	062	143	14
	15	0+	0+	0+	0+	0+	0+	0+	0+	0+	0+	0+	0+	0+	0+	0+	0+	002	007	024	067		15
	16	0+	0+	0+	0+	0+	0+	0+	0+	0+	0+	0+	0+	0+	0+	0+	0+	0+	002	008	026		16
	17	0+	0+	0+	0+	0+	0+	0+	0+	0+	0+	0+	0+	0+	0+	0+	0+	0+	0+	002	008		17
	18	0+	0+	0+	0+	0+	0+	0+	0+	0+	0+	0+	0+	0+	0+	0+	0+	0+	0+	0+	002		18
	19	0+	0+	0+	0+	0+	0+	0+	0+	0+	0+	0+	0+	0+	0+	0+	0+	0+	0+	0+	0+		19
	20	0+	0+	0+	0+	0+	0+	0+	0+	0+	0+	0+	0+	0+	0+	0+	0+	0+	0+	0+	0+		20
	21	0+	0+	0+	0+	0+	0+	0+	0+	0+	0+	0+	0+	0+	0+	0+	0+	0+	0+	0+	0+		21
	22	0+	0+	0+	0+	0+	0+	0+	0+	0+	0+	0+	0+	0+	0+	0+	0+	0+	0+	0+	0+		22
23	0	1	1	1	1	1	1	1	1	1	1	1	1	1	1	1	1	1-	1-	1-	1-	1-	0
	1	206	372	609	693	759	853	911	947	969	976	982	990	994	997	998	999	1-	1-	1-	1-	1-	1
	2	022	077	234	321	405	559	685	781	852	880	902	937	960	975	985	988	997	999	1-	1-	1-	2
	3	002	011	062	105	157	278	408	533	643	692	736	810	867	909	939	951	984	996	999	1-	1-	3
	4	0+	001	012	026	046	107	193	295	405	460	514	615	703	778	838	863	946	982	995	999	1-	4
	5	0+	0+	002	005	011	033	073	133	212	256	303	401	499	593	678	717	864	945	981	995	999	5
	6	0+	0+	0+	001	002	008	023	050	092	119	150	222	305	395	487	532	731	869	946	981	995	6
	7	0+	0+	0+	0+	0+	002	006	015	033	046	062	104	160	227	305	346	560	747	876	949	983	7
	8	0+	0+	0+	0+	0+	0+	001	004	010	015	022	042	072	113	166	196	382	586	763	885	953	8
	9	0+	0+	0+	0+	0+	0+	0+	001	003	004	007	014	027	048	078	096	229	444	612	780	895	9
	10	0+	0+	0+	0+	0+	0+	0+	0+	001	001	002	004	009	017	031	041	120	259	444	636	798	10
	11	0+	0+	0+	0+	0+	0+	0+	0+	0+	0+	0+	001	003	005	011	015	055	142	287	472	661	11
	12	0+	0+	0+	0+	0+	0+	0+	0+	0+	0+	0+	0+	0+	001	003	005	021	068	164	313	500	12
	13	0+	0+	0+	0+	0+	0+	0+	0+	0+	0+	0+	0+	0+	0+	0+	001	005	028	081	184	339	13
	14	0+	0+	0+	0+	0+	0+	0+	0+	0+	0+	0+	0+	0+	0+	0+	0+	002	010	035	094	202	14
	15	0+	0+	0+	0+	0+	0+	0+	0+	0+	0+	0+	0+	0+	0+	0+	0+	001	003	013	041	105	15
	16	0+	0+	0+	0+	0+	0+	0+	0+	0+	0+	0+	0+	0+	0+	0+	0+	0+	001	004	015	047	16
	17	0+	0+	0+	0+	0+	0+	0+	0+	0+	0+	0+	0+	0+	0+	0+	0+	0+	0+	001	005	017	17
	18	0+	0+	0+	0+	0+	0+	0+	0+	0+	0+	0+	0+	0+	0+	0+	0+	0+	0+	0+	001	005	18
	19	0+	0+	0+	0+	0+	0+	0+	0+	0+	0+	0+	0+	0+	0+	0+	0+	0+	0+	0+	0+	001	19
	20	0+	0+	0+	0+	0+	0+	0+	0+	0+	0+	0+	0+	0+	0+	0+	0+	0+	0+	0+	0+	0+	20
	21	0+	0+	0+	0+	0+	0+	0+	0+	0+	0+	0+	0+	0+	0+	0+	0+	0+	0+	0+	0+	0+	21
	22	0+	0+	0+	0+	0+	0+	0+	0+	0+	0+	0+	0+	0+	0+	0+	0+	0+	0+	0+	0+	0+	22
	23	0+	0+	0+	0+	0+	0+	0+	0+	0+	0+	0+	0+	0+	0+	0+	0+	0+	0+	0+	0+	0+	23
24	0	1	1	1	1	1	1	1	1	1	1	1	1	1	1	1	1	1-	1-	1-	1-	1-	0
	1	214	384	625	708	773	865	920	953	973	980	985	991	995	997	999	999	1-	1-	1-	1-	1-	1
	2	024	083	249	339	427	583	708	801	869	894	915	946	967	980	988	991	998	1-	1-	1-	1-	2
	3	002	012	069	116	172	301	436	563	673	720	763	833	885	924	950	960	988	997	999	1-	1-	3
	4	0+	001	014	030	053	121	214	324	439	495	550	650	736	807	862	885	958	987	996	999	1-	4
	5	0+	0+	002	006	013	039	085	153	239	287	337	439	540	634	717	753	889	958	987	996	999	5
	6	0+	0+	0+	001	002	010	028	060	109	139	174	254	344	439	533	578	771	896	960	987	997	6
	7	0+	0+	0+	0+	0+	002	007	019	041	057	076	126	189	264	349	393	611	789	904	964	989	7
	8	0+	0+	0+	0+	0+	0+	002	005	013	020	028	053	089	138	199	234	435	642	808	914	968	8
	9	0+	0+	0+	0+	0+	0+	0+	001	004	006	009	019	036	062	099	121	275	474	672	827	924	9
	10	0+	0+	0+	0+	0+	0+	0+	0+	001	002	002	006	013	024	042	055	153	313	511	701	846	10
	11	0+	0+	0+	0+	0+	0+	0+	0+	0+	0+	001	002	004	008	016	021	074	183	350	546	729	11
	12	0+	0+	0+	0+	0+	0+	0+	0+	0+	0+	0+	0+	001	002	005	007	031	094	213	385	581	12
	13	0+	0+	0+	0+	0+	0+	0+	0+	0+	0+	0+	0+	0+	001	001	002	012	042	114	242	419	13
	14	0+	0+	0+	0+	0+	0+	0+	0+	0+	0+	0+	0+	0+	0+	0+	001	004	016	053	134	271	14

BINOMIAL DISTRIBUTION—CUMULATIVE TERMS (*Concluded*)

Probability of r or more successes in n trials $= \sum_{r}^{n} {}_nC_r p^r q^{n-r}$

n	r	.01	.02	.04	.05	.06	.08	.10	.12	.14	.15	P .16	.18	.20	.22	.24	.25	.30	.35	.40	.45	.50	r
24	15	0+	0+	0+	0+	0+	0+	0+	0+	0+	0+	0+	0+	0+	0+	0+	0+	001	005	022	065	154	15
	16	0+	0+	0+	0+	0+	0+	0+	0+	0+	0+	0+	0+	0+	0+	0+	0+	0+	002	008	027	076	16
	17	0+	0+	0+	0+	0+	0+	0+	0+	0+	0+	0+	0+	0+	0+	0+	0+	0+	002	010	032		17
	18	0+	0+	0+	0+	0+	0+	0+	0+	0+	0+	0+	0+	0+	0+	0+	0+	0+	0+	001	003	011	18
	19	0+	0+	0+	0+	0+	0+	0+	0+	0+	0+	0+	0+	0+	0+	0+	0+	0+	0+	0+	001	003	19
	20	0+	0+	0+	0+	0+	0+	0+	0+	0+	0+	0+	0+	0+	0+	0+	0+	0+	0+	0+	0+	001	20
	21	0+	0+	0+	0+	0+	0+	0+	0+	0+	0+	0+	0+	0+	0+	0+	0+	0+	0+	0+	0+	0+	21
	22	0+	0+	0+	0+	0+	0+	0+	0+	0+	0+	0+	0+	0+	0+	0+	0+	0+	0+	0+	0+	0+	22
	23	0+	0+	0+	0+	0+	0+	0+	0+	0+	0+	0+	0+	0+	0+	0+	0+	0+	0+	0+	0+	0+	23
	24	0+	0+	0+	0+	0+	0+	0+	0+	0+	0+	0+	0+	0+	0+	0+	0+	0+	0+	0+	0+	0+	24
25	0	1	1	1	1	1	1	1	1	1	1	1	1	1	1	1	1	1	1	1	1	1	0
	1	222	397	640	723	787	876	928	959	977	983	987	993	996	998	999	999	1-	1-	1-	1-	1-	1
	2	026	089	264	358	447	605	729	820	883	907	926	955	973	984	991	993	998	1-	1-	1-	1-	2
	3	002	013	076	127	187	323	463	591	700	746	787	853	902	936	959	968	991	998	1-	1-	1-	3
	4	0+	001	017	034	060	135	236	352	471	529	584	683	766	832	883	904	967	990	998	1-	1-	4
	5	0+	0+	003	007	015	045	098	173	267	318	371	477	579	672	752	786	910	968	991	998	1-	5
	6	0+	0+	0+	001	003	012	033	071	127	162	200	288	383	482	577	622	807	917	971	991	998	6
	7	0+	0+	0+	0+	001	003	009	024	051	070	092	149	220	303	393	439	659	827	926	974	993	7
	8	0+	0+	0+	0+	0+	001	002	007	017	025	036	066	109	166	235	273	488	694	846	936	978	8
	9	0+	0+	0+	0+	0+	0+	0+	002	005	008	012	025	047	079	123	149	323	533	726	866	946	9
	10	0+	0+	0+	0+	0+	0+	0+	0+	001	002	003	008	017	033	056	071	189	370	575	758	885	10
	11	0+	0+	0+	0+	0+	0+	0+	0+	0+	0+	001	002	006	012	022	030	098	229	414	616	788	11
	12	0+	0+	0+	0+	0+	0+	0+	0+	0+	0+	0+	001	002	004	008	011	044	125	268	457	655	12
	13	0+	0+	0+	0+	0+	0+	0+	0+	0+	0+	0+	0+	0+	001	002	003	017	060	154	306	500	13
	14	0+	0+	0+	0+	0+	0+	0+	0+	0+	0+	0+	0+	0+	0+	001	001	006	025	078	183	345	14
	15	0+	0+	0+	0+	0+	0+	0+	0+	0+	0+	0+	0+	0+	0+	0+	0+	002	009	034	096	212	15
	16	0+	0+	0+	0+	0+	0+	0+	0+	0+	0+	0+	0+	0+	0+	0+	0+	0+	003	013	044	115	16
	17	0+	0+	0+	0+	0+	0+	0+	0+	0+	0+	0+	0+	0+	0+	0+	0+	0+	001	004	017	054	17
	18	0+	0+	0+	0+	0+	0+	0+	0+	0+	0+	0+	0+	0+	0+	0+	0+	0+	0+	001	006	022	18
	19	0+	0+	0+	0+	0+	0+	0+	0+	0+	0+	0+	0+	0+	0+	0+	0+	0+	0+	0+	002	007	19
	20	0+	0+	0+	0+	0+	0+	0+	0+	0+	0+	0+	0+	0+	0+	0+	0+	0+	0+	0+	0+	002	20
	21	0+	0+	0+	0+	0+	0+	0+	0+	0+	0+	0+	0+	0+	0+	0+	0+	0+	0+	0+	0+	0+	21
	22	0+	0+	0+	0+	0+	0+	0+	0+	0+	0+	0+	0+	0+	0+	0+	0+	0+	0+	0+	0+	0+	22
	23	0+	0+	0+	0+	0+	0+	0+	0+	0+	0+	0+	0+	0+	0+	0+	0+	0+	0+	0+	0+	0+	23
	24	0+	0+	0+	0+	0+	0+	0+	0+	0+	0+	0+	0+	0+	0+	0+	0+	0+	0+	0+	0+	0+	24
	25	0+	0+	0+	0+	0+	0+	0+	0+	0+	0+	0+	0+	0+	0+	0+	0+	0+	0+	0+	0+	0+	25

H. POISSON DISTRIBUTION— INDIVIDUAL TERMS

THE TABLE presents individual Poisson probabilities for the number of occurrences X per unit of measurement for selected values of m, the mean number of occurrences per unit of measurement.

A blank space is left for values less than .0005.

$$f(x) = \frac{e^{-m}m^x}{x!}$$

x	.001	.002	.003	.004	.005	.006	.007	.008	.009	.01	.02	.03	.04	.05	.06	.07	.08	.09	.10	.15	x
0	999	998	997	996	995	994	993	992	991	990	980	970	961	951	942	932	923	914	905	861	0
1	001	002	003	004	005	006	007	008	009	010	020	030	038	048	057	065	074	082	090	129	1
2													001	001	002	002	003	004	005	010	2

m

x	.20	.25	.30	.40	.50	.60	.70	.80	.90	1.0	1.1	1.2	1.3	1.4	1.5	1.6	1.7	1.8	1.9	2.0	x
0	819	779	741	670	607	549	497	449	407	368	333	301	273	247	223	202	183	165	150	135	0
1	164	195	222	268	303	329	348	359	366	368	366	361	354	345	335	323	311	298	284	271	1
2	016	024	033	054	076	099	122	144	165	184	201	217	230	242	251	258	264	268	270	271	2
3	001	002	003	007	013	020	028	038	049	061	074	087	100	113	126	138	150	161	171	180	3
4				001	002	003	005	008	011	015	020	026	032	039	047	055	063	072	081	090	4
5							001	001	002	003	004	006	008	011	014	018	022	026	031	036	5
6										001	001	001	002	003	004	005	006	008	010	012	6
7														001	001	001	001	002	003	003	7
8																			001	001	8

m

x	2.1	2.2	2.3	2.4	2.5	2.6	2.7	2.8	2.9	3.0	3.1	3.2	3.3	3.4	3.5	3.6	3.7	3.8	3.9	4.0	x
0	122	111	100	091	082	074	067	061	055	050	045	041	037	033	030	027	025	022	020	018	0
1	257	244	231	218	205	193	181	170	160	149	140	130	122	113	106	098	091	085	079	073	1
2	270	268	265	261	257	251	245	238	231	224	216	209	201	193	185	177	169	162	154	147	2
3	189	197	203	209	214	218	220	222	224	224	224	223	221	219	216	212	209	205	200	195	3
4	099	108	117	125	134	141	149	156	162	168	173	178	182	186	189	191	193	194	195	195	4
5	042	048	054	060	067	074	080	087	094	101	107	114	120	126	132	138	143	148	152	156	5
6	015	017	021	024	028	032	036	041	045	050	056	061	066	072	077	083	088	094	099	104	6
7	004	005	007	008	010	012	014	016	019	022	025	028	031	035	039	042	047	051	055	060	7
8	001	002	002	003	003	004	005	006	007	008	010	011	013	015	017	019	022	024	027	030	8
9				001	001	001	001	002	002	003	003	004	005	006	007	008	009	010	012	013	9
10									001	001	001	001	002	002	002	003	003	004	005	005	10
11														001	001	001	001	001	002	002	11
12																			001	001	12

POISSON DISTRIBUTION—INDIVIDUAL TERMS (*Continued*)

$$f(x) = \frac{e^{-m}m^x}{x!}$$

m

x	4.1	4.2	4.3	4.4	4.5	4.6	4.7	4.8	4.9	5.0	5.1	5.2	5.3	5.4	5.5	5.6	5.7	5.8	5.9	6.0	x
0	017	015	014	012	011	010	009	008	007	007	006	006	005	005	004	004	003	003	003	002	0
1	068	063	058	054	050	046	043	040	036	034	031	029	026	024	022	021	019	018	016	015	1
2	139	132	125	119	112	106	100	095	089	084	079	075	070	066	062	058	054	051	048	045	2
3	190	185	180	174	169	163	157	152	146	140	135	129	124	119	113	108	103	098	094	089	3
4	195	194	193	192	190	188	185	182	179	175	172	168	164	160	156	152	147	143	138	134	4
5	160	163	166	169	171	173	174	175	175	175	175	175	174	173	171	170	168	166	163	161	5
6	109	114	119	124	128	132	136	140	143	146	149	151	154	156	157	158	159	160	160	161	6
7	064	069	073	078	082	087	091	096	100	104	109	113	116	120	123	127	130	133	135	138	7
8	033	036	039	043	046	050	054	058	061	065	069	073	077	081	085	089	092	096	100	103	8
9	015	017	019	021	023	026	028	031	033	036	039	042	045	049	052	055	059	062	065	069	9
10	006	007	008	009	010	012	013	015	016	018	020	022	024	026	029	031	033	036	039	041	10
11	002	003	003	004	004	005	006	006	007	008	009	010	012	013	014	016	017	019	021	023	11
12	001	001	001	001	002	002	002	003	003	003	004	005	005	006	007	007	008	009	010	011	12
13					001	001	001	001	001	001	002	002	002	002	003	003	004	004	005	005	13
14											001	001	001	001	001	001	001	002	002	002	14
15																001	001	001	001	001	15

m

x	6.1	6.2	6.3	6.4	6.5	6.6	6.7	6.8	6.9	7.0	7.1	7.2	7.3	7.4	7.5	8.0	8.5	9.0	9.5	10.0	x
0	002	002	002	002	002	001	001	001	001	001	001	001	001	001	001						0
1	014	013	012	011	010	009	008	008	007	006	006	005	005	005	004	003	002	001	001		1
2	042	039	036	034	032	030	028	026	024	022	021	019	018	017	016	011	007	005	003	002	2
3	085	081	077	073	069	065	062	058	055	052	049	046	044	041	039	029	021	015	011	008	3
4	129	125	121	116	112	108	103	099	095	091	087	084	080	076	073	057	044	034	025	019	4
5	158	155	152	149	145	142	138	135	131	128	124	120	117	113	109	092	075	061	048	038	5
6	160	160	159	159	157	156	155	153	151	149	147	144	142	139	137	122	107	091	076	063	6
7	140	142	144	145	146	147	148	149	149	149	149	149	148	147	146	140	129	117	104	090	7
8	107	110	113	116	119	121	124	126	128	130	132	134	135	136	137	140	138	132	123	113	8
9	072	076	079	082	086	089	092	095	098	101	104	107	110	112	114	124	130	132	130	125	9
10	044	047	050	053	056	059	062	065	068	071	074	077	080	083	086	099	110	119	124	125	10
11	024	026	029	031	033	035	038	040	043	045	048	050	053	056	059	072	085	097	107	114	11
12	012	014	015	016	018	019	021	023	025	026	028	030	032	034	037	048	060	073	084	095	12
13	006	007	007	008	009	010	011	012	013	014	015	017	018	020	021	030	040	050	062	073	13
14	003	003	003	004	004	005	005	006	006	007	008	009	009	010	011	017	024	032	042	052	14
15	001	001	001	002	002	002	002	003	003	003	004	004	005	005	006	009	014	019	027	035	15
16			001	001	001	001	001	001	001	001	002	002	002	002	003	005	007	011	016	022	16
17									001	001	001	001	001	001	001	002	004	006	009	013	17
18																001	002	003	005	007	18
19																	001	001	002	004	19
20																		001	001	002	20
21																				001	21

I. POISSON DISTRIBUTION— CUMULATIVE TERMS

THE TABLE presents the Poisson probabilities of X *or more* occurrences per unit of measurement for selected values of *m*, the mean number of occurrences per unit of measurement.

The symbol 1 — indicates a value less than 1 but greater than .9995. A blank space is left for values less than .0005.

$$\sum_{x}^{\infty} \frac{e^{-m}m^x}{x!}$$

x	.001	.002	.003	.004	.005	.006	.007	.008	.009	.01	.02	.03	.04	.05	.06	.07	.08	.09	.10	.15	x
0	1	1	1	1	1	1	1	1	1	1	1	1	1	1	1	1	1	1	1	1	0
1	001	002	003	004	005	006	007	008	009	010	020	030	039	049	058	068	077	086	095	139	1
2													001	001	002	002	003	004	005	010	2
3																				001	3

x	.20	.25	.30	.40	.50	.60	.70	.80	.90	1.0	1.1	1.2	1.3	1.4	1.5	1.6	1.7	1.8	1.9	2.0	x
0	1	1	1	1	1	1	1	1	1	1	1	1	1	1	1	1	1	1	1	1	0
1	181	221	259	330	393	451	503	551	593	632	667	699	727	753	777	798	817	835	850	865	1
2	018	026	037	062	090	122	156	191	228	264	301	337	373	408	442	475	507	537	566	594	2
3	001	002	004	008	014	023	034	047	063	080	100	121	143	167	191	217	243	269	296	323	3
4				001	002	003	006	009	013	019	026	034	043	054	066	079	093	109	125	143	4
5							001	001	002	004	005	008	011	014	019	024	030	036	044	053	5
6										001	001	002	002	003	004	006	008	010	013	017	6
7														001	001	001	002	003	003	005	7
8																		001	001	001	8

x	2.1	2.2	2.3	2.4	2.5	2.6	2.7	2.8	2.9	3.0	3.1	3.2	3.3	3.4	3.5	3.6	3.7	3.8	3.9	4.0	x
0	1	1	1	1	1	1	1	1	1	1	1	1	1	1	1	1	1	1	1	1	0
1	878	889	900	909	918	926	933	939	945	950	955	959	963	967	970	973	975	978	980	982	1
2	620	645	669	692	713	733	751	769	785	801	815	829	841	853	864	874	884	893	901	908	2
3	350	377	404	430	456	482	506	531	554	577	599	620	641	660	679	697	715	731	747	762	3
4	161	181	201	221	242	264	286	308	330	353	375	397	420	442	463	485	506	527	547	567	4
5	062	072	084	096	109	123	137	152	168	185	202	219	237	256	275	294	313	332	352	371	5
6	020	025	030	036	042	049	057	065	074	084	094	105	117	129	142	156	170	184	199	215	6
7	006	007	009	012	014	017	021	024	029	034	039	045	051	058	065	073	082	091	101	111	7
8	001	002	003	003	004	005	007	008	010	012	014	017	020	023	027	031	035	040	045	051	8
9					001	001	002	002	003	004	005	006	007	008	010	012	014	016	019	021	9
10									001	001	001	001	001	002	003	003	004	005	007	008	10
11														001	001	001	001	002	002	003	11
12																		001	001	001	12

POISSON DISTRIBUTION—CUMULATIVE TERMS (*Continued*)

$$\sum_{x}^{\infty} \frac{e^{-m}m^{x}}{x!}$$

x	4.1	4.2	4.3	4.4	4.5	4.6	4.7	4.8	4.9	5.0	5.1	5.2	5.3	5.4	5.5	5.6	5.7	5.8	5.9	6.0	x
0	1	1	1	1	1	1	1	1	1	1	1	1	1	1	1	1	1	1	1	1	0
1	983	985	986	988	989	990	991	992	993	993	994	994	995	995	996	997	997	997	997	998	1
2	915	922	928	934	939	944	948	952	956	960	963	966	969	971	973	976	978	979	981	983	2
3	776	790	803	815	826	837	848	857	867	875	884	891	898	905	912	918	923	928	933	938	3
4	586	605	623	641	658	674	690	706	721	735	749	762	775	787	798	809	820	830	840	849	4
5	391	410	430	449	468	487	505	524	542	560	577	594	610	627	642	658	673	687	701	715	5
6	231	247	263	280	297	314	332	349	366	384	402	419	437	454	471	488	505	522	538	554	6
7	121	133	144	156	169	182	195	209	223	238	253	268	283	298	314	330	346	362	378	394	7
8	057	064	071	079	087	095	104	113	123	133	144	155	167	178	191	203	216	229	242	256	8
9	024	028	032	036	040	045	050	056	062	068	075	082	089	097	106	114	123	133	143	153	9
10	010	011	013	015	017	020	022	025	028	032	036	040	044	049	054	059	065	071	077	084	10
11	003	004	005	006	007	008	009	010	012	014	016	018	020	023	025	028	031	035	039	042	11
12	001	001	002	002	002	003	003	004	005	005	006	007	008	010	011	012	014	016	018	020	12
13			001	001	001	001	001	001	002	002	002	003	003	004	004	005	006	007	008	009	13
14									001	001	001	001	001	001	002	002	002	003	003	004	14
15															001	001	001	001	001	001	15
16																				001	16

x	6.1	6.2	6.3	6.4	6.5	6.6	6.7	6.8	6.9	7.0	7.1	7.2	7.3	7.4	7.5	8.0	8.5	9.0	9.5	10.0	x
0	1	1	1	1	1	1	1	1	1	1	1	1	1	1	1	1	1	1	1	1	0
1	998	998	998	998	998	999	999	999	999	999	999	999	999	999	999	1-	1-	1-	1-	1-	1
2	984	985	987	988	989	990	991	991	992	993	993	994	994	995	995	997	998	999	999	1-	2
3	942	946	950	954	957	960	963	966	968	970	973	975	978	980	983	986	991	994	996	997	3
4	857	866	874	881	888	895	901	907	913	918	923	928	933	937	941	958	970	979	985	990	4
5	728	741	753	765	776	787	798	808	818	827	836	844	853	860	868	900	926	945	960	971	5
6	570	586	601	616	631	645	659	673	686	699	712	724	736	747	759	809	850	884	911	933	6
7	410	426	442	458	473	489	505	520	535	550	565	580	594	608	622	687	744	793	835	870	7
8	270	284	298	313	327	342	357	372	386	401	416	431	446	461	475	547	614	676	731	780	8
9	163	174	185	197	208	220	233	245	258	271	284	297	311	324	338	407	477	544	608	667	9
10	091	098	106	114	123	131	140	150	159	170	180	190	201	212	224	283	347	413	478	542	10
11	047	051	056	061	067	073	079	085	092	099	106	113	121	129	138	184	237	294	355	417	11
12	022	025	028	031	034	037	041	045	049	053	058	063	068	074	079	112	151	197	248	303	12
13	010	011	013	014	016	018	020	022	024	027	030	033	036	039	043	064	091	124	164	208	13
14	004	005	005	006	007	008	009	010	011	013	014	016	018	020	022	034	051	074	102	136	14
15	002	002	002	003	003	003	004	004	005	006	006	007	008	009	010	017	027	041	060	083	15
16	001	001	001	001	001	001	002	002	002	002	003	003	004	004	005	008	014	022	033	049	16
17							001	001	001	001	001	001	001	002	002	004	007	011	018	027	17
18												001	001	001	001	002	003	005	009	014	18
19																001	001	002	004	007	19
20																	001	001	002	003	20
21																			001	002	21
22																				001	22

J. VALUES OF e^{-X}

THIS TABLE lists values of e^{-x} for values of X from 0 to 10. Intermediate values can be calculated by making use of the relationship $e^{-(a+b)} = e^{-a} \cdot e^{-b}$. For example, to find $e^{-1.21}$, use $e^{-1.0} = .368$ and $e^{-.21} = .811$; then $e^{-1.21} = (.386)(.811) = .298$.

X	e^{-X}	X	e^{-X}	X	e^{-X}	X	e^{-X}
.00	1.000	.40	.670	.80	.449	3.00	.04979
.01	.990	.41	.664	.81	.445	3.10	.04505
.02	.980	.42	.657	.82	.440	3.20	.04076
.03	.970	.43	.651	.83	.436	3.30	.03688
.04	.961	.44	.644	.84	.432	3.40	.03337
.05	.951	.45	.638	.85	.427	3.50	.03020
.06	.942	.46	.631	.86	.423	3.60	.02732
.07	.932	.47	.625	.87	.419	3.70	.02472
.08	.923	.48	.619	.88	.415	3.80	.02237
.09	.914	.49	.613	.89	.411	3.90	.02024
.10	.905	.50	.607	.90	.407	4.00	.01832
.11	.896	.51	.600	.91	.403	4.10	.01657
.12	.887	.52	.595	.92	.399	4.20	.01500
.13	.878	.53	.589	.93	.395	4.30	.01357
.14	.869	.54	.583	.94	.391	4.40	.01228
.15	.861	.55	.577	.95	.387	4.50	.01111
.16	.852	.56	.571	.96	.383	4.60	.01005
.17	.844	.57	.566	.97	.379	4.70	.00910
.18	.835	.58	.560	.98	.375	4.80	.00823
.19	.827	.59	.554	.99	.372	4.90	.00745
.20	.819	.60	.549	1.00	.368	5.00	.00674
.21	.811	.61	.543	1.10	.333	5.50	.00409
.22	.803	.62	.538	1.20	.301	6.00	.00248
.23	.795	.63	.533	1.30	.273	6.50	.00150
.24	.787	.64	.527	1.40	.247	7.00	.00091
.25	.779	.65	.522	1.50	.223	7.50	.00055
.26	.771	.66	.517	1.60	.202	8.00	.00034
.27	.763	.67	.512	1.70	.183	8.50	.00020
.28	.756	.68	.507	1.80	.165	9.00	.00012
.29	.748	.69	.502	1.90	.150	10.00	.00005
.30	.741	.70	.497	2.00	.135		
.31	.733	.71	.492	2.10	.122		
.32	.726	.72	.487	2.20	.111		
.33	.719	.73	.482	2.30	.100		
.34	.712	.74	.477	2.40	.091		
.35	.705	.75	.472	2.50	.082		
.36	.698	.76	.468	2.60	.074		
.37	.691	.77	.463	2.70	.067		
.38	.684	.78	.458	2.80	.061		
.39	.677	.79	.454	2.90	.055		

K. SUMS OF SQUARES AND FOURTH POWERS USED IN TREND FITTING

THIS TABLE gives the value of Σx^2 and Σx^4 needed to find the constants in secular trend equations fitted by least squares, where the x origin is centered at the midpoint in time. Use the left half of the table for an odd number of years, where the x unit is one year. Use the right half of the table for an even number of years, where the x unit is six months, and the years are numbered 1, 3, 5, $\cdots$ and -1, -3, -5 $\cdots$ from the origin. The sum includes the powers of negative as well as positive values of x. For example, $n = 51$ includes integer values of x from -25 to 25, and $n = 50$ includes odd-numbered values of x from -49 to 49.

FOR ODD NUMBER OF YEARS x UNIT IS 1 YEAR			FOR EVEN NUMBER OF YEARS x UNIT IS 6 MONTHS		
N	Σx^2	Σx^4	N	Σx^2	Σx^4
3	2	2	2	2	2
5	10	34	4	20	164
7	28	196	6	70	1 414
9	60	708	8	168	6 216
11	110	1 958	10	330	19 338
13	182	4 550	12	572	48 620
15	280	9 352	14	910	105 742
17	408	17 544	16	1 360	206 992
19	570	30 666	18	1 938	374 034
21	770	50 666	20	2 660	634 676
23	1 012	79 948	22	3 542	1 023 638
25	1 300	121 420	24	4 600	1 583 320
27	1 638	178 542	26	5 850	2 364 570
29	2 030	255 374	28	7 308	3 427 452
31	2 480	356 624	30	8 990	4 842 014
33	2 992	469 696	32	10 912	6 689 056
35	3 570	654 738	34	13 090	9 060 898
37	4 218	864 690	36	15 540	12 062 148
39	4 940	1 125 332	38	18 278	15 810 470
41	5 740	1 445 332	40	21 320	20 437 352
43	6 622	1 834 294	42	24 682	26 088 874
45	7 590	2 302 806	44	28 380	32 926 476
47	8 628	2 862 488	46	32 430	41 127 726
49	9 800	3 526 040	48	36 848	50 887 088
51	11 050	4 307 290	50	41 650	62 416 690
53	12 402	5 221 242	52	46 852	75 947 092
55	13 860	6 284 124	54	52 470	91 728 054
57	15 428	7 513 436	56	58 520	110 029 304
59	17 110	8 927 998	58	65 018	131 141 306
61	18 910	10 547 998	60	71 980	155 376 028

L. RANDOM NUMBERS

```
05 90 35 89 95    01 61 16 96 94    50 78 13 69 36    37 68 53 37 31    71 26 35 03 71
44 43 80 69 98    46 68 05 14 82    90 78 50 05 62    77 79 13 57 44    59 60 10 39 66
61 81 31 96 82    00 57 25 60 59    46 72 60 18 77    55 66 12 62 11    08 99 55 64 57
42 88 07 10 05    24 98 65 63 21    47 21 61 88 32    27 80 30 21 60    10 92 35 36 12
77 94 30 05 39    28 10 99 00 27    12 73 73 99 12    49 99 57 94 82    96 88 57 17 91

78 83 19 76 16    94 11 68 84 26    23 54 20 86 85    23 86 66 99 07    36 37 34 92 09
87 76 59 61 81    43 63 64 61 61    65 76 36 95 90    18 48 27 45 68    27 23 65 30 72
91 43 05 96 47    55 78 99 95 24    37 55 85 78 78    01 48 41 19 10    35 19 54 07 73
84 97 77 72 73    09 62 06 65 72    87 12 49 03 60    41 15 20 76 27    50 47 02 29 16
87 41 60 76 83    44 88 96 07 80    83 05 83 38 96    73 70 66 81 90    30 56 10 48 59

22 17 68 65 84    68 95 23 92 35    87 02 22 57 51    61 09 43 95 06    58 24 82 03 47
19 36 27 59 46    13 79 93 37 55    39 77 32 77 09    85 52 05 30 62    47 83 51 62 74
16 77 23 02 77    09 61 87 25 21    28 06 24 25 93    16 71 13 59 78    23 05 47 47 25
78 43 76 71 61    20 44 90 32 64    97 67 63 99 61    46 38 03 93 22    69 81 21 99 21
03 28 28 26 08    73 37 32 04 05    69 30 16 09 05    88 69 58 28 99    35 07 44 75 47

93 22 53 64 39    07 10 63 76 35    87 03 04 79 88    08 13 13 85 51    55 34 57 72 69
78 76 58 54 74    92 38 70 96 92    52 06 79 79 45    82 63 18 27 44    69 66 92 19 09
23 68 35 26 00    99 53 93 61 28    52 70 05 48 34    56 65 05 61 86    90 92 10 70 80
15 39 25 70 99    93 86 52 77 65    15 33 59 05 28    22 87 26 07 47    86 96 98 29 06
58 71 96 30 24    18 46 23 34 27    85 13 99 24 44    49 18 09 79 49    74 16 32 23 02

57 35 27 33 72    24 53 63 94 09    41 10 76 47 91    44 04 95 49 66    39 60 04 59 81
48 50 86 54 48    22 06 34 72 52    82 21 15 65 20    33 29 94 71 11    15 91 29 12 03
61 96 48 95 03    07 16 39 33 66    98 56 10 56 79    77 21 30 27 12    90 49 22 23 62
36 93 89 41 26    29 70 83 63 51    99 74 20 52 36    87 09 41 15 09    98 60 16 03 03
18 87 00 42 31    57 90 12 02 07    23 47 37 17 31    54 08 01 88 63    39 41 88 92 10

88 56 53 27 59    33 35 72 67 47    77 34 55 45 70    08 18 27 38 90    16 95 86 70 75
09 72 95 84 29    49 41 31 06 70    42 38 06 45 18    54 84 73 31 65    52 53 37 97 15
12 96 88 17 31    65 19 69 02 83    60 75 86 90 68    24 64 19 35 51    56 61 87 39 12
85 94 57 24 16    92 09 84 38 76    22 00 27 69 85    29 81 94 78 70    21 94 47 90 12
38 64 43 59 98    98 77 87 68 07    91 51 67 62 44    40 98 05 93 78    23 32 65 41 18

53 44 09 42 72    00 41 86 79 79    68 47 22 00 20    35 55 31 51 51    00 83 63 22 55
40 76 66 26 84    57 99 99 90 37    36 63 32 08 58    37 40 13 68 97    87 64 81 07 83
02 17 79 18 05    12 59 52 57 02    22 07 90 47 03    28 14 11 30 79    20 69 22 40 98
95 17 82 06 53    31 51 10 96 46    92 06 88 07 77    56 11 50 81 69    40 23 72 51 39
35 76 22 42 92    96 11 83 44 80    34 68 35 48 77    33 42 40 90 60    73 96 53 97 86

26 29 13 56 41    85 47 04 66 08    34 72 57 59 13    82 43 80 46 15    38 26 61 70 04
77 80 20 75 82    72 82 32 99 90    63 95 73 76 63    89 73 44 99 05    48 67 26 43 18
46 40 66 44 52    91 36 74 43 53    30 82 13 54 00    78 45 63 98 35    55 03 36 67 68
37 56 08 18 09    77 53 84 46 47    31 91 18 95 58    24 16 74 11 53    44 10 13 85 57
61 65 61 68 66    37 27 47 39 19    84 83 70 07 48    53 21 40 06 71    95 06 79 88 54

93 43 69 64 07    34 18 04 52 35    56 27 09 24 86    61 85 53 83 45    19 90 70 99 00
21 96 60 12 99    11 20 99 45 18    48 13 93 55 34    18 37 79 49 90    65 97 38 20 46
95 20 47 97 97    27 37 83 28 71    00 06 41 41 74    45 89 09 39 84    51 67 11 52 49
97 86 21 78 73    10 65 81 92 59    58 76 17 14 97    04 76 62 16 17    17 95 70 45 80
69 92 06 34 13    59 71 74 17 32    27 55 10 24 19    23 71 82 13 74    63 52 52 01 41
```

RANDOM NUMBERS (*Continued*)

```
04 31 17 21 56   33 73 99 19 87   26 72 39 27 67   53 77 57 68 93   60 61 97 22 61
61 06 98 03 91   87 14 77 43 96   43 00 65 98 50   45 60 33 01 07   98 99 46 50 47
85 93 85 86 88   72 87 08 62 40   16 06 10 89 20   23 21 34 74 97   76 38 03 29 63
21 74 32 47 45   73 96 07 94 52   09 65 90 77 47   25 76 16 19 33   53 05 70 53 30
15 69 53 82 80   79 96 23 53 10   65 39 07 16 29   45 33 02 43 70   02 87 40 41 45

02 89 08 04 49   20 21 14 68 86   87 63 93 95 17   11 29 01 95 80   35 14 97 35 33
87 18 15 89 79   85 43 01 72 73   08 61 74 51 69   89 74 39 82 15   94 51 33 41 67
98 83 71 94 22   59 97 50 99 52   08 52 85 08 40   87 80 61 65 31   91 51 80 32 44
10 08 58 21 66   72 68 49 29 31   89 85 84 46 06   59 73 19 85 23   65 09 29 75 63
47 90 56 10 08   88 02 84 27 83   42 29 72 23 19   66 56 45 65 79   20 71 53 20 25

22 85 61 68 90   49 64 92 85 44   16 40 12 89 88   50 14 49 81 06   01 82 77 45 12
67 80 43 79 33   12 83 11 41 16   25 58 19 68 70   77 02 54 00 52   53 43 37 15 26
27 62 50 96 72   79 44 61 40 15   14 53 40 65 39   27 31 58 50 28   11 39 03 34 25
33 78 80 87 15   38 30 06 38 21   14 47 47 07 26   54 96 87 53 32   40 36 40 96 76
13 13 92 66 99   47 24 49 57 74   32 25 43 62 17   10 97 11 69 84   99 63 22 32 98

10 27 53 96 23   71 50 54 36 23   54 31 04 82 98   04 14 12 15 09   26 78 25 47 47
28 41 50 61 88   64 85 27 20 18   83 36 36 05 56   39 71 65 09 62   94 76 62 11 89
34 21 42 57 02   59 19 18 97 48   80 30 03 30 98   05 24 67 70 07   84 97 50 87 46
61 81 77 23 23   82 82 11 54 08   53 28 70 58 96   44 07 39 55 43   42 34 43 39 28
61 15 18 13 54   16 86 20 26 88   90 74 80 55 09   14 53 90 51 17   52 01 63 01 59

91 76 21 64 64   44 91 13 32 97   75 31 62 66 54   84 80 32 75 77   56 08 25 70 29
00 97 79 08 06   37 30 28 59 85   53 56 68 53 40   01 74 39 59 73   30 19 99 85 48
36 46 18 34 94   75 20 80 27 77   78 91 69 16 00   08 43 18 73 68   67 69 61 34 25
88 98 99 60 50   65 95 79 42 94   93 62 40 89 96   43 56 47 71 66   46 76 29 67 02
04 37 59 87 21   05 02 03 24 17   47 97 81 56 51   92 34 86 01 82   55 51 33 12 91

63 62 06 34 41   94 21 78 55 09   72 76 45 16 94   29 95 81 83 83   79 88 01 97 30
78 47 23 53 90   34 41 92 45 71   09 23 70 70 07   12 38 92 79 43   14 85 11 47 23
87 68 62 15 43   53 14 36 59 25   54 47 33 70 15   59 24 48 40 35   50 03 42 99 36
47 60 92 10 77   88 59 53 11 52   66 25 69 07 04   48 68 64 71 06   61 65 70 22 12
56 88 87 59 41   65 28 04 67 53   95 79 88 37 31   50 41 06 94 76   81 83 17 16 33

02 57 45 86 67   73 43 07 34 48   44 26 87 93 29   77 09 61 67 84   06 69 44 77 75
31 54 14 13 17   48 62 11 90 60   68 12 93 64 28   46 24 79 16 76   14 60 25 51 01
28 50 16 43 36   28 97 85 58 99   67 22 52 76 23   24 70 36 54 54   59 28 61 71 96
63 29 62 66 50   02 63 45 52 38   67 63 47 54 75   83 24 78 43 20   92 63 13 47 48
45 65 58 26 51   76 96 59 38 72   86 57 45 71 46   44 67 76 14 55   44 88 01 62 12

39 65 36 63 70   77 45 85 50 51   74 13 39 35 22   30 53 36 02 95   49 34 88 73 61
73 71 98 16 04   29 18 94 51 23   76 51 94 84 86   79 93 96 38 63   08 58 25 58 94
72 20 56 20 11   72 65 71 08 86   79 57 95 13 91   97 48 72 66 48   09 71 17 24 89
75 17 26 99 76   89 37 20 70 01   77 31 61 95 46   26 97 05 73 51   53 33 18 72 87
37 48 60 82 29   81 30 15 39 14   48 38 75 93 29   06 87 37 78 48   45 56 00 84 47

68 08 02 80 72   83 71 46 30 49   89 17 95 88 29   02 39 56 03 46   97 74 06 56 17
14 23 98 61 67   70 52 85 01 50   01 84 02 78 43   10 62 98 19 41   18 83 99 47 99
49 08 96 21 44   25 27 99 41 28   07 41 08 34 66   19 42 74 39 91   41 96 53 78 72
78 37 06 08 43   63 61 62 42 29   39 68 95 10 96   09 24 23 00 62   56 12 80 73 16
37 21 34 17 68   68 96 83 23 56   32 84 60 15 31   44 73 67 34 77   91 15 79 74 58
```

SOURCE: Reprinted with permission from Random Numbers III and IV of Table XXXIII of R. A. Fisher and F. Yates, *Statistical Tables for Biological, Agricultural and Medical Research* (Edinburgh: Oliver & Boyd, Ltd.).

M. VALUES OF t

THE VALUE t describes the sampling distribution of a deviation from the population mean divided by the standard error.

Probabilities in the heading refer to the sum of the two-tailed areas under the curve that lie outside the points $\pm t$. (For a single tail divide the probability by 2.) Degrees of freedom are listed in the first column.

Example: In the distribution of the means of samples of size $n = 10$, $df = n - 1 = 9$; then .05 of the area under the curve falls in the two tails outside the interval $t = \pm2.262$. The last row shows the corresponding areas under the normal curve.

PROBABILITY (P)

df	.20	.10	.05	.02	.01
1	3·078	6·314	12·706	31·821	63·657
2	1·886	2·920	4·303	6·965	9·925
3	1·638	2·353	3·182	4·541	5·841
4	1·533	2·132	2·776	3·747	4·604
5	1·476	2·015	2·571	3·365	4·032
6	1·440	1·943	2·447	3·143	3·707
7	1·415	1·895	2·365	2·998	3·499
8	1·397	1·860	2·306	2·896	3·355
9	1·383	1·833	2·262	2·821	3·250
10	1·372	1·812	2·228	2·764	3·169
11	1·363	1·796	2·201	2·718	3·106
12	1·356	1·782	2·179	2·681	3·055
13	1·350	1·771	2·160	2·650	3·012
14	1·345	1·761	2·145	2·624	2·977
15	1·341	1·753	2·131	2·602	2·947
16	1·337	1·746	2·120	2·583	2·921
17	1·333	1·740	2·110	2·567	2·898
18	1·330	1·734	2·101	2·552	2·878
19	1·328	1·729	2·093	2·539	2·861
20	1·325	1·725	2·086	2·528	2·845
21	1·323	1·721	2·080	2·518	2·831
22	1·321	1·717	2·074	2·508	2·819
23	1·319	1·714	2·069	2·500	2·807
24	1·318	1·711	2·064	2·492	2·797
25	1·316	1·708	2·060	2·485	2·787
26	1·315	1·706	2·056	2·479	2·779
27	1·314	1·703	2·052	2·473	2·771
28	1·313	1·701	2·048	2·467	2·763
29	1·311	1·699	2·045	2·462	2·756
30	1·310	1·697	2·042	2·457	2·750
∞	1·28155	1·64485	1·95996	2·32634	2·57582

Reprinted from Table IV, p. 174, of R. A. Fisher, *Statistical Methods for Research Workers* (11th ed.), published by Oliver and Boyd, Ltd. Edinburgh, by permission of the author and publishers.

N. CHI-SQUARE DISTRIBUTION

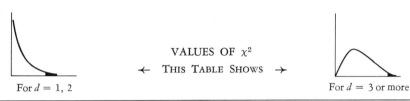

VALUES OF χ^2

← THIS TABLE SHOWS →

For $d = 1, 2$

For $d = 3$ or more

Degrees of Freedom† d	Probability (P)								
	.99	.98	.95	.90	.50	.10	.05	.02	.01
1	.000157	.000628	.00393	.0158	.455	2.706	3.841	5.412	6.635
2	.0201	.0404	.103	.211	1.386	4.605	5.991	7.824	9.210
3	.115	.185	.352	.584	2.366	6.251	7.815	9.837	11.345
4	.297	.429	.711	1.064	3.357	7.779	9.488	11.668	13.277
5	.554	.752	1.145	1.610	4.351	9.236	11.070	13.388	15.086
6	.872	1.134	1.635	2.204	5.348	10.645	12.592	15.033	16.812
7	1.239	1.564	2.167	2.833	6.346	12.017	14.067	16.622	18.475
8	1.646	2.032	2.733	3.490	7.344	13.362	15.507	18.168	20.090
9	2.088	2.532	3.325	4.168	8.343	14.684	16.919	19.679	21.666
10	2.558	3.059	3.940	4.865	9.342	15.987	18.307	21.161	23.209
11	3.053	3.609	4.575	5.578	10.341	17.275	19.675	22.618	24.725
12	3.571	4.178	5.226	6.304	11.340	18.549	21.026	24.054	26.217
13	4.107	4.765	5.892	7.042	12.340	19.812	22.362	25.472	27.688
14	4.660	5.368	6.571	7.790	13.339	21.064	23.685	26.873	29.141
15	5.229	5.985	7.261	8.547	14.339	22.307	24.996	28.259	30.578
16	5.812	6.614	7.962	9.312	15.338	23.542	26.296	29.633	32.000
17	6.408	7.255	8.672	10.085	16.338	24.769	27.587	30.995	33.409
18	7.015	7.906	9.390	10.865	17.338	25.989	28.869	32.346	34.805
19	7.633	8.567	10.117	11.651	18.338	27.204	30.144	33.687	36.191
20	8.260	9.237	10.851	12.443	19.337	28.412	31.410	35.020	37.566
21	8.897	9.915	11.591	13.240	20.337	29.615	32.671	36.343	38.932
22	9.542	10.600	12.338	14.041	21.337	30.813	33.924	37.659	40.289
23	10.196	11.293	13.091	14.848	22.337	32.007	35.172	38.968	41.638
24	10.856	11.992	13.848	15.659	23.337	33.196	36.415	40.270	42.980
25	11.524	12.697	14.611	16.473	24.337	34.382	37.652	41.566	44.314
26	12.198	13.409	15.379	17.292	25.336	35.563	38.885	42.856	45.642
27	12.879	14.125	16.151	18.114	26.336	36.741	40.113	44.140	46.963
28	13.565	14.847	16.928	18.939	27.336	37.916	41.337	45.419	48.278
29	14.256	15.574	17.708	19.768	28.336	39.087	42.557	46.693	49,588
30	14.953	16.306	18.493	20.599	29.336	40.256	43.773	47.962	50.892

Reprinted from Table III of Fisher: *Statistical Methods for Research Workers*, published by Oliver and Boyd Ltd., Edinburgh, by permission of the author and publishers.

† For larger values of degrees of freedom, the quantity $\sqrt{2\chi^2}$ may be assumed to be approximately normally distributed with mean $\sqrt{2d-1}$ and standard deviation 1. Thus, the statistic, $\sqrt{2\chi^2} - \sqrt{2d-1}$, may be taken to have the standard normal distribution.

O. THE F DISTRIBUTION

VALUES OF F

d_2	d_1 = Degrees of Freedom for Numerator											
	1	2	3	4	5	6	7	8	9	10	11	12
1	161	200	216	225	230	234	237	239	241	242	243	244
	4,052	**4,999**	**5,403**	**5,625**	**5,764**	**5,859**	**5,928**	**5,981**	**6,022**	**6,056**	**6,082**	**6,106**
2	18.51	19.00	19.16	19.25	19.30	19.33	19.36	19.37	19.38	19.39	19.40	19.41
	98.49	**99.01**	**99.17**	**99.25**	**99.30**	**99.33**	**99.34**	**99.36**	**99.38**	**99.40**	**99.41**	**99.42**
3	10.13	9.55	9.28	9.12	9.01	8.94	8.88	8.84	8.81	8.78	8.76	8.74
	34.12	**30.81**	**29.46**	**28.71**	**28.24**	**27.91**	**27.67**	**27.49**	**27.34**	**27.23**	**27.13**	**27.05**
4	7.71	6.94	6.59	6.39	6.26	6.16	6.09	6.04	6.00	5.96	5.93	5.91
	21.20	**18.00**	**16.69**	**15.98**	**15.52**	**15.21**	**14.98**	**14.80**	**14.66**	**14.54**	**14.45**	**14.37**
5	6.61	5.79	5.41	5.19	5.05	4.95	4.88	4.82	4.78	4.74	4.70	4.68
	16.26	**13.27**	**12.06**	**11.39**	**10.97**	**10.67**	**10.45**	**10.27**	**10.15**	**10.05**	**9.96**	**9.89**
6	5.99	5.14	4.76	4.53	4.39	4.28	4.21	4.15	4.10	4.06	4.03	4.00
	13.74	**10.92**	**9.78**	**9.15**	**8.75**	**8.47**	**8.26**	**8.10**	**7.98**	**7.87**	**7.79**	**7.72**
7	5.59	4.74	4.35	4.12	3.97	3.87	3.79	3.73	3.68	3.63	3.60	3.57
	12.25	**9.55**	**8.45**	**7.85**	**7.46**	**7.19**	**7.00**	**6.84**	**6.71**	**6.62**	**6.54**	**6.47**
8	5.32	4.46	4.07	3.84	3.69	3.58	3.50	3.44	3.39	3.34	3.31	3.28
	11.26	**8.65**	**7.59**	**7.01**	**6.63**	**6.37**	**6.19**	**6.03**	**5.91**	**5.82**	**5.74**	**5.67**
9	5.12	4.26	3.86	3.63	3.48	3.37	3.29	3.23	3.18	3.13	3.10	3.07
	10.56	**8.02**	**6.99**	**6.42**	**6.06**	**5.80**	**5.62**	**5.47**	**5.35**	**5.26**	**5.18**	**5.11**
10	4.96	4.10	3.71	3.48	3.33	3.22	3.14	3.07	3.02	2.97	2.94	2.91
	10.04	**7.56**	**6.55**	**5.99**	**5.64**	**5.39**	**5.21**	**5.06**	**4.95**	**4.85**	**4.78**	**4.71**
11	4.84	3.98	3.59	3.36	3.20	3.09	3.01	2.95	2.90	2.86	2.82	2.79
	9.65	**7.20**	**6.22**	**5.67**	**5.32**	**5.07**	**4.88**	**4.74**	**4.63**	**4.54**	**4.46**	**4.40**
12	4.75	3.88	3.49	3.26	3.11	3.00	2.92	2.85	2.80	2.76	2.72	2.69
	9.33	**6.93**	**5.95**	**5.41**	**5.06**	**4.82**	**4.65**	**4.50**	**4.39**	**4.30**	**4.22**	**4.16**
13	4.67	3.80	3.41	3.18	3.02	2.92	2.84	2.77	2.72	2.67	2.63	2.60
	9.07	**6.70**	**5.74**	**5.20**	**4.86**	**4.62**	**4.44**	**4.30**	**4.19**	**4.10**	**4.02**	**3.96**
14	4.60	3.74	3.34	3.11	2.96	2.85	2.77	2.70	2.65	2.60	2.56	2.53
	8.86	**6.51**	**5.56**	**5.03**	**4.69**	**4.46**	**4.28**	**4.14**	**4.03**	**3.94**	**3.86**	**3.80**
15	4.54	3.68	3.29	3.06	2.90	2.79	2.70	2.64	2.59	2.55	2.51	2.48
	8.68	**6.36**	**5.42**	**4.89**	**4.56**	**4.32**	**4.14**	**4.00**	**3.89**	**3.80**	**3.73**	**3.67**
16	4.49	3.63	3.24	3.01	2.85	2.74	2.66	2.59	2.54	2.49	2.45	2.42
	8.53	**6.23**	**5.29**	**4.77**	**4.44**	**4.20**	**4.03**	**3.89**	**3.78**	**3.69**	**3.61**	**3.55**
17	4.45	3.59	3.20	2.96	2.81	2.70	2.62	2.55	2.50	2.45	2.41	2.38
	8.40	**6.11**	**5.18**	**4.67**	**4.34**	**4.10**	**3.93**	**3.79**	**3.68**	**3.59**	**3.52**	**3.45**
18	4.41	3.55	3.16	2.93	2.77	2.66	2.58	2.51	2.46	2.41	2.37	2.34
	8.28	**6.01**	**5.09**	**4.58**	**4.25**	**4.01**	**3.85**	**3.71**	**3.60**	**3.51**	**3.44**	**3.37**
19	4.38	3.52	3.13	2.90	2.74	2.63	2.55	2.48	2.43	2.38	2.34	2.31
	8.18	**5.93**	**5.01**	**4.50**	**4.17**	**3.94**	**3.77**	**3.63**	**3.52**	**3.43**	**3.36**	**3.30**
20	4.35	3.49	3.10	2.87	2.71	2.60	2.52	2.45	2.40	2.35	2.31	2.28
	8.10	**5.85**	**4.94**	**4.43**	**4.10**	**3.87**	**3.71**	**3.56**	**3.45**	**3.37**	**3.30**	**3.23**
21	4.32	3.47	3.07	2.84	2.68	2.57	2.49	2.42	2.37	2.32	2.28	2.25
	8.02	**5.78**	**4.87**	**4.37**	**4.04**	**3.81**	**3.65**	**3.51**	**3.40**	**3.31**	**3.24**	**3.17**
22	4.30	3.44	3.05	2.82	2.66	2.55	2.47	2.40	2.35	2.30	2.26	2.23
	7.94	**5.72**	**4.82**	**4.31**	**3.99**	**3.76**	**3.59**	**3.45**	**3.35**	**3.26**	**3.18**	**3.12**
23	4.28	3.42	3.03	2.80	2.64	2.53	2.45	2.38	2.32	2.28	2.24	2.20
	7.88	**5.66**	**4.76**	**4.26**	**3.94**	**3.71**	**3.54**	**3.41**	**3.30**	**3.21**	**3.14**	**3.07**
24	4.26	3.40	3.01	2.78	2.62	2.51	2.43	2.36	2.30	2.26	2.22	2.18
	7.82	**5.61**	**4.72**	**4.22**	**3.90**	**3.67**	**3.50**	**3.36**	**3.25**	**3.17**	**3.09**	**3.03**
25	4.24	3.38	2.99	2.76	2.60	2.49	2.41	2.34	2.28	2.24	2.20	2.16
	7.77	**5.57**	**4.68**	**4.18**	**3.86**	**3.63**	**3.46**	**3.32**	**3.21**	**3.13**	**3.05**	**2.99**
26	4.22	3.37	2.98	2.74	2.59	2.47	2.39	2.32	2.27	2.22	2.18	2.15
	7.72	**5.53**	**4.64**	**4.14**	**3.82**	**3.59**	**3.42**	**3.29**	**3.17**	**3.09**	**3.02**	**2.96**

d_2 = Degrees of Freedom for Denominator

* This table is reprinted by permission from George W. Snedecor, *Statistical Methods* (5th ed.; Iowa City: Iowa State University Press, Copyright 1956).

THE F DISTRIBUTION (*Continued*)

d_1 = Degrees of Freedom for Numerator

14	16	20	24	30	40	50	75	100	200	500	∞	d_2
245 / 6,142	246 / 6,169	248 / 6,208	249 / 6,234	250 / 6,258	251 / 6,286	252 / 6,302	253 / 6,323	253 / 6,334	254 / 6,352	254 / 6,361	254 / 6,366	1
19.42 / 99.43	19.43 / 99.44	19.44 / 99.45	19.45 / 99.46	19.46 / 99.47	19.47 / 99.48	19.47 / 99.48	19.48 / 99.49	19.49 / 99.49	19.49 / 99.49	19.50 / 99.50	19.50 / 99.50	2
8.71 / 26.92	8.69 / 26.83	8.66 / 26.69	8.64 / 26.60	8.62 / 26.50	8.60 / 26.41	8.58 / 26.35	8.57 / 26.27	8.56 / 26.23	8.54 / 26.18	8.54 / 26.14	8.53 / 26.12	3
5.87 / 14.24	5.84 / 14.15	5.80 / 14.02	5.77 / 13.93	5.74 / 13.83	5.71 / 13.74	5.70 / 13.69	5.68 / 13.61	5.66 / 13.57	5.65 / 13.52	5.64 / 13.48	5.63 / 13.46	4
4.64 / 9.77	4.60 / 9.68	4.56 / 9.55	4.53 / 9.47	4.50 / 9.38	4.46 / 9.29	4.44 / 9.24	4.42 / 9.17	4.40 / 9.13	4.38 / 9.07	4.37 / 9.04	4.36 / 9.02	5
3.96 / 7.60	3.92 / 7.52	3.87 / 7.39	3.84 / 7.31	3.81 / 7.23	3.77 / 7.14	3.75 / 7.09	3.72 / 7.02	3.71 / 6.99	3.69 / 6.94	3.68 / 6.90	3.67 / 6.88	6
3.52 / 6.35	3.49 / 6.27	3.44 / 6.15	3.41 / 6.07	3.38 / 5.98	3.34 / 5.90	3.32 / 5.85	3.29 / 5.78	3.28 / 5.75	3.25 / 5.70	3.24 / 5.67	3.23 / 5.65	7
3.23 / 5.56	3.20 / 5.48	3.15 / 5.36	3.12 / 5.28	3.08 / 5.20	3.05 / 5.11	3.03 / 5.06	3.00 / 5.00	2.98 / 4.96	2.96 / 4.91	2.94 / 4.88	2.93 / 4.86	8
3.02 / 5.00	2.98 / 4.92	2.93 / 4.80	2.90 / 4.73	2.86 / 4.64	2.82 / 4.56	2.80 / 4.51	2.77 / 4.45	2.76 / 4.41	2.73 / 4.36	2.72 / 4.33	2.71 / 4.31	9
2.86 / 4.60	2.82 / 4.52	2.77 / 4.41	2.74 / 4.33	2.70 / 4.25	2.67 / 4.17	2.64 / 4.12	2.61 / 4.05	2.59 / 4.01	2.56 / 3.96	2.55 / 3.93	2.54 / 3.91	10
2.74 / 4.29	2.70 / 4.21	2.65 / 4.10	2.61 / 4.02	2.57 / 3.94	2.53 / 3.86	2.50 / 3.80	2.47 / 3.74	2.45 / 3.70	2.42 / 3.66	2.41 / 3.62	2.40 / 3.60	11
2.64 / 4.05	2.60 / 3.98	2.54 / 3.86	2.50 / 3.78	2.46 / 3.70	2.42 / 3.61	2.40 / 3.56	2.36 / 3.49	2.35 / 3.46	2.32 / 3.41	2.31 / 3.38	2.30 / 3.36	12
2.55 / 3.85	2.51 / 3.78	2.46 / 3.67	2.42 / 3.59	2.38 / 3.51	2.34 / 3.42	2.32 / 3.37	2.28 / 3.30	2.26 / 3.27	2.24 / 3.21	2.22 / 3.18	2.21 / 3.16	13
2.48 / 3.70	2.44 / 3.62	2.39 / 3.51	2.35 / 3.43	2.31 / 3.34	2.27 / 3.26	2.24 / 3.21	2.21 / 3.14	2.19 / 3.11	2.16 / 3.06	2.14 / 3.02	2.13 / 3.00	14
2.43 / 3.56	2.39 / 3.48	2.33 / 3.36	2.29 / 3.29	2.25 / 3.20	2.21 / 3.12	2.18 / 3.07	2.15 / 3.00	2.12 / 2.97	2.10 / 2.92	2.08 / 2.89	2.07 / 2.87	15
2.37 / 3.45	2.33 / 3.37	2.28 / 3.25	2.24 / 3.18	2.20 / 3.10	2.16 / 3.01	2.13 / 2.96	2.09 / 2.89	2.07 / 2.86	2.04 / 2.80	2.02 / 2.77	2.01 / 2.75	16
2.33 / 3.35	2.29 / 3.27	2.23 / 3.16	2.19 / 3.08	2.15 / 3.00	2.11 / 2.92	2.08 / 2.86	2.04 / 2.79	2.02 / 2.76	1.99 / 2.70	1.97 / 2.67	1.96 / 2.65	17
2.29 / 3.27	2.25 / 3.19	2.19 / 3.07	2.15 / 3.00	2.11 / 2.91	2.07 / 2.83	2.04 / 2.78	2.00 / 2.71	1.98 / 2.68	1.95 / 2.62	1.93 / 2.59	1.92 / 2.57	18
2.26 / 3.19	2.21 / 3.12	2.15 / 3.00	2.11 / 2.92	2.07 / 2.84	2.02 / 2.76	2.00 / 2.70	1.96 / 2.63	1.94 / 2.60	1.91 / 2.54	1.90 / 2.51	1.88 / 2.49	19
2.23 / 3.13	2.18 / 3.05	2.12 / 2.94	2.08 / 2.86	2.04 / 2.77	1.99 / 2.69	1.96 / 2.63	1.92 / 2.56	1.90 / 2.53	1.87 / 2.47	1.85 / 2.44	1.84 / 2.42	20
2.20 / 3.07	2.15 / 2.99	2.09 / 2.88	2.05 / 2.80	2.00 / 2.72	1.96 / 2.63	1.93 / 2.58	1.89 / 2.51	1.87 / 2.47	1.84 / 2.42	1.82 / 2.38	1.81 / 2.36	21
2.18 / 3.02	2.13 / 2.94	2.07 / 2.83	2.03 / 2.75	1.98 / 2.67	1.93 / 2.58	1.91 / 2.53	1.87 / 2.46	1.84 / 2.42	1.81 / 2.37	1.80 / 2.33	1.78 / 2.31	22
2.14 / 2.97	2.10 / 2.89	2.04 / 2.78	2.00 / 2.70	1.96 / 2.62	1.91 / 2.53	1.88 / 2.48	1.84 / 2.41	1.82 / 2.37	1.79 / 2.32	1.77 / 2.28	1.76 / 2.26	23
2.13 / 2.93	2.09 / 2.85	2.02 / 2.74	1.98 / 2.66	1.94 / 2.58	1.89 / 2.49	1.86 / 2.44	1.82 / 2.36	1.80 / 2.33	1.76 / 2.27	1.74 / 2.23	1.73 / 2.21	24
2.11 / 2.89	2.06 / 2.81	2.00 / 2.70	1.96 / 2.62	1.92 / 2.54	1.87 / 2.45	1.84 / 2.40	1.80 / 2.32	1.77 / 2.29	1.74 / 2.23	1.72 / 2.19	1.71 / 2.17	25
2.10 / 2.86	2.05 / 2.77	1.99 / 2.66	1.95 / 2.58	1.90 / 2.50	1.85 / 2.41	1.82 / 2.36	1.78 / 2.28	1.76 / 2.25	1.72 / 2.19	1.70 / 2.15	1.69 / 2.13	26

d_2 = Degrees of Freedom for Denominator

THE F DISTRIBUTION (*Continued*)

		d_1 = Degrees of Freedom for Numerator										
d_2	1	2	3	4	5	6	7	8	9	10	11	12
27	4.21	3.35	2.96	2.73	2.57	2.46	2.37	2.30	2.25	2.20	2.16	2.13
	7.68	5.49	4.60	4.11	3.79	3.56	3.39	3.26	3.14	3.06	2.98	2.93
28	4.20	3.34	2.95	2.71	2.56	2.44	2.36	2.29	2.24	2.19	2.15	2.12
	7.64	5.45	4.57	4.07	3.76	3.53	3.36	3.23	3.11	3.03	2.95	2.90
29	4.18	3.33	2.93	2.70	2.54	2.43	2.35	2.28	2.22	2.18	2.14	2.10
	7.60	5.42	4.54	4.04	3.73	3.50	3.33	3.20	3.08	3.00	2.92	2.87
30	4.17	3.32	2.92	2.69	2.53	2.42	2.34	2.27	2.21	2.16	2.12	2.09
	7.56	5.39	4.51	4.02	3.70	3.47	3.30	3.17	3.06	2.98	2.90	2.84
32	4.15	3.30	2.90	2.67	2.51	2.40	2.32	2.25	2.19	2.14	2.10	2.07
	7.50	5.34	4.46	3.97	3.66	3.42	3.25	3.12	3.01	2.94	2.86	2.80
34	4.13	3.28	2.88	2.65	2.49	2.38	2.30	2.23	2.17	2.12	2.08	2.05
	7.44	5.29	4.42	3.93	3.61	3.38	3.21	3.08	2.97	2.89	2.82	2.76
36	4.11	3.26	2.86	2.63	2.48	2.36	2.28	2.21	2.15	2.10	2.06	2.03
	7.39	5.25	4.38	3.89	3.58	3.35	3.18	3.04	2.94	2.86	2.78	2.72
38	4.10	3.25	2.85	2.62	2.46	2.35	2.26	2.19	2.14	2.09	2.05	2.02
	7.35	5.21	4.34	3.86	3.54	3.32	3.15	3.02	2.91	2.82	2.75	2.69
40	4.08	3.23	2.84	2.61	2.45	2.34	2.25	2.18	2.12	2.07	2.04	2.00
	7.31	5.18	4.31	3.83	3.51	3.29	3.12	2.99	2.88	2.80	2.73	2.66
42	4.07	3.22	2.83	2.59	2.44	2.32	2.24	2.17	2.11	2.06	2.02	1.99
	7.27	5.15	4.29	3.80	3.49	3.26	3.10	2.96	2.86	2.77	2.70	2.64
44	4.06	3.21	2.82	2.58	2.43	2.31	2.23	2.16	2.10	2.05	2.01	1.98
	7.24	5.12	4.26	3.78	3.46	3.24	3.07	2.94	2.84	2.75	2.68	2.62
46	4.05	3.20	2.81	2.57	2.42	2.30	2.22	2.14	2.09	2.04	2.00	1.97
	7.21	5.10	4.24	3.76	3.44	3.22	3.05	2.92	2.82	2.73	2.66	2.60
48	4.04	3.19	2.80	2.56	2.41	2.30	2.21	2.14	2.08	2.03	1.99	1.96
	7.19	5.08	4.22	3.74	3.42	3.20	3.04	2.90	2.80	2.71	2.64	2.58
50	4.03	3.18	2.79	2.56	2.40	2.29	2.20	2.13	2.07	2.02	1.98	1.95
	7.17	5.06	4.20	3.72	3.41	3.18	3.02	2.88	2.78	2.70	2.62	2.56
55	4.02	3.17	2.78	2.54	2.38	2.27	2.18	2.11	2.05	2.00	1.97	1.93
	7.12	5.01	4.16	3.68	3.37	3.15	2.98	2.85	2.75	2.66	2.59	2.53
60	4.00	3.15	2.76	2.52	2.37	2.25	2.17	2.10	2.04	1.99	1.95	1.92
	7.08	4.98	4.13	3.65	3.34	3.12	2.95	2.82	2.72	2.63	2.56	2.50
65	3.99	3.14	2.75	2.51	2.36	2.24	2.15	2.08	2.02	1.98	1.94	1.90
	7.04	4.95	4.10	3.62	3.31	3.09	2.93	2.79	2.70	2.61	2.54	2.47
70	3.98	3.13	2.74	2.50	2.35	2.23	2.14	2.07	2.01	1.97	1.93	1.89
	7.01	4.92	4.08	3.60	3.29	3.07	2.91	2.77	2.67	2.59	2.51	2.45
80	3.96	3.11	2.72	2.48	2.33	2.21	2.12	2.05	1.99	1.95	1.91	1.88
	6.96	4.88	4.04	3.56	3.25	3.04	2.87	2.74	2.64	2.55	2.48	2.41
100	3.94	3.09	2.70	2.46	2.30	2.19	2.10	2.03	1.97	1.92	1.88	1.85
	6.90	4.82	3.98	3.51	3.20	2.99	2.82	2.69	2.59	2.51	2.43	2.36
125	3.92	3.07	2.68	2.44	2.29	2.17	2.08	2.01	1.95	1.90	1.86	1.83
	6.84	4.78	3.94	3.47	3.17	2.95	2.79	2.65	2.56	2.47	2.40	2.33
150	3.91	3.06	2.67	2.43	2.27	2.16	2.07	2.00	1.94	1.89	1.85	1.82
	6.81	4.75	3.91	3.44	3.14	2.92	2.76	2.62	2.53	2.44	2.37	2.30
200	3.89	3.04	2.65	2.41	2.26	2.14	2.05	1.98	1.92	1.87	1.83	1.80
	6.76	4.71	3.88	3.41	3.11	2.90	2.73	2.60	2.50	2.41	2.34	2.28
400	3.86	3.02	2.62	2.39	2.23	2.12	2.03	1.96	1.90	1.85	1.81	1.78
	6.70	4.66	3.83	3.36	3.06	2.85	2.69	2.55	2.46	2.37	2.29	2.23
1,000	3.85	3.00	2.61	2.38	2.22	2.10	2.02	1.95	1.89	1.84	1.80	1.76
	6.66	4.62	3.80	3.34	3.04	2.82	2.66	2.53	2.43	2.34	2.26	2.20
∞	3.84	2.99	2.60	2.37	2.21	2.09	2.01	1.94	1.88	1.83	1.79	1.75
	6.64	4.60	3.78	3.32	3.02	2.80	2.64	2.51	2.41	2.32	2.24	2.18

d_2 = Degrees of Freedom for Denominator

THE F DISTRIBUTION (Concluded)

d_1 = Degrees of Freedom for Numerator												
14	16	20	24	30	40	50	75	100	200	500	∞	d_2
2.08	2.03	1.97	1.93	1.88	1.84	1.80	1.76	1.74	1.71	1.68	1.67	27
2.83	2.74	2.63	2.55	2.47	2.38	2.33	2.25	2.21	2.16	2.12	2.10	
2.06	2.02	1.96	1.91	1.87	1.81	1.78	1.75	1.72	1.69	1.67	1.65	28
2.80	2.71	2.60	2.52	2.44	2.35	2.30	2.22	2.18	2.13	2.09	2.06	
2.05	2.00	1.94	1.90	1.85	1.80	1.77	1.73	1.71	1.68	1.65	1.64	29
2.77	2.68	2.57	2.49	2.41	2.32	2.27	2.19	2.15	2.10	2.06	2.03	
2.04	1.99	1.93	1.89	1.84	1.79	1.76	1.72	1.69	1.66	1.64	1.62	30
2.74	2.66	2.55	2.47	2.38	2.29	2.24	2.16	2.13	2.07	2.03	2.01	
2.02	1.97	1.91	1.86	1.82	1.76	1.74	1.69	1.67	1.64	1.61	1.59	32
2.70	2.62	2.51	2.42	2.34	2.25	2.20	2.12	2.08	2.02	1.98	1.96	
2.00	1.95	1.89	1.84	1.80	1.74	1.71	1.67	1.64	1.61	1.59	1.57	34
2.66	2.58	2.47	2.38	2.30	2.21	2.15	2.08	2.04	1.98	1.94	1.91	
1.98	1.93	1.87	1.82	1.78	1.72	1.69	1.65	1.62	1.59	1.56	1.55	36
2.62	2.54	2.43	2.35	2.26	2.17	2.12	2.04	2.00	1.94	1.90	1.87	
1.96	1.92	1.85	1.80	1.76	1.71	1.67	1.63	1.60	1.57	1.54	1.53	38
2.59	2.51	2.40	2.32	2.22	2.14	2.08	2.00	1.97	1.90	1.86	1.84	
1.95	1.90	1.84	1.79	1.74	1.69	1.66	1.61	1.59	1.55	1.53	1.51	40
2.56	2.49	2.37	2.29	2.20	2.11	2.05	1.97	1.94	1.88	1.84	1.81	
1.94	1.89	1.82	1.78	1.73	1.68	1.64	1.60	1.57	1.54	1.51	1.49	42
2.54	2.46	2.35	2.26	2.17	2.08	2.02	1.94	1.91	1.85	1.80	1.78	
1.92	1.88	1.81	1.76	1.72	1.66	1.63	1.58	1.56	1.52	1.50	1.48	44
2.52	2.44	2.32	2.24	2.15	2.06	2.00	1.92	1.88	1.82	1.78	1.75	
1.91	1.87	1.80	1.75	1.71	1.65	1.62	1.57	1.54	1.51	1.48	1.46	46
2.50	2.42	2.30	2.22	2.13	2.04	1.98	1.90	1.86	1.80	1.76	1.72	
1.90	1.86	1.79	1.74	1.70	1.64	1.61	1.56	1.53	1.50	1.47	1.45	48
2.48	2.40	2.28	2.20	2.11	2.02	1.96	1.88	1.84	1.78	1.73	1.70	
1.90	1.85	1.78	1.74	1.69	1.63	1.60	1.55	1.52	1.48	1.46	1.44	50
2.46	2.39	2.26	2.18	2.10	2.00	1.94	1.86	1.82	1.76	1.71	1.68	
1.88	1.83	1.76	1.72	1.67	1.61	1.58	1.52	1.50	1.46	1.43	1.41	55
2.43	2.35	2.23	2.15	2.06	1.96	1.90	1.82	1.78	1.71	1.66	1.64	
1.86	1.81	1.75	1.70	1.65	1.59	1.56	1.50	1.48	1.44	1.41	1.39	60
2.40	2.32	2.20	2.12	2.03	1.93	1.87	1.79	1.74	1.68	1.63	1.60	
1.85	1.80	1.73	1.68	1.63	1.57	1.54	1.49	1.46	1.42	1.39	1.37	65
2.37	2.30	2.18	2.09	2.00	1.90	1.84	1.76	1.71	1.64	1.60	1.56	
1.84	1.79	1.72	1.67	1.62	1.56	1.53	1.47	1.45	1.40	1.37	1.35	70
2.35	2.28	2.15	2.07	1.98	1.88	1.82	1.74	1.69	1.62	1.56	1.53	
1.82	1.77	1.70	1.65	1.60	1.54	1.51	1.45	1.42	1.38	1.35	1.32	80
2.32	2.24	2.11	2.03	1.94	1.84	1.78	1.70	1.65	1.57	1.52	1.49	
1.79	1.75	1.68	1.63	1.57	1.51	1.48	1.42	1.39	1.34	1.30	1.28	100
2.26	2.19	2.06	1.98	1.89	1.79	1.73	1.64	1.59	1.51	1.46	1.43	
1.77	1.72	1.65	1.60	1.55	1.49	1.45	1.39	1.36	1.31	1.27	1.25	125
2.23	2.15	2.03	1.94	1.85	1.75	1.68	1.59	1.54	1.46	1.40	1.37	
1.76	1.71	1.64	1.59	1.54	1.47	1.44	1.37	1.34	1.29	1.25	1.22	150
2.20	2.12	2.00	1.91	1.83	1.72	1.66	1.56	1.51	1.43	1.37	1.33	
1.74	1.69	1.62	1.57	1.52	1.45	1.42	1.35	1.32	1.26	1.22	1.19	200
2.17	2.09	1.97	1.88	1.79	1.69	1.62	1.53	1.48	1.39	1.33	1.28	
1.72	1.67	1.60	1.54	1.49	1.42	1.38	1.32	1.28	1.22	1.16	1.13	400
2.12	2.04	1.92	1.84	1.74	1.64	1.57	1.47	1.42	1.32	1.24	1.19	
1.70	1.65	1.58	1.53	1.47	1.41	1.36	1.30	1.26	1.19	1.13	1.08	1,000
2.09	2.01	1.89	1.81	1.71	1.61	1.54	1.44	1.38	1.28	1.19	1.11	
1.69	1.64	1.57	1.52	1.46	1.40	1.35	1.28	1.24	1.17	1.11	1.00	∞
2.07	1.99	1.87	1.79	1.69	1.59	1.52	1.41	1.36	1.25	1.15	1.00	

d_2 = Degrees of Freedom for Denominator

P. RANK-SUM TEST

CRITICAL VALUES FOR T_1

FOR THE RANK-SUM TEST FOR TWO INDEPENDENT SAMPLES

T_1 = Sum of ranks in one group, n_1 and n_2 are numbers in each group. If $T_1 \leq 1$ (lower limit) or $T_1 \geq u$ (upper limit), reject hypothesis that groups are equal, at .05 or .01 level of significance shown in subhead. (When n_1 or n_2 = one, and for blanks in table, hypothesis cannot be rejected).

n_1		$n_2 \rightarrow$ 2		3		4		5		6		7		8		9		10	
		.05	.01	.05	.01	.05	.01	.05	.01	.05	.01	.05	.01	.05	.01	.05	.01	.05	.01
2	1													3		3		3	
	u													19		21		23	
3	1							6		7		7		8		8	6	9	6
	u							21		23		26		28		31	33	33	36
4	1					10		11		12	10	13	10	14	11	14	11	15	12
	u					26		29		32	34	35	38	38	41	42	45	45	48
5	1			15		16		17	15	18	16	20	16	21	17	22	18	23	19
	u			30		34		38	40	42	44	45	49	49	53	53	57	57	61
6	1			22		23	21	24	22	26	23	27	24	29	25	31	26	32	27
	u			38		43	45	48	50	52	55	57	60	61	65	65	70	70	75
7	1			29		31	28	33	29	34	31	36	32	38	34	40	35	42	37
	u			48		53	56	58	62	64	67	69	73	74	78	79	84	84	89
8	1	36		38		40	37	42	38	44	40	46	42	49	43	51	45	53	47
	u	52		58		64	67	70	74	76	80	82	86	87	93	93	99	99	105
9	1	45		47	45	49	46	52	48	55	50	57	52	60	54	62	56	65	58
	u	63		70	72	77	80	83	87	89	94	96	101	102	108	109	115	115	122
10	1	55		58	55	60	57	63	59	66	61	69	64	72	66	75	68	78	71
	u	75		82	85	90	93	97	101	104	109	111	116	118	124	125	132	132	139

INDEX

This book has been set in 12 point and 10 point Garamond #3, leaded 1 point. Chapter numbers and titles are in 18 point Lydian Bold. The size of the type page is 27 × 46½ picas.